SET THEORY

STUDIES IN LOGIC

AND

THE FOUNDATIONS OF MATHEMATICS

VOLUME 102

ELSEVIER

AMSTERDAM • BOSTON • HEIDELBERG • LONDON • NEW YORK • OXFORD
PARIS • SAN DIEGO • SAN FRANCISCO • SINGAPORE • SYDNEY • TOKYO

SET THEORY

An Introduction to Independence Proofs

Kenneth KUNEN
University of Texas, Austin, U.S.A.

ELSEVIER

AMSTERDAM • BOSTON • HEIDELBERG • LONDON • NEW YORK • OXFORD
PARIS • SAN DIEGO • SAN FRANCISCO • SINGAPORE • SYDNEY • TOKYO

ELSEVIER B.V.
Radarweg 29
P.O. Box 211, 1000 AE
Amsterdam, The Netherlands

ELSEVIER Inc.
525 B Street
Suite 1900, San Diego
CA 92101-4495, USA

ELSEVIER Ltd
The Boulevard
Langford Lane, Kidlington,
Oxford OX5 1GB, UK

ELSEVIER Ltd
84 Theobalds Road
London WC1X 8RR
UK

First edition 1980
Second impression 1983
Third impression 1988
Fourth impression 1990
Fifth impression 1992
Sixth impression 1995
Seventh impression 1999
Eighth impression 2004
Ninth impression 2005
Tenth impression 2006

Library of Congress Cataloging in Publication Data
Kunen, Kenneth.
Set theory.

(Studies in logic and the foundation of mathematics; v.102)
Bibliography: p. Includes indexes.
1. Axiomatic set theory. I. Title II. Series
QA248.K75 510.3'22 80-20375

ISBN: 0-444-86839-9

Transferred to digital printing 2006

To Anne, Isaac, and Adam

PREFACE

This book provides an introduction to relative consistency proofs in axiomatic set theory, and is intended to be used as a text in beginning graduate courses in that subject. It is hoped that this treatment will make the subject accessible to those mathematicians whose research is sensitive to axiomatics. We assume that the reader has had the equivalent of an undergraduate course on cardinals and ordinals, but no specific training in logic is necessary.

The author is grateful to the large number of people who have suggested improvements in the original manuscript for this book. In particular we would like to thank John Baldwin, Eric van Douwen, Peter Nyikos, and Dan Velleman. Special thanks are due to Jon Barwise, who tried out the manuscript in a course at the University of Wisconsin.

CONTENTS

INTRODUCTION

Set theory is the foundation of mathematics. All mathematical concepts are defined in terms of the primitive notions of set and membership. In axiomatic set theory we formulate a few simple axioms about these primitive notions in an attempt to capture the basic "obviously true" set-theoretic principles. From such axioms, all known mathematics may be derived. However, there are some questions which the axioms fail to settle, and that failure is the subject of this book.

§1. Consistency results

The specific axiom system we discuss is ZFC, or Zermelo–Frankel set theory with the Axiom of Choice. We say that a statement ϕ is independent of ZFC if neither ϕ nor $\neg\phi$ (the negation of ϕ) is provable from ZFC; this is equivalent to saying that both ZFC + $\neg\phi$ and ZFC + ϕ are consistent. The most famous example of such a ϕ is the Continuum Hypothesis (CH), but within the past few years, a large number of statements, coming from various branches of mathematics, have been shown to be independent of ZFC.

In this book, we study the techniques for showing that a statement ϕ is consistent with ZFC. ϕ will be shown to be independent if we can successfully apply these techniques to ϕ and to $\neg\phi$; this will always involve two separate arguments. There are also many statements which have been shown to be consistent but whose independence has remained unsettled.

Some of the statements known to be consistent with ZFC are "quotable" principles of abstract set theory, such as CH, or $\neg$CH, or Martin's Axiom, or Suslin's Hypothesis, or $\diamondsuit$. Workers in the more abstract areas of analysis and topology are well aware of these principles and often apply them. Since any consequence of a consistent statement is also consistent, this provides a source for many consistency proofs in mathematics. In addition, those mathematicians with a background in set theory often return to the basic methods to prove consistency results for specific mathematical statements which do not follow from one of the known "quotable" principles.

The purpose of this book is to explain the basic techniques for proving statements consistent with ZFC. We include consistency proofs for many of the "quotable" principles. More importantly, we hope to enable mathematicians to produce new consistency proofs of their own, as needed.

§2. Prerequisites

We assume that the reader has seen a development of axiomatic set theory through the basic properties of von Neumann ordinals and cardinals. This material is contained in set theory texts such as [Enderton 1977] or [Halmos 1960], as well as in appendices to books in other areas of mathematics which use set theory, such as [Chang–Keisler 1973] or [Kelley 1955]. This material is also reviewed in Chapter I.

It is not necessary for the reader to have seen the particular axiom system ZFC. There are other systems which differ from ZFC in the formal way proper classes are handled (see I §12). A reader familiar with one of these should have no trouble with ZFC, but should bear in mind that in ZFC proper classes have no formal existence, and all variables range over sets.

The reader need not be knowledgeable about very picky axiomatic questions—such as which axioms of ZFC are used to prove which theorems. In those few cases where such questions are of any importance, they are reviewed quite extensively in Chapter I. However, we do presume some sophistication in the way set theory is handled in its mathematical applications, as one would see in a course in general topology or measure theory.

Our prerequisite in formal logic is elastic. A book whose main results involve consistency of axiomatic systems cannot avoid logic entirely. We have included a sketch of background material on formal logic to enable readers with no training in the subject to understand independence proofs, but such readers might be suspicious about the complete mathematical rigor of our methods. A good undergraduate course in logic would dispel that suspicion. On a higher level, there are many foundational questions raised by our subject which are of interest to the student of logic per se, and we have collected such material in appendices to the various chapters. In these appendices, we have felt free to assume as much logical sophistication as is needed for the particular argument at hand.

§3. Outline

Chapter I contains some logical background and a sketch of the development of the axioms of ZFC, excluding Foundation (Regularity). Since this material is partly a review, we have omitted many proofs. We have been fairly pedantic about the fact that for many of the theorems, certain axioms, especially Choice and Power Set, are not needed, and we have indicated explicitly where these axioms are used; such considerations are not important for the development of mathematics within ZFC, but will be useful when we get to independence proofs.

Chapter II covers some special topics in combinatorial set theory. In part, this chapter provides some combinatorial lemmas needed in Chapters VI–VIII, but its main purpose is to introduce the reader to the vast array of set-theoretic questions that one might try to prove independent of ZFC.

We have departed from tradition in basing our treatment of forcing in Chapter VII upon the discussion of Martin's Axiom in Chapter II. This has the advantage of separating the mathematical difficulties in handling forcing from the metamathematical ones. It has the disadvantage of requiring those readers (if there are any) who wish to learn forcing without learning Martin's Axiom to do some extra work.

The Axiom of Foundation is discussed in Chapter III. This axiom is never used in mathematics, but it leads to a much clearer picture of the set theoretic universe.

Chapter IV develops the basic methods used in producing consistency proofs, including inner models, relativization, and absoluteness. We also discuss the Reflection Theorem and related results.

Chapter V discusses the formalization of the logical notion of definability within ZFC. These ideas are used in defining the class **L** of constructible sets in Chapter VI. In Chapter VI we establish the consistency of the Generalized Continuum Hypothesis by showing that it holds in **L**. We also show that the combinatorial principles $\diamondsuit$ and $\diamondsuit^+$ are true in **L**.

Chapter VII introduces forcing and uses it to prove the consistency of $\neg$CH and various related statements of cardinal arithmetic. Chapter VIII covers iterated forcing and the consistency of Martin's Axiom with $\neg$CH.

§4. How to use this book

In internal cross referencing, chapters are denoted by Roman numerals and § denotes section number. Thus, VII §5 is the fifth section of Chapter Seven, and VII 5.16 is the sixteenth numbered enunciation in that section.

The exercises range from routine verifications to additional development of the material in the chapter. The more difficult ones are starred. The exercises are not necessary for understanding later material in the text, although they are sometimes required for later exercises. There are probably more exercises in some of the chapters than most readers will want to do.

It is not necessary to read the book straight through from cover to cover. In particular, the material in Chapter II is not used at all until the end of

Chapter VI, so the reader may simply skip Chapter II and refer back to it as needed. Also, a knowledge of constructibility is not necessary to understand forcing, so it is possible to read Chapters VII and VIII without reading Chapters V and VI, although the reader doing this would have to take on faith the existence of models of GCH. Furthermore, the appendices of all chapters may be omitted without loss of continuity.

§5. What has been omitted

We have two goals in writing this book. First, we hope to bridge the gap between the current literature and the elementary texts on cardinals and ordinals. Second, we hope to emphasize the interplay between classical combinatorial set theory and modern independence proofs. Much important material in set theory which is secondary to these goals has been omitted.

Specifically, topics which are already well covered in the literature by texts or survey articles have often been omitted. We have little here on large cardinals: the interested reader may consult [Drake 1974] or [Solovay–Reinhardt–Kanamori 1978]. Likewise, we do not treat the fine-structure methods in **L**; see [Devlin 1973] for this.

We have also avoided topics which require some sophistication in logic. In particular, we do not discuss model-theoretic applications of large cardinals (see [Drake 1974]), or results in descriptive set theory, or the relationship between these fields (see [Martin 1977] or [Moschovakis 1980]).

This book gives short shrift to the Axiom of Choice (AC). We consider AC to be one of the basic axioms of set theory, although we do indicate proofs that it is neither provable (see VII Exercise E4) nor refutable (see V 2.14 and VI 4.9) from the other axioms. For more on set theory without AC, see [Jech 1973].

§6. On references

Since this is a text and not a research monograph, we have not attempted to give references to the literature for every theorem we prove. Our bibliography is intended primarily as suggestions for further reading, and not as a source for establishing priority. We apologize to those mathematicians who are chagrined at not seeing their name mentioned more often. Aside from a few trivial exercises, none of the results in this book are due to the author.

§7. The axioms

For reference, we list here the axioms of ZFC and of some related theories; these are explained in much greater detail in Chapters I and III. After each axiom we list the section in Chapters I or III where it first occurs.

AXIOM 0. *Set Existence* (I §5).

$$\exists x\,(x = x).$$

AXIOM 1. *Extensionality* (I §5).

$$\forall x\,\forall y\,\big(\forall z\,(z \in x \leftrightarrow z \in y) \rightarrow x = y\big).$$

AXIOM 2. *Foundation* (III §4).

$$\forall x\,[\exists y\,(y \in x) \rightarrow \exists y\,(y \in x \wedge \neg\exists z\,(z \in x \wedge z \in y))].$$

AXIOM 3. *Comprehension Scheme* (I §5). For each formula ϕ with free variables among $x, z, w_1, \ldots, w_n$,

$$\forall z\,\forall w_1, \ldots, w_n\,\exists y\,\forall x\,(x \in y \leftrightarrow x \in z \wedge \phi).$$

AXIOM 4. *Pairing* (I §6).

$$\forall x\,\forall y\,\exists z\,(x \in z \wedge y \in z).$$

AXIOM 5. *Union* (I §6).

$$\forall \mathscr{F}\,\exists A\,\forall Y\,\forall x\,(x \in Y \wedge Y \in \mathscr{F} \rightarrow x \in A).$$

AXIOM 6. *Replacement Scheme* (I §6). For each formula ϕ with free variables among $x, y, A, w_1, \ldots, w_n$,

$$\forall A\,\forall w_1, \ldots, w_n\,[\forall x \in A\,\exists! y\,\phi \rightarrow \exists Y\,\forall x \in A\,\exists y \in Y\,\phi].$$

On the basis of Axioms 0, 1, 3, 4, 5 and 6, one may define $\subset$ (subset), 0 (empty set), S (ordinal successor; $S(x) = x \cup \{x\}$), and the notion of well-ordering. The following axioms are then defined.

AXIOM 7. *Infinity* (I §7).

$$\exists x\,\big(0 \in x \wedge \forall y \in x\,(S(y) \in x)\big).$$

AXIOM 8. *Power Set* (I §10).

$$\forall x\,\exists y\,\forall z\,(z \subset x \rightarrow z \in y).$$

AXIOM 9. *Choice* (I §6).

$$\forall A\, \exists R\, (R \text{ well-orders } A).$$

ZFC is the system of Axioms 0–9.

For technical reasons, it will sometimes be important to know that some of the results which we prove from ZFC do not in fact require all the axioms of ZFC; the reason for this is discussed at the end of I §4. We list here some abbreviations for commonly used subtheories of ZFC. ZF consists of Axioms 0–8, ZF − P consists of Axioms 0–7, and ZFC − P consists of Axioms 0–7 plus Axiom 9. By ZFC^-, ZF^-, $\text{ZF}^- - \text{P}$, and $\text{ZFC}^- - \text{P}$, we mean the respective theory (ZFC, ZF, ZF − P, and ZFC − P) with Axiom 2 (Foundation) deleted. Other abbreviations for weakenings of ZFC are usually self-explanatory. For example, $\text{ZF}^- - \text{P} - \text{Inf}$ is $\text{ZF}^- - \text{P}$ with the Axiom of Infinity deleted.

CHAPTER I

THE FOUNDATIONS OF SET THEORY

It is assumed that the reader has seen a development of mathematics based on some principles roughly resembling the axioms listed in §7 of the Introduction. In this chapter we review such a development, stressing some foundational points which will be important for later work.

§1. Why axioms?

Most mathematicians have little need for a precise codification of the set theory they use. It is generally understood which principles are correct beyond any doubt, and which are subject to question. For example, it is generally agreed that the Continuum Hypothesis (CH) is not a basic principle, but rather an open conjecture, and we are all able, without the benefit of any formal axiomatization, to tell which of our theorems we have proved absolutely and which depend upon the (as yet undecided) truth or falsity of CH.

However, in this book we are concerned with establishing results like: "CH is neither provable nor refutable from ordinary set-theoretic principles". In order to make that statement precise, we must say exactly what those principles are; in this book, we have defined them to be the axioms of ZFC listed in the Introduction. The assertion: "CH is neither provable nor refutable from ZFC" is now a well-defined statement which we shall establish in Chapters VI and VII.

The question remains as to whether the axioms of ZFC do embody all the "ordinary set-theoretic principles". In this chapter we shall develop them far enough to be able to see how one can derive from them all of current conventional mathematics. Of course, future generations of mathematicians may come to realize some "obviously true" set-theoretic principles which do not follow from ZFC. Conceivably, CH could be then settled using those principles.

Even at the present, there are several ways besides ZFC for handling the axiomatization of currently accepted set-theoretic principles (see §12). The methods of this book are easily modified to handle those systems as well, although the technical details are slightly simpler for ZFC.

§2. Why formal logic?

The idea of setting down one's axioms harks back to Euclid, and is hardly revolutionary. Usually in mathematics the axioms are stated in an informal language, such as Greek or English. But here we shall state our axioms in a formal, or artificial, language, called the first-order predicate calculus. The feature of a formal language is that there are precise rules of formation for linguistic objects. There are two main reasons for this approach.

Reason 1. Formal logic is needed to state the axioms of set theory precisely. For example, ZFC has a Comprehension Axiom asserting that sets of the form

$$\{x \in A : P(x)\}$$

exist, where A is a given set and $P(x)$ is any property of x. But what is a property? Intuitively, it is any well-defined English assertion about the variable x. Is "x is happy" a property? It is clear that we need a rigorous definition of which properties we are to admit. We shall require that $P(x)$ be expressible in our formal language, which will be capable of expressing mathematical notions, but not non-mathematical ones. The fact that an imprecise notion of property can lead to trouble is illustrated by the following "paradox" in ordinary reasoning: Let n be the least positive integer not definable by an English expression using forty words or less. But I have just defined it in forty words or less.

Reason 2. Even after we have defined ZFC, what does it mean to say that CH is not provable from ZFC? Intuitively, it means that there is no way of deriving CH from ZFC using legitimate rules of inference. This intuitive notion can be made precise using the concept of a *formal deduction.*

We shall only sketch here the development of formal logic, referring the reader to a text on the subject, such as [Enderton 1972], [Kleene 1952], or [Shoenfield 1967], for a more detailed treatment. We shall give a precise definition of the formal language, as this is easy to do and is necessary for stating the axioms of ZFC. We shall only hint at the rules of formal deduction; these are also not hard to define, but it then takes some work to see that the standard mathematical arguments can all be formalized within the prescribed rules.

The basic symbols of our formal language are $\wedge$, $\neg$, $\exists$, (,), $\in$, $=$, and

v_j for each natural number j. Intuitively, $\wedge$ means "and," $\neg$ means "not," $\exists$ means "there exists," $\in$ denotes membership, $=$ denotes equality, $v_0, v_1, \ldots$ are variables, and the parentheses are used for phrasing. An *expression* is any finite sequence of basic symbols, such as $)\exists\exists\neg)$. The intuitive interpretation of the symbols indicates which expressions are meaningful; these are called formulas. More precisely, we define *a formula* to be any expression constructed by the rules:

(1) $v_i \in v_j$, $v_i = v_j$ are formulas for any i, j.

(2) If ϕ and ψ are formulas, so are $(\phi) \wedge (\psi)$, $\neg(\phi)$, and $\exists v_i(\phi)$ for any i.

So, for example, $\exists v_0 (\exists v_1 ((v_0 \in v_1) \wedge (v_1 \in v_0)))$ is a formula.

Our formal definition departs somewhat from intuition in that, in an effort to make the definition simple, the use of parentheses was prescribed very restrictively. For example,

$$v_0 \in v_1 \ \wedge \neg(v_1 \in v_0)$$

is not a formula. Another seeming drawback of our formal language is that it seems to lack the ability to express certain very basic logical notions, like, e.g. $\forall$ (for all). However, this is not really a problem, since $\neg(\exists v_i(\neg(\phi)))$ expresses $\forall v_i(\phi)$. Similar remarks hold for $\vee$ (or), $\rightarrow$ (implies) and $\leftrightarrow$ (iff), which may all be expressed using $\wedge$ and $\neg$. To save ourselves the work of always writing these longer expressions, we agree at the outset to use the following abbreviations.

(1) $\forall v_i(\phi)$ abbreviates $\neg(\exists v_i(\neg(\phi)))$.

(2) $(\phi) \vee (\psi)$ abbreviates $\neg((\neg(\phi)) \wedge (\neg(\psi)))$.

(3) $(\phi) \rightarrow (\psi)$ abbreviates $(\neg(\phi)) \vee (\psi)$.

(4) $(\phi) \leftrightarrow (\psi)$ abbreviates $((\phi) \rightarrow (\psi)) \wedge ((\psi) \rightarrow (\phi))$.

(5) Parentheses are dropped if it is clear from the context how to put them in.

(6) $v_i \neq v_j$ abbreviates $\neg(v_i = v_j)$ and $v_i \notin v_j$ abbreviates $\neg(v_i \in v_j)$.

(7) Other letters and subscripted letters from the English, Greek, and Hebrew alphabet are used for variables.

We shall explain (7) in more detail later.

There are many abbreviations other than these seven. Actually, in this book we shall very rarely see a formula. We shall follow standard mathematical usage of writing expressions mostly in English, augmented by logical symbols when this seems useful. For example, we might say "there are sets x, y, z such that $x \in y \wedge y \in z$", rather than

$$\exists v_0 (\exists v_1 (\exists v_2 ((v_0 \in v_1) \wedge (v_1 \in v_2)))).$$

The Comprehension Axiom (see Reason (1) above) will be made into a precise statement by requiring that properties $P(x)$ occurring in it be expressible in the formal language, but it will not be necessary to write out the formula expressing $P(x)$ each time the Comprehension Axiom is used.

A *subformula* of ϕ is a consecutive sequence of symbols of ϕ which form a formula. For example, the 5 subformulae of

$$(\exists v_0\,(v_0 \in v_1)) \wedge (\exists v_1\,(v_2 \in v_1)) \tag{1}$$

are $v_0 \in v_1$, $\exists v_0\,(v_0 \in v_1)$, $v_2 \in v_1$, $\exists v_1\,(v_2 \in v_1)$, and the formula (1) itself. The *scope* of an occurrence of a quantifier $\exists v_i$ is the (unique) subformula beginning with that $\exists v_i$. For example, the scope of the $\exists v_0$ in (1) is $\exists v_0\,(v_0 \in v_1)$. An occurrence of a variable in a formula is called *bound* iff it lies in the scope of a quantifier acting on that variable, otherwise it is called *free*. For example, in (1) the first occurrence of v_1 is free, but the second is bound, whereas v_0 is bound at its occurrences and v_2 is free at its occurrence.

Intuitively, a formula expresses a property of its free variables, whereas the bound or dummy variables are used just to make existential statements which could be made equally well with different bound variables. Thus, formula (1) means the same as

$$(\exists v_4\,(v_4 \in v_1)) \wedge (\exists v_4\,(v_2 \in v_4)).$$

Note that since $\forall v_i$ is an abbreviation for $\neg \exists v_i \neg$, it also binds its variable v_i, whereas the abbreviations $\vee$, $\rightarrow$, $\leftrightarrow$ are defined in terms of other propositional connections and do not bind variables.

Often in a discussion, we present a formula and call it $\phi(x_1, \ldots, x_n)$ to emphasize its dependence on $x_1, \ldots, x_n$. Then, later, if $y_1, \ldots, y_n$ are other variables, $\phi(y_1, \ldots, y_n)$ will denote the formula resulting from substituting a y_i for each free occurrence of x_i. Such a substitution is called *free*, or *legitimate* iff no free occurrence of an x_i is in the scope of a quantifier $\exists y_i$. The idea is that $\phi(y_1, \ldots, y_n)$ says about $y_1, \ldots, y_n$ what $\phi(x_1, \ldots, x_n)$ said about $x_1, \ldots, x_n$, but this will not be the case if the substitution is not free and some y_i gets bound by a quantifier of ϕ. In general, we shall always assume that our substitutions are free. The use of the notation $\phi(x_1, \ldots, x_n)$ does not imply that each x_i actually occurs free in $\phi(x_1, \ldots, x_n)$; also, $\phi(x_1, \ldots, x_n)$ may have other free variables which in the particular discussion we are not emphasizing.

For example, let $\phi(v_1, v_3)$ be formula (1). Then $\phi(v_2, v_8)$ is

$$(\exists v_0\,(v_0 \in v_2)) \wedge (\exists v_1\,(v_2 \in v_1)),$$

and $\phi(v_0, v_8)$ is

$$(\exists v_0\,(v_0 \in v_0)) \wedge (\exists v_1\,(v_2 \in v_1)). \tag{2}$$

But this latter substitution is not free, and has perverted the meaning of ϕ. The assertion in $\phi(v_1, v_3)$ that "v_1 has an element" became "some set is an element of itself" in $\phi(v_0, v_8)$.

A *sentence* is a formula with no free variables; intuitively, it states an assertion which is either true or false. ZFC is a certain set of sentences. If S is any set of sentences and ϕ is a sentence, then $S \vdash \phi$ means intuitively that ϕ is provable from S by a purely logical argument that may quote sentences in S as axioms but may not refer to the intended "meaning" of $\in$. Formally, we define $S \vdash \phi$ iff there is a *formal deduction* of ϕ from S; this is a finite sequence $\phi_1, \ldots, \phi_n$ of formulas such that ϕ_n is ϕ and for each i, either ϕ_i is in S or ϕ_i is a logical axiom or ϕ_i follows from $\phi_1, \ldots, \phi_{i-1}$ by certain rules of inference. Notions such as "logical axiom" and "rule of inference" are defined purely syntactically.

If $S \vdash \phi$ where S is the empty set of sentences, we write $\vdash \phi$ and say that ϕ is *logically valid*. If $\vdash(\phi \leftrightarrow \psi)$, we say ϕ and ψ are *logically equivalent*.

We do not dwell here on the precise definition of $\vdash$, but merely remark on some of its properties. See §4 for further discussion.

If ϕ is a formula, a *universal closure* of ϕ is a sentence obtained by universally quantifying all free variables of ϕ. For example, if ϕ is

$$x = y \rightarrow \forall z (z \in x \leftrightarrow z \in y),$$

then $\forall x \, \forall y \, \phi$ and $\forall y \, \forall x \, \phi$ are universal closures of ϕ. All universal closures of a formula are logically equivalent. In common parlance, when one asserts ϕ, one means to assert its universal closure. Formally, if S is a set of sentences and ϕ is a formula, we define $S \vdash \phi$ to mean that the universal closure of ϕ is derivable from S. The meaning of $S \vdash \phi$ when elements of S are not sentences is not the same in all presentations of logic and will not be discussed here.

We extend to formulas our notions of logical validity and logical equivalence. Thus, we call ϕ logically valid iff its universal closure is logically valid; and we say ϕ and ψ are logically equivalent iff $\phi \leftrightarrow \psi$ is logically valid.

Using the notion of logical equivalence, we may make precise the idea that bound variables are dummy variables. If $\phi(x_1, \ldots, x_n)$ is a formula with only $x_1, \ldots, x_n$ free and $\phi'(x_1, \ldots, x_n)$ results from replacing the bound variables of ϕ with other variables, then ϕ and ϕ' are logically equivalent. This enables us to be sloppy in our use of the various English, Greek, and Hebrew letters to stand for the official variables $v_0, v_1, v_2, \ldots$. For example, we have stated the Pairing Axiom as

$$\forall x \, \forall y \, \exists z (x \in z \wedge y \in z).$$

Formally, we should have chosen some distinct i, j, k and written the axiom as

$$\forall v_i \, \forall v_j \, \exists v_k (v_i \in v_k \wedge v_j \in v_k).$$

However, all such choices of i, j, k yield logically equivalent axioms.

Likewise, when we use other abbreviations, we can be vague about which of a number of logically equivalent unabbreviations is intended. For example, $\phi \wedge \psi \wedge \chi$ could abbreviate either $\phi \wedge (\psi \wedge \chi)$ or $(\phi \wedge \psi) \wedge \chi$, but since these two formulas are logically equivalent, it usually does not matter which of the two sentences we choose officially to represent $\phi \wedge \psi \wedge \chi$.

If S is a set of sentences, S is *consistent* (Con(S)) iff for no ϕ does $S \vdash \phi$ and $S \vdash \neg\phi$. If S is inconsistent, then $S \vdash \psi$ for all ψ and S is thus of no interest. By formalizing reductio ad absurdum, one proves for any sentence ϕ that $S \vdash \phi$ iff $S \cup \{\neg\phi\}$ is inconsistent and $S \vdash \neg\phi$ iff $S \cup \{\phi\}$ is inconsistent. Thus, ZFC $\nvdash$ CH is equivalent to Con(ZFC + $\neg$CH), (i.e., Con(ZFC $\cup \{\neg$CH$\}$)).

Intuitively, $x = y$ means that x and y are the same object. This is reflected formally in the fact that basic properties of equality are logically valid and need not be stated explicitly as axioms of ZFC. For example,

$$\vdash x = y \rightarrow \forall z\,(z \in x \leftrightarrow z \in y),$$

whereas the converse is not logically valid, although it is a theorem of ZFC since its universal closure is an axiom (Extensionality):

$$\forall x\, \forall y\, (\forall z\,(z \in x \leftrightarrow z \in y) \rightarrow x = y).$$

The fact that formal deductions from S are finite objects means that they can only mention a finite number of sentences in S even if S infinite. Thus, the following theorem holds.

2.1. THEOREM. (a) *If* $S \vdash \phi$, *then there is a finite* $S_0 \subset S$ *such that* $S_0 \vdash \phi$.
(b) *If* S *is inconsistent, there is a finite* $S_0 \subset S$ *such that* S_0 *is inconsistent.* □

This will be important since ZFC is an infinite set of axioms.

§3. The philosophy of mathematics

This section presents a caricature of some extremes of mathematical thought. For a more serious discussion, see [Fraenkel–Bar-Hillel–Lévy 1973], [Kleene 1952], or [Kreisel–Krivine 1967].

A Platonist believes that the set-theoretic universe has an existence outside of ourselves, so that CH is in fact either true or false (although at present we do not know which). From this point of view, the axioms of ZFC are merely certain obviously true set-theoretic principles. The fact that these axioms neither prove nor refute CH says nothing about its truth or falsity and does not preclude the possibility of our eventually being able to decide CH using some other obviously true principles which we forgot to list in ZFC. But a Platonist should still be interested in this independence result, since it says that we are wasting our time trying to decide CH unless we can recognize some new valid principle outside of ZFC.

A Finitist believes only in finite objects; one is not justified in forming the set of rational numbers, let alone the set of real numbers, so CH is a meaningless statement. There is some merit in the Finitist's position, since all objects in known physical reality are finite, so that infinite sets may be discarded as figments of the mathematician's imagination. Unfortunately, this point of view also discards much of modern mathematics.

The Formalist can hedge his bets. The formal development of ZFC makes sense from a strictly finitistic point of view: the axioms of ZFC do not say anything, but are merely certain finite sequences of symbols. The assertion ZFC $\vdash \phi$ means that there is a certain kind of finite sequence of finite sequences of symbols—namely, a formal proof of ϕ. Even though ZFC contains infinitely many axioms, notions like ZFC $\vdash \phi$ will make sense, since one can recognize when a particular sentence is an axiom of ZFC. A Formalist can thus do his mathematics just like a Platonist, but if challenged about the validity of handling infinite objects, he can reply that all he is really doing is juggling finite sequences of symbols.

Pedagogically, it is much easier to develop ZFC from a platonistic point of view, and we shall do so throughout this book. Thus, to establish that ZFC $\vdash \phi$, we shall simply produce an argument that ϕ is true based on the assumption that the axioms of ZFC are true. Those readers who are Formalists and are skilled in formal logic will then see how to produce a formal proof of ϕ from ZFC. In some cases, when the formalistic interpretation of the material in a chapter is not immediately apparent, we have elaborated on this in an appendix to the chapter.

It is important to make a distinction between the formal theory and the metatheory. If we are discussing ZFC, then the *formal theory* is ZFC, and a *formal theorem* is a sentence in the formal language provable from ZFC. If we announce, in our development of the formal theory:

3.1. THEOREM. *There are uncountably many real numbers.* □

Then we mean that the sentence of the formal language which expresses Theorem 3.1 is a formal theorem of ZFC.

The *metatheory* consists of what is really true. This distinction is somewhat easier for the Finitist to make, since he must view the metatheory as being strictly finitistic. An example of a result in the metatheory is

$$\mathrm{Con}(\mathrm{ZFC}) \rightarrow \mathrm{Con}(\mathrm{ZFC} + \mathrm{CH}).$$

This is a statement about the formal theories which we shall establish in

VI. The proof will provide an explicit constructive procedure which, when applied to an inconsistency in ZFC + CH would produce one in ZFC.

The distinction between formal theory and metatheory is a little trickier for a Platonist. The Finitist can interpret Theorem 3.1 only as a formal theorem, since it talks about infinite objects and is therefore not really meaningful. To the Platonist, Theorem 3.1 also represents a true statement about the real world, but the fact that it is also a formal theorem means that it can be established on the basis of ZFC only. The Platonists of 2100 may know whether CH is true, but neither CH nor $\neg$CH will ever be a formal theorem of ZFC. Likewise, many Platonists of today believe that inaccessible cardinals exist, even though the statement that they exist is not a formal theorem of ZFC (see IV 6.9). Since what is "really true" for some Platonists may not be so for others, one cannot specify precisely what the platonistic metatheory is. Fortunately, in this book we need only assume that the metatheory contains all finitistic reasoning.

§4. What we are describing

We present here an informal discussion of the intended interpretation of the axioms of ZFC. The fact that there are other possible interpretations is the basis for all our independence proofs.

An *interpretation* of the language of set theory is defined by specifying a non-empty *domain of discourse*, over which the variables are intended to vary, together with a binary relation on that domain, which is the interpretation of $\in$. If ϕ is any sentence in the language of set theory, ϕ is either true or false under a specified interpretation. As a frivolous example, we may let the domain of discourse be the set $\mathbb{Z}$ of integers, and interpret $x \in y$ as $x < y$. This is a legitimate interpretation for the *language* of set theory, even though the sentence

$$\forall x \, \exists y \, (y \in x)$$

is true under this interpretation but refutable from ZFC. Of course, not all the axioms of ZFC are true under this interpretation.

In the *intended* interpretation, under which the axioms of ZFC are presumed true, $x \in y$ is interpreted to mean that x is a member of y, but the domain of discourse is somewhat harder to describe. In accordance with the belief that set theory is the foundation of mathematics, we should be able to capture all of mathematics by just talking about sets, so our variables should not range over objects like cows and pigs. But if C is a cow, $\{C\}$ is a set, but not a legitimate mathematical object. More generally, since we wish to talk only about sets but also should be able to talk about any element of a set in our domain of discourse, all the elements of such a set should be sets also. Repeating this, we shall understand that our domain of discourse consists of those x such that

$$x \text{ is a set}, \quad \text{and}$$

$$\forall y\,(y \in x \rightarrow y \text{ is a set}), \quad \text{and}$$

$$\forall z\,\forall y\,(y \in x \wedge z \in y \rightarrow z \text{ is a set}), \quad \text{and etc.}$$

We say such an x is *hereditarily* a set. Examples of such sets are 0 (the empty set), $\{0\}$, $\{0, \{\{0\}\}\}$, etc.

The following is a more set-theoretic way of looking at the hereditary sets. Let $\bigcup x$ be the union of all the sets in x. Let $\bigcup^0 x = x$ and $\bigcup^{n+1} x = \bigcup(\bigcup^n x)$. Then x is an hereditary set iff x is a set and for each $n = 0, 1, 2, \ldots$, all elements of $\bigcup^n x$ are sets.

An important feature of our domain of discourse is that every element of an hereditary set is an hereditary set. This is needed to see the truth of the Axiom of Extensionality (see §5).

The *intended* interpretation of set theory will be further discussed in III §4. We turn now to ad hoc interpretations; this is the basis of all consistency proofs in this book. If S is any set of sentences, we may show S is consistent by producing any interpretation under which all sentences of S are true. Usually, $\in$ will be still interpreted as membership, but the domain of discourse will be some sub-domain of the hereditary sets. Thus, we shall produce one interpretation for ZFC + CH and another for ZFC + $\neg$CH without ever deciding whether CH is true in the intended interpretation.

The justification for this method of producing consistency proofs is the easy direction of the Gödel Completeness Theorem; that if S holds in some interpretation, then S is consistent. The reason this theorem holds is that the rules of formal deduction are set up so that if $S \vdash \phi$, then ϕ must be true under any interpretation which makes all sentences in S true. If we fix an interpretation in which S holds, then any sentence false in that interpretation is not provable from S. Since $\neg\phi$ and ϕ cannot both hold in a given interpretation, S cannot prove both ϕ and $\neg\phi$; thus, S is consistent.

The non-trivial direction of the Gödel Completeness Theorem is that if S is consistent, then S holds in some interpretation, whose domain of discourse may be taken to be a countable set (but we may not be able to interpret $\in$ as real membership). We do not need this result in our work, but it is of interest, since it shows that the notion of consistency is not tied to a particular development of formal derivability. In fact, if we allow infinitistic methods in the metatheory, we may dispense entirely with formal proofs and *define* S to be consistent iff S holds in some interpretation, and *define* $S \vdash \phi$ iff ϕ is true in every interpretation which makes all sentences of S true. It is then much easier to see when $S \vdash \phi$. In this approach, the Compactness Theorem (2.1) becomes a deep result rather than a trivial remark.

We now explain why it is of interest that some elementary set theory can be developed without the full strength of ZFC. When we define an interpretation for, say, ZFC + CH, it will not be trivial to verify immediately that all axioms of ZFC do indeed hold in our interpretation. We shall first check that a weak theory, such as ZF − P − Inf holds; this will imply that some simple set-theoretic facts are true, which will make it easier to understand what the axioms of Choice, Power Set, and Infinity mean under this interpretation.

§5. Extensionality and Comprehension

We begin to list and discuss the axiom of ZFC.

AXIOM 0. *Set Existence.*

$$\exists x\,(x = x). \quad \square$$

This axiom says that our universe is non-void. Under most developments of formal logic, this is derivable from the logical axioms and thus redundant to state here, but we do so for emphasis.

AXIOM 1. *Extensionality.*

$$\forall x\,\forall y\,(\forall z\,(z \in x \leftrightarrow z \in y) \rightarrow x = y). \quad \square$$

This says that a set is determined by its members. To recognize (informally) the truth of this axiom, it is important to note, by our discussion in §4, that the variables x, y, z range only over the hereditary sets. Given hereditary sets x and y, $\forall z\,(z \in x \leftrightarrow z \in y)$ means that x and y have the same hereditary sets as members; but all members of x and y must be hereditary sets, so x and y have the same members and are hence the same set.

The Comprehension Axiom is intended to formalize the construction of sets of the form $\{x : P(x)\}$ where $P(x)$ denotes some property of x. Since the notion of property is made rigorous via formulas, it is tempting to set forth as axioms statements of the form

$$\exists y\,\forall x\,(x \in y \leftrightarrow \phi),$$

where ϕ is a formula. Unfortunately, such a scheme is inconsistent by the famous Russell paradox: If ϕ is $x \notin x$, then this axiom gives us a y such that

$$\forall x\,(x \in y \leftrightarrow x \notin x),$$

whence $y \in y \leftrightarrow y \notin y$. Fortunately in mathematical applications it is sufficient to be able to use a property $P(x)$ to define a subset of a given set, so we postulate Comprehension as follows.

AXIOM 3. *Comprehension Scheme.* For each formula ϕ without y free, the universal closure of the following is an axiom:

$$\exists y\, \forall x\, (x \in y \leftrightarrow x \in z \wedge \phi). \quad \square$$

ϕ may have any number of other variables free. The y asserted to exist is unique by Extensionality, and we denote this y by

$$\{x : x \in z \wedge \phi\} \quad \text{or} \quad \{x \in z : \phi\}.$$

Variables other than x which are free in ϕ are considered parameters in this definition of a subset of z.

Our restriction on y not being free in ϕ eliminates self-referential definitions of sets, for example,

$$\exists y\, \forall x\, (x \in y \leftrightarrow x \in z \wedge x \notin y)$$

would be inconsistent with the existence of a non-empty z.

Note that the Comprehension Scheme, although it expresses one idea, yields an infinite collection of axioms—one for each ϕ.

If z is any set, we may form, by Comprehension, $\{x \in z : x \neq x\}$, which is then a set with no members. By Axiom 0, some set z exists, so there is a set with no members. By Extensionality, such a set is unique. We are thus justified in making:

5.1. DEFINITION. 0 is the unique set y such that $\forall x\, (x \notin y)$. $\square$

We can also prove that there is no universal set.

5.2. THEOREM.

$$\neg \exists z\, \forall x\, (x \in z).$$

PROOF. If $\forall x\, (x \in z)$, then, by Comprehension, form $\{x \in z : x \notin x\} = \{x : x \notin x\}$, which would yield a contradiction by the Russell paradox discussed above. $\square$

We let $A \subset B$ abbreviate $\forall x\, (x \in A \to x \in B)$. So, $A \subset A$ and $0 \subset A$.

0 is the only set which can be proved to exist from Axioms 0, 1, and 3. To see this, consider the interpretation whose domain of discourse contains only the empty set, with $\in$ interpreted as the (vacuous) membership relation. Axioms 0, 1, and 3 hold in this interpretation, but so does $\forall y\, (y = 0)$, so Axioms 0, 1, and 3 cannot refute $\forall y\, (y = 0)$ (see IV 2.8 for a more formal presentation of this argument). Of course, we need more axioms.

§6. Relations, functions, and well-ordering

The following intuitive picture should emerge from §5. For a given $\phi(x)$, there need not necessarily exist a set $\{x: \phi(x)\}$; this collection (or class) may be too big to form a set. In some cases, for example with $\{x: x = x\}$, the collection is provably too big. Comprehension says that if the collection is a sub-collection of a given set, then it does exist. In certain other cases, e.g. where the collection is finite or is not too much bigger in cardinality than a given set, it should exist but the axioms of §5 are not strong enough to prove that it does. We begin this section with a few more axioms saying that certain sets which should exist do, and then sketch the development of some basic set-theoretic notions using these axioms.

Axioms 4–8 of ZFC all say that certain collections do form sets. We actually state these axioms in the (apparently) weaker form that the desired collection is a subcollection of a set, since we may then apply Comprehension to prove that the desired set exists. Stating Axioms 4–8 in this way will make it fairly easy to verify them in the various interpretations considered in Chapters VI and VII.

AXIOM 4. *Pairing.*

$$\forall x\, \forall y\, \exists z\, (x \in z \ \wedge\ y \in z). \qquad \square$$

AXIOM 5. *Union.*

$$\forall \mathscr{F}\, \exists A\, \forall Y\, \forall x\, (x \in Y \ \wedge\ Y \in \mathscr{F} \rightarrow x \in A). \qquad \square$$

AXIOM 6. *Replacement Scheme.* For each formula ϕ without Y free, the universal closure of the following is an axiom:

$$\forall x \in A\, \exists ! y\, \phi(x, y) \rightarrow \exists Y\, \forall x \in A\, \exists y \in Y\, \phi(x, y). \qquad \square$$

By Pairing, for a given x and y we may let z be any set such that $x \in z \wedge y \in z$; then $\{v \in z: v = x \vee v = y\}$ is the (unique by Extensionality) set whose elements are precisely x and y; we call this set $\{x, y\}$. $\{x\} = \{x, x\}$ is the set whose unique element is x. $\langle x, y\rangle = \{\{x\}, \{x, y\}\}$ is the *ordered pair* of x and y. One must check that

$$\forall x\, \forall y\, \forall x'\, \forall y'\, (\langle x, y\rangle = \langle x', y'\rangle \rightarrow x = x' \ \wedge\ y = y').$$

In the Union Axiom, we are thinking of $\mathscr{F}$ as a family of sets and postulate the existence of a set A such that each member Y of $\mathscr{F}$ is a subset of A. This justifies our defining the *union* of the family $\mathscr{F}$, or $\bigcup \mathscr{F}$, by

$$\bigcup \mathscr{F} = \{x: \exists Y \in \mathscr{F}\, (x \in Y)\};$$

this set exists since it is also

$$\{x \in A : \exists Y \in \mathscr{F} \, (x \in Y)\}.$$

When $\mathscr{F} \neq 0$, we let

$$\bigcap \mathscr{F} = \{x : \forall Y \in \mathscr{F} \, (x \in Y)\};$$

this set exists since, for any $B \in \mathscr{F}$, it is equal to

$$\{x \in B : \forall Y \in \mathscr{F} \, (x \in Y)\}$$

(so we do not appeal to the Union Axiom here). If $\mathscr{F} = 0$, then $\bigcup \mathscr{F} = 0$ and $\bigcap \mathscr{F}$ "should be" the set of all sets, which does not exist. Finally, we set $A \cap B = \bigcap \{A, B\}$, $A \cup B = \bigcup \{A, B\}$, and $A \setminus B = \{x \in A : x \notin B\}$.

The Replacement Scheme, like Comprehension, yields an infinite collection of axioms—one for each ϕ. The justification of Replacement is: assuming $\forall x \in A \, \exists ! y \, \phi(x, y)$, we can try to let $Y = \{y : \exists x \in A \, \phi(x, y)\}$; Y should be small enough to exist as a set, since its cardinality is $\leq$ that of the set A. Of course, by Replacement (and Comprehension),

$$\{y : \exists x \in A \, \phi(x, y)\}$$

does exist, since it is also $\{y \in Y : \exists x \in A \, \phi(x, y)\}$ for any Y such that $\forall x \in A \, \exists y \in Y \, \phi(x, y)$.

For any A and B, we define the *cartesian product*

$$A \times B = \{\langle x, y \rangle : x \in A \wedge y \in B\}.$$

To justify this definition, we must apply Replacement twice. First, for any $y \in B$, we have

$$\forall x \in A \, \exists ! z \, (z = \langle x, y \rangle),$$

so by Replacement (and Comprehension) we may define

$$\text{prod}(A, y) = \{z : \exists x \in A \, (z = \langle x, y \rangle)\}.$$

Now,

$$\forall y \in B \, \exists ! z \, (z = \text{prod}(A, y)),$$

so by Replacement we may define

$$\text{prod}'(A, B) = \{\text{prod}(A, y) : y \in B\}.$$

Finally, we define $A \times B = \bigcup \text{prod}'(A, B)$.

We now review some other notions which may be developed on the basis of the Axioms 0, 1, 3, 4, 5, and 6. A *relation* is a set R all of whose elements are ordered pairs.

$$\text{dom}(R) = \{x : \exists y (\langle x, y \rangle \in R)\}$$

and

$$\operatorname{ran}(R) = \{y : \exists x(\langle x, y\rangle \in R)\}.$$

These definitions make sense for any set R, but are usually used only when R is a relation, in which case $R \subset \operatorname{dom}(R) \times \operatorname{ran}(R)$. We define $R^{-1} = \{\langle x, y\rangle : \langle y, x\rangle \in R\}$, so $(R^{-1})^{-1} = R$ if R is a relation.

f is a *function* iff f is a relation and

$$\forall x \in \operatorname{dom}(f)\, \exists! y \in \operatorname{ran}(f)(\langle x, y\rangle \in f).$$

$f: A \to B$ means f is a function, $A = \operatorname{dom}(f)$, and $\operatorname{ran}(f) \subset B$. If $f: A \to B$ and $x \in R$, $f(x)$ is the unique y such that $\langle x, y\rangle \in f$; if $C \subset A$, $f \restriction C = f \cap C \times B$ is the *restriction* of f to C, and $f''C = \operatorname{ran}(f \restriction C) = \{f(x): x \in C\}$. Many people use $f(C)$ for $f''C$, but the notation would cause confusion in this book since often elements of A will be subsets of A as well.

$f: A \to B$ is 1–1, or an *injection*, iff f^{-1} is a function, and f is onto, or a *surjection*, iff $\operatorname{ran}(f) = B$. $f: A \to B$ is a *bijection* iff f is both 1–1 and onto.

A *total ordering* (sometimes called a *strict* total ordering) is a pair $\langle A, R\rangle$ such that R totally orders A—that is, A is a set, R is a relation, R is transitive on A:

$$\forall x, y, z \in A\,(xRy \wedge yRz \to xRz),$$

trichotomy holds:

$$\forall x, y \in A\,(x = y \vee xRy \vee yRx),$$

and R is irreflexive:

$$\forall x \in A\,(\neg(xRx)).$$

As usual, we write xRy for $\langle x, y\rangle \in R$. Note that our definition does not assume $R \subset A \times A$, so if $\langle A, R\rangle$ is a total ordering so is $\langle B, R\rangle$ whenever $B \subset A$.

Whenever R and S are relations, and A, B are sets, we say $\langle A, R\rangle \cong \langle B, S\rangle$ iff there is a bijection f: $A \to B$ such that $\forall x, y \in A\,(xRy \leftrightarrow f(x)\,Sf(y))$. f is called an *isomorphism* from $\langle A, R\rangle$ to $\langle B, S\rangle$.

We say R *well-orders* A, or $\langle A, R\rangle$ is a *well-ordering* iff $\langle A, R\rangle$ is a total ordering and every non-0 subset of A has an R-least element.

If $x \in A$, let $\operatorname{pred}(A, x, R) = \{y \in A: yRx\}$. This notation is used mainly when dealing with ordering. The basic rigidity properties of well-ordering are given as follows.

6.1. LEMMA. *If $\langle A, R\rangle$ is a well-ordering, then for all $x \in A$, $\langle A, R\rangle \not\cong \langle \operatorname{pred}(A, x, R), R\rangle$.*

PROOF. If $f: A \to \text{pred}(A, x, R)$ were an isomorphism, derive a contradiction by considering the R-least element of $\{y \in A: f(y) \neq y\}$. □

6.2. LEMMA. *If $\langle A, R \rangle$ and $\langle B, S \rangle$ are isomorphic well-orderings, then the isomorphism between them is unique.*

PROOF. If f and g were different isomorphisms, derive a contradiction by considering the R-least $y \in A$ such that $f(y) \neq g(y)$. □

The proofs of Lemmas 6.1 and 6.2 are examples of proofs by transfinite induction.

A basic fact about well-orderings is that any two are comparable:

6.3. THEOREM. *Let $\langle A, R \rangle$, $\langle B, S \rangle$ be two well-orderings. Then exactly one of the following holds:*

(a) $\langle A, R \rangle \cong \langle B, S \rangle$;

(b) $\exists y \in B(\langle A, R \rangle \cong \langle \text{pred}(B, y, S), S \rangle)$;

(c) $\exists x \in A(\langle \text{pred}(A, x, R), R \rangle \cong \langle B, S \rangle)$.

PROOF. Let

$$f = \{\langle v, w \rangle : v \in A \wedge w \in B \wedge \langle \text{pred}(A, v, R), R \rangle \cong \langle \text{pred}(B, w, S), S \rangle\};$$

note that f is an isomorphism from some initial segment of A onto some initial segment of B, and that these initial segments cannot both be proper. □

The notion of well-ordering gives us a convenient way of stating the Axiom of Choice (AC).

Axiom 9. *Choice.*

$$\forall A \, \exists R \, (R \text{ well-orders } A). \qquad \square$$

There are many equivalent versions of AC. See, e.g., [Jech 1973], [Rubin–Rubin 1963], or Exercises 9–11.

This book is concerned mainly with set theory with AC. However, it is of some interest that much of the elementary development of set theory does not need AC, so in this chapter we shall explicitly indicate which results have used AC in their proofs. AC is not provable in ZF; see [Jech 1973], or VII Exercise E3.

§7. Ordinals

The basics of the von Neumann theory of ordinals can be developed using the axioms so far presented, but at some point an axiom postulating the existence of a limit ordinal must be introduced.

7.1. DEFINITION. A set x is *transitive* iff every element of x is a subset of x. □

In 6 we tried to use different type for different "kinds" of sets, $x, y, \ldots$ for elements, $A, B, \ldots$ for sets, and $\mathscr{F}$ for families of sets. But in the light of Definition 7.1, it is impossible to maintain this distinction.

Examples of transitive sets are 0, $\{0\}$, $\{0, \{0\}\}$, and $\{\{\{0\}\}, \{0\}, 0\}$. $\{\{0\}\}$ is not transitive. If $x = \{x\}$, then x is transitive; for more on such pathological sets see III.

7.2. DEFINITION. x is an *ordinal* iff x is transitive and well-ordered by $\in$. □

More formally, the assertion that x is well-ordered by $\in$ means that $\langle x, \in_x\rangle$ is a well-ordering, where $\in_x = \{\langle y, z\rangle \in x \times x : y \in z\}$. Examples of ordinals are 0, $\{0\}$, $\{0, \{0\}\}$, whereas $\{\{\{0\}\}, \{0\}, 0\}$ is not an ordinal. If $x = \{x\}$, then x is not ordinal since we have defined orderings to be strict.

We shall often drop explicit mention of $\in_x$ in discussing an ordinal x. Thus, we write $x \cong \langle A, R\rangle$ for $\langle x, \in_x\rangle \cong \langle A, R\rangle$ and, when $y \in x$, $\mathrm{pred}(x, y)$ for $\mathrm{pred}(x, y, \in_x)$.

7.3. THEOREM. (1) *If x is an ordinal and $y \in x$, then y is an ordinal and $y = \mathrm{pred}(x, y)$.*

(2) *If x and y are ordinals and $x \cong y$, then $x = y$.*

(3) *If x and y are ordinals, then exactly one of the following is true: $x = y$, $x \in y$, $y \in x$.*

(4) *If x, y, and z are ordinals, $x \in y$, and $y \in z$, then $x \in z$.*

(5) *If C is a non-empty set of ordinals, then $\exists x \in C\, \forall y \in C\, (x \in y \vee x = y)$.*

PROOF. For (3), use (1), (2) and Theorem 6.3 to show that at least one of of the three conditions holds. That no more than one holds follows from the fact that no ordinal can be a member of itself, since $x \in x$ would imply that $\langle x, \in_x\rangle$ is not a (strict) total ordering (since $x \in_x x$). For (5), note that the conclusion is, by (3), equivalent to $\exists x \in C\, (x \cap C = 0)$. Let $x \in C$ be arbitrary. If $x \cap C \neq 0$, then, since x is well-ordered by $\in$, there is an $\in$-least element, x' of $x \cap C$; then $x' \cap C = 0$. □

Theorem 7.3 implies that the set of all ordinals, if it existed, would be an ordinal, and thus cannot exist. More precisely, the following holds.

7.4. THEOREM. $\neg \exists z \forall x$ (*x is an ordinal* $\rightarrow x \in z$).

PROOF. If there were such a z then we would have a set ON such that

$$ON = \{x : x \text{ is an ordinal}\}.$$

Then ON is transitive by (1) of Theorem 7.3 and well-ordered by $\in$ (by (3), (4), (5)), so ON is an ordinal, so $ON \in ON$; but, as pointed out in the proof of Theorem 7.3, no ordinal is a member of itself. □

This so-called Burali–Forti paradox indicates (as did the Russell paradox, Theorem 5.1) that one must exercise some care when forming the set of elements satisfying a given property.

Any proper initial segment of the non-existent ON is an ordinal.

7.5. LEMMA. *If A is a set of ordinals and $\forall x \in A\, \forall y \in x\, (y \in A)$, then A is an ordinal.* □

7.6. THEOREM. *If $\langle A, R\rangle$ is a well-ordering, then there is a unique ordinal C such that $\langle A, R\rangle \cong C$.*

PROOF. Uniqueness follows from Theorem 7.3 (2). To prove existence, let $B = \{a \in A : \exists x\,(x \text{ is an ordinal } \wedge \langle \text{pred}(A, a, R), R\rangle \cong x)\}$. Let f be the function with domain B such that for $a \in B$, $f(a) =$ the (unique) ordinal x such that $\langle \text{pred}(A, a, R), R\rangle \cong x$, and let $C = \text{ran}(f)$. Now check that C is an ordinal (using Lemma 7.5), that f is an isomorphism from $\langle B, R\rangle$ onto C, and that either $B = A$ (in which case we are done), or $B = \text{pred}(A, b, R)$ for some $b \in A$ (in which case we would have $b \in B$ and hence a contradiction). □

Note that the proof of Theorem 7.6 used the Axiom of Replacement in an essential way to justify the existence of the set f. More formally, we let $\phi(a, x)$ be the formula asserting

$$\langle \text{pred}(A, a, R), R\rangle \cong x.$$

Then $\forall a \in B\, \exists! x\, \phi(a, x)$, so by Replacement (and Comprehension) one can form $C = \{x : \exists a \in B\, \phi(a, x)\}$, and then use Comprehension to define $f \subset B \times C$. If one drops Replacement from ZFC, one can still develop most of "ordinary" mathematics, but one cannot prove Theorem 7.6; see IV Exercise 9 for more details.

Theorem 7.6 implies that one may use ordinals as representatives of order types.

7.7. DEFINITION. If $\langle A, R\rangle$ is a well-ordering, type(A, R) is the unique ordinal C such that $\langle A, R\rangle \cong C$. □

From now on we use Greek letters α, β, γ, ... to vary over ordinals. We may thus say, e.g., $\forall\alpha \cdots$ instead of $\forall x\,(x$ is an ordinal $\rightarrow \cdots)$. Since $\in$ orders the ordinals, we write $\alpha < \beta$ for $\alpha \in \beta$ and use the standard conventions with order; e.g., $\alpha \geq \beta$ means $\beta \in \alpha \vee \beta = \alpha$.

7.8. DEFINITION. If X is a set of ordinals, $\sup(X) = \bigcup X$, and if $X \neq 0$, $\min(X) = \bigcap X$. □

7.9. LEMMA. (1) $\forall\alpha, \beta\,(\alpha \leq \beta \leftrightarrow \alpha \subset \beta)$.

(2) *If X is a set of ordinals,* $\sup(X)$ *is the least ordinal* $\geq$ *all elements of* X, *and, if* $X \neq 0$, $\min(X)$ *is the least ordinal in* X. □

The first few ordinals are the natural numbers. We use natural numbers to count finite sets. The importance of ordinals in set theory is that, assuming AC, every set can be counted by an ordinal (see §10).

Many of the standard arithmetic operations on natural numbers can be defined on all the ordinals. We begin with successor.

7.10. DEFINITION. $S(\alpha) = \alpha \cup \{\alpha\}$. □

7.11. LEMMA. *For any* α, $S(\alpha)$ *is an ordinal,* $\alpha < S(\alpha)$, *and* $\forall\beta\,(\beta < S(\alpha) \leftrightarrow \beta \leq \alpha)$. □

7.12. DEFINITION. α is a *successor ordinal* iff $\exists\beta\,(\alpha = S(\beta))$. α is a *limit ordinal* iff $\alpha \neq 0$ and α is not a successor ordinal. □

7.13. DEFINITION. $1 = S(0)$, $2 = S(1)$, $3 = S(2)$, $4 = S(3)$, etc. □

So, 0 is the empty set, $1 = \{0\}$, $2 = \{0, 1\}$, $3 = \{0, 1, 2\}$, $4 = \{0, 1, 2, 3\}$, etc.

7.14. DEFINITION. α *is a natural number* iff $\forall\beta \leq \alpha\,(\beta = 0 \vee \beta$ is a successor ordinal$)$. □

It is immediate from the definition that the natural numbers form an initial segment of the ordinals. Intuitively, they are those ordinals obtained by applying S to 0 a finite number of times, since if β is the least ordinal not so obtained, β could not be a successor ordinal, so that β and all larger α would not satisfy Definition 7.14. Formally, the notion of "finite" has not yet been defined; it will be defined, by using the concept of "natural number," in §10.

Many mathematical arguments involve operations with the *set* of natural numbers, but one cannot prove on the basis of the axioms so far presented that there is such a set (see IV 3.12). We thus need a new axiom:

AXIOM 7. *Infinity.*

$$\exists x\,(0 \in x \;\wedge\; \forall y \in x\,(S(y) \in x)). \quad \square$$

If x satisfies the Axiom of Infinity, then "by induction", x contains all natural numbers. More rigorously, suppose n is a natural number and $n \notin x$. $n \neq 0$, so $n = S(m)$ for some m; then $m < n$, m is a natural number and $m \notin x$; so $n \smallsetminus x \neq 0$. Let n' be the least element of $n \smallsetminus x$; but applying the above argument to n' produces an $m' < n'$ with $m' \notin x$, which is a contradiction.

Now, by Comprehension, there is a set of natural numbers. The usual principle of induction (7.16 (4)) is stated in terms of this set, and in the future will replace awkward arguments in the style of the previous paragraph.

7.15. DEFINITION. ω is the set of natural numbers. $\square$

ω is an ordinal (by Lemma 7.5) and all smaller ordinals (i.e., all its elements) are successor ordinals or 0. So ω is a limit ordinal (since if not it would be a natural number), and hence ω is the least limit ordinal. Actually, the Axiom of Infinity is equivalent to postulating the existence of a limit ordinal, since any limit ordinal satisfies the axiom.

It is a philosophical quibble whether the elements of ω are the *real* natural numbers (whatever that means). The important thing is that they satisfy the Peano Postulates, namely

7.16. THEOREM. *The Peano Postulates.*

(1) $0 \in \omega$.
(2) $\forall n \in \omega\,(S(n) \in \omega)$.
(3) $\forall n, m \in \omega\,(n \neq m \rightarrow S(n) \neq S(m))$.
(4) (*Induction*) $\forall X \subset \omega\,[(0 \in X \;\wedge\; \forall n \in X\,(S(n) \in X)) \rightarrow X = \omega]$.

PROOF. For (4), if $X \neq \omega$, let γ be the least element of $\omega \smallsetminus X$, and show that γ is a limit ordinal $< \omega$. $\square$

Given the natural numbers with the Peano Postulates, one may temporarily forget about ordinals and proceed to develop elementary mathematics directly, constructing the integers and the rationals, and then introducing the Power Set Axiom and constructing the set of real numbers. The first step would be to define $+$ and $\cdot$ on ω. We do not take this approach here, since we wish to discuss $+$ and $\cdot$ on all ordinals. The approach we take actually defines $+$ and $\cdot$ without using the Axiom of Infinity, although this fact is not of great importance here.

To define $+$, note that $2 + 3 = 5$ means that if I lay out 2 apples in a row followed by 3 bananas, I will have a row of 5 pieces of fruit. We thus define $\alpha + \beta$ as follows.

7.17. DEFINITION. $\alpha + \beta = \text{type}(\alpha \times \{0\} \cup \beta \times \{1\}, R)$, where

$$R = \{\langle\langle\xi, 0\rangle, \langle\eta, 0\rangle\rangle : \xi < \eta < \alpha\} \cup$$

$$\{\langle\langle\xi, 1\rangle, \langle\eta, 1\rangle\rangle : \xi < \eta < \beta\} \cup [(\alpha \times \{0\}) \times (\beta \times \{1\})]. \quad \square$$

7.18. LEMMA. *For any* α, β, γ,

(1) $\alpha + (\beta + \gamma) = (\alpha + \beta) + \gamma$.
(2) $\alpha + 0 = \alpha$.
(3) $\alpha + 1 = S(\alpha)$.
(4) $\alpha + S(\beta) = S(\alpha + \beta)$.
(5) *If* β *is a limit ordinal,* $\alpha + \beta = \sup\{\alpha + \xi : \xi < \beta\}$.

PROOF. Directly from the definition of $+$. For example, to check (1), note that both $\alpha + (\beta + \gamma)$ and $(\alpha + \beta) + \gamma$ are isomorphic to α apples followed by β bananas and then γ grapes (i.e., $\alpha \times \{0\} \cup \beta \times \{1\} \cup \gamma \times \{2\}$ ordered in the obvious way). $\square$

$+$ is not commutative. For example $1 + \omega = \omega \neq \omega + 1$. $+$ is commutative on the natural numbers (see §10).

We compute $\alpha \cdot \beta$ by counting out α apples β times. Thus, for example, $\omega \cdot 2 = \omega + \omega$.

7.19. DEFINITION. $\alpha \cdot \beta = \text{type}(\beta \times \alpha, R)$, where R is lexicographic order on $\beta \times \alpha$:

$$\langle\xi, \eta\rangle R \langle\xi', \eta'\rangle \leftrightarrow (\xi < \xi' \vee (\xi = \xi' \wedge \eta < \eta')). \quad \square$$

Again, we check from the definition the basic properties of $\cdot$:

7.20. LEMMA. *For any* α, β, γ,

(1) $\alpha \cdot (\beta \cdot \gamma) = (\alpha \cdot \beta) \cdot \gamma$.
(2) $\alpha \cdot 0 = 0$.
(3) $\alpha \cdot 1 = \alpha$.

(4) $\alpha \cdot S(\beta) = \alpha \cdot \beta + \alpha$.
(5) *If* β *is a limit ordinal*, $\alpha \cdot \beta = \sup\{\alpha \cdot \xi : \xi < \beta\}$.
(6) $\alpha \cdot (\beta + \gamma) = \alpha \cdot \beta + \alpha \cdot \gamma$. □

Multiplication is not commutative, since $2 \cdot \omega = \omega \neq \omega \cdot 2$. The distributive law, (6), fails for multiplication on the right, since $(1 + 1) \cdot \omega = \omega \neq 1 \cdot \omega + 1 \cdot \omega$. On the natural numbers, $\cdot$ is commutative (see §10).

Natural numbers give us a way of handling finite sequences.

7.21. Definition. (a) A^n is the set of functions from n into A.
(b) $A^{<\omega} = \bigcup \{A^n : n \in \omega\}$. □

Under this definition, A^2 and $A \times A$ are not the same, but there is an obvious 1–1 correspondence between them.

It is not completely trivial to prove that Definition 7.21 makes sense without using the Power Set Axiom (to be introduced in §10). Let $\phi(n, y)$ say that

$$\forall s\,(s \in y \leftrightarrow s \text{ is a function from } n \text{ into } A).$$

By induction on n (i.e., by Peano Postulate 7.16 (4)), show $\forall n \in \omega\, \exists y\, \phi(n, y)$; the induction step uses Replacement plus the identification of A^{n+1} with $A^n \times A$. By Extensionality, $\forall n \in \omega\, \exists! y\, \phi(n, y)$, so by Replacement we may form $\{y : \exists n \in \omega\, \phi(n, y)\} = \{A^n : n \in \omega\}$, whence by the Union Axiom, $A^{<\omega}$ exists.

We often think that of the elements of A^n as the sequences from A of length n.

7.22. Definition. For each n, $\langle x_0, \ldots, x_{n-1} \rangle$ is the function s with domain n such that $s(0) = x_0$, $s(1) = x_1, \ldots, s(n-1) = x_{n-1}$. □

In the case $n = 2$, this definition of $\langle x, y \rangle$ is inconsistent with the definition of ordered pair in §6. The more elementary definition, $\langle x, y \rangle = \{\{x\}, \{x, y\}\}$ is convenient while developing basic properties of functions and relations, while Definition 7.22 becomes more useful when we wish to handle finite sequences of various finite lengths. In those few cases when it makes a difference which definition of $\langle x, y \rangle$ is intended, we shall say so explicitly.

In general, if s is a function with $\mathrm{dom}(s) = I$, we may think of I as an index set and s as a sequence indexed by I. In this case, we often write s_i for $s(i)$. Variants of Definition 7.22 are then used to explicitly define such functions; for example, $\langle i \cup \{x\} : i \in I \rangle$ is the function s with $\mathrm{dom}(s) = I$ and $s(i) = i \cup \{x\}$ for all $i \in I$.

When $\mathrm{dom}(s)$ is an ordinal α, we may think of s as a sequence of length α.

If $\mathrm{dom}(t) = \beta$, we may *concatenate* the sequences s and t to form a sequence $s^\frown t$ of length $\alpha + \beta$.

7.23. DEFINITION. If s and t are functions with $\mathrm{dom}(s) = \alpha$ and $\mathrm{dom}(t) = \beta$, the function $s^\frown t$ with domain $\alpha + \beta$ is defined by: $(s^\frown t)\restriction \alpha = s$ and $(s^\frown t)(\alpha + \xi) = t(\xi)$ for all $\xi < \beta$. □

§8. Remarks on defined notions

In the previous few sections, we have introduced a large number of set-theoretic definitions, starting with 0 and $\subset$ in §5. We now address the question of whether our handling of these is justified.

From a Platonistic point of view, this question might seem to be a mere quibble, since it is standard mathematical practice to enlarge one's vocabulary as one introduces new concepts. However, in our axiomatic treatment, we explicitly stated our Comprehension and Replacement Axioms to apply only with properties expressible in the original vocabulary as defined in §2. Yet, we have frequently quoted these axioms with properties defined using new vocabulary. It would thus appear that we need our language and axioms to be elastic, and to expand with time, although such an approach seems highly inelegant.

To avoid this difficulty, we take the official position that our formal language never changes, and that any new symbols introduced merely represent new ways of abbreviating formulas. The reason for these new abbreviations is the same as that for using $\forall x$ to abbreviate $\neg \exists x \neg$: to save space and promote clarity. It is important to note that the properties we express with these new symbols can, if need be, be expressed without them, so that our original Comprehension and Replacement Axioms apply.

To elaborate on this further, there are two kinds of defined notions: defined relations and defined operations.

When we introduce a defined relation between objects, we are merely defining a way of abbreviating a formula using a new symbol. For example, $x \subset y$ abbreviates $\forall z\,(z \in x \rightarrow z \in y)$. The "new symbol" may be a fragment of English; "x is empty" abbreviates $\forall z\,(z \notin x)$.

The introduction of defined operations (e.g., $x \cap y$, $\{x\}$, $\alpha + \beta$) or constants (e.g., 0, ω) is slightly more tricky, since this can be done only when our axiom system can prove that the definition makes sense. More formally, if $\phi(x_1, \ldots, x_n, y)$ is a formula with no variables except $x_1, \ldots, x_n, y$ free, and S is a set of axioms such that

$$S \vdash \forall x_1, \ldots, \forall x_n\, \exists ! y\, \phi(x_1, \ldots, x_n, y),$$

then we may, when arguing from S, "define" $F(x_1, \ldots, x_n)$ to be the y such that $\phi(x_1, \ldots, x_n, y)$. Then expressions in which F occurs are to be considered abbreviations for expressions in which F does not occur. The introduction of constants is the special case when $n = 0$.

For example, let $\phi(x, y, z)$ be $\forall v\,(v \in z \leftrightarrow v \in x \;\wedge v \in y)$. If S contains the instance of the Comprehension Axiom used to prove that $\forall x\, \forall y\, \exists ! z\, \phi(x, y, z)$, then we may use the symbol $x \cap y$ when arguing from S. A formula such as $x \cap y \in A \cap B$ can be expressed without the symbol $\cap$ in several ways—for example

$$\exists z\, \exists C\, \big(\phi(x, y, z) \wedge \phi(A, B, C) \wedge z \in C\big), \tag{1}$$

or

$$\forall z\, \forall C\, \big(\phi(x, y, z) \wedge \phi(A, B, C) \rightarrow z \in C\big). \tag{2}$$

Since $S \vdash \forall x\, \forall y\, \exists ! z\, \phi(x, y, z)$, formulas (1) and (2) are provably equivalent from S.

Note that the $\{: \cdots\cdots\cdots\}$ notation may be viewed similarly as a way of introducing defined operations. Thus, $\{x: \phi(x, y_1, \ldots, y_n)\}$ is the unique z such that

$$\forall x\, \big(x \in z \leftrightarrow \phi(x, y_1, \ldots, y_n)\big).$$

This notation is only used when S contains Extensionality and

$$S \vdash \forall y_1, \ldots, \forall y_n\, \exists z\, \forall x\, \big(x \in z \leftrightarrow \phi(x, y_1, \ldots, y_n)\big).$$

We consider partially defined operations (e.g., $\alpha + \beta$), to be defined to be 0 outside their natural domains. So, $x + y$ is defined for all x, y and is 0 unless x and y are both ordinals.

We refer the reader to § 13 for a more formal treatment of defined symbols.

§ 9. Classes and recursion

We have seen that there need not exist a set of the form $\{x: \phi(x)\}$; the simplest example being $\{x: x = x\}$. There is nothing wrong with thinking about such collections, and they sometimes provide useful motivation; but since they are outside the domain of discourse described by the axioms, they must never appear in a formal proof. See § 12 for remarks on set theories in which such collections do appear formally.

Informally, we call any collection of the form $\{x: \phi(x)\}$ a *class*. We allow ϕ to have free variables other than x, which are thought of as parameters upon which the class depends. A *proper class* is a class which does not form a set (because it is "too big"). The Comprehension Axiom says that any subclass of a set is a set. We use boldface letters to denote classes. Two useful classes, which are proper by Theorems 5.2 and 7.4, are given by the following.

9.1. DEFINITION.

$$\mathbf{V} = \{x : x = x\}$$

$$\mathbf{ON} = \{x : x \text{ is an ordinal}\}. \quad \square$$

Formally, proper classes do not exist, and expressions involving them must be thought of as abbreviations for expressions not involving them. Thus, $x \in \mathbf{ON}$ abbreviates the formula expressing that x is an ordinal, and $\mathbf{ON} = \mathbf{V}$ abbreviates the (false) sentence (abbreviated by)

$$\forall x\,(x \text{ is an ordinal} \leftrightarrow x = x).$$

There is, in fact, no formal distinction between a formula and a class; the distinction is only in the informal presentation. We could think of 9.1 as defining $\mathbf{ON}(x)$ to abbreviate the formula "x is an ordinal," but it is useful to think of $\mathbf{ON}$ as a class if we wish to write expressions like $\mathbf{ON} \cap y$ (to abbreviate $\{x \in y : x \text{ is an ordinal}\}$). Any of our defined predicates and functions might be thought of as a class. For example, we could think of the union operation as defining a class $\mathbf{UN} = \{\langle\langle x, y\rangle, z\rangle : z = x \cup y\}$. Intuitively, $\mathbf{UN}: \mathbf{V} \times \mathbf{V} \to \mathbf{V}$, and this motivates using an abbreviation like $\mathbf{UN} \upharpoonright (a \times b)$ for

$$\{\langle\langle x, y\rangle, z\rangle : z = x \cup y \wedge x \in a \wedge y \in b\}$$

The abbreviations obtained with the class become very useful when discussing general properties of classes. Asserting that a statement is true of *all classes* is equivalent to asserting a *theorem schema.* As an example of this, we state the principles of induction and recursion on **ON**.

9.2. THEOREM. *Transfinite Induction on* **ON**. *If* $\mathbf{C} \subset \mathbf{ON}$ *and* $\mathbf{C} \neq 0$ *then* **C** *has a least element.*

PROOF. Exactly like Theorem 7.3(5), which asserted the same thing when **C** is a set. Fix $\alpha \in \mathbf{C}$. If α is not the least element of **C**, let β be the least element of $\alpha \cap \mathbf{C}$. Then β is the least element of **C**. $\square$

Mathematically, Theorems 7.3 (5) and 9.2 are very similar. But formally there is a great difference. Theorem 7.3 (5) is an abbreviation for one sentence which is provable, whereas 9.2 is a theorem schema, which represents an infinite collection of theorems. To state Theorem 9.2 without classes, we would have to say: for each formula $\mathbf{C}(x, z_1, \ldots, z_n)$, the following is a theorem:

$$\forall z_1, \ldots, z_n \{[\forall x\,(\mathbf{C} \to x \text{ is an ordinal}) \wedge \exists x\, \mathbf{C}] \to$$
$$\to [\exists x\,(\mathbf{C} \wedge \forall y\,(\mathbf{C}(y, z_1, \ldots, z_n) \to y \geq x))]\}. \quad (1)$$

Note that here we are thinking of $\mathbf{C}$ as defining $\{x: \mathbf{C}(x, z_1, \dots, z_n)\}$, with $z_1, \dots, z_n$ as parameters.

The fact that we may use parameters in the definition of classes implies that a schema such as 9.2 about all classes has, as one special case, the universal statement about all sets. Thus, if $\mathbf{C}(x, z)$ is $x \in z$, then (1) is equivalent to

$$\forall z\,\{[z \text{ is a non-0 set of ordinals}] \rightarrow [\exists x \in z\, \forall y \in z\, (y \geq x)]\},$$

which is Theorem 7.3 (5).

A "proof by transfinite induction on α" establishes $\forall \alpha\, \psi(\alpha)$ by showing, for each α, that

$$(\forall \beta < \alpha\, \psi(\beta)) \rightarrow \psi(\alpha).$$

Then $\forall \alpha\, \psi(\alpha)$ follows, since $\exists \alpha\, \neg \psi(\alpha)$, the least α such that $\neg \psi(\alpha)$ would lead to a contradiction.

A similar result says that one can define a function of α recursively from information about the function below α.

9.3. THEOREM. *Transfinite Recursion on* **ON**. *If* $\mathbf{F}: \mathbf{V} \rightarrow \mathbf{V}$, *then there is a unique* $\mathbf{G}: \mathbf{ON} \rightarrow \mathbf{V}$ *such that*

$$\forall \alpha\, [\mathbf{G}(\alpha) = \mathbf{F}(\mathbf{G} \restriction \alpha)]. \tag{2}$$

PROOF. For uniqueness, if $\mathbf{G}_1$ and $\mathbf{G}_2$ both satisfied (2), one proves $\forall \alpha\, (\mathbf{G}_1(\alpha) = \mathbf{G}_2(\alpha))$ by transfinite induction on α.

To establish existence, call g a *δ-approximation* iff g is a function with domain δ and

$$\forall \alpha < \delta\, [g(\alpha) = \mathbf{F}(g \restriction \alpha)].$$

As in the uniqueness proof, if g is a δ-approximation and g' is a δ'-approximation, then $g \restriction (\delta \cap \delta') = g' \restriction (\delta \cap \delta')$. Next, by transfinite induction on δ, show that for each δ there is a δ-approximation (which is then unique). Now, define $\mathbf{G}(\alpha)$ to be the value $g(\alpha)$, where g is the δ-approximation for some (any) $\delta > \alpha$. □

Theorem 9.3 may be stated more verbosely without using classes. Given a formula $\mathbf{F}(x, y)$ (with possibly other free variables), one can explicitly define a formula $\mathbf{G}(v, y)$ (the way $\mathbf{G}$ was defined above) so that

$$\forall x\, \exists! y\, \mathbf{F}(x, y) \rightarrow [\forall \alpha\, \exists! y\, \mathbf{G}(\alpha, y) \wedge \forall \alpha\, \exists x\, \exists y (\mathbf{G}(\alpha, y) \wedge \mathbf{F}(x, y) \wedge x = \mathbf{G} \restriction \alpha)] \tag{3}$$

is a theorem, where $x = \mathbf{G} \restriction \alpha$ abbreviates

$$x \text{ is a function} \wedge \operatorname{dom} x = \alpha \wedge \forall \beta \in \operatorname{dom} x\, \mathbf{G}(\beta, x(\beta)).$$

To express the uniqueness statement in Theorem 9.3, let $\mathbf{G}'(v, y)$ be any other formula and let (3)′ be like (3) but with $\mathbf{G}'$ instead of $\mathbf{G}$. Then the following is also a theorem:

$$[\forall x\, \exists ! y\, \mathbf{F}(x, y) \wedge (3)'] \rightarrow \forall \alpha\, \forall y\, (\mathbf{G}(\alpha, y) \leftrightarrow \mathbf{G}'(\alpha, y)).$$

Fortunately, it is rarely necessary in mathematical arguments about classes to translate away the classes; it is, however, important to know that this can be done in principle.

One may think of the operations $\alpha + \beta$ and $\alpha \cdot \beta$ as being defined by transfinite recursion on β, with α as a fixed parameter, although it is easier to check their properties directly from the definitions given in §7. We dwell somewhat on the recursive definition of $\alpha + \beta$ to illustrate how the usual informal manner of presenting recursive definitions can, if desired, be reduced to Theorem 9.3.

Informally, we could have defined $\alpha + \beta$ by recursion on β via the clauses:

$$\alpha + 0 = \alpha.$$

$$\alpha + S(\beta) = S(\alpha + \beta).$$

$$\alpha + \beta = \sup\{\alpha + \xi : \xi < \beta\} \text{ when } \beta \text{ is a limit ordinal.}$$

Since the function $+$ as defined in 7.17. satisfies these clauses (by 7.18 (2), (4), (5)), the two definitions are equivalent.

More formally, we interject first a definition.

9.4. Definition. $\beta - 1$ is β if β is a limit or 0, and γ if $\beta = S(\gamma)$. □

Now, for each α, define $\mathbf{F}_\alpha : \mathbf{V} \rightarrow \mathbf{V}$ so that $\mathbf{F}_\alpha(x)$ is 0 unless x is a function with domain some ordinal β, in which case $\mathbf{F}_\alpha(x)$ is α if $\beta = 0$, $S(x(\beta - 1))$ if β is a successor, and $\bigcup\{x(\xi) : \xi < \beta\}$ if β is a limit. Then Theorem 9.3 yields a unique $\mathbf{G}_\alpha : \mathbf{ON} \rightarrow \mathbf{V}$ such that $\forall \beta\, [\mathbf{G}_\alpha(\beta) = \mathbf{F}_\alpha(\mathbf{G}_\alpha \restriction \beta)]$. The uniqueness implies, using Lemma 7.18 (2), (4), (5), that $\forall \alpha \beta\, [\mathbf{G}_\alpha(\beta) = \alpha + \beta]$.

More formally still, the subscripted α becomes an additional free variable in the formula $\mathbf{F}$ occurring in our official explication of Theorem 9.3.

One can also define $\cdot$ by recursion using the clauses in Lemma 7.20 (2), (4), (5). A more useful application of recursion is in defining ordinal exponentiation, since a direct combinatorial definition of α^β is slightly complicated (see Exercise 7).

9.5. Definition. α^β is defined by recursion on β by

(1) $\alpha^0 = 1$.
(2) $\alpha^{\beta+1} = \alpha^\beta \cdot \alpha$.
(3) If β is a limit, $\alpha^\beta = \sup\{\alpha^\xi : \xi < \beta\}$. □

Note that $2^\omega = \omega$; this should not be confused with cardinal exponentiation (see §10).

A minor variant of Theorem 9.3 is transfinite recursion on an ordinal, δ. If $\mathbf{F}: \mathbf{V} \to \mathbf{V}$, there is a unique function g with domain δ such that $\forall \alpha < \delta\, [g(\alpha) = \mathbf{F}(g \restriction \alpha)]$; to see this, let $\mathbf{G}: \mathbf{ON} \to \mathbf{V}$ be the function satisfying (2), and let $g = \mathbf{G} \restriction \delta$. g is a set by the Axiom of Replacement.

An important special case, when $\delta = \omega$, is often used in arithmetic. For example, we define $n!$ by the clauses:

$$0! = 1.$$

$$(n+1)! = n! \cdot (n+1).$$

This may be cast more formally in the form of Theorem 9.3 as in the discussion of $\alpha + \beta$ above. Here there are only two clauses, as there are no limit ordinals $< \omega$.

§10. Cardinals

We use 1–1 functions to compare the size of sets.

10.1. Definition. (1) $A \preccurlyeq B$ iff there is a 1–1 function from A into B.

(2) $A \approx B$ iff there is a 1–1 function from A onto B.

(3) $A \prec B$ iff $A \preccurlyeq B$ and $B \not\preccurlyeq A$. □

It is easily seen that $\preccurlyeq$ is transitive and that $\approx$ is an equivalence relation. A much deeper result is given in the following theorem.

10.2. Theorem. *Schröder–Bernstein.*

$$A \preccurlyeq B, B \preccurlyeq A \to A \approx B.$$

Proof. See Exercise 8. □

One determines the size of a finite set by counting it. More generally, if A can be well-ordered, then $A \approx \alpha$ for some α (Theorem 7.6), and there is then a least such α, which we call the *cardinality* of A.

10.3. Definition. If A can be well-ordered, $|A|$ is the least α such that $\alpha \approx A$. □

If we write a statement involving $|A|$, such as $|A| < \alpha$, we take it to imply that A can be well-ordered.

Under AC, $|A|$ is defined for every A. Since $A \approx B \to |A| = |B|$ and $|A| \approx A$, the operation $|A|$ picks, under AC, a unique representative of each $\approx$-equivalence class.

Regardless of AC, $|\alpha|$ is defined and $\leq \alpha$ for all α.

10.4. DEFINITION. α is a *cardinal* iff $\alpha = |\alpha|$. □

Equivalently, α is a cardinal iff $\forall \beta < \alpha (\beta \not\approx \alpha)$. We use κ and λ to range over cardinals.

10.5. LEMMA. *If* $|\alpha| \leq \beta \leq \alpha$, *then* $|\beta| = |\alpha|$.

PROOF. $\beta \subset \alpha$ so $\beta \preccurlyeq \alpha$, and $\alpha \approx |\alpha| \subset \beta$ so $\alpha \preccurlyeq \beta$. Thus, by Theorem 10.2, $\alpha \approx \beta$. □

10.6. LEMMA. *If* $n \in \omega$, *then*

(1) $n \not\approx n + 1$.

(2) $\forall \alpha (\alpha \approx n \rightarrow \alpha = n)$.

PROOF. (1) is by induction on n. (2) follows using Lemma 10.5. □

10.7. COROLLARY. ω *is a cardinal and each* $n \in \omega$ *is a cardinal.* □

10.8. DEFINITION. A is *finite* iff $|A| < \omega$. A is *countable* iff $|A| \leq \omega$. *Infinite* means not finite. *Uncountable* means not countable. □

One cannot prove on the basis of the axioms so far given that uncountable sets exist (see IV 6.7).

Cardinal multiplication and addition must be distinguished from ordinal multiplication.

10.9. DEFINITION. (1) $\kappa \oplus \lambda = |\kappa \times \{0\} \cup \lambda \times \{1\}|$.

(2) $\kappa \otimes \lambda = |\kappa \times \lambda|$. □

Unlike the ordinal operations, $\oplus$ and $\otimes$ are commutative, as is easily checked from their definitions. Also, the definitions of $+$ and $\cdot$ (7.17 and 7.19) imply that $|\kappa + \lambda| = |\lambda + \kappa| = \kappa \oplus \lambda$ and $|\kappa \cdot \lambda| = |\lambda \cdot \kappa| = \kappa \otimes \lambda$. Thus, e.g., $\omega \oplus 1 = |1 + \omega| = \omega < \omega + 1$ and $\omega \otimes 2 = |2 \cdot \omega| = \omega < \omega \cdot 2$.

10.10. LEMMA. *For* $n, m \in \omega$, $n \oplus m = n + m < \omega$ *and* $n \otimes m = n \cdot m < \omega$.

PROOF. First show $n + m < \omega$ by induction on m. Then show $n \cdot m < \omega$ by induction on m. The rest follows by 10.6 (2). □

We now consider $\oplus$ and $\otimes$ on infinite cardinals.

10.11. LEMMA. *Every infinite cardinal is a limit ordinal.*

PROOF. If $\kappa = \alpha + 1$, then since $1 + \alpha = \alpha$, $\kappa = |\kappa| = |1 + \alpha| = |\alpha|$, a contradiction. □

We remark that the principle of transfinite induction (Theorem 9.2) can be applied to prove results about cardinals, since every class of cardinals is a class of ordinals. This is illustrated by the following Theorem.

10.12. THEOREM. *If κ is an infinite cardinal, $\kappa \otimes \kappa = \kappa$.*

PROOF. By transfinite induction on κ. Assume this holds for smaller cardinals. Then for $\alpha < \kappa$, $|\alpha \times \alpha| = |\alpha| \otimes |\alpha| < \kappa$ (applying Lemma 10.10 when α is finite). Define a well-ordering $\lhd$ on $\kappa \times \kappa$ by $\langle \alpha, \beta \rangle \lhd \langle \gamma, \delta \rangle$ iff

$$\max(\alpha, \beta) < \max(\gamma, \delta) \vee [\max(\alpha, \beta) = \max(\gamma, \delta)$$
$$\wedge \langle \alpha, \beta \rangle \text{ precedes } \langle \gamma, \delta \rangle \text{ lexicographically}].$$

Each $\langle \alpha, \beta \rangle \in \kappa \times \kappa$ has no more than $|(\max(\alpha, \beta) + 1) \times (\max(\alpha, \beta) + 1)| < \kappa$ predecessors in $\lhd$, so $\text{type}(\kappa \times \kappa, \lhd) \leq \kappa$, whence $|\kappa \times \kappa| \leq \kappa$. Since clearly $|\kappa \times \kappa| \geq \kappa$, $|\kappa \times \kappa| = \kappa$. □

10.13. COROLLARY. *Let κ, λ be infinite cardinals, then*

(1) $\kappa \oplus \lambda = \kappa \otimes \lambda = \max(\kappa, \lambda)$.

(2) $|\kappa^{<\omega}| = \kappa$ *(see Definition 7.21).*

PROOF. For (2), use the proof of Theorem 10.12 to define, by induction on n, a 1–1 map $f_n : \kappa^n \to \kappa$. This yields a 1–1 map $f : \bigcup_n \kappa^n \to \omega \times \kappa$, whence $|\kappa^{<\omega}| \leq \omega \otimes \kappa = \kappa$. □

It is consistent with the axioms so far presented ($\text{ZFC}^- - \text{P}$) that the only infinite cardinal is ω (see IV 6.7).

AXIOM 8. *Power Set.*

$$\forall x \, \exists y \, \forall z \, (z \subset x \to z \in y). \quad \square$$

10.14. DEFINITION. $\mathscr{P}(x) = \{z : z \subset x\}$. □

This definition is justified by the Power Set and Comprehension Axioms. The operation $\mathscr{P}$ gives us a way of constructing sets of larger and larger cardinalities.

10.15. THEOREM. *Cantor.* $x \prec \mathscr{P}(x)$. □

Under AC, it is immediate from 10.15 that there is a cardinal $>\omega$, namely $|\mathscr{P}(\omega)|$, but in fact AC is not needed here.

10.16. **Theorem.** $\forall\alpha\,\exists\kappa\,(\kappa > \alpha$ *and* κ *is a cardinal*).

Proof. Assume $\alpha \geq \omega$. Let $W = \{R \in \mathscr{P}(\alpha \times \alpha)\colon R$ well-orders $\alpha\}$. Let $S = \{\text{type}(\langle \alpha, R\rangle)\colon R \in W\}$ (S exists by Replacement). Then $\sup(S)$ is a cardinal $>\alpha$. □

10.17. **Definition.** α^+ is the least cardinal $>\alpha$. κ is a successor cardinal iff $\kappa = \alpha^+$ for some α. κ is a limit cardinal iff $\kappa > \omega$ and is not a successor cardinal. □

10.18. **Definition.** $\aleph_\alpha = \omega_\alpha$ is defined by transfinite recursion on α by:

(1) $\omega_0 = \omega$.

(2) $\omega_{\alpha+1} = (\omega_\alpha)^+$.

(3) For γ a limit, $\omega_\gamma = \sup\{\omega_\alpha\colon \alpha < \gamma\}$. □

10.19. **Lemma.** (1) *Each* ω_α *is a cardinal.*

(2) *Every infinite cardinal is equal to* ω_α *for some* α.

(3) $\alpha < \beta \rightarrow \omega_\alpha < \omega_\beta$.

(4) ω_α *is a limit cardinal iff* α *is a limit ordinal.* ω_α *is a successor cardinal iff* α *is a successor ordinal.* □

Many of the basic properties of cardinals need AC. See [Jech 1973] for a discussion of what can happen if AC is dropped.

10.20. **Lemma** (AC). *If there is a function* f *from* X *onto* Y, *then* $|Y| \leq |X|$.

Proof. Let R well-order X, and define $g\colon Y \rightarrow X$ so that $g(y)$ is the R-least element of $f^{-1}(\{y\})$. Then g is 1–1, so $Y \preccurlyeq X$. □

As in Theorem 10.16, one can prove without AC that there is a map from $\mathscr{P}(\omega)$ onto ω_1, but one cannot produce a 1–1 map from ω_1 into $\mathscr{P}(\omega)$.

10.21. **Lemma** (AC). *If* $\kappa \geq \omega$ *and* $|X_\alpha| \leq \kappa$ *for all* $\alpha < \kappa$, *then* $|\bigcup_{\alpha<\kappa} X_\alpha| \leq \kappa$.

Proof. For each α, pick a 1–1 map f_α from X_α into κ. Use these to define a 1–1 map from $\bigcup_{\alpha<\kappa} X_\alpha$ into $\kappa \times \kappa$. The f_α are picked using a well-ordering of $\mathscr{P}(\bigcup_\alpha X_\alpha \times \kappa)$. □

Lévy showed that is consistent with ZF that $\mathscr{P}(\omega)$ and ω_1 are countable unions of countable sets.

A very important modification of Lemma 10.21 is the downward Löwenheim–Skolem–Tarski theorem of model theory, which is frequently applied in set theory (see, e.g., IV 7.8). 10.23 is a purely combinatorial version of this theorem.

10.22. DEFINITION. An *n-ary function* on A is an $f: A^n \to A$ if $n > 0$, or an element of A if $n = 0$. If $B \subset A$, B is closed under f iff $f''B^n \subset B$ (or $f \in B$ when $n = 0$). A *finitary function* is an n-ary function for some n. If $\mathscr{S}$ is a set of finitary functions and $B \subset A$, the *closure* of B under $\mathscr{S}$ is the least $C \subset A$ such that $B \subset C$ and C is closed under all the functions in $\mathscr{S}$. □

Note that there is a least C, namely $\bigcap \{D: B \subset D \subset A \wedge D \text{ is closed under } \mathscr{S}\}$.

10.23. THEOREM (AC). *Let κ be an infinite cardinal. Suppose $B \subset A$, $|B| \leq \kappa$, and $\mathscr{S}$ is a set of $\leq \kappa$ finitary functions on A. Then the closure of B under $\mathscr{S}$ has cardinality $\leq \kappa$.*

PROOF. If $f \in \mathscr{S}$ and $D \subset A$, let $f * D$ be $f''(D^n)$ if f is n-place, or $\{f\}$ if f is 0-place. Note that $|D| \leq \kappa \to |f * D| \leq \kappa$. Let $C_0 = B$ and $C_{n+1} = C_n \cup \bigcup \{f * C_n : f \in \mathscr{S}\}$. By Lemma 10.21 and induction on n, $|C_n| \leq \kappa$ for all n. Let $C_\omega = \bigcup_n C_n$. Then C_ω is the closure of B under $\mathscr{S}$ and, by 10.21 again, $|C_\omega| \leq \kappa$. □

A simple illustration of Theorem 10.23 is the fact that every infinite group, G, has a countably infinite subgroup. To see this, let $B \subset G$ be arbitrary such that $|B| = \omega$, and apply 10.23 with $\mathscr{S}$ consisting of the 2-ary group multiplication and the 1-ary group inverse.

Our intended application of 10.23 is not with groups, but with models of set theory.

We turn now to cardinal exponentiation.

10.24. DEFINITION. $A^B = {}^B A = \{f : f \text{ is a function } \wedge \operatorname{dom}(f) = B \wedge \operatorname{ran}(f) \subset A\}$. □

$A^B \subset \mathscr{P}(B \times A)$, so A^B exists by the Power Set Axiom.

10.25. DEFINITION (AC). $\kappa^\lambda = |{}^\lambda \kappa|$. □

The notations A^B and ${}^B A$ are both common in the literature. When discussing cardinal exponentiation, one can avoid confusion by using κ^λ for the cardinal and ${}^\lambda \kappa$ for the set of functions.

10.26. LEMMA. *If $\lambda \geq \omega$ and $2 \leq \kappa \leq \lambda$, then ${}^{\lambda}\kappa \approx {}^{\lambda}2 \approx \mathscr{P}(\lambda)$.*

PROOF. ${}^{\lambda}2 \approx \mathscr{P}(\lambda)$ follows by identifying sets with their characteristic functions, then

$$ {}^{\lambda}2 \preccurlyeq {}^{\lambda}\kappa \preccurlyeq {}^{\lambda}\lambda \preccurlyeq \mathscr{P}(\lambda \times \lambda) \approx \mathscr{P}(\lambda) \approx {}^{\lambda}2. \quad \square $$

Cardinal exponentiation is not the same as ordinal exponentiation (Definition 9.5). The ordinal 2^{ω} is ω, but the cardinal $2^{\omega} = |\mathscr{P}(\omega)| > \omega$. In this book, ordinal exponentiation is rarely used, and κ^{λ} denotes cardinal exponentiation unless otherwise stated.

If $n, m \in \omega$, the ordinal and cardinal exponentiations n^m are equal (Exercise 13).

The familiar laws for handling exponents for finite cardinals are true in general.

10.27. LEMMA (AC). *If κ, λ, σ are any cardinals,*

$$ \kappa^{\lambda \oplus \sigma} = \kappa^{\lambda} \otimes \kappa^{\sigma} \quad \textit{and} \quad (\kappa^{\lambda})^{\sigma} = \kappa^{\lambda \otimes \sigma}. $$

PROOF. One easily checks without AC that

$$ {}^{(B \cup C)}A \approx {}^{B}A \times {}^{C}A \quad (\text{if} \quad B \cap C = 0), $$

and

$$ {}^{C}({}^{B}A) \approx {}^{C \times B}A. \quad \square $$

Since Cantor could show that $2^{\omega_\alpha} \geqq \omega_{\alpha+1}$ (Theorem 10.15), and had no way of producing cardinals between ω_α and 2^{ω_α}, he conjectured that $2^{\omega_\alpha} = \omega_{\alpha+1}$.

10.28. DEFINITION (AC). CH (the Continuum Hypothesis) is the statement $2^{\omega} = \omega_1$. GCH (the Generalized Continuum Hypothesis) is the statement $\forall \alpha (2^{\omega_\alpha} = \omega_{\alpha+1})$. $\square$

Under GCH, κ^{λ} can be easily computed, but one must first introduce the notion of cofinality.

10.29. DEFINITION. If $f : \alpha \to \beta$, f maps α *cofinally* iff $\operatorname{ran}(f)$ is unbounded in β. $\square$

10.30. DEFINITION. The cofinality of β $(\operatorname{cf}(\beta))$ is the least α such that there is a map from α cofinally into β. $\square$

So $\operatorname{cf}(\beta) \leq \beta$. If β is a successor, $\operatorname{cf}(\beta) = 1$.

10.31. LEMMA. *There is a cofinal map* $f : \mathrm{cf}(\beta) \to \beta$ *which is strictly increasing* ($\xi < \eta \to f(\xi) < f(\eta)$).

PROOF. Let $g : \mathrm{cf}(\beta) \to \beta$ be any cofinal map, and define f recursively by

$$f(\eta) = \max(g(\eta), \sup\{f(\xi) + 1 : \xi < \eta\}). \quad \square$$

10.32. LEMMA. *If* α *is a limit ordinal and* $f : \alpha \to \beta$ *is a strictly increasing cofinal map, then* $\mathrm{cf}(\alpha) = \mathrm{cf}(\beta)$.

PROOF. $\mathrm{cf}(\beta) \le \mathrm{cf}(\alpha)$ follows by composing a cofinal map from $\mathrm{cf}(\alpha)$ into α with f. To see $\mathrm{cf}(\alpha) \le \mathrm{cf}(\beta)$, let $g : \mathrm{cf}(\beta) \to \beta$ be a cofinal map, and let $h(\xi)$ be the least η such that $f(\eta) > g(\xi)$; then $h : \mathrm{cf}(\beta) \to \alpha$ is a cofinal map. $\square$

10.33. COROLLARY. $\mathrm{cf}(\mathrm{cf}(\beta)) = \mathrm{cf}(\beta)$.

PROOF. Apply Lemma 10.32 to the strictly increasing cofinal map $f : \mathrm{cf}(\beta) \to \beta$ guaranteed by Lemma 10.31. $\square$

10.34. DEFINITION. β is *regular* iff β is a limit ordinal and $\mathrm{cf}(\beta) = \beta$. $\square$

So, by Corollary 10.33, $\mathrm{cf}(\beta)$ is regular for all limit ordinals β.

10.35. LEMMA. *If* β *is regular then* β *is a cardinal.* $\square$

10.36. LEMMA. ω *is regular.* $\square$

10.37. LEMMA (AC). κ^+ *is regular.*

PROOF. If f mapped α cofinally into κ^+ where $\alpha < \kappa^+$, then

$$\kappa^+ = \bigcup \{f(\xi) : \xi < \alpha\},$$

but a union of $\le\kappa$ sets each of cardinality $\le\kappa$ must have cardinality $\le\kappa$ by Lemma 10.21. $\square$

Without AC, it is consistent that $\mathrm{cf}(\omega_1) = \omega$. It is unknown whether one can prove in ZF that there exists a cardinal of cofinality $> \omega$.

Limit cardinals often fail to be regular. For example, $\mathrm{cf}(\omega_\omega) = \omega$. More generally, the following holds.

10.38. LEMMA. *If* α *is a limit ordinal, then* $\mathrm{cf}(\omega_\alpha) = \mathrm{cf}(\alpha)$.

PROOF. By Lemma 10.32. $\square$

Thus, if ω_α is a regular limit cardinal, then $\omega_\alpha = \alpha$. But the condition $\omega_\alpha = \alpha$ is not sufficient. For example, let $\sigma_0 = \omega$, $\sigma_{n+1} = \omega_{\sigma_n}$, and $\alpha = \sup\{\sigma_n : n \in \omega\}$. Then α is the first ordinal to satisfy $\omega_\alpha = \alpha$ but $\mathrm{cf}(\alpha) = \omega$. Thus, the first regular limit cardinal is rather large.

10.39. DEFINITION. (1) κ is *weakly inaccessible* iff κ is a regular limit cardinal.

(2) (AC) κ is *strongly inaccessible* iff $\kappa > \omega$, κ is regular, and

$$\forall \lambda < \kappa (2^\lambda < \kappa). \quad \square$$

So, strong inaccessibles are weak inaccessibles, and under GCH the notions coincide. It is consistent that 2^ω is weakly inaccessible or that it is larger than the first weak inaccessible (see VII 5.16). One cannot prove in ZFC that weak inaccessibles exist (see VI 4.13).

A modification of Cantor's diagonal argument yields that $(\omega_\omega)^\omega > \omega_\omega$. More generally, the following holds.

10.40. LEMMA (AC). *König. If κ is infinite and* $\mathrm{cf}(\kappa) \le \lambda$, *then* $\kappa^\lambda > \kappa$.

PROOF. Fix any cofinal map $f : \lambda \to \kappa$. Let $G : \kappa \to {}^\lambda\kappa$. We show that G cannot be onto. Define $h : \lambda \to \kappa$ so that $h(\alpha)$ is the least element of

$$\kappa \smallsetminus \{(G(\mu))(\alpha) : \mu < f(\alpha)\}.$$

Then $h \notin \mathrm{ran}\, G$. $\square$

10.41. COROLLARY (AC). *If* $\lambda \ge \omega$, $\mathrm{cf}(2^\lambda) > \lambda$.

PROOF. $(2^\lambda)^\lambda = 2^{\lambda \otimes \lambda} = 2^\lambda$, so apply Lemma 10.40 with $\kappa = 2^\lambda$. $\square$

10.42. LEMMA (AC + GCH). *Assume that* $\kappa, \lambda \ge 2$ *and at least one of them is infinite, then*

(1) $\kappa \le \lambda \to \kappa^\lambda = \lambda^+$.
(2) $\kappa > \lambda \ge \mathrm{cf}(\kappa) \to \kappa^\lambda = \kappa^+$.
(3) $\lambda < \mathrm{cf}(\kappa) \to \kappa^\lambda = \kappa$.

PROOF. (1) is by Lemma 10.26. For (2), $\kappa^\lambda > \kappa$ by Lemma 10.40, but $\kappa^\lambda \le \kappa^\kappa = 2^\kappa = \kappa^+$. For (3), $\lambda < \mathrm{cf}(\kappa)$ implies that ${}^\lambda\kappa = \bigcup\{{}^\lambda\alpha : \alpha < \kappa\}$, and each $|{}^\lambda\alpha| \le \max(\alpha, \lambda)^+ \le \kappa$. $\square$

The following definitions are sometimes useful.

10.43. DEFINITION (AC). (a) ${}^{<\beta}A = A^{<\beta} = \bigcup\{{}^\alpha A : \alpha < \beta\}$.

(b) $\kappa^{<\lambda} = |{}^{<\lambda}\kappa|$. $\square$

When $\kappa \geq \omega$, $\kappa^{<\omega} = \kappa$ (10.13 (2)), and $\kappa^{<\lambda} = \sup\{\kappa^\theta : \theta < \lambda \wedge \theta$ is a cardinal$\}$ (Exercise 15), so 10.43 (b) is used mainly when λ is a limit cardinal.

10.44. Definition (AC). $\beth_\alpha$ is defined by transfinite recursion on α by:
(1) $\beth_0 = \omega$.
(2) $\beth_{\alpha+1} = 2^{\beth_\alpha}$,
(3) For γ a limit, $\beth_\gamma = \sup\{\beth_\alpha : \alpha < \gamma\}$. □

Thus, GCH is equivalent to the statement $\forall \alpha (\beth_\alpha = \omega_\alpha)$.

§11. The real numbers

11.1. Definition. $\mathbb{Z}$ is the ring of integers, $\mathbb{Q}$ is the field of rational numbers, $\mathbb{R}$ is the field of real numbers, and $\mathbb{C}$ is the field of complex numbers. □

Any reasonable way of defining these from the natural numbers will do, but for definiteness we take $\mathbb{Z} = \omega \times \omega / \sim$, where $\langle n, m \rangle$ is intended to represent $n - m$, the equivalence relation $\sim$ is defined appropriately, $\mathbb{Z}$ is the set of equivalence classes, and operations $+$ and $\cdot$ are defined appropriately. $\mathbb{Q} = (\mathbb{Z} \times (\mathbb{Z} \smallsetminus \{0\}))/\simeq$ where $\langle x, y \rangle$ is intended to represent x/y.

$$\mathbb{R} = \{X \in \mathscr{P}(\mathbb{Q}) : X \neq 0 \wedge X \neq \mathbb{Q} \wedge \forall x \in X\, \forall y \in \mathbb{Q}(y < x \rightarrow y \in X)\}.$$

So $\mathbb{R}$ is the set of left sides of Dedekind cuts. $\mathbb{C} = \mathbb{R} \times \mathbb{R}$, with field operations defined in the usual way.

§12. Appendix 1: Other set theories

We discuss briefly two other systems of set theory which differ from ZF in that they give classes a formal existence. In both, all sets are classes, but not all classes are sets. Let us temporarily use capital letters to range over classes. We define X to be a set iff $\exists Y (X \in Y)$, and we use lower case letters to range sets. In both systems, the sets satisfy the usual ZF axioms, and the intersection of a class with a set is a set.

The system NBG (von Neumann–Bernays–Gödel, see [Gödel 1940]) has as a class comprehension axiom, the universal closure of

$$\exists X\, \forall y\, (y \in X \leftrightarrow \phi(y)),$$

where ϕ may have other free set and class variables, but the bound variables of ϕ may only range over sets. NBG is a conservative extension of ZF; that is, if ψ is a sentence with only set variables, NBG $\vdash \psi$ iff ZF $\vdash \psi$ (see [Wang 1949], [Shoenfield 1954]). Unlike ZF, NBG is finitely axiomatizable.

The system MK (Morse–Kelley; see the Appendix of [Kelley 1955]) strengthens NBG by allowing an arbitrary ϕ to appear in the class comprehension axiom. MK is not a conservative extension of ZF and is not finitely axiomatizable (see Exercises 25, 26).

None of the three theories, ZF, NBG, and MK, can claim to be the "right" one. ZF seems inelegant, since it forces us to treat classes, as we did in §9, via a circumlocution in the metatheory. Once we give classes a formal existence, it is hard to justify the restriction in NBG on the ϕ occurring in the class comprehension axiom, so MK seems like the right theory. However, once we have decided to give classes their full rights, it is natural to consider various properties of classes, and to try to form super-classes, such as

$$\{\mathbf{R} \subset \mathbf{ON} \times \mathbf{ON}\colon \mathbf{R} \text{ well-orders } \mathbf{ON}\}.$$

In MK, such objects can be handled only via an inelegant circumlocution in the metatheory.

ZF, NBG, and MK are all founded on the same basic concepts, and the proofs in this book for ZF are easily adapted to the other two theories, although ZF is technically slightly easier to deal with.

A set theory which is different in principle from ZF is Quine's New Foundations, NF, see [Quine 1937], [Quine 1951], and [Rosser 1953]. Like ZF, NF has only sets, but a universal set exists: $\exists v\, \forall x\, (x \in v)$. Unlike ZF, NF makes the restriction on forming $\{x\colon \phi(x)\}$ not one of size, but of the syntactical form of ϕ; ϕ must be *stratified*, which means roughly that it can be obtained from a formula of type theory by erasing the types. $x = x$ is stratified, but $x \in x$ and $x \notin x$ are not, so one avoids Russell's paradox. It is unknown whether NF is consistent, even assuming Con(ZF). It is known that NF $\vdash \neg$AC [Specker 1953]. NF is usually rejected as a foundation of mathematics for this reason and because we do not have a "clear" picture of the objects it describes as we get with ZF using Foundation (see III).

The reader is referred to [Fraenkel–Bar-Hillel–Lévy 1973] for a more detailed discussion of the various axiomatizations of set theory.

§13. Appendix 2: Eliminating defined notions

One aspect of our development of ZF which seems to be lacking in rigor is our treatment in §8 of defined notions. This is usually handled by the following general discussion in the metatheory. We specify a formal language by defining its set, $\mathscr{L}$, of non-logical symbols; a symbol of $\mathscr{L}$ may be an n-place predicate symbol or an n-place function symbol for some finite n; 0-place function symbols are called constant symbols. $=$ is considered a logical symbol, so for the language of set theory, $\mathscr{L} = \{\in\}$.

Let S be a set of sentences of $\mathscr{L}$. We consider extensions of S by definitions. If $\mathscr{L}' = \mathscr{L} \cup \{P\}$, where P is an n-place predicate symbol, $P \notin \mathscr{L}$, and $\phi(x_i, \ldots, x_n)$ is a formula of $\mathscr{L}$ with at most $x_1, \ldots, x_n$ free, the axioms S plus

$$\forall x_1, \ldots, x_n\, [\phi(x_1, \ldots, x_n) \leftrightarrow P(x_1, \ldots, x_n)]$$

form a *one-step extension* of S by adding a defined predicate. If $\mathscr{L}' = \mathscr{L} \cup \{f\}$, where f is an n-place function symbol, $f \notin \mathscr{L}$, $\phi(x_1, \ldots, x_n, y)$ is a formula of $\mathscr{L}$ with at most $x_1, \ldots, x_n, y$ free, and

$$S \vdash \forall x_1, \ldots, x_n\, \exists ! y\, \phi(x_1, \ldots, x_n, y),$$

then the axioms S plus

$$\forall x_1, \ldots, x_n, y\, [y = f(x_1, \ldots, x_n) \leftrightarrow \phi(x_1, \ldots, x_n, y)]$$

form a *one-step extension* of S by adding a defined function. Finally, if $\mathscr{L} = \mathscr{L}_0 \subset \mathscr{L}_1 \subset \mathscr{L}_2 \subset \cdots \subset \mathscr{L}_n = \mathscr{L}'$, and $S = S_0 \subset S_1 \subset S_2 \subset \cdots \subset S_n = S'$, where each $\mathscr{L}_{i+1}$ is $\mathscr{L}_i$ plus one new symbol and each S_{i+1} is a one-step extension of S_i by adding that defined symbol, then we call the set of sentences S' of $\mathscr{L}'$ an *extension of S by definitions*.

Thus, at any point in this book, we are operating in some extension of ZF (or of ZFC or ZF^- etc.) by a finite number of definitions. The following general theorem of logic is tacitly being applied.

13.1. THEOREM. *If S' in $\mathscr{L}'$ is an extension by definitions of S in $\mathscr{L}$, then for each formula $\psi'(x_1, \ldots, x_n)$ of $\mathscr{L}'$ there is a formula $\psi(x_1, \ldots, x_n)$ of $\mathscr{L}$ such that*

$$S' \vdash \forall x_1, \ldots, x_n\, (\psi(x_1, \ldots, x_n) \leftrightarrow \psi'(x_1, \ldots, x_n)). \quad \square$$

This is relevant in set theory, where S is ZF and $\mathscr{L} = \{\in\}$, when we wish to apply the Comprehension or Replacement Axioms with a formula ψ' of $\mathscr{L}'$; the existence of ψ shows that on the basis of S', these axioms in the original language are sufficient. But, we then need a result saying that S' adds nothing essentially new.

13.2. THEOREM. *If S' in $\mathscr{L}'$ is an extension by definitions of S in $\mathscr{L}$, then S' is a conservative extension; i.e., if $S' \vdash \phi$, where ϕ is a sentence of $\mathscr{L}$, then $S \vdash \phi$.* $\square$

Theorem 13.1 is easily proved by induction on the number of steps in the extension, and, for one-step extensions, by induction on ψ'. Theorem 13.2 is easy model-theoretically, since any model for S has an expansion satisfying S'. A finitistic proof of 13.2 is rather tricky, see [Shoenfield 1967].

The theorem on defining functions by transfinite recursion may be viewed formally as a schema for introducing defined functions. Thus, if F is an $n+1$-place function symbol in some extension S of ZF (or $\mathrm{ZF}^- - \mathrm{P}$) by definitions, the theorem asserts that we may pass to a one-step extension S' of S by adding a new $n+1$-place function symbol G so that

$$S' \vdash \forall x_1, \ldots, x_n \forall \alpha \left[G(x_1, \ldots, x_n, \alpha) = F(x_1, \ldots, x_n, G \restriction \alpha)\right],$$

where $G \restriction \alpha$ means

$$\{\langle \xi, G(x_1, \ldots, x_n, \xi)\rangle : \xi < \alpha\}.$$

The formula of $\mathscr{L}$ defining G may be obtained by examining the proof of Theorem 9.3.

§14. Appendix 3: Formalizing the metatheory

Another point in our development of ZF which should be cleared up is the confusion between objects in the metatheory and formal objects. A careful examination of this point leads to Gödel's incompleteness theorems and Tarski's theorem on non-definability of truth.

The discussion of these theorems requires the notion of a *recursive* (or *decidable*) set. Informally, we say (in the metatheory) that a set R of natural numbers is recursive iff we may write a computer program which inputs a natural number n, and outputs "yes" if $n \in R$ and "no" if $n \notin R$. For example, the set of even numbers is recursive. This notion extends naturally to other sets of finitistic objects. For example, the set of formulas of predicate calculus is recursive because we may program a computer to read a string of symbols and tell us whether or not that string is a formula. Likewise, the set of axioms of ZF is recursive. For a more formal discussion, see [Enderton 1972], [Kleene 1952], or [Shoenfield 1967].

One very important result on these notions is that, assuming Con(ZF), the set of theorems of ZF is *not* recursive. More generally, if T is any consistent set of axioms extending ZF, then $\{\varphi : T \vdash \varphi\}$ is not recursive; for a proof, see one of the above three references or Exercise 24. A consequence of this is Gödel's *First Incompleteness Theorem*—namely, that if such a T is recursive, then it is *incomplete* in the sense that there is a sentence φ such that $T \nvdash \varphi$ and $T \nvdash \neg\varphi$. The proof is: if there were no such φ, then for every φ, either $T \vdash \varphi$ or $T \vdash \neg\varphi$, and not both. But then we could program a computer to decide whether $T \vdash \varphi$ as follows: start listing all formal deductions from T, and stop when a deduction of either φ or of $\neg\varphi$ is found.

Of course, if T is ZF itself, then, as we shall see in this book, there is a very explicit φ which is neither provable nor refutable from T—namely AC. Likewise, if T is ZFC, then we may take φ to be CH. But the importance of the First Incompleteness Theorem is that no matter how we extend ZF to a recursive consistent T, there will always be sentences which are not decided by T. Thus, there is no recursive axiomatization for "all that is true".

Gödel's Second Incompleteness Theorem says roughly that there is no elementary proof of Con(ZF); in fact, there is no proof at all by methods formalizable within ZF. In order to understand this result and Tarski's theorem better, we must look more closely at the relationship between objects in the metatheory and objects defined within the formal theory.

As an example, there are two different ways of using the number 1. One is as a concept in the metatheory, as in "$x \in x$ has 1 free variable." The other is as a defined notion, $1 = \{0\}$, within the formal theory, ZF.

In those cases where these two usages might cause confusion, we shall use the following sort of notation (called Quine's corner convention) for separating them: If Ob is any finitistic object in the metatheory, we shall use $\ulcorner \mathrm{Ob} \urcorner$ for a constant symbol denoting Ob in an extension of ZF by definitions.

Specifically, we should use 0 only to denote the informal concept, and $\ulcorner 0 \urcorner$ the constant symbol introduced by the defining axiom

$$\forall y\,(y = \ulcorner 0 \urcorner \leftrightarrow \forall x\,(x \notin y)).$$

Likewise, our definition of 1, 2, 3, ... within ZF (see Definition 7.13) should really be a schema in the metatheory for introducing new defined symbols $\ulcorner 1 \urcorner, \ulcorner 2 \urcorner, \ulcorner 3 \urcorner$, via the axioms

$$\forall y\,(y = \ulcorner 1 \urcorner \leftrightarrow y = S(\ulcorner 0 \urcorner)),$$
$$\forall y\,(y = \ulcorner 2 \urcorner \leftrightarrow y = S(\ulcorner 1 \urcorner)),$$
etc.

Another finitistic object is a finite sequence of natural numbers, which is formalized within ZF as an element of $\omega^{<\omega}$ (see Definition 7.21). If s is a finite sequence of natural numbers in the metatheory, we again use $\ulcorner s \urcorner$ for the formal object. For example, $\ulcorner \langle 8, 1, 5 \rangle \urcorner$ is introduced by the definition,

$$\forall y\,[y = \ulcorner \langle 8,1,5 \rangle \urcorner \leftrightarrow (y \text{ is a function } \wedge \operatorname{dom}(y) = \ulcorner 3 \urcorner \ \wedge y(\ulcorner 0 \urcorner) =$$
$$= \ulcorner 8 \urcorner \ \wedge y(\ulcorner 1 \urcorner) = \ulcorner 1 \urcorner\, y(\ulcorner 2 \urcorner) = \ulcorner 5 \urcorner)].$$

A formula in the language of set theory is a finite sequence of symbols such as $\wedge$, $\neg$, etc. (see §2). But what is a symbol? We sidestep this question by *defining* (in the metatheory) $\wedge$ to be the number 1 (whatever that is); likewise $\neg$ is 3, $\exists$ is 5, and (,), $\in$, $=$ are 7, 9, 11, 13, respectively, while v_i is the number $2i$. So $v_3 = v_3$ *is* the sequence $\langle 6, 13, 6 \rangle$. Thus, any formula ϕ is a finite sequence of numbers, and $\ulcorner \phi \urcorner$ is defined as above.

Likewise, if Υ is a finite sequence of finite sequences of natural numbers (for example, a formal deduction), we may introduce $\ulcorner\Upsilon\urcorner$ in the obvious way.

Now that we have succeeded in naming in the formal theory various finite objects in the metatheory, we may ask how properties of these objects are reflected by formal theorems. There are two levels to this investigation.

As an example of Level 1, we may let $\chi_{\text{even}}(x)$ be the formula $\exists y \in \omega\,(x = \ulcorner 2\urcorner \cdot y)$. Then we should be able to check that $\text{ZF} \vdash \chi_{\text{even}}(\ulcorner 8\urcorner)$ and $\text{ZF} \vdash \neg\chi_{\text{even}}(\ulcorner 7\urcorner)$. More generally, the following holds.

14.1. THEOREM. *Given any recursive set, R of natural numbers, there is a formula $\chi_R(x)$ which represents R in the sense that for all n,*

$$n \in R \rightarrow (\text{ZF} \vdash \chi_R(\ulcorner n\urcorner)) \quad \textit{and} \quad n \notin R \rightarrow (\text{ZF} \vdash \neg\chi_R(\ulcorner n\urcorner)).$$

Recursive sets of finite sequences and recursive predicates in several variables are likewise representable. □

Theorem 14.1 is easily proved (in the metatheory) using any one of the usual definitions of recursive.

As an example of Level 2,

$$\text{ZF} \vdash \forall x \in \omega\,(\chi_{\text{even}}(x) \vee \chi_{\text{even}}(x + \ulcorner 1\urcorner)).$$

More generally, if A is any assertion in the metatheory about recursive predicates, we can, using our representing formulas, write a corresponding sentence A^* in the language of set theory. We would expect that if we can prove A by a finitistic argument, then $\text{ZF} \vdash A^*$, since ZF should incorporate all finitistic methods, and much more (this presupposes a "reasonable" choice of the representing formulas—see Exercise 22). Unlike Level 1, Level 2 does not lend itself to a precise theorem, since the notion of finitistic is not rigorously defined, but we may verify Level 2 assertions individually as needed. In particular, Gödel's Second Incompleteness Theorem requires checking that all the basic syntactical results of elementary logic may be proved within ZF.

We may now state the basic result behind the Second Incompleteness Theorem and Tarski's theorem on undefinability of truth.

14.2. THEOREM. *Gödel. If $\phi(x)$ is any formula in one free variable, x, then there is a sentence ψ such that*

$$\text{ZF} \vdash \psi \leftrightarrow \phi(\ulcorner\psi\urcorner).$$

PROOF. Let $\sigma(v)$ be $\phi(v(\ulcorner v\urcorner))$. Then for each formula θ in one free variable, $\text{ZF} \vdash \sigma(\ulcorner\theta\urcorner) \leftrightarrow \phi(\ulcorner\theta(\ulcorner\theta\urcorner)\urcorner)$. In particular,

$$\text{ZF} \vdash \sigma(\ulcorner\sigma\urcorner) \leftrightarrow \phi(\ulcorner\sigma(\ulcorner\sigma\urcorner)\urcorner),$$

so let ψ be the sentence $\sigma(\ulcorner\sigma\urcorner)$.

The definition of σ might require some additional explanation. For each θ, we may add a defined constant $\ulcorner\theta\urcorner$ as above, but we shall consider the sentence $\theta(\ulcorner\theta\urcorner)$ to be in the original language of set theory, so we obtain $\theta(\ulcorner\theta\urcorner)$ by using Theorem 13.1 to eliminate the defined constant. The map $\theta \mapsto \theta(\ulcorner\theta\urcorner)$ is a recursive map of finite sequences, so is represented by a formula $\chi(v, w)$. Then for any sentence τ, τ is $\theta(\ulcorner\theta\urcorner)$ iff $\mathrm{ZF} \vdash \chi(\ulcorner\theta\urcorner, \ulcorner\tau\urcorner)$, so $\mathrm{ZF} \vdash \chi(\ulcorner\theta\urcorner, \ulcorner\theta(\ulcorner\theta\urcorner)\urcorner)$. Also, $\mathrm{ZF} \vdash \forall v\, \exists! w\, \chi(v, w)$ (this is an example of Level 2), so for each θ,

$$\mathrm{ZF} \vdash \forall w\,(\chi(\ulcorner\theta\urcorner, w) \leftrightarrow w = \ulcorner\theta(\ulcorner\theta\urcorner)\urcorner).$$

Let $\sigma(v)$ be $\exists w\,(\chi(v, w) \wedge \phi(w))$. Then for each θ,

$$\mathrm{ZF} \vdash \sigma(\ulcorner\theta\urcorner) \leftrightarrow \exists w\,(\chi(\ulcorner\theta\urcorner, w) \wedge \phi(w)) \quad \text{so that}$$

$$\mathrm{ZF} \vdash \sigma(\ulcorner\theta\urcorner) \leftrightarrow \phi(\ulcorner\theta(\ulcorner\theta\urcorner)\urcorner). \quad \square$$

If $\chi(x)$ is any formula in one free variable, we may apply 14.2 with $\neg\chi(x)$ to get Tarski's theorem on non-definability of truth; namely, for some sentence ψ,

$$\mathrm{ZF} \vdash \psi \leftrightarrow \neg\chi(\ulcorner\psi\urcorner).$$

The platonistic interpretation of this is that no formula $\chi(x)$ can say "x is a true sentence", since there is always a sentence ψ which is true iff $\chi(\ulcorner\psi\urcorner)$ is false.

We now discuss the Second Incompleteness Theorem. Fix a recursive extension, T, of ZF. Let $\chi^T_{\mathrm{pf}}(v, w)$ represent the predicate "Υ is a formal proof of ψ from T," so that for each ψ and Υ, $\mathrm{ZF} \vdash \chi^T_{\mathrm{pf}}(\ulcorner\Upsilon\urcorner, \ulcorner\psi\urcorner)$ iff Υ is a formal proof of ψ from T. Let $\varphi^T(w)$ be $\neg\exists v\, \chi^T_{\mathrm{pf}}(v, w)$; then $\varphi^T(w)$ says, "w is not provable from T". Finally, let CON^T be $\varphi^T(\ulcorner\exists v\,(v \neq v)\urcorner)$; then CON^T asserts that a specific logically refutable sentence is not provable from T; equivalently, that T is consistent.

14.3. THEOREM. *Gödel's Second Incompleteness Theorem. If T is a recursive consistent extension of* ZF, *then* $T \nvdash \mathrm{CON}^T$.

PROOF. Apply Theorem 14.2 to produce a ψ such that $\mathrm{ZF} \vdash (\psi \leftrightarrow \varphi^T(\ulcorner\psi\urcorner))$. Note first that

$$\mathrm{Con}(T) \rightarrow T \nvdash \psi. \tag{$*$}$$

To see this, suppose $T \vdash \psi$, and let Υ be a formal proof of ψ from T. Then $\mathrm{ZF} \vdash \chi^T_{\mathrm{pf}}(\ulcorner\Upsilon\urcorner, \ulcorner\psi\urcorner)$, so $\mathrm{ZF} \vdash \neg\varphi^T(\ulcorner\psi\urcorner)$, so $\mathrm{ZF} \vdash \neg\psi$, so $T \vdash \neg\psi$, so T is inconsistent.

The proof of (*) took place in the metatheory, but one can, by a careful Level 2 analysis, formalize it in ZF to yield

$$\mathrm{ZF} \vdash \mathrm{CON}^T \rightarrow \varphi^T(\ulcorner \psi \urcorner).$$

But $\varphi^T(\ulcorner \psi \urcorner)$ and ψ are provably equivalent in ZF and T is a stronger theory, so

$$T \vdash \mathrm{CON}^T \rightarrow \psi.$$

It follows that we may replace ψ by CON^T in (*) to yield

$$\mathrm{Con}(T) \rightarrow (T \nvdash \mathrm{CON}^T). \qquad \square$$

Actually, in the above,

$$T \vdash \mathrm{CON}^T \leftrightarrow \psi,$$

since, within T, we may argue that $\varphi^T(\ulcorner \psi \urcorner)$, which is equivalent to ψ, implies that some sentence is not provable from T, so that T must be consistent. Any sentence ψ which asserts its own non-provability from T is called a Gödel sentence for T. All Gödel sentences for T are provably from T (actually, from ZF) equivalent to CON^T and hence to each other.

Platonistically, the Second Incompleteness Theorem, like the First, shows that no recursive set of axioms can capture all truth. If T is a set of axioms and we can recognize that all the axioms of T are true, then all statements provable from T are true; thus, T is consistent, so CON^T is a true statement which is not provable from T. Actually, it is not necessary for T to extend all of ZF here; it is only required that within T one can develop enough finite combinatorics to prove basic facts about first-order predicate calculus (see Exercise 23).

EXERCISES

Work in ZFC unless otherwise indicated. Foundation is never relevant, but AC occasionally is.

(1) Write a formula expressing $z = \langle\langle x, y\rangle, \langle v, w\rangle\rangle$ using just $\in$ and $=$.

(2) Show that $\alpha < \beta$ implies that $\gamma + \alpha < \gamma + \beta$ and $\alpha + \gamma \leq \beta + \gamma$. Give an example to show that the "$\leq$" cannot be replaced by "$<$". Also, show:

$$\alpha \leq \beta \rightarrow \exists! \delta (\alpha + \delta = \beta).$$

(3) Show that if $\gamma > 0$, then $\alpha < \beta$ implies that $\gamma \cdot \alpha < \gamma \cdot \beta$ and $\alpha \cdot \gamma \leq \beta \cdot \gamma$. Give an example to show that the "$\leq$" cannot be replaced by "$<$". Also,

show:

$$(\alpha \le \beta \wedge \alpha > 0) \rightarrow \exists! \delta, \xi (\xi < \alpha \wedge \alpha \cdot \delta + \xi = \beta).$$

(4) Verify that ordinal exponentiation satisfies:

$$\alpha^{\beta + \gamma} = \alpha^{\beta} \cdot \alpha^{\gamma} \quad \text{and} \quad (\alpha^{\beta})^{\gamma} = \alpha^{\beta \cdot \gamma}.$$

(5) Let α be a limit ordinal. Show that the following are equivalent:
(a) $\forall \beta, \gamma < \alpha (\beta + \gamma < \alpha)$.
(b) $\forall \beta < \alpha (\beta + \alpha = \alpha)$.
(c) $\forall X \subset \alpha (\text{type}(X) = \alpha \vee \text{type}(\alpha \smallsetminus X) = \alpha)$.
(d) $\exists \delta (\alpha = \omega^{\delta})$ (ordinal exponentiation).
Such α are called *indecomposable*.

(6) Prove the Cantor Normal Form Theorem for ordinals: Every non-0 ordinal α may be represented in the form:

$$\alpha = \omega^{\beta_1} \cdot l_1 + \cdots + \omega^{\beta_n} \cdot l_n,$$

where $1 \le n < \omega$, $\alpha \ge \beta_1 > \cdots > \beta_n$, and $1 \le l_i < \omega$ for $i = 1, \ldots, n$. Furthermore, this representation is unique. α is called an *epsilon number* iff $n = 1$, $l_1 = 1$, and $\beta_1 = \alpha$ (i.e., $\omega^{\alpha} = \alpha$). Show that if κ is an uncountable cardinal, then κ is an epsilon number and there are κ epsilon numbers below κ; in particular, the first epsilon number, called ϵ_0, is countable. All exponentiation is ordinal exponentiation in this exercise.

(7) Prove that the following definition of ordinal exponentiation is equivalent to Definition 9.5: Let

$$F(\alpha, \beta) = \{ f \in {}^{\beta}\alpha : |\{\xi : f(\xi) \ne 0\}| < \omega \}.$$

If $f, g \in F(\alpha, \beta)$ and $f \ne g$, say $f \lhd g$ iff $f(\xi) < g(\xi)$, where ξ is the largest ordinal such that $f(\xi) \ne g(\xi)$. Then $\alpha^{\beta} = \text{type}(\langle F(\alpha, \beta), \lhd \rangle)$.

(8) IN ZF^-, prove the Schröder–Bernstein Theorem (10.2).
Hint. Assume $f : A \rightarrow B$ and $g : B \rightarrow A$, where f and g are 1–1. Let $A_0 = A$, $B_0 = B$, $A_{n+1} = g''B_n$, $B_{n+1} = f''A_n$, $A_{\infty} = \bigcap_n A_n$, and $B_{\infty} = \bigcap_n B_n$. Let $h(x)$ be $f(x)$ if $x \in A_{\infty} \cup \bigcup_n (A_{2n} \smallsetminus A_{2n+1})$; otherwise $h(x) = g^{-1}(x)$. Then $h : A \rightarrow B$ and h is 1–1 and onto.

(9) Show in ZF^- that for any set X the following are equivalent.
(a) X can be well-ordered.
(b) There is a $C : (\mathscr{P}(X) \smallsetminus \{0\}) \rightarrow X$ such that $\forall Y \subset X (Y \ne 0 \rightarrow C(Y) \in Y)$.
Hint for (b) $\rightarrow$ (a). Fix $p \ne X$, and let $C(Y) = p$ if $Y \notin \mathscr{P}(X) \smallsetminus \{0\}$. Define,

by transfinite recursion,

$$F(\alpha) = C(X \setminus \{F(\xi); \xi < \alpha\}).$$

(10) Show, in ZF^-, that the following are equivalent.

(a) AC.

(b) $\forall \mathscr{S} \, [0 \notin \mathscr{S} \rightarrow \exists C (C : \mathscr{S} \rightarrow \bigcup \mathscr{S} \wedge \forall Y \in \mathscr{S} (C(Y) \in Y))]$.

(c) The cartesian product of non-empty sets is non-empty.

(d) The Tychonov Theorem.

Hint for (d) $\rightarrow$ (c) (Kelley). Let $A_i \, (i \in I)$ be non-0 sets. Fix $p \notin \bigcup_i A_i$. Let $X_i = A_i \cup \{p\}$, where neighborhoods of p in X_i are cofinite. Use compactness of $\prod_i X_i$ to prove that $\prod_i A_i \neq 0$. *Remark*. The Tychonov Theorem for compact Hausdorff spaces does not imply AC (Halpern–Lévy; see [Jech 1973]).

(11) $\mathscr{F} \subset \mathscr{P}(A)$ is of *finite character* iff for all $X \subset A$,

$$X \in \mathscr{F} \leftrightarrow \forall Y \subset X \, (|Y| < \omega \rightarrow Y \in \mathscr{F}).$$

Tuckey's Lemma (a form of *Zorn's Lemma*) says that whenever $\mathscr{F}$ is of finite character,

$$\forall X \in \mathscr{F} \, \exists Y \in \mathscr{F} \, (X \subset Y \wedge \forall Z \in \mathscr{F} \, (Y \subset Z \rightarrow Y = Z)).$$

Show, in ZF^-, that AC $\leftrightarrow$ Tuckey's Lemma.
Hint. For $\rightarrow$, let $A = \{a_\xi : \xi < \kappa\}$, and put $a_\xi \in Y$ iff

$$(X \cup \{a_\eta : \eta < \xi \wedge a_\eta \in Y\} \cup \{a_\xi\}) \in \mathscr{F}.$$

For $\leftarrow$, see 10(b); let C be a maximal partial choice function on $\mathscr{S}$.

(12) Define, in ZF^-, $\aleph(X) = \sup(\{\alpha : \exists f \in {}^{\alpha}X \, (f \text{ is } 1\text{–}1)\})$ (Hartogg's aleph function). Show

*(a) $\aleph(X) < \aleph(\mathscr{P}(\mathscr{P}(\mathscr{P}(X))))$ ($\aleph(X) < \mathscr{P}^4(X)$ is easier).

(b) There is no sequence $\langle X_n : n \in \omega \rangle$ such that $\forall n \, (\mathscr{P}(X_{n+1}) \preccurlyeq X_n)$.

(c) AC implies that $\aleph(X) = |X|^+$ whenever X is infinite.

(13) Show that for $n, m \in \omega$, the ordinal and cardinal exponentiations n^m are equal. *Hint*. Use induction on m.

(14) For infinite cardinals $\lambda \leq \kappa$, show $|\{X \subset \lambda : |X| = \lambda\}| = \kappa^\lambda$.

(15) When λ is an infinite cardinal and κ is any cardinal, show

$$\kappa^{<\lambda} = \sup\{\kappa^\theta : \theta < \lambda \wedge \theta \text{ is a cardinal}\}.$$

*(16) Assume CH but don't assume GCH. Show that $(\omega_n)^\omega = \omega_n$ whenever $1 \leq n < \omega$.

(17) Show that the following sets have cardinality 2^ω:
(a) $\mathbb{R}$.
(b) $\{f \in \mathbb{R}^{\mathbb{R}} : f \text{ is continuous}\}$.
*(c) $\{X \subset \mathbb{R} : X \text{ is Borel}\}$.

(18) If κ_i are cardinals for $i \in I$, define

$$\textstyle\sum_i \kappa_i = \left|\bigcup \{\{i\} \times \kappa_i : i \in I\}\right|$$

and

$$\textstyle\prod_i \kappa_i = \left|\{f : \operatorname{dom}(f) = I \wedge \forall i\,(f(i) \in \kappa_i)\}\right|.$$

(a) Show that

$$\forall i \in I(\theta_i < \kappa_i) \rightarrow \textstyle\sum_i \theta_i < \prod_i \kappa_i.$$

(b) Derive König's Lemma (10.40) directly from (a).
Hint. For (a), generalize the proof of 10.40. For (b), each $\kappa_i = \kappa$, $I = \lambda$, and $\{\theta_i ; i \in \lambda\}$ is unbounded in κ.

*(19) Let κ be an infinite cardinal and $\lhd$ any well-ordering of κ. Show that there is an $X \subset \kappa$ such that: $|X| = \kappa$, and $\lhd$ and $<$ agree on X.

*(20) Prove the Rado–Milner Paradox: If $\kappa \leq \alpha < \kappa^+$, then there are $X_n \subset \alpha$ $(n \in \omega)$ such that $\alpha = \bigcup_n X_n$ and $\operatorname{type}(X_n) \leq \kappa^n$.
Hint. Use induction on α.

(21) Algebraic topology and homological algebra rely on categories (which are proper classes) to prove theorems about sets (spaces, rings, etc.). Show how to formalize these subjects within ZFC.

(22) Assume Con(ZF). Show that there is a formula $\psi(x)$ such that ψ represents $\{n : n \text{ is even}\}$ (in the sense of Theorem 14.1) but

$$\mathrm{ZF} \nvdash \forall x \in \omega\,(\psi(x) \vee \psi(x + \ulcorner 1 \urcorner)).$$

(23) Show that Gödel Incompleteness Theorems and Tarski's theorem on undefinability of truth go through for any recursive extension of $\mathrm{ZF}^- - \mathrm{P} - \mathrm{Inf}$.

(24) Let T be any consistent set of axioms extending ZF. Show that $\{\psi: T \vdash \psi\}$ is not recursive. *Hint.* If it were recursive, then, by Theorem 14.1, there would be a formula $\chi(x)$ such that

$$(T \vdash \psi) \rightarrow (\mathrm{ZF} \vdash \chi(\ulcorner \psi \urcorner))$$

and

$$(T \nvdash \psi) \rightarrow (\mathrm{ZF} \vdash \neg \chi(\ulcorner \psi \urcorner))$$

for any ψ. Now, fix ψ (by Theorem 14.2) such that $\mathrm{ZF} \vdash \psi \leftrightarrow \neg \chi(\ulcorner \psi \urcorner)$, and show that T is inconsistent,

(25) Show that MK (see §12) proves the consistency of ZF, and is thus not a conservative extension of ZF. *Hint.* Formalize in MK definition of satisfaction for formulas with set variables.

(26) Show that MK is not finitely axiomatizable. *Hint.* Show that MK proves the consistency of any finite subtheory of itself.

CHAPTER II

INFINITARY COMBINATORICS

By infinitary combinatorics we mean the field that used to be called set theory before there were independence proofs. Elementary examples of this subject are the results in cardinal arithmetic discussed in I §10, but those are only the beginning.

There is a two-way interplay between combinatorics and independence proofs. On the one hand, the answers to many combinatorial questions which had been raised classically are now known to be independent of ZFC by the methods of Chapters VI–VIII. On the other hand, classical combinatorial facts are often used in carrying out the independence proofs. This interplay of ideas has also resulted in the creation of new combinatorial concepts, such as $\diamondsuit$ (§7) and MA (§2), which could have been discovered classically but in fact were not.

The material covered in this chapter has been selected for its relevance to Chapters VI–VIII, so many important topics in combinatorics per se have been omitted. For more on the subject, see [Erdös–Hajnal–Máté–Rado 1900] or [Kunen 1977].

The reader who is not particularly interested in combinatorics may skip this entire chapter, referring back to it as needed. The only part of this chapter needed later in a fundamental way is §2, on MA, since that forms the basis for our treatment of forcing in VII; and even there the technical details about measure theory and topology may be omitted.

We may think of this entire chapter as based on the axioms ZFC$^-$. The Axiom of Foundation is, as always in mathematics, totally irrelevant.

§ 1. Almost disjoint and quasi-disjoint sets

1.1. DEFINITION. Let κ be an infinite cardinal. If $x, y \subset \kappa$, x and y are *almost disjoint* (a.d.) iff $|x \cap y| < \kappa$. An a.d. *family* is an $\mathscr{A} \subset \mathscr{P}(\kappa)$ such that $\forall x \in \mathscr{A}(|x| = \kappa)$ and any two distinct elements of $\mathscr{A}$ are a.d. A *maximal* a.d. *family* (m.a.d.f.) is an a.d. family $\mathscr{A}$ with no a.d. family $\mathscr{B}$ properly containing it. $\square$

Almost disjointness is similar to disjointness in name only. It is clear that there can be no family of more than κ disjoint subsets of κ; but, since $|\kappa| = |\kappa \times \kappa|$, there is a family of κ disjoint subsets whose union is all of κ, so that this family is a maximal disjoint family. However, the following theorem holds.

1.2. THEOREM. *Let $\kappa \geq \omega$ be a regular cardinal, then:*

(a) *If $\mathscr{A} \subset \mathscr{P}(\kappa)$ is an* a.d. *family and $|\mathscr{A}| = \kappa$, then $\mathscr{A}$ is not maximal.*

(b) *There is a* m.a.d.f. *$\mathscr{B} \subset \mathscr{P}(\kappa)$ of cardinality $\geq \kappa^+$.*

PROOF. (b) is immediate from (a) and Zorn's Lemma: Let $\mathscr{A} \subset \mathscr{P}(\kappa)$ be any disjoint (or a.d.) family of size κ, and let $\mathscr{B} \supset \mathscr{A}$ be a m.a.d.f:, then $|\mathscr{B}| > \kappa$ by (a).

(a) is proved by a diagonal argument. Let $\mathscr{A} = \{A_\xi : \xi < \kappa\}$. Let $B_\xi = A_\xi \setminus \bigcup_{\eta<\xi} A_\eta$. $B_\xi \neq 0$, since $B_\xi = A_\xi \setminus \bigcup_{\eta<\xi}(A_\xi \cap A_\eta)$, $|A_\xi| = \kappa$, and $|\bigcup_{\eta<\xi}(A_\xi \cap A_\eta)| < \kappa$ by regularity of κ. Pick $\beta_\xi \in B_\xi$. The β_ξ are distinct since the B_ξ are disjoint, so $D = \{\beta_\xi : \xi < \kappa\}$ has cardinality κ. $\beta_\xi \in A_\eta \to \eta \geq \xi$, so $D \cap A_\eta \subset \{\beta_\xi : \xi \leq \eta\}$, so D and A_η are a.d. for each η. $\square$

If one does not assume GCH, Theorem 1.2 suggests a number of questions. First, is there an a.d. family of 2^κ subsets of κ? The answer is yes if $\kappa = \omega$. It is still yes if $\kappa = \omega_1$ if one assumes CH (but 2^{ω_1} can be anything). More generally, the following holds.

1.3. THEOREM. *If $\kappa \geq \omega$ and $2^{<\kappa} = \kappa$, then there is an* a.d. *family $\mathscr{A} \subset \mathscr{P}(\kappa)$ with $|\mathscr{A}| = 2^\kappa$.*

PROOF. Let $I = \{x \subset \kappa : \sup(x) < \kappa\}$. Since $2^{<\kappa} = \kappa$, $|I| = \kappa$. If $X \subset \kappa$, let $A_X = \{X \cap \alpha : \alpha < \kappa\}$. If $|X| = \kappa$, then $|A_X| = \kappa$. If $X \neq Y$, then $|A_X \cap A_Y| < \kappa$, since if we fix β such that $\neg(\beta \in X \leftrightarrow \beta \in Y)$, then

$$A_X \cap A_Y \subset \{X \cap \alpha : \alpha \leq \beta\}.$$

Let $\mathscr{A} = \{A_X : X \subset \kappa \wedge |X| = \kappa\}$; then $|\mathscr{A}| = 2^\kappa$ and is an a.d. family of subsets of I. If we let f be a 1–1 function from I onto κ, $\{f''A : A \in \mathscr{A}\}$ is an a.d. family of 2^κ subsets of κ. $\square$

The hypothesis $2^{<\kappa} = \kappa$ cannot be dropped; the existence of an a.d. family of 2^{ω_1} subsets of ω_1 can be neither proved nor refuted from $2^\omega = 2^{\omega_1} = \omega_3$ (see VIII Exercise B5).

Another question when $2^\kappa > \kappa^+$ is: is there a m.a.d.f. of cardinality κ^+? The answer to this question is independent of the axioms of set theory (see §2 and VIII 2.3).

Another property with "disjoint" in its name is "quasi-disjoint".

1.4. DEFINITION. A family $\mathscr{A}$ of sets is called a *Δ-system*, or a *quasi-disjoint* family iff there is a fixed set r, called the *root* of the Δ-system, such that $a \cap b = r$ whenever a and b are distinct members of $\mathscr{A}$ (see Figure 1.1). □

The main result on Δ-systems is the so-called "Δ-system lemma", which we state first in the special case most often quoted.

1.5. THEOREM. *If $\mathscr{A}$ is any uncountable family of finite sets, there is an uncountable $\mathscr{B} \subset \mathscr{A}$ which forms a Δ-system.* □

This is an immediate corollary of the following theorem when $\kappa = \omega$ and $\theta = \omega_1$. For an easier direct proof of 1.5, see Exercise 1.

1.6. THEOREM. *Let κ be any infinite cardinal. Let $\theta > \kappa$ be regular and satisfy $\forall \alpha < \theta(|\alpha^{<\kappa}| < \theta)$. Assume $|\mathscr{A}| \geq \theta$ and $\forall x \in \mathscr{A}(|x| < \kappa)$, then there is a $\mathscr{B} \subset \mathscr{A}$, such that $|\mathscr{B}| = \theta$ and $\mathscr{B}$ forms a Δ-system.*

PROOF. By shrinking $\mathscr{A}$ if necessary, we may assume $|\mathscr{A}| = \theta$. Then $|\bigcup \mathscr{A}| \leq \theta$. Since what the elements of $\mathscr{A}$ are as individuals is irrelevant, we may assume $\bigcup \mathscr{A} \subset \theta$. Then each $x \in \mathscr{A}$ has some order type $<\kappa$ as a subset of θ. Since θ is regular and $\theta > \kappa$, there is some $\rho < \kappa$, such that $\mathscr{A}_1 = \{x \in \mathscr{A} : x \text{ has type } \rho\}$ has cardinality θ. We now fix such a ρ and deal only with $\mathscr{A}_1$.

For each $\alpha < \theta$, $|\alpha^{<\kappa}| < \theta$ implies that less than θ elements of $\mathscr{A}_1$ are subsets of α. Thus, $\bigcup \mathscr{A}_1$ is unbounded in θ. If $x \in \mathscr{A}_1$ and $\xi < \rho$, let $x(\xi)$ be the ξ-th element of x. Since θ is regular, there is some ξ such that $\{x(\xi) : x \in \mathscr{A}_1\}$ is unbounded in θ. Now fix ξ_0 to be the least such ξ (ξ_0 may be 0). Let

$$\alpha_0 = \sup\{x(\eta) + 1 : x \in \mathscr{A}_1 \wedge \eta < \xi_0\};$$

then $\alpha_0 < \theta$ and $x(\eta) < \alpha_0$ for all $x \in \mathscr{A}_1$ and all $\eta < \xi_0$.

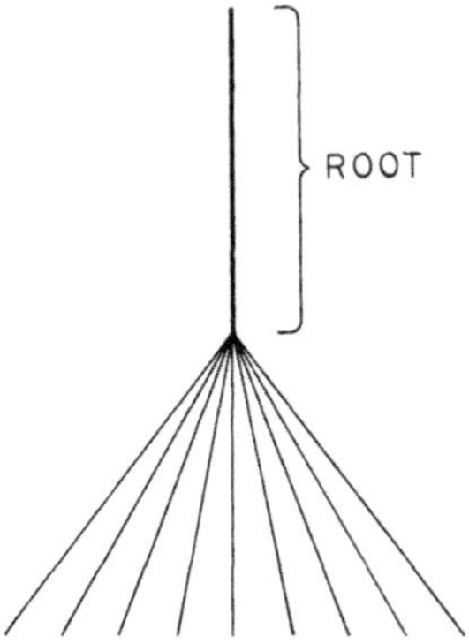

Figure 1.1. A Δ-System.

By transfinite recursion on $\mu < \theta$, pick $x_\mu \in \mathscr{A}_1$ so that $x_\mu(\xi_0) > \alpha_0$ and $x_\mu(\xi_0)$ is above all elements of earlier x_ν; i.e.,

$$x_\mu(\xi_0) > \max(\alpha_0, \sup\{x_\nu(\eta): \eta < \rho \wedge \nu < \mu\}).$$

Let $\mathscr{A}_2 = \{x_\mu: \mu < \theta\}$. Then $|\mathscr{A}_2| = \theta$ and $x \cap y \subset \alpha_0$ whenever x and y are distinct elements of $\mathscr{A}_2$. Since $|\alpha_0^{<\kappa}| < \theta$, there is an $r \subset \alpha_0$ and a $\mathscr{B} \subset \mathscr{A}_2$ with $|\mathscr{B}| = \theta$ and $\forall x \in \mathscr{B}(x \cap \alpha_0 = r)$, whence $\mathscr{B}$ forms a Δ-system with root r. □

We conclude this section with an application of Theorem 1.5 to topology.

1.7. Definition. A topological space X has the *countable chain condition* (c.c.c.) iff there is no uncountable family of pairwise disjoint non-empty open subsets of X. □

The word "chain" usually refers to a total ordering; for an equivalent of c.c.c. in terms of chains, see Exercise 6.

If X is countable, X is trivially c.c.c. If X is uncountable then X may fail to be c.c.c.; for example, if X has the discrete topology, then the singletons form an uncountable disjoint family of open sets.

A non-trivial example of a c.c.c. space is the real numbers, $\mathbb{R}$. $\mathbb{R}$ has c.c.c. since it is *separable*; that is, it has a countable dense subspace. In general, the following holds.

1.8. Lemma. *If X is separable, then X has* c.c.c.

Proof. Let $D \subset X$ be dense and countable. If U_α $(\alpha < \omega_1)$ were open, non-empty, and disjoint, then we could pick $d_\alpha \in U_\alpha \cap D$. Then the d_α would all be distinct, a contradiction. □

A natural question to ask about a topological property is whether it is preserved under products. For example, the famous Tychonov theorem says that compactness is preserved under arbitrary products. The product of two separable spaces is separable since if D is dense in X and E is dense in Y then $D \times E$ is dense in $X \times Y$. However, separability is not preserved under arbitrary products (it is if there are $\leq 2^\omega$ factors, but not if there are $(2^\omega)^+$ factors; see Exercises 3 and 4).

The question of whether the product of any two c.c.c. spaces must be c.c.c. is independent of ZFC; this is true under MA + $\neg$CH (see 2.24), but it fails if CH holds (see VIII Exercise 8) or if Suslin's Hypothesis fails (see Lemma 4.3). The c.c.c. has the strange property that if it is preserved by products with two factors, it is preserved by arbitrary products. To see this, first note that if the product of any two c.c.c. spaces is c.c.c., then by induction we see that the product of any n c.c.c. spaces is c.c.c. ($n \in \omega$). We now pass immediately to arbitrary products by the following.

1.9. THEOREM. *Suppose X_i ($i \in I$) are spaces such that for every finite $r \subset I$, $\prod_{i\in r} X_i$ is c.c.c. Then $\prod_{i\in I} X_i$ is c.c.c.*

PROOF. Suppose U_α for $\alpha < \omega_1$ were pairwise disjoint non-empty sets in $\prod_{i\in I} X_i$. By shrinking the U_α if necessary, we may assume that each U_α is a basic open set. Then U_α depends on a finite set of coordinates, $a_\alpha \subset I$. By the Δ-system lemma there is an uncountable $A \subset \omega_1$ such that $\{a_\alpha : \alpha \in A\}$ forms a Δ-system with some root, r. r cannot be empty, since $a_\alpha \cap a_\beta = 0$ implies $U_\alpha \cap U_\beta \neq 0$. Let $\pi(U_\alpha)$ be the projection of U_α onto $\prod_{i\in r} X_i$. Then the $\pi(U_\alpha)$ for $\alpha \in A$ form an uncountable disjoint family in $\prod_{i\in r} X_i$. □

A simple example of Theorem 1.9 is that ${}^{\kappa}2$ is c.c.c. for any κ, where 2 is the space $\{0, 1\}$ with the discrete topology. When $\kappa > 2^\omega$, this is an example of a c.c.c. space which is not separable (by Exercise 4).

§ 2. Martin's Axiom

If CH fails, there arises a large number of questions about the various infinite cardinals $\kappa < 2^\omega$. Some of these questions are of a purely combinatorial nature, such as

Question 1. If $\kappa < 2^\omega$, does $2^\kappa = 2^\omega$?

Question 2. If $\kappa < 2^\omega$, does every a.d. family $\mathscr{A} \subset \mathscr{P}(\omega)$ of size κ fail to be maximal?

Other questions come from measure theory, such as

Question 3. If $\kappa < 2^\omega$, does the union of κ subsets of $\mathbb{R}$, each of Lebesgue measure 0, have measure 0?

There is also the analogous question about category:

Question 4. If $\kappa < 2^\omega$, does the union of κ first-category subsets of $\mathbb{R}$ have first category?

Since these questions all clearly have answer "yes" when $\kappa = \omega$, they are only of interest for $\omega < \kappa < 2^\omega$. For such κ, it is known by the method of forcing that none of the questions can be settled under the axioms ZFC + $\neg$CH.

It is commonly felt that the "yes" answer to Questions 1–4 is the reasonable one for all $\kappa < 2^\omega$; i.e., that all infinite cardinals $< 2^\omega$ behave similarly to ω. In the case of Question 1, the consistency of "yes" with $\neg$CH follows from the original construction by Cohen (VII §5). For Questions 2–4, the consistency of "yes" with $\neg$CH is much more difficult, but was proved by Solovay by applying the technique of iterated forcing developed by him and Tennenbaum. It was then noticed by Martin that one could amalgamate all these "yes" answers into one axiom, now called Martin's Axiom (MA). One may apply iterated forcing once to prove the consistency of MA + $\neg$CH (see VIII), and then derive "yes" answers to Questions 1–4 from MA by purely combinatorial arguments.

Since MA itself can be understood without any knowledge of forcing, many people have worked on it, and MA is known to have a large number of important consequences, primarily in combinatorics and set-theoretic topology, but also in algebra and analysis. For examples other than those given here, see [Martin–Solovay 1970], [Rudin 1977], or [Shoenfield 1975].

Unlike the basic axioms of ZFC, MA does not pretend to be an "intuitively evident" principle, and in fact at first sight it seems strange and ill-motivated. Its original motivation grew out of the technical details of certain forcing arguments, although in this book we shall attempt to motivate forcing by our treatment of MA.

MA can easily be defined topologically as the assertion that no compact Hausdorff space with the c.c.c. is the union of $< 2^\omega$ closed nowhere dense sets. Unfortunately, this form of MA is hard to apply directly. Our official version of MA, in terms of partial orders, is harder to comprehend but easier to use once understood. For a proof that the two versions are equivalent, see Theorem 3.4.

We begin by fixing our notation for partial orders. It will be more convenient here to take the $\leq$, rather than the $<$, as basic.

2.1. Definition. (a) A *partial order* is a pair $\langle \mathbb{P}, \leq \rangle$ such that $\mathbb{P} \neq 0$ and $\leq$ is a relation on $\mathbb{P}$ which is transitive and reflexive $(\forall p \in \mathbb{P}(p \leq p))$. $p \leq q$ is read "*p extends q*". Elements of $\mathbb{P}$ are called *conditions*.

(b) $\langle \mathbb{P}, \leq \rangle$ is a partial order in the *strict sense* iff it in addition satisfies

$$\forall p, q\,(p \leq q \wedge q \leq p \rightarrow p = q).$$

In that case, define $p < q$ iff $p \leq q \wedge p \neq q$. $\square$

We often abuse notation by referring to "the partial order $\mathbb{P}$" or "the partial order $\leq$" if $\leq$ or $\mathbb{P}$ is clear from context.

A trivial example of a partial order on $\mathbb{P}$ is $\mathbb{P} \times \mathbb{P}$; i.e., $\forall p, q\,(p \leq q)$. However, most of our partial orders will be partial orders in the strict sense;

in that case, observe that $<$ is transitive and irreflexive $(\forall p(p \not< p))$, and $p \leq q$ iff $p = q \vee p < q$. It will eventually be useful (see VIII §5) to have developed the theory of general (non-strict) partial orders, and that additional generality will cause us no additional work here.

2.2. DEFINITION. Let $\langle \mathbb{P}, \leq \rangle$ be a partial order. A *chain* in $\mathbb{P}$ is a set $C \subset \mathbb{P}$ such that $\forall p, q \in C (p \leq q \cup q \leq p)$. p and q are *compatible* iff

$$\exists r \in \mathbb{P} (r \leq p \wedge r \leq q);$$

they are *incompatible* $(p \perp q)$ iff $\neg \exists r \in \mathbb{P} (r \leq p \wedge r \leq q)$. An *antichain* in $\mathbb{P}$ is a subset $A \subset \mathbb{P}$ such that $\forall p, q \in A (p \neq q \rightarrow p \perp q)$. □

2.3. DEFINITION. A partial order $\langle \mathbb{P}, \leq \rangle$ has the *countable chain condition* (c.c.c.) iff every antichain in $\mathbb{P}$ is countable. □

Example 1. $\mathbb{P} = \omega_1$ with the usual order. Every subset of $\mathbb{P}$ is a chain (i.e., is totally ordered), but every antichain has cardinality ≤ 1, so $\mathbb{P}$ has the c.c.c.

Example 2. Let X be any non-empty set and $\mathbb{P} = \mathscr{P}(X) \setminus \{0\}$, with $p \leq q \leftrightarrow p \subset q$; then $p \perp q$ iff $p \cap q = 0$. $A \subset \mathbb{P}$ is an antichain iff the elements of A are pairwise disjoint, so $\mathbb{P}$ has the c.c.c. iff $|X| \leq \omega$.

Example 3. Let X be any topological space and $\mathbb{P} = \{p \subset X : p \text{ is open} \wedge p \neq 0\}$, with $p \leq q \leftrightarrow p \subset q$. As in Example 2, $p \perp q$ if $p \cap q = 0$, so $\mathbb{P}$ has c.c.c. iff X does (Definition 1.7).

Example 4. Let $\mathscr{B}$ be any Boolean algebra, and $\mathbb{P} = \mathscr{B} \setminus \{0\}$, with the same order as in $\mathscr{B}$; then $p \perp q$ iff $p \wedge q = 0$.

Since the notion of c.c.c. involves antichains rather than chains, it should really be called c.a.c., and is in some of the literature. In Examples 3 and 4, c.c.c. is related to *well-ordered* chains (see Exercises 5 and 7).

Examples 1–4 were intended primarily to illustrate Definitions 2.1–2.3, although Examples 3 and 4 become important in the more abstract discussion of MA, beginning with Theorem 2.22. We shall give a fifth example, which has more of the flavor of typical MA applications, after we say what MA is.

2.4. DEFINITION. Let $\langle \mathbb{P}, \leq \rangle$ be a partial order. $D \subset \mathbb{P}$ is *dense* in $\mathbb{P}$ iff $\forall p \in \mathbb{P}\, \exists q \leq p (q \in D)$. $G \subset \mathbb{P}$ is a *filter* in $\mathbb{P}$ iff

(a) $\forall p, q \in G\, \exists r \in G (r \leq p \wedge r \leq q)$, and

(b) $\forall p \in G\, \forall q \in \mathbb{P} (q \leq p \rightarrow q \in G)$. □

2.5. Definition. MA(κ) is the statement: Whenever $\langle \mathbb{P}, \leq \rangle$ is a non-empty c.c.c. partial order, and $\mathscr{D}$ is a family of $\leq \kappa$ dense subsets of $\mathbb{P}$, then there is a filter G in $\mathbb{P}$ such that $\forall D \in \mathscr{D} (G \cap D \neq 0)$. MA is the statement $\forall \kappa < 2^\omega (\text{MA}(\kappa))$. □

Intuitively, the conditions (elements of $\mathbb{P}$) say something about G or some object we plan to construct directly from G. If p extends q then p says more than q. The following example illustrates these ideas.

Example 5. Let $\mathbb{P}$ be the set of finite partial functions from ω to 2; formally,

$$\mathbb{P} = \{p : p \subset \omega \times 2 \wedge |p| < \omega \wedge p \text{ is a function}\}.$$

Let $p \leq q$ iff $q \subset p$ (iff p extends q as a function). p and q are compatible iff they agree on $\text{dom}(p) \cap \text{dom}(q)$, in which case $p \cup q$ is a common extension of both p and q. $\mathbb{P}$ clearly has c.c.c., since $|\mathbb{P}| = \omega$. If G is a filter in $\mathbb{P}$, then the elements of G are pairwise compatible, so if we let $f = f_G = \bigcup G$, then f_G is a function with $\text{dom}(f_G) \subset \omega$. In this example, we are more interested in f than G. If $p \in \mathbb{P}$, we think of p as a finite approximation to f, and we say intuitively that p forces "$p \subset f$" in the sense that if $p \in G$, then $p \subset f_G$. So p says what f is restricted to $\text{dom}(p)$. If $q \leq p$, then q says more about f than p does.

$\text{dom}(f_G)$ could be a very small subset of ω; for example, 0 is a filter and $\bigcup 0 = 0$, the empty function. However, by requiring that G intersect many dense sets, we can avoid this pathology and make f very "generic" (or representative of typical functions). For $n \in \omega$, let $D_n = \{p \in \mathbb{P} : n \in \text{dom}(p)\}$. Since any $p \in \mathbb{P}$ can be extended to a condition with n in its domain, D_n is dense. If $\forall n \in \omega (G \cap D_n \neq 0)$, then f_G has as domain all of ω. Also, the "typical" function from ω to 2 would probably not be the constant function 0. Let $E = \{p \in \mathbb{P} : \exists n \in \text{dom}(p)(p(n) = 1)\}$. E is dense; and if $G \cap E \neq 0$, then f_G takes value 1 somewhere.

Likewise, a "typical" function from ω to 2 should not equal any function given in advance, a fact which leads to a contradiction from MA(2^ω). Thus, when $h : \omega \to 2$, let $E_h = \{p \in \mathbb{P} : \exists n \in \text{dom}(p)(p(n) \neq h(n))\}$. E_h is dense; and if $G \cap E_h \neq 0$, then $f_G \neq h$. Let $\mathscr{D} = \{D_n : n \in \omega\} \cup \{E_h : h \in {}^\omega 2\}$; then $|\mathscr{D}| = 2^\omega$. If G is a filter and $G \cap D \neq 0$ for all $D \in \mathscr{D}$, then f_G is a function from ω to 2 which differs from every function from ω to 2, which is impossible.

Leaving now this specific example, we come to the following.

2.6. Lemma. (a) *If* $\kappa < \kappa'$, *then* MA(κ') $\to$ MA(κ).

(b) MA(2^ω) *is false.*

(c) MA(ω) *is true.*

PROOF. (a) is clear and (b) was just done. For (c), let $\mathscr{D} = \{D_n : n \in \omega\}$ and define, by induction on n, $p_n \in \mathbb{P}$ so that p_0 is an arbitrary element of $\mathbb{P}$ (since $\mathbb{P} \neq 0$), and p_{n+1} is any extension of p_n such that $p_{n+1} \in D_n$; this is possible since D_n is dense. So $p_0 \geq p_1 \geq p_2 \geq \cdots$. Let G be the filter generated by $\{p_n : n \in \omega\}$; i.e., $G = \{q \in \mathbb{P} : \exists n(q \geq p_n)\}$; then G is a filter and $\forall n (G \cap D_n \neq 0)$. $\square$

By (a) and (b), MA—that is, $\forall \kappa < 2^\omega$ MA(κ)—is the strongest assertion of this type which is not outright inconsistent. By (c), MA follows from CH, a fact which is not surprising if we view MA as saying that all infinite cardinals $< 2^\omega$ have properties similar to ω. We shall show in VIII §6 that, roughly, MA is consistent with 2^ω being any regular cardinal, such as ω_2, ω_5, $\omega_{\omega+1}$, or the first weak inaccessible. MA does imply that 2^ω is regular (see Corollary 2.19).

Since the proof of Lemma 2.6 (c) did not need that $\mathbb{P}$ was c.c.c., one might attempt to strengthen MA(κ) by dropping this requirement. But, for $\kappa > \omega$, this strengthening becomes inconsistent by our next example.

Example 6. Let

$$\mathbb{P} = \{p : p \subset \omega \times \omega_1 \wedge |p| < \omega \wedge p \text{ is a function}\}.$$

This is like Example 5, but now our conditions talk about a function from ω to ω_1, and a suitably generic function would be onto, which is impossible. More formally, if G is a filter in $\mathbb{P}$, then as before, $\bigcup G$ is a function with $\text{dom}(\bigcup G) \subset \omega$ and $\text{ran}(\bigcup G) \subset \omega_1$. If $\alpha < \omega_1$, let $D_\alpha = \{p \in \mathbb{P} : \alpha \in \text{ran}(p)\}$; then D_α is dense. No filter G could intersect D_α for all $\alpha < \omega_1$, since that would mean $\text{ran}(\bigcup G) = \omega_1$. Of course $\mathbb{P}$ is not c.c.c., since the conditions $\{\langle 0, \alpha \rangle\}$ for $\alpha \in \omega_1$ are pairwise incompatible.

That concludes the elementary discussion of MA. We now proceed to show how MA is applied, and in particular, how it answers Questions 1–4 in the beginning of this section. Questions 1, 2, and 4 may all be answered with the same partial order, although the order we define will, at first sight, seem to be relevant only to Question 2.

2.7. DEFINITION. Let $\mathscr{A} \subset \mathscr{P}(\omega)$. The *almost disjoint sets partial order*, $\mathbb{P}_{\mathscr{A}}$, is

$$\{\langle s, F \rangle : s \subset \omega \wedge |s| < \omega \wedge F \subset \mathscr{A} \wedge |F| < \omega\},$$

where $\langle s', F' \rangle \leq \langle s, F \rangle$ iff

$$s \subset s' \wedge F \subset F' \wedge \forall x \in F (x \cap s' \subset s). \quad \square$$

Intuitively, the conditions $\langle s, F \rangle \in \mathbb{P}_{\mathscr{A}}$ are intended to describe a $d \subset \omega$

which is almost disjoint from the elements of $\mathscr{A}$. $\langle s, F\rangle$ "forces" that $s \subset d$ and $\forall x \in F\,(d \cap x \subset s)$, so $\langle s, F\rangle$ "forces" $n \notin d$ whenever $n \in x \smallsetminus s$ for some $x \in F$. This explains our definition of $\leq$. We extend $\langle s, F\rangle$ to $\langle s', F'\rangle$ by expanding s and F, but we are not allowed to put a number n into s' which was already "forced" by $\langle s, F\rangle$ not to be in d; so we require, for each $x \in F$, that $\forall n \in x \smallsetminus s\,(n \notin s')$, or $x \cap s' \subset s$.

2.8. Lemma. *In* $\mathbb{P}_{\mathscr{A}}$, $\langle s_1, F_1\rangle$ *and* $\langle s_2, F_2\rangle$ *are compatible iff*

$$\forall x \in F_1\,(x \cap s_2 \subset s_1) \wedge \forall x \in F_2\,(x \cap s_1 \subset s_2),$$

in which case $\langle s_1 \cup s_2, F_1 \bigcup F_2\rangle$ *is a common extension.* □

Lemma 2.8 is easily checked directly from the definition of $\leq$. The condition for compatibility is equivalent to

$$\forall x \in F_1\,\forall n \in x \smallsetminus s_1\,(n \notin s_2) \wedge \forall x \in F_2\,\forall n \in x \smallsetminus s_2\,(n \notin s_1);$$

i.e., no number is "forced" to be not in d by one of the conditions and in d by the other. We now make precise our intuitive discussion of what a condition "forces".

2.9. Definition. If G is a filter in $\mathbb{P}_{\mathscr{A}}$, $d_G = \bigcup \{s\colon \exists F(\langle s, F\rangle \in G)\}$. □

2.10. Lemma. *If* G *is a filter in* $\mathbb{P}_{\mathscr{A}}$ *and* $\langle s, F\rangle \in G$, *then* $\forall x \in F\,(x \cap d_G \subset s)$.

Proof. If $\langle s', F'\rangle \in G$, then $\langle s', F'\rangle$ and $\langle s, F\rangle$ are compatible, so by Lemma 2.8, $x \cap s' \subset s$ for all $x \in F$. □

2.11. Definition. If $x \in \mathscr{A}$, then $D_x = \{\langle s, F\rangle \in \mathbb{P}_{\mathscr{A}} : x \in F\}$. □

2.12. Lemma. *If* G *is a filter in* $\mathbb{P}_{\mathscr{A}}$ *and* $G \cap D_x \neq 0$, *then* $|x \cap d_G| < \omega$.

Proof. By Lemma 2.10. □

2.13. Lemma. *If* $x \in \mathscr{A}$, *then* D_x *is dense in* $\mathbb{P}_{\mathscr{A}}$.

Proof. For any $\langle s, F\rangle \in \mathbb{P}_{\mathscr{A}}$, $\langle s, F \bigcup \{x\}\rangle \leq \langle s, F\rangle$. □

2.14. Lemma. $\mathbb{P}_{\mathscr{A}}$ *has the* c.c.c.

Proof. Suppose $\langle s_\xi, F_\xi\rangle$, for $\xi < \omega_1$, were pairwise incompatible. By Lemma 2.8, the s_ξ would all be distinct, which is impossible. □

We now are almost set up for an application of MA. If $|\mathscr{A}| \leq \kappa$ and MA(κ) holds, then there is a filter G intersecting D_x for each $x \in \mathscr{A}$, whence d_G will be almost disjoint from each element of $\mathscr{A}$. However, d_G might be finite, or even empty. In fact, if there were some finite $F \subset \mathscr{A}$ such that $\bigcup F$ were cofinite, then no infinite d could be almost disjoint from all the elements of F. We shall show that if there is no such F, then we can make d infinite. More generally, we may make d have infinite intersection with any subset of ω which is not almost covered by a finite union from $\mathscr{A}$.

2.15. THEOREM. *Assume* MA(κ). *Let* $\mathscr{A}, \mathscr{C} \subset \mathscr{P}(\omega)$, *where* $|\mathscr{A}| \leq \kappa$, $|\mathscr{C}| \leq \kappa$, *and assume that for all* $y \in \mathscr{C}$ *and all finite* $F \subset \mathscr{A}$, $|y \setminus \bigcup F| = \omega$; *then there is* $d \subset \omega$, *such that* $\forall x \in \mathscr{A}\,(|d \cap x| < \omega)$ *and* $\forall y \in \mathscr{C}\,(|d \cap y| = \omega)$.

PROOF. For $y \in \mathscr{C}$ and $n \in \omega$, let

$$E_n^y = \{\langle s, F\rangle \in \mathbb{P}_{\mathscr{A}} : s \cap y \not\subset n\}.$$

E_n^y is dense in $\mathbb{P}_{\mathscr{A}}$, since for each $\langle s, F\rangle \in \mathbb{P}_{\mathscr{A}}$, $|y \setminus \bigcup F| = \omega$; if we pick $m \in y \setminus \bigcup F$ with $m > n$, then $\langle s \cup \{m\}, F\rangle$ is an extension of $\langle s, F\rangle$ in E_n^y.

By MA(κ), there is a filter G in $\mathbb{P}_{\mathscr{A}}$ intersecting all the dense sets in

$$\{D_x : x \in \mathscr{A}\} \cup \{E_n^y : y \in \mathscr{C} \wedge n \in \omega\}.$$

We have just seen that $d_G \cap x$ is finite for $x \in \mathscr{A}$. If $y \in \mathscr{C}$, then $d_G \cap y \not\subset n$ for all n, so $d_G \cap y$ is infinite. □

This answers Question 2 directly.

2.16. COROLLARY. *Let* $\mathscr{A} \subset \mathscr{P}(\omega)$ *be an almost disjoint family of cardinality* κ, *where* $\omega \leq \kappa < 2^{\omega}$. *Assume* MA($\kappa$). *Then* $\mathscr{A}$ *is not maximal.*

PROOF. Let $\mathscr{C} = \{\omega\}$. $|\omega \setminus \bigcup F| = \omega$ for all finite $F \subset \mathscr{A}$, since $\mathscr{A}$ is a.d. and infinite. Thus, Theorem 2.15 applies to yield an infinite d a.d. from each member of $\mathscr{A}$. □

We now settle Question 1.

2.17. LEMMA. *Let* $\mathscr{B} \subset \mathscr{P}(\omega)$ *be an almost disjoint family of size* κ, *where* $\omega \leq \kappa < 2^{\omega}$. *Let* $\mathscr{A} \subset \mathscr{B}$. *If* MA($\kappa$), *then there is a* $d \subset \omega$ *such that* $\forall x \in \mathscr{A}\,(|d \cap x| < \omega)$ *and* $\forall x \in \mathscr{B} \setminus \mathscr{A}\,(|d \cap x| = \omega)$.

PROOF. Apply Theorem 2.15 with $\mathscr{C} = \mathscr{B} \setminus \mathscr{A}$. □

2.15–2.17 are due to Solovay, who used Lemma 2.17 as a means of encoding subsets of κ by subsets of ω.

2.18. THEOREM. $\mathrm{MA}(\kappa) \to 2^\kappa = 2^\omega$.

PROOF. Fix any a.d. family $\mathscr{B}$ of size κ ($\mathscr{B}$ exists by Theorem 1.3). Define $\Phi: \mathscr{P}(\omega) \to \mathscr{P}(\mathscr{B})$ by $\Phi(d) = \{x \in \mathscr{B}: |d \cap x| < \omega\}$. Lemma 2.17 says Φ is onto, so $2^\kappa = |\mathscr{P}(\mathscr{B})| \leq |\mathscr{P}(\omega)| = 2^\omega$. □

2.19. COROLLARY. MA $\to 2^\omega$ *is regular.*

PROOF. If $\kappa < 2^\omega$, then $2^\kappa = 2^\omega$, so, by König's Theorem (I 10.41), $\mathrm{cf}(2^\omega) > \kappa$. □

It is consistent with ZFC that 2^ω is a singular cardinal, such as ω_{ω_1} (see VII 5.16).

One might attempt to improve Lemma 2.17 by demanding that d almost contain the elements of $\mathscr{B} \setminus \mathscr{A}$; that is,

$$\forall x \in \mathscr{B} \setminus \mathscr{A} \, (|x \setminus d| < \omega).$$

However, this improvement is inconsistent unless $\mathscr{A}$ or $\mathscr{B} \setminus \mathscr{A}$ is countable (see Exercises 9, 10).

We now answer Question 4.

2.20. THEOREM. *Assume* $\mathrm{MA}(\kappa)$. *Let* M_α, *for* $\alpha < \kappa$, *be first category subsets of* $\mathbb{R}$; *then* $\bigcup_{\alpha<\kappa} M_\alpha$ *is of first category.*

PROOF. Recall that a set $M \subset \mathbb{R}$ is of first category iff there are closed nowhere dense sets K_n ($n \in \omega$), such that $M \subset \bigcup_n K_n$. In particular, since each of our M_α is contained in a countable union of closed nowhere dense sets, it is sufficient to show that whenever we have κ closed nowhere dense sets, K_α for $\alpha < \kappa$, we can find countably many closed nowhere dense sets, H_n ($n \in \omega$) such that $\bigcup_{\alpha<\kappa} K_\alpha \subset \bigcup_{n<\omega} H_n$.

The complement of a closed nowhere dense set is a dense open set. Replacing the K_α and the H_n by their complements, we see that Theorem 2.20 will be proved if we can show that whenever $U_\alpha \subset \mathbb{R}$ is dense and open for $\alpha < \kappa$, we can find dense open V_n for $n < \omega$ such that

$$\bigcap_{n<\omega} V_n \subset \bigcap_{\alpha<\kappa} U_\alpha.$$

Let B_i ($i < \omega$) enumerate all the non-empty open intervals in $\mathbb{R}$ with rational endpoints. These form a base for the topology of $\mathbb{R}$; that is, for every open W,

$$W = \bigcup \{B_i : B_i \subset W\}.$$

For a suitably chosen $d \subset \omega$, we shall set $V_n = \bigcup \{B_i : i \in d \wedge i > n\}$. d will be chosen by applying Theorem 2.15.

Let $c_j = \{i \in \omega \colon B_i \subset B_j\}$. If $|d \cap c_j| = \omega$, then for each n there is an $i > n$, such that $i \in d$ and $B_i \subset B_j$, so $V_n \cap B_j \neq 0$. Thus, if $|d \cap c_j| = \omega$ for all j, then $\forall n \, \forall j \, (V_n \cap B_j \neq 0)$, so each V_n is dense.

For each $\alpha < \kappa$, let $a_\alpha = \{i \in \omega \colon B_i \not\subset U_\alpha\}$. If $|d \cap a_\alpha| < \omega$, then for some n, $d \cap a_\alpha \subset n$, so $\forall i > n \, (i \in d \to B_i \subset U_\alpha)$, whence $V_n \subset U_\alpha$. Thus, if $|d \cap a_\alpha| < \omega$ for all α, then

$$\bigcap_{n<\omega} V_n \subset \bigcap_{\alpha<\kappa} U_\alpha .$$

We now quote Theorem 2.15 directly with $\mathscr{C} = \{c_j \colon j < \omega\}$ and $\mathscr{A} = \{a_\alpha \colon \alpha < \kappa\}$. We are thus done if we can check that Theorem 2.15 applies; i.e., that whenever F is a finite subset of κ and $j < \omega$, $|c_j \smallsetminus \bigcup_{\alpha \in F} a_\alpha| = \omega$. But

$$c_j \smallsetminus \bigcup_{\alpha \in F} a_\alpha = \{i \in \omega \colon B_i \subset (B_j \cap \bigcap_{\alpha \in F} U_\alpha)\},$$

which is infinite since $B_j \cap \bigcap_{\alpha \in F} U_\alpha$ is open and non-empty (since $B_j \neq 0$ and $\bigcap_{\alpha \in F} U_\alpha$ is dense). □

We now introduce a new partial order and answer Question 3.

2.21. THEOREM. *Assume* MA(κ). *Let M_α, for $\alpha < \kappa$, be subsets of* $\mathbb{R}$, *each of Lebesgue measure* 0. *Then $\bigcup_{\alpha<\kappa} M_\alpha$ has Lebesgue measure* 0.

PROOF. Let μ be Lebesgue measure. Recall that a subset $M \subset \mathbb{R}$ has measure 0 iff for all $\varepsilon > 0$, there is an open $U \subset \mathbb{R}$ such that $M \subset U$ and $\mu(U) \leq \varepsilon$. Fix $\varepsilon > 0$. We shall find an open U with $\mu(U) \leq \varepsilon$ and $\bigcup_{\alpha<\kappa} M_\alpha \subset U$. Let

$$\mathbb{P} = \{p \subset \mathbb{R} \colon p \text{ is open} \wedge \mu(p) < \varepsilon\}.$$

Define $p \leq q$ iff $q \subset p$, then p and q are compatible iff $\mu(p \cup q) < \varepsilon$, in which case $p \cup q$ is a common extension of p and q.

Intuitively, p forces "$p \subset U$". Formally, if G is a filter in $\mathbb{P}$, let $U_G = \bigcup G$. U_G is clearly open. We must also check that $\mu(U_G) \leq \varepsilon$. First note that if $p, q \in G$, they have a common extension $r \in G$; and since $r \leq p \cup q$, we have $p \cup q \in G$. Next, by induction on n, if $p_1, \ldots, p_n \in G$, then $p_1 \cup \cdots \cup p_n$ is in G and hence in $\mathbb{P}$, so $\mu(p_1 \cup \cdots \cup p_n) < \varepsilon$. Thus, by the countable additivity of μ, whenever A is a countable subset of G, $\mu(\bigcup A) \leq \varepsilon$. G is uncountable, but we now show that $\bigcup A = \bigcup G$ for some countable $A \subset G$; this will follow from the fact that the topology of $\mathbb{R}$ has a countable base. Let $\mathscr{B}$ be the countable set of open intervals with rational endpoints. If $x \in p$ for some $p \in G$, then there is a $q \in \mathscr{B}$ with $x \in q$ and $q \subset p$. Then $q \geq p$, so $q \in G$ since G is a filter. Thus, if $A = G \cap \mathscr{B}$, then $\bigcup A = \bigcup G = U_G$, so $\mu(U_G) \leq \varepsilon$.

We now check that $\mathbb{P}$ has the c.c.c. This uses the separability of the measure space of $\mathbb{R}$ (or of $L^1(\mathbb{R})$). More precisely, let $\mathscr{C}$ be the set of all finite unions of elements of $\mathscr{B}$. Then whenever V is open (or just measurable) and $\delta > 0$, there is a $C \in \mathscr{C}$ such that $\mu(C \triangle V) \leq \delta$ ($\triangle$ is symmetric difference). Now, suppose $\{p_\alpha : \alpha < \omega_1\}$ is a family of pairwise incompatible conditions. Since $\mu(p_\alpha) < \varepsilon$, there is a fixed $\delta > 0$ such that $X = \{\alpha < \omega_1 : \mu(p_\alpha) \leq \varepsilon - 3\delta\}$ is uncountable. For $\alpha \in X$, choose $C_\alpha \in \mathscr{C}$ such that $\mu(p_\alpha \triangle C_\alpha) \leq \delta$. If α and β are distinct elements of X, then $p_\alpha \perp p_\beta$, so $\mu(p_\alpha \cup p_\beta) \geq \varepsilon$; since $\mu(p_\alpha \cap p_\beta) \leq \varepsilon - 3\delta$, we have $\mu(p_\alpha \triangle p_\beta) \geq 3\delta$; since $\mu(p_\alpha \triangle C_\alpha) \leq \delta$ and $\mu(p_\beta \triangle C_\beta) \leq \delta$, we have $\mu(C_\alpha \triangle C_\beta) \geq \delta$, whence $C_\alpha \neq C_\beta$. Thus, the C_α for $\alpha \in X$ are distinct elements of $\mathscr{C}$, a contradiction.

Finally, we come to the dense sets. For $\alpha < \kappa$, let $D_\alpha = \{p : M_\alpha \subset p\}$. To see that D_α is dense, fix $q \in \mathbb{P}$. $\mu(q) < \varepsilon$, so there is an open V with $M_\alpha \subset V$ and $\mu(V) < \varepsilon - \mu(q)$. Then $p = q \cup V$ has measure $< \varepsilon$, so $p \in \mathbb{P}$. Thus, p is an extension of q in D_α.

If $G \cap D_\alpha \neq 0$, then $M_\alpha \subset U_G$. By MA(κ), there is a G such that $G \cap D_\alpha \neq 0$ for all α, whence U_G is an open set of measure $\leq \varepsilon$ such that $\bigcup_{\alpha<\kappa} M_\alpha \subset U_G$. □

We now leave applications of MA to ω and the continuum, and turn to some applications in general topology. The following easy consequence of MA(κ) is in fact equivalent to MA(κ) by a much harder argument (see Theorem 3.4).

2.22. Theorem. *Assume* MA(κ). *Let X be a compact* c.c.c. *Hausdorff space and U_α dense open subsets of X for $\alpha < \kappa$. Then $\bigcap_{\alpha<\kappa} U_\alpha \neq 0$.*

Proof. Let $\mathbb{P} = \{p \subset X : p \text{ is open} \wedge p \neq 0\}$, with $p \leq q \leftrightarrow p \subset q$ (see Example 3); then $p \perp q \leftrightarrow p \cap q = 0$, so $\mathbb{P}$ has c.c.c. since X has. If G is a filter in $\mathbb{P}$, then G has the finite intersection property, so by compactness $\bigcap\{\bar{p} : p \in G\} \neq 0$. For each α, let $D_\alpha = \{p \in \mathbb{P} : \bar{p} \subset U_\alpha\}$; D_α is dense in the ordering $\mathbb{P}$ since U_α is dense in the space X and X is regular. If we take G so that $G \bigcap D_\alpha \neq 0$ for all $\alpha < \kappa$, then $\bigcap\{\bar{p} : p \in G\}$ is a non-empty set contained in $\bigcap_{\alpha<\kappa} U_\alpha$. □

If $\kappa = \omega$, Theorem 2.22 is just the Baire category theorem and does not require X to be c.c.c., just as the proof in ZFC of MA(ω) (Lemma 2.6) did not require $\mathbb{P}$ to be c.c.c. However, Theorem 2.22 is simply false when $\kappa = \omega_1$ if c.c.c. is dropped; a counter-example may be obtained from $X = (\omega_1 + 1)^\omega$ (see Exercise 11).

Theorem 2.22 should be compared with Theorem 2.20. Replacing U_α by $X \smallsetminus U_\alpha$ in Theorem 2.22, we see that X is not the union of κ closed nowhere dense sets. We cannot conclude as we did in Theorem 2.20 that such a union is of first category. A counter-example is provided by $X = [0,1]^{\omega_1}$, which is c.c.c. by Theorem 1.9 (see Exercise 12).

One can easily conclude directly from Theorem 2.22 the (apparently) stronger result that whenever X is locally compact Hausdorff and c.c.c., the intersection of κ dense open subsets of X is dense.

We shall now show that under $\mathrm{MA}(\omega_1)$, any product of c.c.c. spaces is c.c.c.; see §1 for a discussion of this question.

2.23. LEMMA. *Assume* $\mathrm{MA}(\omega_1)$. *Let X be* c.c.c. *and* $\{U_\alpha : \alpha < \omega_1\}$ *a family of non-empty open subsets of X, then there is an uncountable $A \subset \omega_1$, such that $\{U_\alpha : \alpha \in A\}$ has the finite intersection property.*

PROOF. Let $V_\alpha = \bigcup_{\gamma > \alpha} U_\gamma$. Then $\alpha < \beta$ implies $V_\beta \subseteq V_\alpha$. We check first that there is an α, such that

$$\forall \beta > \alpha (\overline{V}_\beta = \overline{V}_\alpha). \tag{$*$}$$

If there were no such α, we could find an increasing sequence of ordinals $\alpha_\xi (\xi < \omega_1)$, such that for each ξ $\overline{V}_{\alpha_{\xi+1}} \neq \overline{V}_{\alpha_\xi}$, whence $V_{\alpha_\xi} \smallsetminus \overline{V}_{\alpha_{\xi+1}} \neq 0$. These sets are open and pairwise disjoint, contradicting the fact that X has c.c.c.

Now, fix α satisfying ($*$), and let $\mathbb{P} = \{p \subset V_\alpha : p \text{ is open} \wedge p \neq 0\}$. $\mathbb{P}$ is c.c.c. because X is. If G is a filter in $\mathbb{P}$, then G has the finite intersection property, so if

$$A = \{\gamma < \omega_1 : \exists p \in G (p \subset U_\gamma)\},$$

then $\{U_\gamma : \gamma \in A\}$ has the finite intersection property. If G is suitably generic, A will be unbounded in ω_1, and hence uncountable. More formally, for each β let

$$D_\beta = \{p \in \mathbb{P} : \exists \gamma > \beta (p \subset U_\gamma)\}.$$

To see that D_β is dense, note that by ($*$), $\overline{V}_\alpha \subset \overline{V}_\beta$, so if $p \in \mathbb{P}$, then $p \cap V_\beta \neq 0$, so $p \cap U_\gamma \neq 0$ for some $\gamma > \beta$; thus $p \cap U_\gamma$ is an extension of p in D_β. Now, if $G \cap D_\beta \neq 0$, A will contain a $\gamma > \beta$, so if $G \cap D_\beta \neq 0$ for all β, A will be unbounded in ω_1. □

2.24. THEOREM. *Assume* $\mathrm{MA}(\omega_1)$; *then any product of* c.c.c. *spaces is* c.c.c.

PROOF. It is sufficient to prove that the product of two c.c.c. spaces is c.c.c., since that will immediately imply the theorem for finite products, and hence for arbitrary products by Theorem 1.9.

Let X and Y be c.c.c., and suppose $X \times Y$ were not. Let $\{W_\alpha : \alpha < \omega_1\}$ be a family of pairwise disjoint non-0 open subsets of $X \times Y$. For each α, pick a non-0 open box, $U_\alpha \times V_\alpha \subset W_\alpha$. By Lemma 2.23, let $A \subset \omega_1$ be uncountable such that $\{U_\alpha : \alpha \in A\}$ has the finite intersection property. If $\alpha, \beta \in A$ and $\alpha \neq \beta$, then $U_\alpha \cap U_\beta \neq 0$ but $U_\alpha \times V_\alpha \cap U_\beta \times V_\beta = 0$, so $V_\alpha \cap V_\beta = 0$. Thus, $\{V_\alpha : \alpha \in A\}$ contradicts the fact that Y has c.c.c. □

§3. Equivalents of MA

Our first equivalent of MA (Lemma 3.1) will be of importance in VIII §6 when we prove the consistency of MA + ¬CH, since it shows that we need only consider partial orders of size $<2^{\omega}$. The other equivalents are of some theoretical interest, and are important for understanding the connection between forcing and Boolean-valued models (see VII §7).

3.1. LEMMA. MA(κ) *is equivalent to* MA(κ) *restricted to partial orders of cardinality* $\leq \kappa$.

PROOF. Assume the restricted form of MA(κ), and let $\langle \mathbb{Q}, < \rangle$ be a c.c.c. partial order of arbitrary cardinality, and $\mathscr{D}$ a family of $\leq \kappa$ dense subsets of $\mathbb{Q}$. We shall find a filter H in $\mathbb{Q}$ intersecting each $D \in \mathscr{D}$ by applying the restricted form of MA(κ) to a suitable sub-order $\mathbb{P} \subset \mathbb{Q}$ with $|\mathbb{P}| \leq \kappa$.

We first show that we can find $\mathbb{P} \subset \mathbb{Q}$ such that

(1) $|\mathbb{P}| \leq \kappa$.

(2) $D \cap \mathbb{P}$ is dense in $\mathbb{P}$ for each $D \in \mathscr{D}$.

(3) If $p, q \in \mathbb{P}$, p and q are compatible in $\mathbb{P}$ iff they are compatible in $\mathbb{Q}$.

The model-theorist will realize that this is immediate by the downward Löwenheim–Skolem–Tarski theorem applied to $\langle \mathbb{Q}; <, D \rangle_{D \in \mathscr{D}}$, but we give a direct proof. For $D \in \mathscr{D}$, let $f_D : \mathbb{Q} \to \mathbb{Q}$ be such that

$$\forall p \in \mathbb{Q} \left(f_D(p) \in D \wedge f_D(p) \leq p \right),$$

and let $g : \mathbb{Q} \times \mathbb{Q} \to \mathbb{Q}$ be such that

$$\forall p, q \in \mathbb{Q} \left(p, q \text{ compatible} \to g(p, q) \leq p \wedge g(p, q) \leq q \right).$$

Let $\mathbb{P} \subset \mathbb{Q}$ be such that $|\mathbb{P}| \leq \kappa$ and $\mathbb{P}$ is closed under g and each f_D (this is possible by I 10.23). $\mathbb{P}$ satisfies (3) by closure under g and (2) by closure under the f_D.

(3) implies that $\mathbb{P}$ has the c.c.c., so by applying the restricted MA(κ) to $\mathbb{P}$, let G be a filter in $\mathbb{P}$ such that

$$\forall D \in \mathscr{D} \, (G \cap D \cap \mathbb{P} \neq 0).$$

If H is the filter generated by G, i.e.,

$$H = \{ q \in \mathbb{Q} : \exists p \in G \, (p \leq q) \},$$

then H is a filter in $\mathbb{Q}$ intersecting each $D \in \mathscr{D}$. □

We remark that in many applications of MA(κ), the "natural" partial order to use has cardinality $>\kappa$ (see 2.21, 2.22, 2.23).

We now turn to the proof that the conclusion of Theorem 2.22 implies MA(κ). It is possible to give a direct proof, which would be very much like the proof of the Stone representation theorem of Boolean algebra, only more complicated. This direct proof would only serve to obscure the very important link between MA and Boolean algebra (and later between forcing and Boolean algebra). For that reason, we shall give a proof using Boolean algebra, although this will require that the reader be familiar with the elements of this subject (see [Halmos 1963]).

We review some basic notation on Boolean algebras. If $\mathscr{B}$ is a Boolean algebra, we always use $<$ for the order on $\mathscr{B}$. If $a, b \in \mathscr{B}$, $a \vee b$ is the least upper bound of a and b, and $a \wedge b$ is their greatest lower bound. a' is the complement of a. $\mathbb{1}$ is the largest element of $\mathscr{B}$ and $\mathbb{0}$ is the smallest element.

$\mathscr{B}$ is called *complete* iff every subset $S \subset \mathscr{B}$ has both a supremum (written $\bigvee S$) and an infinum (written $\bigwedge S$).

It is actually $\mathscr{B} \smallsetminus \{\mathbb{0}\}$ that is relevant to MA applications (see §2, Example 4). We shall abuse notation and say that an antichain in $\mathscr{B}$ is a set $A \subset \mathscr{B} \smallsetminus \{\mathbb{0}\}$ such that

$$\forall a, b \in A\,(a \neq b \rightarrow a \wedge b = \mathbb{0}).$$

(A is really an antichain $\mathscr{B} \smallsetminus \{\mathbb{0}\}$). We say $\mathscr{B}$ has c.c.c. iff all antichains in $\mathscr{B}$ are countable. By a filter on $\mathscr{B}$, we mean a subset $G \subset \mathscr{B} \smallsetminus \{\mathbb{0}\}$ which is a filter in the sense of Definition 2.4, but this is easily seen to coincide with the usual notion of a filter on a Boolean algebra. By MA restricted to complete Boolean algebras, we mean MA restricted to those partial orders of the form $\mathscr{B} \smallsetminus \{\mathbb{0}\}$ for some complete c.c.c. Boolean algebra $\mathscr{B}$. Note that none of the partial orders so far considered came from Boolean algebras. Nevertheless, the following holds.

3.2. THEOREM. *For any $\kappa \geq \omega$,* MA(κ) *is equivalent to* MA(κ) *restricted to complete Boolean algebras.*

To prove this, we shall first describe a general procedure for associating a complete Boolean algebra with a given partial order.

3.3. LEMMA. *Let $\mathbb{P}$ be a partial order. Then there is a complete Boolean algebra $\mathscr{B}$ and a map $i : \mathbb{P} \rightarrow \mathscr{B} \smallsetminus \{\mathbb{0}\}$ such that:*

(1) *$i''\mathbb{P}$ is dense in $\mathscr{B} \smallsetminus \{\mathbb{0}\}$.*

(2) $\forall p, q \in \mathbb{P}\,(p \leq q \rightarrow i(p) \leq i(q))$.

(3) $\forall p, q \in \mathbb{P}\,(p \perp q \leftrightarrow i(p) \wedge i(q) = \mathbb{0})$.

In many cases of interest, i will be 1–1 and $\forall p, q \in \mathbb{P}\,(p \leq q \leftrightarrow i(p) \leq i(q))$, so we may identify $\mathbb{P}$ as a sub-order of $\mathscr{B}$ (see Exercise 15). However, if, for example, all elements of $\mathbb{P}$ are compatible, then $\mathscr{B}$ is the 2-element algebra and $i(p) = \mathbb{1}$ for all $p \in \mathbb{P}$. $\mathscr{B}$ is called *the completion* of $\mathbb{P}$; it is unique up to isomorphism (see Exercise 18). We may think of the elements of $\mathscr{B}$ as formed by taking suprema of various subsets of $\mathbb{P}$, and one may prove Lemma 3.3 by constructing $\mathscr{B}$ as the algebra of "formal suprema" (see Exercise 19). Our proof here will be topological.

Recall that if X is any topological space, we may define the regular open algebra of X, $\mathrm{ro}(X)$. The elements of $\mathrm{ro}(X)$ are the regular open subsets $b \subset X$. (b is regular iff $b = \mathrm{int\,cl}(b)$). $b \leq c$ iff $b \subset c$. The algebraic operations in $\mathrm{ro}(X)$ are as follows: $b \wedge c = b \cap c$; $b \vee c = \mathrm{int\,cl}(b \cup c)$; $b' = \mathrm{int}(X \smallsetminus b)$. $\mathrm{ro}(X)$ is complete, and if $S \subset \mathrm{ro}(X)$, $\bigvee S = \mathrm{int\,cl}(\bigcup S)$ and $\bigwedge S = \mathrm{int}(\bigcap S)$.

Proof of Lemma 3.3. Define a topology on $\mathbb{P}$ as follows. If $p \in \mathbb{P}$, let $N_p = \{q \in \mathbb{P}: q \leq p\}$. Give $\mathbb{P}$ the topology whose base is $\{N_p: p \in \mathbb{P}\}$. The N_p form a base since $q \in N_p \rightarrow N_q \subset N_p$. This topology is usually not Hausdorff, or even T_1. For each p, N_p is the smallest open set containing p.

Let $\mathscr{B} = \mathrm{ro}(\mathbb{P})$ and let $i(p) = \mathrm{int\,cl}(N_p)$. To check (1), let b any non-0 regular open set. Fix $p \in b$, then $N_p \subset b$, so $i(p) = \mathrm{int\,cl}(N_p) \subset \mathrm{int\,cl}(b) = b$. (2) is obvious. For (3), suppose first that p and q are compatible, and fix r with $r \leq p$ and $r \leq q$. By (2), $i(r) \leq i(p)$ and $i(r) \leq i(q)$, so $i(p) \wedge i(q) \neq 0$. Conversely, suppose $p \perp q$; then $N_p \cap N_q = 0$. Since N_q is open, $\mathrm{cl}(N_p) \cap N_q = 0$, so $i(p) \cap N_q = \mathrm{int\,cl}(N_p) \cap N_q = 0$. Since $i(p)$ is open, the same argument applied to q yields $i(p) \wedge i(q) = i(p) \cap i(q) = 0$. □

Proof of 3.2. Assume MA(κ) restricted to complete Boolean algebras. Let $\mathbb{P}$ be an arbitrary c.c.c. partial order and $\mathscr{D}$ a family of $\leq\kappa$ dense subsets of $\mathbb{P}$. Let $\mathscr{B}$ and i be as in 3.3. $\mathscr{B}$ has c.c.c., since if $\{b_\alpha: \alpha < \omega_1\}$ were an antichain in $\mathscr{B}$, we could, by (1), find p_α with $i(p_\alpha) \leq b_\alpha$; then, by (2), $\{p_\alpha: \alpha < \omega_1\}$ would be antichain in $\mathbb{P}$. Also, if $D \in \mathscr{D}$, then $i''D$ is dense, since if $b \in \mathscr{B} \smallsetminus \{\mathbb{0}\}$, there is a p with $i(p) \leq b$ and then a $q \in D$ with $q \leq p$; so $i(q) \leq b$ and $i(q) \in i''D$. Now, by the restricted MA, there is a filter G in $\mathscr{B}$ such that $G \cap i''D \neq 0$ for each $D \in \mathscr{D}$. Let $H = i^{-1}(G)$; then $H \cap D \neq 0$ for each $D \in \mathscr{D}$.

Unfortunately, H may fail to be a filter. It is clear, by (2), that H satisfies

(b) $\forall p \in H\, \forall q \in \mathbb{P}\, (q \geq p \rightarrow q \in H)$.

The difficulty is with

(a) $\forall p, q \in H\, \exists r \in H\, (r \leq p \wedge r \leq q)$

If $p, q \in H$, then p and q are compatible by (3), but they may not have a common extension in H. To remedy this, let

$$D_{pq} = \{r \in \mathbb{P}: (r \leq p \wedge r \leq q) \vee r \perp p \vee r \perp q\}.$$

D_{pq} is dense for every $p, q \in \mathbb{P}$. To see this, fix $r_0 \in \mathbb{P}$. If r_0 has an extension incompatible with either p or q, then such an extension is in D_{pq}. If not, then in particular r_0 is compatible with p, so fix r_1, such that $r_1 \leq r_0$ and $r_1 \leq p$; since r_1 is compatible with q, fix r_2, such that $r_2 \leq r_1$ and $r_2 \leq q$. Since also $r_2 \leq p$, r_2 is an extension of r_0 in D_{pq}.

Now, *if* $|\mathbb{P}| \leq \kappa$, we may assume that each $D_{pq} \in \mathscr{D}$, since adding $\{D_{pq}: p, q \in \mathbb{P}\}$ to $\mathscr{D}$ will not change $|\mathscr{D}| \leq \kappa$. In this case, if $p, q \in H$, fix $r \in H \cap D_{pq}$. Since elements of H are pairwise compatible, we cannot have $r \perp p$ or $r \perp q$, so $r \leq p$ and $r \leq q$. Thus, (a) is satisfied and H is a filter.

We have thus shown that MA(κ) restricted to complete Boolean algebras implies MA(κ) restricted to partial orders of cardinality $\leq\kappa$, but this implies MA(κ) by Lemma 3.1. □

Finally, we show that MA is equivalent to a purely topological statement. The following summarizes our results.

3.4. THEOREM. *For any* $\kappa \geq \omega$, *the following are equivalent*:

(a) MA(κ).

(b) MA(κ) *restricted to partial orders of cardinality* $\leq\kappa$.

(c) MA(κ) *restricted to complete Boolean algebras.*

(d) *If X is any compact* c.c.c. *Hausdorff space and U_α are dense open sets for $\alpha < \kappa$, then* $\bigcap_\alpha U_\alpha \neq 0$.

PROOF (a), (b), and (c) are equivalent by 3.1 and 3.2, and (a) → (d) by 2.14. We show (d) → (c). Let $\mathscr{B}$ be a c.c.c. Boolean algebra and $\mathscr{D}$ a family of $\leq\kappa$ dense subsets of $\mathscr{B} \setminus \{\mathbb{0}\}$. Assuming (d), we shall produce a filter $G \subset \mathscr{B}$ intersecting each element of $\mathscr{D}$. It is not necessary here that $\mathscr{B}$ be complete.

Let X be the Stone space of $\mathscr{B}$. Thus, the elements of X are the ultrafilters on $\mathscr{B}$, and, if $b \in \mathscr{B}$, a basic open set in X is given by

$$N_b = \{G \in X: b \in G\}.$$

X is a compact Hausdorff space. $N_b \bigcap . N_c = 0$ iff $b \wedge c = 0$, so X has the c.c.c. For each $D \in \mathscr{D}$, let

$$W_D = \bigcup \{N_b: b \in D\}.$$

W_D is clearly open. W_D is dense (in the topological sense), since if N_c were disjoint from W_D, then $c \wedge b = 0$ for all $b \in D$, which is impossible since D is dense (in the order sense) and thus contains an extension of c. By (d), let $G \in \bigcap \{W_D: D \in \mathscr{D}\}$. Then G is a filter (in fact an ultrafilter), and for each $D \in \mathscr{D}$, $G \in W_D$, so $\exists b \in D\,(G \in N_b)$, or $\exists b \in D\,(b \in G)$, or $G \bigcap D \neq 0$. □

§4. The Suslin problem

We discussed in §1 the relationship between separability and the c.c.c. in topological spaces. In particular, we remarked that every separable space is c.c.c., but not every c.c.c. space is separable. However, if we consider these properties for ordered spaces, the question of whether they are equivalent is independent of ZFC.

4.1. Definition. A *Suslin line* is a total ordering, $\langle X, <\rangle$ such that in the order topology, X is c.c.c. but not separable. Suslin's Hypothesis (SH) is the statement, "there are no Suslin lines." □

SH arose naturally in an attempt to characterize the order type of the real numbers, $\langle \mathbb{R}, <\rangle$. It was well-known (see Exercise 29) that any total ordering $\langle X, <\rangle$ satisfying

(a) X has no first or last element,
(b) X is connected in the order topology, and
(c) X is separable in the order topology

is isomorphic to $\langle \mathbb{R}, <\rangle$. [Suslin 1920] asked whether (c) may be replaced by

(c′) X is c.c.c. in the order topology.

Clearly, under SH, (c) and (c′) are equivalent, and one can show (Exercise 30) that if there is a Suslin line, then there is one satisfying (a) and (b). Thus, SH is equivalent to the statement that (a), (b), and (c′) characterizes the ordering $\langle \mathbb{R}, <\rangle$.

SH turns out to be independent of ZFC. SH is consistent since, as we shall presently see, it follows from MA + $\neg$CH, which is consistent (see VIII §6). In fact, Jensen has shown that SH is consistent with GCH (see [Devlin–Johnsbråten 1974]). But $\neg$SH follows from $\diamondsuit$ (see §7), which is consistent with ZFC + GCH (see VI).

4.2. Theorem. $\mathrm{MA}(\omega_1) \to$ SH.

Proof. $\mathrm{MA}(\omega_1)$ implies that the product of c.c.c. spaces is c.c.c. (Theorem 2.24), so the theorem is immediate from the following lemma. □

4.3. Lemma. *If X is a Suslin line, X^2 is not* c.c.c.

Proof. If $a, b \in X$ and $a < b$, let (a, b) denote the open interval, $\{x \in X: a < x < b\}$. (a, b) could be empty if a and b are adjacent.

By induction on $\alpha < \omega_1$, we shall find $a_\alpha, b_\alpha, c_\alpha \in X$ so that

(1) $a_\alpha < b_\alpha < c_\alpha$.

(2) $(a_\alpha, b_\alpha) \neq 0$ and $(b_\alpha, c_\alpha) \neq 0$.
(3) $(a_\alpha, c_\alpha) \cap \{b_\xi : \xi < \alpha\} = 0$.

Assuming we can do this, let $U_\alpha = (a_\alpha, b_\alpha) \times (b_\alpha, c_\alpha)$. By (2), $U_\alpha \neq 0$. If $\xi < \alpha$, then $U_\xi \cap U_\alpha = 0$, since, by (3), either $b_\xi \leq a_\alpha$, in which case $(a_\xi, b_\xi) \cap (a_\alpha, b_\alpha) = 0$, or $b_\xi \geq c_\alpha$, in which case $(b_\xi, c_\xi) \cap (b_\alpha, c_\alpha) = 0$. Thus, $\{U_\alpha : \alpha < \omega_1\}$ refutes the c.c.c. of X^2.

To find a_α, b_α, c_α, we first let W be the set all of isolated points of X. Since an isolated point is open and X has c.c.c., $|W| \leq \omega$. Now, assume we have picked a_ξ, b_ξ, c_ξ for $\xi < \alpha$. Since X is not separable, $X \setminus \mathrm{cl}(W \cup \{b_\xi : \xi < \alpha\})$ is a non-0 open set, and thus contains a non-empty open interval, (a_α, c_α). Since (a_α, c_α) contains no isolated points, it is infinite, so we may choose $b_\alpha \in (a_\alpha, c_\alpha)$ such that $(a_\alpha, b_\alpha) \neq 0$ and $(b_\alpha, c_\alpha) \neq 0$. □

Our definition of a Suslin line allowed for lines which could be very bad; for example, X could have gaps in it, or isolated points. We show now how to manipulate a line into nicer form.

4.4. THEOREM. *If there is a Suslin line, then there is a Suslin line X such that*
(1) *X is dense in itself (i.e., $a < b \rightarrow (a, b) \neq 0$), and*
(2) *no non-empty open subset of X is separable.*

PROOF. Let Y be any Suslin line. Define an equivalence relation $\sim$ on Y by setting $x \sim y$ iff the interval between them ((x, y) if $x < y$, or (y, x) if $x > y$) is separable. Let X be the set of $\sim$-equivalence classes. If $I \in X$, then I is convex; i.e., $x, y \in I \wedge x < y \rightarrow (x, y) \subset I$. We totally order X by setting $I < J$ iff some (any) element of I is less than some (any) element of J.

Note that each $I \in X$ is separable. To see this, let $\mathcal{M}$ be a maximal disjoint collection of non-0 open intervals of the form (x, y) with $x, y \in I$. $\mathcal{M}$ is countable since Y has the c.c.c., so let $\mathcal{M} = \{(x_n, y_n) : n \in \omega\}$. Since $x_n \sim y_n$, let D_n be a countable dense subset of (x_n, y_n). Let $D = \bigcup_n D_n$; then D is dense in $\bigcup_n (x_n, y_n)$. If $z \in I$ and $z \in (x, y) \subset I$, then (x, y) intersects some (x_n, y_n) by maximality of $\mathcal{M}$; this implies $z \in \bar{D}$ unless z is the first or last element of I. Thus, D together with the first and last elements of I (if I has a first or last element) forms a countable dense subset of I.

To see that X is dense in itself, suppose $I < J$ but $(I, J) = 0$. Pick $x \in I$ and $y \in J$, then $(x, y) \subset I \cup J$, which is separable, so $x \sim y$, a contradiction.

To verify (2), it is sufficient to see that (I, J) is not separable whenever $I < J$. Suppose it were. Let $\{K_n : 2 \leq n < \omega\}$ be dense in (I, J), and let $K_0 = I$, $K_1 = J$. In Y, let D_n be a countable dense subset of K_n, then $\bigcup_n D_n$ is dense in $\bigcup \{L : I \leq L \leq J\}$, so points of I are equivalent to points of J, a contradiction.

Finally, to see that X is c.c.c., suppose (I_α, J_α) were, for $\alpha < \omega_1$, disjoint open intervals in X. Pick $x_\alpha \in I_\alpha$ and $y_\alpha \in J_\alpha$, then the (x_α, y_α) would be disjoint and non-0 in Y. □

Further manipulation on X will make it still nicer; see Exercise 30.

§5. Trees

Suslin's Hypothesis, as well as a number of other combinatorial statements, are best understood as assertions about trees.

5.1. DEFINITION. A *tree* is a partial order in the strict sense (see Definition 2.1), $\langle T, \leq \rangle$, such that for each $x \in T$, $\{y \in T: y < x\}$ is well-ordered by $<$. □

As usual, we shall abuse notation and refer to T when we mean $\langle T, \leq \rangle$.

5.2. DEFINITION. Let T be a tree

(a) If $x \in T$, the *height* of x in T, or $\mathrm{ht}(x, T)$, is $\mathrm{type}(\{y \in T: y < x\})$.

(b) For each ordinal α, the α-th *level* of T, or $\mathrm{Lev}_\alpha(T)$, is

$$\{x \in T: \mathrm{ht}(x, T) = \alpha\}.$$

(c) The *height* of T, or $\mathrm{ht}(T)$, is the least α such that $\mathrm{Lev}_\alpha(T) = 0$.

(d) A *sub-tree* of T is a subset $T' \subset T$ with the induced order such that

$$\forall x \in T' \, \forall y \in T \, (y < x \rightarrow y \in T'). \quad \square$$

$\mathrm{ht}(T)$ is also equal to $\sup\{\mathrm{ht}(x, T) + 1: x \in T\}$. If T' is a sub-tree of T, then, for $x \in T'$, $\mathrm{ht}(x, T) = \mathrm{ht}(x, T')$.

One trivial example of a tree is any set, T, with $<$ being the empty order 0; then $\mathrm{ht}(x, T) = 0$ for all $x \in T$ and $\mathrm{ht}(T) = 1$. Another trivial example is any ordinal δ with the usual order. $\mathrm{ht}(\alpha, \delta) = \alpha$, and $\mathrm{ht}(\delta) = \delta$.

A more useful example of a tree is ${}^{<\delta}I = \bigcup \{{}^{\alpha}I: \alpha < \delta\}$, the *complete I-ary tree* of height δ. We think of elements of ${}^{\alpha}I$ as sequences of elements of I of length α. In ${}^{<\delta}I$, we define $s \leq t$ iff $s \subset t$ iff the sequence t extends s. If $\alpha < \delta$, then $\mathrm{Lev}_\alpha({}^{<\delta}I) = {}^{\alpha}I$, and $\mathrm{ht}({}^{<\delta}I) = \delta$. When $I = 2$, we refer to ${}^{<\delta}2$ as the *complete binary tree* of height δ.

We now borrow some terminology from our discussion of MA.

5.3. DEFINITION. Let T be a tree. A *chain* in T is a set $C \subset T$ which is totally ordered by $<$. An *antichain* in T is a set $A \subset T$, such that

$$\forall x, y \in A \, (x \neq y \rightarrow (x \nleq y \wedge y \nleq x)). \quad \square$$

If $\mathbb{P} = \langle T, \geq \rangle$ (the reverse order of T), then our notation coincides precisely with Definition 2.2. Note that x and y are *compatible* in $\mathbb{P}$ $(\exists z (z \geq x \wedge z \geq y))$ iff they are *comparable* $(x \leq y \vee y \leq x)$ since the predecessors of any z in T are totally ordered.

The tree analogue to a Suslin line is a Suslin tree.

5.4. DEFINITION. For any infinite cardinal κ, a *κ-Suslin tree* is a tree T, such that $|T| = \kappa$ and every chain and every antichain of T has cardinality $<\kappa$. □

Suslin trees were introduced by Kurepa (see, e.g., [Kurepa 1936]), who showed that there is an ω_1-Suslin tree iff there is a Suslin line (see Theorem 5.13). We first continue with a general discussion of κ-Suslin trees and related concepts. We confine our attention to the case when κ is regular; when κ is singular, κ-Suslin trees exist (Exercise 33) but are of little interest. We first try to get a rough picture of the shape of a κ-Suslin tree.

5.5. DEFINITION. For any regular κ, a κ-tree is a tree T of height κ such that $\forall \alpha < \kappa (|\mathrm{Lev}_\alpha(T)| < \kappa)$. □

5.6. LEMMA. *For any regular κ, every κ-Suslin tree is a κ-tree.*

PROOF. $\mathrm{Lev}_\kappa(T) = 0$, since if $x \in \mathrm{Lev}_\kappa(T)$, $\{y: y < x\}$ would be a chain of cardinality κ. Thus, $\mathrm{ht}(T) \leq \kappa$. Since each $\mathrm{Lev}_\alpha(T)$ is an antichain, $|\mathrm{Lev}_\alpha(T)| < \kappa$. Since $|T| = \kappa$ and $T = \bigcup \{\mathrm{Lev}_\alpha(T): \alpha < \mathrm{ht}(T)\}$, regularity of κ implies $\mathrm{ht}(T) = \kappa$. □

We now show that there are no ω-Suslin trees.

5.7. LEMMA. *König. If T is an ω-tree, then T has an infinite chain.*

PROOF. Pick $x_0 \in \mathrm{Lev}_0(T)$ such that $\{y \in T: y \geq x_0\}$ is infinite. This is possible since $\mathrm{Lev}_0(T)$ is finite, T is infinite, and every element of T is $\geq$ some element of $\mathrm{Lev}_0(T)$. By a similar argument, we may inductively pick $x_n \in \mathrm{Lev}_n(T)$ so that for each n, $x_{n+1} > x_n$ and $\{y \in T: y \geq x_{n+1}\}$ is infinite. Then $\{x_n: n \in \omega\}$ is an infinite chain in T. □

Since Lemma 5.7 did not mention antichains, it established a stronger result than the non-existence of ω-Suslin trees.

5.8. DEFINITION. For any regular κ, a *κ-Aronszajn tree* is a κ-tree such that every chain in T is of cardinality $<\kappa$. □

Thus, every κ-Suslin tree is a κ-Aronszajn tree, and Lemma 5.7 says that there are no ω-Aronszajn trees. At ω_1, the situation is different. The existence of an ω_1-Suslin tree is independent of ZFC, but there is always an ω_1-Aronszajn tree.

5.9. THEOREM. *There is an ω_1-Aronszajn tree.*

PROOF. Let

$$T = \{s \in {}^{<\omega_1}\omega \colon s \text{ is } 1\text{–}1\}.$$

Thus, T is a sub-tree of ${}^{<\omega_1}\omega$. $\mathrm{ht}(T) = \omega_1$, since for every $\alpha < \omega_1$, there is a 1–1 function from α into ω. If C were an uncountable chain in T, then $\bigcup C$ would be a 1–1 function from ω_1 into ω; thus, every chain in T is countable. Unfortunately, T is not Aronszajn, since T is not an ω_1-tree; $\mathrm{Lev}_\alpha(T)$ is uncountable for $\omega \le \alpha < \omega_1$. However, we shall define a sub-tree of T which is Aronszajn.

If $s, t \in {}^\alpha\omega$, define $s \sim t$ iff $\{\xi < \alpha \colon s(\xi) \neq t(\xi)\}$ is finite. We shall find s_α for $\alpha < \omega_1$ such that

(i) $s_\alpha \in {}^\alpha\omega$ and s_α is 1–1,

(ii) $\alpha < \beta \to s_\alpha \sim s_\beta \restriction \alpha$, and

(iii) $\omega \setminus \mathrm{ran}(s_\alpha)$ is infinite.

Assuming such s_α may be found, let

$$T^* = \bigcup_{\alpha<\omega_1} \{t \in \mathrm{Lev}_\alpha(T) \colon t \sim s_\alpha\}.$$

By (ii), T^* is a sub-tree of T. By (i), each $s_\alpha \in T^*$, so $\mathrm{Lev}_\alpha(T) \neq 0$. Unlike T, T^* is an ω_1-tree, since $\{t \in {}^\alpha\omega \colon t \sim s_\alpha\}$ is countable. Thus, T^* is an ω_1-Aronszajn tree.

We pick the s_α by induction. Given s_α, take any $n \in \omega \setminus \mathrm{ran}(s_\alpha)$ and let $s_{\alpha+1} = s_\alpha \cup \{\langle \alpha, n\rangle\}$; it is here that (iii) is used. Now, suppose we have s_α for $\alpha < \gamma$, where γ is a limit. Fix α_n for $n < \omega$ so that $\alpha_0 < \alpha_1 < \alpha_2 < \cdots$ and $\sup_n \alpha_n = \gamma$. Let $t_0 = s_{\alpha_0}$, and inductively define $t_n \colon \alpha_n \to \omega$ so that t_n is 1–1, $t_n \sim s_{\alpha_n}$, and $t_{n+1} \restriction \alpha_n = t_n$. Let $t = \bigcup_n t_n$. $t \in {}^\gamma\omega$ and t is 1–1; if we set $s_\gamma = t$, then (i) would hold for $\alpha = \gamma$ and (ii) would hold for $\alpha < \beta = \gamma$, but (iii) might fail. To fix this, we define $s_\gamma(\alpha_n) = t(\alpha_{2n})$ and $s_\gamma(\xi) = t(\xi)$ for $\xi \notin \{\alpha_n \colon n \in \omega\}$. Then

$$\{t(\alpha_{2n+1}) \colon n \in \omega\} \subset (\omega \setminus \mathrm{ran}(s_\gamma)),$$

so (iii) holds as well. □

A tree constructed as in Theorem 5.9 can never be a Suslin tree (see Exercise 39).

We now survey the situation for $\kappa > \omega_1$. If $\kappa = \lambda^+$, λ is regular, and

$2^{<\lambda} = \lambda$, then there is a κ-Aronszajn tree (Exercise 37). Thus, under GCH, there is a κ-Aronszajn tree for every regular $\kappa \geq \omega_1$ except possibly when κ is inaccessible or the successor of a singular cardinal. It is unknown whether GCH implies that there must be a κ-Aronszajn tree for κ the successor of a singular cardinal, although Jensen showed that it is consistent with GCH that there is for all such κ, and in fact this follows from Gödel's Axiom of Constructibility ($\mathbf{V} = \mathbf{L}$) discussed in VI (see [Devlin 1973] for the proof). For κ strongly inaccessible, then (without assuming GCH) there is a κ-Aronszajn tree unless κ is weakly compact, a property which implies that κ is the κ-th inaccessible and much more (see Exercise 49 or [Kunen 1977]).

Without GCH, there need not even be an ω_2-Aronszajn tree. The existence of such a tree is consistent with but independent of ZFC + $2^{\omega} = \omega_2$ ([Mitchell 1972]).

We turn now to Suslin trees. Jensen showed that if $\mathbf{V} = \mathbf{L}$, then there is a κ-Suslin tree for every regular $\kappa \geq \omega_1$ which is not weakly compact (see [Devlin 1973]). $\mathbf{V} = \mathbf{L}$ implies GCH (see VI 4.7); it is unknown whether it is consistent with GCH that there is a regular non-weakly compact $\kappa > \omega_1$ for which there is no κ-Suslin tree. Specifically, it is unknown whether the non-existence of an ω_2-Suslin tree is consistent with GCH; it is consistent with CH [Laver–Shelah 1900], although CH implies that there is an ω_2-Aronszajn tree (Exercise 37). Jensen showed that the non-existence of an ω_1-Suslin tree is consistent with GCH (see [Devlin–Johnsbråten 1974]).

We now give the proof that there is a Suslin line iff there is an ω_1-Suslin tree. First, some preliminary remarks.

It is considered good practice to prune a tree, removing branches that are sickly or do not produce fruit. At the same time, we may ensure that the tree has a single trunk.

5.10. Definition. A *well-pruned* κ-tree is a κ-tree T, such that $|\mathrm{Lev}_0(T)| = 1$ and

$$\forall x \in T\, \forall \alpha (\mathrm{ht}(x, T) < \alpha < \kappa \rightarrow \exists y \in \mathrm{Lev}_\alpha(T)(x < y)). \quad \square \qquad (*)$$

5.11. Lemma. *If κ is regular and T is a κ-tree, then T has a well-pruned κ-sub-tree.*

Proof. Let T' be the set of $x \in T$, such that

$$|\{z \in T\colon z > x\}| = \kappa.$$

T' is clearly a sub-tree of T. To verify (*) for T', fix $x \in T'$ and α such that $\mathrm{ht}(x, T) < \alpha < \kappa$. Let $Y = \{y \in \mathrm{Lev}_\alpha(T)\colon x < y\}$. By definition of T' and the fact that each $|\mathrm{Lev}_\beta(T)| < \kappa$, $\{z \in T\colon z > x \wedge \mathrm{ht}(z, T) > \alpha\}$ has cardinality κ, and each element of this set is above some element of Y. Since $|Y| < \kappa$, there is a $y \in Y$, such that $|\{z \in T: z > y\}| = \kappa$, and this y is in T'. A similar argument shows that $\mathrm{Lev}_0(T') \neq 0$, so $T' \neq 0$.

Now for every $x \in \mathrm{Lev}_0(T')$, $\{y \in T': y \geq x\}$ is a well-pruned sub-tree of T. □

Pruning a tree tends to make the remaining branches more bushy.

5.12. LEMMA. *If κ is regular, T is a well-pruned κ-Aronszajn tree and $x \in T$, then*

$$\forall n < \omega\, \exists \alpha > \mathrm{ht}(x, T)\big(|\{y \in \mathrm{Lev}_\alpha(T): y > x\}| \geq n\big).$$

PROOF. For $n = 2$, this follows from the fact that $\{y: y > x\}$ meets all levels above x and cannot form a chain. For $n > 2$, we proceed by induction. If the lemma holds for n, fix $\alpha > \mathrm{ht}(x, T)$ and distinct $y_1, \ldots, y_n \in \mathrm{Lev}_\alpha(T)$ with each $y_i > x$. Now, let $\beta > \alpha$ be such there are distinct $z_n, z_{n+1} \in \mathrm{Lev}_\beta(T)$ with $z_n, z_{n+1} > y_n$. For $i < n$, there are $z_i \in \mathrm{Lev}_\beta(T)$ with $z_i > y_i$. Then $\{z_1, \ldots, z_{n+1}\}$ establishes the lemma for $n + 1$. □

Actually, Lemma 5.12 holds for any cardinal $\lambda < \kappa$ in place of n (see Exercise 38).

We may now prove that SH is equivalent to the non-existence of an ω_1-Suslin tree.

5.13. THEOREM. *There is an ω_1-Suslin tree iff there is a Suslin line.*

PROOF. First, let T be an ω_1-Suslin tree. By Lemma 5.11, we may assume that T is well-pruned. Let

$$L = \{C \subset T: C \text{ is a maximal chain in } T\}.$$

If $C \in L$, then there is an ordinal $h(C)$ such that C contains exactly one element from $\mathrm{Lev}_\alpha(T)$ for $\alpha < h(C)$ and no elements from $\mathrm{Lev}_\alpha(T)$ for $a \geq h(C)$. Since T is Aronszajn, $h(C) < \omega_1$. Since T is well-pruned, a maximal chain cannot have a largest element, so each $h(C)$ is a limit ordinal. For $\alpha < h(C)$, Let $C(\alpha)$ be the element of C on level α.

We order L as follows. Fix an arbitrary total order $\prec$ of T. If $C, D \in L$, and $C \neq D$, let $d(C, D)$ be the least α such that $C(\alpha) \neq D(\alpha)$. Observe that $d(C, D) < \min(h(C), h(D))$. Let $C \lhd D$ iff $C(d(C, D)) \prec D(d(C, D))$. We have thus used $\prec$ to define a kind of lexicographic order on L. It is easily verified that it is indeed a total order of L. We now show that $\langle L, \lhd\rangle$ is a Suslin line.

First, to show that L has the c.c.c, suppose $\{(C_\xi, D_\xi): \xi < \omega_1\}$ is a family of disjoint non-empty open intervals. Pick $E_\xi \in (C_\xi, D_\xi)$, and pick α_ξ, so that

$$\max\big(d(C_\xi, E_\xi), d(E_\xi, D_\xi)\big) < \alpha_\xi < h(E_\xi);$$

then $\{E_\xi(\alpha_\xi): \xi < \omega_1\}$ forms an antichain in T, contradicting that T is Suslin.

To show that L is not separable, it is sufficient to see that for each $\delta < \omega_1$, $\{C: h(C) < \delta\}$ is not dense in L. Fix $x \in \mathrm{Lev}_\delta(T)$. By Lemma 5.12, there is an $\alpha > \delta$ with 3 distinct elements, $y, z, \omega \in \mathrm{Lev}_\alpha(T)$ above x. Let D, E, F be elements of L containing y, z, w respectively. Say they are ordered, $D \lhd E \lhd F$, then (D, F) is a non-empty interval, but since $x \in D \cap F$, (D, F) contains no $C \in L$ with $h(C) < \delta$.

Conversely, suppose we are given a Suslin line, $\langle L, \lhd \rangle$. By Theorem 4.4, we may assume that L is dense in itself and that no non-empty open subset of L is separable. Let $\mathscr{J}$ be the set of all non-empty open intervals of L; so elements of $\mathscr{J}$ are of the form (a, b), where $a \lhd b$. $\mathscr{J}$ is partially ordered by reverse inclusion: $I \leq J$ iff $J \subset I$. We shall define a subset $T \subset \mathscr{J}$ so that $\leq$ is a Suslin tree ordering on T.

To find T, we first find $\mathscr{I}_\beta \subset \mathscr{J}$ for $\beta < \omega_1$ so that for each β,

(1) the elements of $\mathscr{I}_\beta$ are pairwise disjoint,

(2) $\bigcup \mathscr{I}_\beta$ is dense in L, and

(3) if $\alpha < \beta$, $I \in \mathscr{I}_\alpha$, and $J \in \mathscr{I}_\beta$, then either

  (a) $I \cap J = 0$, or

  (b) $J \subset I$ and $I \setminus \mathrm{cl}(J) \neq 0$.

Assuming this can be done, we let $T = \bigcup_\beta \mathscr{I}_\beta$. By (1)–(3), T is a tree and each $\mathscr{I}_\beta = \mathrm{Lev}_\beta(T)$. If $A \subset T$ is an antichain, then the elements of A are pairwise disjoint, so $|A| \leq \omega$. T can have no uncountable chains, since if $\{I_\xi: \xi < \omega_1\}$ were such a chain, with $\xi < \eta \to I_\xi \leq I_\eta$, then by (3b),

$$\xi < \eta \to \big(I_\eta \subset I_\xi \wedge I_\xi \setminus \mathrm{cl}(I_\eta) \neq 0\big),$$

so $\{I_\xi \setminus \mathrm{cl}(I_{\xi+1}): \xi < \omega_1\}$ would contradict the c.c.c. of L. Finally, $|T| = \omega_1$, since (2) implies in particular that each $\mathscr{I}_\beta \neq 0$. Thus, T is Suslin.

We now construct the $\mathscr{I}_\beta$ by induction. $\mathscr{I}_0$ is any maximal disjoint subfamily of $\mathscr{J}$; maximality implies $\bigcup \mathscr{I}_0$ is dense. Given $\mathscr{I}_\alpha$, we define $\mathscr{I}_{\alpha+1}$ as follows: For $I \in \mathscr{I}_\alpha$, let $\mathscr{K}_I$ be a maximal disjoint subfamily of

$$\{K \in \mathscr{J}: K \subset I \wedge I \setminus \mathrm{cl}(J) \neq 0\}.$$

Let $\mathscr{I}_{\alpha+1} = \bigcup \{\mathscr{K}_I: I \in \mathscr{I}_\alpha\}$.

Finally, assume γ is a limit and we have defined the $\mathscr{I}_\alpha$ for $\alpha < \gamma$ satisfying (1)–(3) for $\alpha < \beta < \gamma$. Let

$$\mathscr{K} = \{K \in \mathscr{J}: \forall \alpha < \gamma\, \forall I \in \mathscr{I}_\alpha [I \cap K = 0 \vee (K \subset I \wedge I \setminus \mathrm{cl}(K) \neq 0)]\},$$

and let $\mathscr{I}_\gamma$ be a maximal disjoint subfamily of $\mathscr{K}$. Then (1) and (3) now hold for all $\alpha < \beta \leq \gamma$. (2) for $\beta = \gamma$ says that no $J \in \mathscr{J}$ is disjoint from all members of $\mathscr{I}_\gamma$; this will follow by maximality of $\mathscr{I}_\gamma$ if we can show that for each $J \in \mathscr{J}$, $\exists K \in \mathscr{K}$ $(K \subset J)$. Let E be the set of all left and right endpoints of all intervals in $\bigcup_{\alpha<\gamma} \mathscr{I}_\alpha$. E is countable and J is not separable, so fix $K_1 \in \mathscr{J}$ with $K_1 \subset J$ and $K_1 \cap E = 0$. If $I \in \bigcup_{\alpha<\gamma} \mathscr{I}_\alpha$, then K_1 does not contain the endpoints of I, so $I \cap K_1 = 0$ or $K_1 \subset I$. Now take $K \in \mathscr{J}$ with $K \subset K_1$ and $K_1 \smallsetminus \mathrm{cl}(K) \neq 0$; then $K \subset \mathscr{J}$ and $K \in \mathscr{K}$. □

It follows now that $\mathrm{MA}(\omega_1)$ implies that there are no ω_1-Suslin trees, but the argument we have obtained for this (using Theorems 5.13 and 4.2) is rather indirect. There is a much simpler proof, using the tree itself to prove its own non-existence.

5.14. Lemma. $\mathrm{MA}(\omega_1)$ *implies that there are no* ω_1*-Suslin trees.*

Proof. Let $\langle T, \leq \rangle$ be an ω_1-Suslin tree, and let $\mathbb{P} = \langle T, \geq \rangle$, the reverse order of T. Since T has no uncountable antichains, $\mathbb{P}$ has c.c.c. By Lemma 5.11, we may assume T is well-pruned, in which case $D_\alpha = \{x \in T : \mathrm{ht}(x, T) > \alpha\}$ is dense in $\mathbb{P}$. By $\mathrm{MA}(\omega_1)$, there is a filter G intersecting each D_α. Then G is an uncountable chain, contradicting the fact that T is Suslin. □

We now take up a new kind of tree.

5.15. Definition. If T is a κ-tree, a *path* through T is a chain C which intersects $\mathrm{Lev}_\alpha(T)$ for each $\alpha < \kappa$. □

Equivalently, a path is a maximal chain of cardinality κ.

It is easy to find examples of κ-trees which are not Aronszajn; for example, the ordinal κ itself. However, the natural examples have few paths. κ has precisely one. For a less trivial example, if

$$T = \{s \in {}^{<\kappa}2 : |\{\alpha \in \mathrm{dom}(s) : s(\alpha) = 1\}| < \omega\},$$

then T has precisely κ paths.

5.16. Definition. For any regular κ, a *κ-Kurepa tree* is a κ-tree with at least κ^+ paths. κ-KH is the statement, "there is a κ-Kurepa tree." KH, or *Kurepa's Hypothesis*, is ω_1-KH. □

Note that KH says there *is* an ω_1-Kurepa tree, whereas SH says that there is *no* ω_1-Suslin tree.

KH is independent of ZFC + GCH. KH follows from $\diamondsuit^+$ (see §7), which in turn follows from $\mathbf{V} = \mathbf{L}$ (see VI §5). In VIII §3, we show $\neg$KH is consistent with GCH.

Unlike SH, κ-SH for successor cardinals κ other than ω_1 can be seen to be independent by minor modifications of the argument for ω_1 (see VIII, Exercises F1–F6). However, for $\kappa = \omega$ or κ strongly inaccessible, κ-KH is trivial, since the complete binary tree of height κ is a κ-Kurepa tree.

KH is equivalent to a simple combinatorial statement not involving trees.

5.17. DEFINITION. A κ-Kurepa family is an $\mathscr{F} \subset \mathscr{P}(\kappa)$ such that $|\mathscr{F}| \geq \kappa^+$ and

$$\forall \alpha < \kappa (|\{A \cap \alpha : A \in \mathscr{F}\}| < \kappa). \quad \square$$

Observe that the notion of a κ-Kurepa family does not really depend on the ordering of κ. In general, if $\mathscr{F} \subset \mathscr{P}(I)$ and $X \subset I$, let

$$\mathscr{F}_X = \{A \cap X : A \in \mathscr{F}\}.$$

If $X \subset Y \subset I$, we may map $\mathscr{F}_Y$ onto $\mathscr{F}_X$ by taking $A \cap Y$ to $A \cap X$, so $|\mathscr{F}_X| \leq |\mathscr{F}_Y|$. In particular, if $I = \kappa$ and κ is regular, then $\forall \alpha < \kappa (|\mathscr{F}_\alpha| < \kappa)$ iff $\forall X \subset \kappa (|X| < \kappa \to |\mathscr{F}_X| < \kappa)$. Thus, for κ regular, there is a κ-Kurepa family iff for some (or any) I with $|I| = \kappa$, there is an $\mathscr{F} \subset \mathscr{P}(I)$ with $|\mathscr{F}| \geq \kappa^+$ and

$$\forall X \subset I (|X| < \kappa \to |\mathscr{F}_X| < \kappa).$$

5.18. THEOREM. *For any regular κ, there is a κ-Kurepa family iff there is a κ-Kurepa tree.*

PROOF. If T is a κ-Kurepa tree, let $\mathscr{F}$ be the set of all paths through T; then $|\mathscr{F}| \geq \kappa^+$. $|T| = \kappa$, so, as we have just observed, we will have a Kurepa family if we can show that for each $X \subset T$ with $|X| < \kappa$, $|\mathscr{F}_X| < \kappa$. Fix such an X. Since κ is regular, there is an $\alpha < \kappa$ such that $\forall x \in X\,(\text{ht}(x, T) < \alpha)$. Since every $C \in \mathscr{F}$ intersects $\text{Lev}_\alpha(T)$, each element of $\mathscr{F}_X$ is of the form $\{x \in X : x < z\}$ for some $z \in \text{Lev}_\alpha(T)$. Thus, $|\mathscr{F}_X| \leq |\text{Lev}_\alpha(T)| < \kappa$.

Conversely, suppose $\mathscr{F} \subset \mathscr{P}(\kappa)$ is a κ-Kurepa family. For $B \in \mathscr{F}_\alpha$, let let $\chi_B \in {}^\alpha 2$ be its characteristic function; then

$$\bigcup_{\alpha < \kappa} \{\chi_B : B \in \mathscr{F}_\alpha\} \subset {}^{<\kappa}2$$

is a κ-Kurepa tree. $\square$

§6. The c.u.b. filter

In §1, we discussed almost disjoint subsets of κ, where "almost" meant "$<\kappa$". There are other notions of "almost"; the general concept involved is that of a filter.

6.1. Definition. For any non-empty set A, a *filter* on A is a subset $\mathscr{F} \subset \mathscr{P}(A)$ such that:

(a) $A \in \mathscr{F}$ and $0 \notin \mathscr{F}$.

(b) $\forall X, Y \in \mathscr{F}\,(X \cap Y \in \mathscr{F})$.

(c) $\forall X \in \mathscr{F}\, \forall Y \subset A(X \subset Y \to Y \in \mathscr{F})$. □

6.2. Definition. For any non-empty set A, an *ideal* on A is a subset $\mathscr{I} \subset \mathscr{P}(A)$ such that:

(a) $0 \in \mathscr{I}$ and $A \notin \mathscr{I}$.

(b) $\forall X, Y \in \mathscr{I}\,(X \cup Y \in \mathscr{I})$.

(c) $\forall X \in \mathscr{I}\, \forall Y \subset A\,(X \supset Y \to Y \in \mathscr{I})$. □

6.3. Definition. If $\mathscr{I}$ is an ideal on A, the *dual filter*, $\mathscr{I}^*$, is

$$\{X \subset A : A \setminus X \in \mathscr{I}\}.$$

If $\mathscr{F}$ is a filter on A, the *dual ideal*, $\mathscr{F}^*$, is $\{X \subset A : A \setminus X \in \mathscr{F}\}$. □

6.4. Lemma. *Let $\mathscr{F}$ be a filter and $\mathscr{I}$ an ideal on A, then $\mathscr{F}^*$ is an ideal, $\mathscr{I}^*$ is a filter, $\mathscr{F}^{**} = \mathscr{F}$, and $\mathscr{I}^{**} = \mathscr{I}$.* □

We remark that a filter on A is a filter in $\mathscr{P}(A) \setminus \{0\}$ in the sense of Definition 2.4, where the partial order is set-theoretic inclusion.

Example 1. For any infinite A, $\{X \subset A : |X| < \omega\}$ is an ideal on A.

Example 2. Let $A = [0, 1] \subset \mathbb{R}$. Let μ be Lebesgue measure; "$\mu(X) = r$" means "X is Lebesgue measurable and $\mu(X) = r$"; then $\mathscr{F} = \{X \subset [0, 1]: \mu(X) = 1\}$ is a filter, and $\mathscr{F}^* = \{X \subset [0, 1]: \mu(X) = 0\}$. We say a property $\phi(x)$ holds *almost everywhere* in $[0, 1]$ iff $\{x \in [0, 1]: \phi(x)\} \in \mathscr{F}$. This measure-theoretic use of "almost" is frequently borrowed in discussions of filters in general.

Example 3. Let κ be an infinite cardinal, and $\mathscr{I} = \{X \subset \kappa: |X| < \kappa\}$, then an almost disjoint family (Definition 1.1) is an $\mathscr{A} \subset (\mathscr{P}(\kappa) \setminus \mathscr{I})$, such that $X \cap Y \in \mathscr{I}$ whenever X and Y are distinct members of $\mathscr{A}$.

It is immediate from the definition that any ideal is closed under finite unions, but ideals such as the ones in Examples 2 and 3 can be closed under bigger unions as well.

6.5. DEFINITION. An ideal $\mathscr{I}$ is *κ-complete* iff

$$\forall \mathscr{A} \subset \mathscr{I}(|\mathscr{A}| < \kappa \rightarrow \bigcup \mathscr{A} \in \mathscr{I}).$$

A filter $\mathscr{F}$ is *κ-complete* iff

$$\forall \mathscr{A} \subset \mathscr{F}(|\mathscr{A}| < \kappa \rightarrow \bigcap \mathscr{A} \in \mathscr{F}). \quad \square$$

Thus, an ideal is κ-complete iff its dual filter is κ-complete. Every ideal and filter is ω-complete. If $\mathscr{I}$ is an ideal on A and $\{a\} \in \mathscr{I}$ for each $a \in A$, then $\mathscr{I}$ is not $|A|^+$-complete, since

$$A = \bigcup \{\{a\} : a \in A\} \notin \mathscr{I}.$$

$\mathscr{I}$ could be $|A|$-complete; for example $\{X \subset \kappa: |X| < \kappa\}$ is κ-complete iff κ is regular.

The measure ideal of Example 2 is ω_1-complete but not $(2^\omega)^+$-complete. Under MA, the measure ideal is 2^ω-complete (see 2.21), as is the category ideal,

$$\{X \subset [0, 1] : X \text{ is first category}\}$$

(see Theorem 2.20).

We now introduce the closed unbounded (c.u.b.) ideal and filter.

6.6. DEFINITION. For any limit ordinal μ, a set $C \subset \mu$ is *closed* iff for all limit $\delta < \mu$, if $C \cap \delta$ is unbounded in δ then $\delta \in C$. C is c.u.b. iff C is closed and unbounded in μ. $\square$

We remark that being closed is equivalent to being closed in the order topology. Examples of closed sets are $\{\gamma < \mu: \gamma \text{ is a limit ordinal}\}$ and $\{\gamma < \mu: \gamma \text{ is a limit of limits}\}$. If μ is a cardinal $> \omega$, these sets are c.u.b. in μ. If μ is a limit cardinal, then $\{\gamma < \mu : \gamma \text{ is a cardinal}\}$ is c.u.b. in μ. If $\text{cf}(\mu) = \omega$, then any cofinal ω-sequence in μ is c.u.b.

6.7. DEFINITION. If $\text{cf}(\mu) > \omega$, the c.u.b. *filter* on μ, $\text{Cub}(\mu)$ is

$$\{X \subset \mu: \exists C \subset X (C \text{ is c.u.b. in } \mu)\}. \quad \square$$

For example, let

$$C = \{\gamma < \omega_1 : \gamma \text{ is a limit and } \gamma > \omega\},$$

and let $X = C \cup \omega$, then $X \in \text{Cub}(\omega_1)$ but is not c.u.b. in ω_1. This example shows that the c.u.b. sets themselves do not form a filter, since condition (c) of Definition 6.1 is violated. $\mathrm{Cub}(\mu)$ has at least a chance of being a filter, since conditions (a) and (c) are easily verified for it. Condition (b) requires some further comment; note that it would fail if we allowed $\mathrm{cf}(\mu)$ to be ω, since there would be disjoint cofinal ω-sequences in μ.

6.8. LEMMA. *If* $\mathrm{cf}(\mu) > \omega$, *then*

(a) *the intersection of any family of less than* $\mathrm{cf}(\mu)$ c.u.b. *subsets of* μ *is* c.u.b.

(b) $\mathrm{Cub}(\mu)$ *is a* $\mathrm{cf}(\mu)$*-complete filter.*

PROOF. For (a), let C_α be c.u.b. in μ for $\alpha < \lambda$, where $\lambda < \mathrm{cf}(\mu)$, and let $D = \bigcap_{\alpha<\lambda} C_\alpha$. D is easily seen to be closed. To show that D is unbounded, first let $f_\alpha(\xi)$ be the least element of C_α greater than ξ. So $f_\alpha\colon \mu \to \mu$. Let $g(\xi) = \sup\{f_\alpha(\xi)\colon \alpha < \lambda\}$; then $\xi < g(\xi) < \mu$ since $\mathrm{cf}(\mu) > \lambda$. Let $g^0(\xi) = \xi$, $g^{n+1}(\xi) = g(g^n(\xi))$ and $g^\omega(\xi) = \sup\{g^n(\xi)\colon n < \omega\}$; then $\xi < g^\omega(\xi) < \mu$ since $\mathrm{cf}(\mu) > \omega$. For each α, C_α is unbounded in $g^\omega(\xi)$, so $g^\omega(\xi) \in C_\alpha$; so $g^\omega(\xi) \in \bigcap_\alpha C_\alpha$. Thus, for each ξ, $g^\omega(\xi)$ is an element of D greater than ξ.

To prove (b), let $X_\alpha \in \mathrm{Cub}(\mu)$ for $\alpha < \lambda$, where $\lambda < \mathrm{cf}(\mu)$. Pick c.u.b. $C_\alpha \subset X_\alpha$; then $\bigcap_\alpha C_\alpha \subset \bigcap_\alpha X_\alpha$, so $\bigcap_\alpha X_\alpha \in \mathrm{Cub}(\mu)$ by (a). □

$\mathrm{Cub}(\mu)$ is most often used when μ is regular, in which case it is μ-complete and $\mathrm{Cub}^*(\mu) \supset \{X \subset \mu\colon |X| < \mu\}$.

We present some auxiliary notation.

6.9. DEFINITION. If $\mathrm{cf}(\mu) > \omega$, $X \subset \mu$ is *stationary* iff $X \notin \mathrm{Cub}^*(\mu)$, and X is *non-stationary* iff $X \in \mathrm{Cub}^*(\mu)$. □

Equivalently, X is stationary iff $X \cap C \neq 0$ for all c.u.b. C. It is immediate from Lemma 6.8 that the union of any family of less than $\mathrm{cf}(\mu)$ non-stationary subsets of μ is non-stationary, and that if $\mu = \bigcup_{\alpha<\lambda} X_\alpha$, where $\lambda < \mathrm{cf}(\mu)$, then some X_α is stationary. An example of stationary sets is given by

6.10. LEMMA. *If* $\mathrm{cf}(\mu) > \lambda$, *where* λ *is regular, then* $\{\gamma < \mu\colon \mathrm{cf}(\gamma) = \lambda\}$ *is stationary in* μ.

PROOF. If C is c.u.b. in μ, then the λ-th element of C has cofinality λ. □

Thus, if $\mathrm{cf}(\mu) \geq \omega_2$, we can find two disjoint stationary subsets of μ—namely $\{\gamma < \mu\colon \mathrm{cf}(\gamma) = \omega\}$ and $\{\gamma < \mu\colon \mathrm{cf}(\gamma) = \omega_1\}$ (equivalently, $\mathrm{Cub}(\mu)$ is not an ultrafilter). If $\mathrm{cf}(\mu) = \omega_1$, this is not so obvious, but still true, by a classical result of Ulam. We confine our attention from now on to regular cardinals, since the general case is easily reduced to it (Exercise 43).

6.11. THEOREM. *Ulam. Let κ be a successor cardinal and $\mathscr{I}$ a κ-complete ideal on κ such that each singleton is in $\mathscr{I}$. Then there are disjoint $X_\alpha \subset \kappa$ for $\alpha < \kappa$, such that each $X_\alpha \notin \mathscr{I}$.*

PROOF. Let $\kappa = \lambda^+$. Note that every subset of κ of size $\le\lambda$ is in $\mathscr{I}$. Let $\mathscr{F} = \mathscr{I}^*$.

For each $\rho < \kappa$, let f_ρ be a 1–1 function from ρ into λ. Now, for each $\alpha < \kappa$ and $\xi < \lambda$, let

$$X_\alpha^\xi = \{\rho > \alpha : f_\rho(\alpha) = \xi\} \subset \kappa;$$

then $\alpha \neq \beta \to X_\alpha^\xi \cap X_\beta^\xi = 0$ for any $\xi < \lambda$, since each f_ρ is 1–1. Furthermore, for each $\alpha < \kappa$,

$$\bigcup_\xi X_\alpha^\xi = \{\rho < \kappa : \rho > \alpha\} \in \mathscr{F}.$$

Since $\mathscr{I}$ is κ-complete, we cannot have $\forall\xi < \lambda\,(X_\alpha^\xi \in \mathscr{I})$, so let $h(\alpha) < \lambda$ be such that $X_\alpha^{h(\alpha)} \notin \mathscr{I}$. Since $h: \kappa \to \lambda$, there is a fixed $\xi < \lambda$, such that $|h^{-1}\{\xi\}| = \kappa$; then $\{X_\alpha^\xi : h(\alpha) = \xi\}$ satisfies the theorem. □

The $\lambda \times \kappa$ matrix of sets X_α^ξ is known as an *Ulam matrix.*

6.12. COROLLARY. *For any regular $\kappa > \omega$, there is a family of κ disjoint stationary subsets of κ.*

PROOF. If κ is a successor, apply Theorem 6.11 to $\mathrm{Cub}^*(\kappa)$. If κ is a limit, and hence weakly inaccessible, then there are κ regular cardinals $\lambda < \kappa$, and for each such λ, $\{\gamma < \kappa : \mathrm{cf}(\gamma) = \lambda\}$ is stationary by Lemma 6.10. □

If κ is a successor cardinal, then whenever S is any stationary subset of κ, S may be partitioned into κ disjoint stationary subsets, since one may apply Theorem 6.11 to the ideal, $\{X \subset \kappa : X \cap S \text{ is nonstationary}\}$. This fact is also true for κ weakly inaccessible by a much harder proof [Solovay 1971].

It is unknown whether there can be a regular $\kappa > \omega$ such that there is no almost disjoint family of κ^+ stationary sets (see also Exercise 52).

We close with a few additional facts about c.u.b. sets whose proofs are modifications of the κ-completeness of $\mathrm{Cub}(\kappa)$ (Lemma 6.8). The first is a version of the downward Löwenheim–Skolem–Tarski theorem (I 10.23).

6.13. LEMMA. *Let $\kappa > \omega$ be regular and let $\mathscr{A}$ be a set of less than κ finitary functions on κ; then $C = \{\gamma < \kappa : \gamma \text{ is closed under } \mathscr{A}\}$ is* c.u.b. *in κ.*

PROOF. C is easily seen to be closed. To see that it is unbounded, first let $G(\xi)$ be the closure of ξ under $\mathscr{A}$; then $\xi \subset G(\xi) \subset \kappa$, and $|G(\xi)| < \kappa$ by

I 10.23. Since κ is regular, we may pick $g(\xi)$ so that $\xi < g(\xi) < \kappa$ and $G(\xi) \subset g(\xi)$. As in the proof of Lemma 6.8, let g^n be the n-th iterate of g and $g^\omega(\xi) = \sup_n g^n(\xi)$; then $g^\omega(\xi)$ is an element of C greater than ξ. $\square$

6.14. Lemma. *Let $\kappa > \omega$ be regular, and let C_α be* c.u.b. *in κ for all $\alpha < \kappa$. Then*

$$D = \{\gamma \colon \forall \alpha < \gamma\,(\gamma \in C_\alpha)\}$$

is c.u.b. *in κ.*

Proof. D is easily seen to be closed. To see that D is unbounded, let $g(\xi)$ be some element of $\bigcap_{\alpha<\xi} C_\alpha$ which is larger than ξ (note that $\bigcap_{\alpha<\xi} C_\alpha$ is unbounded by Lemma 6.8), then $g^\omega(\xi)$ (defined as in Lemma 6.13) is an element of D greater than ξ. $\square$

D is called the *diagonal intersection* of the C_α.

The following consequence of Lemma 6.14, due to Fodor, has frequent applications.

6.15. Lemma. *The Pressing-Down Lemma. Let $\kappa > \omega$ be regular, S a stationary subset of κ, and $f: S \to \kappa$ such that $\forall \gamma \in S\,(f(\gamma) < \gamma)$; then for some $\alpha < \kappa$, $f^{-1}\{\alpha\}$ is stationary.*

Proof. If not, then for each α pick a c.u.b. C_α with $C_\alpha \cap f^{-1}\{\alpha\} = 0$. Let $D = \{\gamma \colon \forall \alpha < \gamma\,(\gamma \in C_\alpha)\}$. D is c.u.b. by Lemma 6.14. But also $D \cap S = 0$, since if $\gamma \in D \cap S$, $f(\gamma) \neq \alpha$ for all $\alpha < \gamma$, contradicting $f(\gamma) < \gamma$. Thus, S could not have been stationary. $\square$

§7. $\diamondsuit$ and $\diamondsuit^+$

7.1. Definition. $\diamondsuit$ is the statement: There are sets $A_\alpha \subset \alpha$ for $\alpha < \omega_1$ such that

$$\forall A \subset \omega_1\,(\{\alpha < \omega_1 : A \cap \alpha = A_\alpha\} \text{ is stationary}).$$

The sequence $\langle A_\alpha \colon \alpha < \omega_1 \rangle$ is called a *$\diamondsuit$-sequence.* $\square$

7.2. Lemma. $\diamondsuit \to$ CH.

Proof. If $\langle A_\alpha \colon \alpha < \omega_1 \rangle$ is a $\diamondsuit$-sequence, then

$$\forall A \subset \omega\, \exists \alpha > \omega (A = A_\alpha),$$

so $\{A_\alpha \colon A_\alpha \subset \omega\} = \mathscr{P}(\omega)$. $\square$

$\diamondsuit$ may be viewed as a strengthening of CH. From a $\diamondsuit$-sequence, we easily read off a listing of $\mathscr{P}(\omega)$, but the $\diamondsuit$-sequence "captures" all subsets of ω_1 as well. $\diamondsuit$ may be used to construct various objects of size ω_1 which have some property involving all their subsets. We shall illustrate this by showing that $\diamondsuit \rightarrow \neg$SH; i.e., we use $\diamondsuit$ to construct an ω_1-Suslin tree. It is known that CH, or even GCH does not imply the existence of an ω_1-Suslin tree (see [Devlin–Johnsbråten 1974]). In VI §5 we shall show that $\diamondsuit$ is consistent with GCH.

Recall from 5.4–5.6 that an ω_1-Suslin tree is an ω_1-tree in which all chains and all antichains are countable. If we avoid certain trivialities in our construction of the tree, we need only worry about antichains, and in fact just maximal ones.

7.3. Definition. A tree, $\langle T, \leq \rangle$ is *ever-branching* iff for all $x \in T$, $\{y \in T: y > x\}$ is not totally ordered by $<$. □

7.4. Lemma. *Suppose $\langle T, \leq \rangle$ is an ever-branching ω_1-tree in which every maximal antichain is countable; then T is an ω_1-Suslin tree.*

Proof. By Zorn's Lemma, every antichain is contained in a maximal one, so every antichain is countable. Suppose B were an uncountable chain. We may assume B is maximal; so B intersects every level of T. Since T is ever-branching, there is, for each $x \in T$, an $f(x) > x$ such that $f(x) \notin B$. Now, inductively pick $x_\alpha \in B$ for $\alpha < \omega_1$ so that $\mathrm{ht}(x_\alpha, T) > \sup\{\mathrm{ht}(f(x_\beta), T): \beta < \alpha\}$; then $\{f(x_\alpha): \alpha < \omega_1\}$ would be an uncountable antichain. □

Since $\diamondsuit$ talks about subsets of ω_1, it is natural to try to build a Suslin tree of the form $\langle \omega_1, \lhd \rangle$; i.e., we shall construct some Suslin tree order $\lhd$ on ω_1. We first note that by Löwenheim–Skolem arguments, elementary properties of our tree will reflect to a c.u.b. set of countable ordinals. More specifically, the following applies.

7.5. Definition. For any tree T, let $T_\alpha = \bigcup\{\mathrm{Lev}_\beta(T): \beta < \alpha\}$. □

Thus, T_α is the sub-tree of T below level α.

7.6. Lemma. *Let $T = \langle \omega_1, \lhd \rangle$ be an ω_1-tree. Then*

(a) $\{\alpha < \omega_1 : T_\alpha = \alpha\}$ *is* c.u.b. *in* ω_1.

(b) *If $A \subset \omega_1$ is a maximal antichain in T, then* $\{\alpha < \omega_1 : T_\alpha = \alpha \wedge A \cap T_\alpha$ *is a maximal antichain in* $T_\alpha\}$ *is* c.u.b. *in* ω_1.

Proof. For (a), the set is clearly closed. To see that it is unbounded, define $f(\xi) = \mathrm{ht}_T(\xi)$ and $g(\xi) = \sup\{\eta: \eta \in \mathrm{Lev}_\xi(T)\}$. By Lemma 6.13, the set of α closed under f and g forms a c.u.b., and $T_\alpha = \alpha$ for any such α.

For (b), note that A is maximal in T iff all $x \in T \setminus A$ are comparable with some element of A. Again, the set described is easily seen to be closed. To see that it is unbounded, let $h(\xi)$ be some element of A comparable with ξ (so $\xi \in A \to h(\xi) = \xi$). If α is closed under f, g, and h, then $T_\alpha = \alpha$ and $A \cap T_\alpha$ is maximal in T_α. □

We now state precisely how $\diamondsuit$ is used in the construction of a Suslin tree.

7.7. LEMMA. *Let $T = \langle \omega_1, \lhd \rangle$ be an ever-branching ω_1-tree, and $\langle A_\alpha : \alpha < \omega_1 \rangle$ a $\diamondsuit$-sequence. Suppose that for all limit $\alpha < \omega_1$,*

$$(T_\alpha = \alpha \wedge A_\alpha \text{ is a maximal antichain in } \alpha) \to \forall x \in \mathrm{Lev}_\alpha(T)\, \exists y \in A_\alpha (y \lhd x). \tag{$*$}$$

Then T is an ω_1-Suslin tree.

PROOF. By Lemma 7.4, it is sufficient to check that every maximal antichain, A, is countable. By Lemma 7.6,

$$C = \{\alpha < \omega_1 : \alpha \text{ is a limit} \wedge T_\alpha = \alpha \wedge A \cap T_\alpha \text{ is maximal in } T_\alpha\}$$

is c.u.b. in ω_1. Since $\{\alpha : A \cap \alpha = A_\alpha\}$ is stationary, we may fix an $\alpha \in C$, such that $A \cap \alpha = A_\alpha$. By $(*)$, if $z \in T$ and $\mathrm{ht}_T(z) \geq \alpha$, then z is above some element of $A_\alpha = A \cap \alpha$, so $z \notin A$. Thus, $A = A_\alpha$, so A is countable. □

It is now an easy matter to build T by transfinite induction to make $(*)$ hold.

7.8. THEOREM. *$\diamondsuit$ implies that there is an ω_1-Suslin tree.*

PROOF. T will be $\langle \omega_1, \lhd \rangle$. Let $I_\beta = \{\omega \cdot \beta + n : n \in \omega\}$. Fix a $\diamondsuit$-sequence $\langle A_\alpha : \alpha < \omega_1 \rangle$. $\lhd$ will be inductively constructed so that

(1) $\lhd$ is a tree order on ω_1 and for each $\beta < \omega_1$, $\mathrm{Lev}_\beta(T) = I_\beta$.

(2) For each $\beta < \omega_1$ and $n < \omega$, $(\omega \cdot \beta + n) \lhd (\omega \cdot (\beta + 1) + 2n)$, and $(\omega \cdot \beta + n) \lhd (\omega \cdot (\beta + 1) + 2n + 1)$.

(3) If $\beta < \alpha < \omega_1$ and $x \in I_\beta$, then $\exists y \in I_\alpha (x \lhd y)$.

(4) $(*)$ of Lemma 7.7 holds.

Assuming that $\lhd$ can be so constructed, (1) and (2) guarantee that T is an ever-branching ω_1-tree, so that (4) implies that T is Suslin. Condition (3) will facilitate the construction of T. Incidentally, note that now $T_\alpha = \omega \cdot \alpha$.

To construct $\lhd$ inductively, we assume that $\lhd$ has been defined on the elements of $\omega \cdot \alpha$ so that (1)–(4) hold below α, and we describe how to extend $\lhd$ to the elements of $\omega \cdot \alpha \cup I_\alpha$. If $\alpha = \beta + 1$, then condition (2) specifies the construction: if $x \in \omega \cdot \alpha$, then $x \lhd (\omega \cdot \alpha + 2n)$ iff $x = (\omega \cdot \beta + n)$ or $x \lhd (\omega \cdot \beta + n)$; likewise for $x \lhd (\omega \cdot \alpha + 2n + 1)$. This preserves (1) and (3), and (4) says nothing for successor α.

From now on, assume α is a limit. For each $x \in T_\alpha = \omega \cdot \alpha$, let $B(x)$ be a chain in T_α, such that $x \in B(x)$ and $B(x)$ intersects $I_\eta = \mathrm{Lev}_\eta(T_\alpha)$ for each $\eta < \alpha$. To find such a $B(x)$, first choose $\xi_m = \xi_m(x)$ for $m < \omega$, such that $\mathrm{ht}(x) < \xi_0 < \xi_1 < \cdots$ and $\sup\{\xi_m : m \in \omega\} = \alpha$; then inductively choose $y_m = y_m(x) \in I_{\xi_m}$ such that $x \lhd y_0 \lhd y_1 \lhd \cdots$ (this is possible by condition (3)); then set

$$B(x) = \{z \in T_\alpha : \exists n (z \lhd y_n(x))\}.$$

Now, let $\omega \cdot \alpha = \{x_n : n \in \omega\}$, and define, for $z \in \omega \cdot \alpha$, $z \lhd (\omega \cdot \alpha + n)$ iff $z \in B(x_n)$. The fact that $B(x_n)$ intersects each level of T_α implies that $\omega \cdot \alpha + n$ indeed has height α in T; the rest of (1)–(3) is now easily seen to be preserved.

Finally. condition (4) at level α is only a problem if $\omega \cdot \alpha = \alpha$ and A_α is a maximal antichain in T_α, so assume that this is the case. Then, modify the construction of $B(x)$ for $x \in T_\alpha$ by first choosing $y_0(x)$ so that $x \lhd y_0(x)$ and $\exists z \in A_\alpha (z \lhd y_0(x))$; this is possible since x is comparable with some element of A_α. Then $\xi_0(x) = \mathrm{ht}(y_0(x))$. Now choose $\xi_m(x)(1 \le m < \omega)$ and $y_m(x)(1 \le m < \omega)$ as before. Then each $B(x)$ intersects A_α, so (*) holds. □

To simplify the ordinal arithmetic, we have made $\mathrm{Lev}_\alpha(T)$ uniformly $\{\omega \cdot \alpha + n : n \in \omega\}$, so that $\mathrm{Lev}_0(T) = \omega$. It is easy to modify the construction to make T everywhere binary, so that $\mathrm{Lev}_0(T) = \{0\}$, $\mathrm{Lev}_1(T) = \{1, 2\}$, $\mathrm{Lev}_2(T) = \{3, 4, 5, 6\}$, etc., and $\mathrm{Lev}_{\omega+\alpha}(T) = \{\omega \cdot (1 + \alpha) + n : n \in \omega\}$; then T would be embeddable as a subtree of $2^{<\omega_1}$.

One might now attempt to construct a Kurepa tree, but a principle stronger than $\diamondsuit$ is needed, since GCH + $\diamondsuit$ + $\neg$KH is consistent (see VIII Exercise J6). An obvious strengthening of $\diamondsuit$ is the principle $\diamondsuit!$ obtained by replacing "stationary" by "c.u.b." in Definition 7.1. $\diamondsuit! \to$ KH, but also $\diamondsuit! \to 0 = 1$, since if A and B are distinct subsets of ω_1,

$$\{\alpha < \omega_1 : A \cap \alpha = A_\alpha\} \cap \{\alpha < \omega_1 : B \cap \alpha = A_\alpha\}$$

is countable. The principle $\diamondsuit^+$ is a weakening of $\diamondsuit!$ which is consistent (see VI §5), and which still implies KH. In $\diamondsuit^+$, we capture sets on c.u.b. sets, but we allow ourselves, for each α, ω chances to do the capturing.

7.9. **Definition.** $\diamondsuit^+$ is the statement: There are sets $\mathscr{A}_\alpha \subset \mathscr{P}(\alpha)$ for $\alpha < \omega_1$, such that each $|\mathscr{A}_\alpha| \le \omega$ and for each $A \subset \omega_1$, there is a c.u.b. $C \subset \omega_1$, such that

(a) $\forall \alpha \in C (A \cap \alpha \in \mathscr{A}_\alpha)$, and

(b) $\forall \alpha \in C (C \cap \alpha \in \mathscr{A}_\alpha)$.

The sequence $\langle \mathscr{A}_\alpha : \alpha < \omega_1 \rangle$ is called a $\diamondsuit^+$-sequence. □

It is true, but not immediately obvious, that $\diamondsuit^+ \to \diamondsuit$, as we shall check after constructing a Kurepa tree.

Actually, we shall construct a Kurepa family, and retrieve the tree from it (see 5.17 and 5.18).

7.10. THEOREM. *$\diamondsuit^+$ implies that there is a family $\mathscr{F} \subset \mathscr{P}(\omega_1)$ such that*
(1) $\forall \beta < \omega_1 (|\{X \cap \beta : X \in \mathscr{F}\}| \leq \omega)$, *and*
(2) $\forall A \subset \omega_1 (|A| = \omega_1 \to \exists X \in \mathscr{F} (|X| = \omega_1 \wedge X \subset A))$.

7.11. COROLLARY. *$\diamondsuit^+$ implies that there is an ω_1-Kurepa tree with 2^{ω_1} paths.*

7.12. COROLLARY. *$\diamondsuit^+$ implies that there is a Kurepa family $\mathscr{G}$ which is a maximal almost disjoint family of subsets of ω_1.*

PROOFS. We show first that Theorem 7.10 implies Corollary 7.11. Observe that $\diamondsuit^+ \to$ CH, since, as in the proof of $\diamondsuit \to$ CH, if $\langle \mathscr{A}_\alpha : \alpha < \omega_1 \rangle$ is a $\diamondsuit^+$-sequence, $\mathscr{P}(\omega) \subset \bigcup_{\alpha < \omega_1} \mathscr{A}_\alpha$ (or, $\diamondsuit^+ \to \diamondsuit$ by Theorem 7.14, and $\diamondsuit \to$ CH). By CH, there is an almost disjoint family, $\mathscr{B} \subset \mathscr{P}(\omega_1)$, of cardinality 2^{ω_1} (see Theorem 1.3). For each $B \in \mathscr{B}$, there is an $X_B \in \mathscr{F}$ with $X_B \subset B$ and $|X_B| = \omega_1$. These X_B are distinct, so $|\mathscr{F}| = 2^{\omega_1}$. It follows that if T is the Kurepa tree constructed from $\mathscr{F}$ by identifying sets with their characteristic functions (see Theorem 5.18), T has 2^{ω_1} paths.

We now show that Theorem 7.10 implies Corollary 7.12. For $\mathscr{B}$ as above, let $\mathscr{G}_1 = \{X_B : B \in \mathscr{B}\}$; then $\mathscr{G}_1 \subset \mathscr{F}$, $|\mathscr{G}_1| = 2^{\omega_1}$, and $\mathscr{G}_1$ is an almost disjoint family. Let $\mathscr{G}$ be such that $\mathscr{G}_1 \subset \mathscr{G}$,
(a) the elements of $\mathscr{G}$ are uncountable and pairwise almost disjoint,
(b) $\mathscr{G} \subset \mathscr{F}$,

and $\mathscr{G}$ is maximal with respect to (a) and (b). By (2) of Theorem 7.10, $\mathscr{G}$ is a maximal almost disjoint family. Since $\mathscr{G} \subset \mathscr{F}$, (1) of Theorem 7.10 implies that

$$\forall \beta < \omega_1 (|\{X \cap \beta : X \in \mathscr{G}\}| \leq \omega).$$

Since $|\mathscr{G}| \geq |\mathscr{G}_1| \geq \omega_2$, $\mathscr{G}$ is a Kurepa family.

Finally, we prove Theorem 7.10. If $C \subset \omega_1$ and $\xi < \omega_1$, define $s(C, \xi) = \sup(C \cup \{0\}) \cap (\xi + 1)$; so if C is closed, then $s(C, \xi)$ is the largest element of $C \cup \{0\}$ less than or equal to ξ. If $A \subset \omega_1$, let

$$X(A, C) = \{\xi \in A : \neg \exists \eta \in A (s(C, \xi) \leq \eta < \xi)\};$$

so, for each $\beta \in C$, $X(A, C)$ contains the least $\xi \in A$, such that $s(C, \xi) = \beta$, if there is such a ξ. Observe that $X(A, C) \subset A$, and that if $|A| = \omega_1$ and C is c.u.b., then $|X(A, C) = \omega_1|$. Let $\langle \mathscr{A}_\alpha : \alpha < \omega_1 \rangle$ be a $\diamondsuit^+$-sequence. Let $\mathscr{F}$ be the set of all $X(A, C)$, such that

(a) $\forall\alpha\in C\,(A\cap\alpha\in\mathscr{A}_\alpha)$,
(b) $\forall\alpha\in C\,(C\cap\alpha\in\mathscr{A}_\alpha)$, and
(c) $A\subset\omega_1$, $C\subset\omega_1$, $|A|=\omega_1$, and C is c.u.b.

Then (2) of Theorem 7.10 is immediate. (1) will follow if we can show that for each A and C satisfying (a)–(c) and each $\beta<\omega_1$,

$$|X(A,C)\cap\beta|\leq 1 \text{ or}$$
$$\exists\alpha\leq\beta\,\exists x\subset\beta\,\exists B,D\in\mathscr{A}_\alpha(X(A,C)\cap\beta=X(B,D)\cup x \text{ and } |x|\leq 1). \quad (*)$$

To prove (*), let $\alpha=s(C,\beta)$. If $\alpha>0$, then $\alpha\in C$, so let $B=A\cap\alpha$ and $D=C\cap\alpha$, and let ξ be the least element of $A\smallsetminus\alpha$; then $X(A,C)\cap\beta$ equals $X(B,D)$ if $\xi\geq\beta$ and $X(B,D)\cup\{\xi\}$ if $\xi<\beta$. Likewise, if $\alpha=0$, then $X(A,C)\cap\beta$ is either 0 or a singleton. □

Finally, we show that $\Diamond^+\rightarrow\Diamond$. It is convenient to interpolate here a statement which is obviously a weakening of $\Diamond^+$.

7.13. Definition. $\Diamond^-$ is the statement: There are sets $\mathscr{A}_\alpha\subset\mathscr{P}(\alpha)$ for $\alpha<\omega_1$, such that each $|\mathscr{A}_\alpha|\leq\omega$ and for each $A\subset\omega_1$,

$$\{\alpha<\omega_1: A\cap\alpha\in\mathscr{A}_\alpha\}$$

is stationary. □

Clearly $\Diamond^+\rightarrow\Diamond^-$, so the fact that $\Diamond^+\rightarrow\Diamond$ follows from

7.14. Theorem. $\Diamond^-\leftrightarrow\Diamond$.

Proof. $\Diamond\rightarrow\Diamond^-$ is trivial, so we assume $\Diamond^-$ and conclude $\Diamond$.

Let $\langle\mathscr{A}_\alpha:\alpha<\omega_1\rangle$ be a $\Diamond^-$-sequence, i.e., satisfy Definition 7.13. We show first that we may alter the sequence to capture subsets of $\omega\times\omega_1$ rather than ω_1. To do this, let f be any 1–1 function from ω_1 onto $\omega\times\omega_1$. If $A\subset\omega_1$, let $A'=f''A$. If $B\subset\omega\times\omega_1$, let $B^*=f^{-1}B$. Then $A'^*=A$ and $B^{*\prime}=B$. Let

$$C=\{\alpha<\omega_1:\alpha\geq\omega\wedge f^{-1}(\omega\times\alpha)\subset\alpha\wedge f''(\alpha)\subset\omega\times\alpha\}.$$

C is c.u.b. (see Lemma 6.13). If $\alpha\in C$, $A\subset\alpha$, and $B\subset\omega\times\alpha$, then $A'\subset\omega\times\alpha$ and $B^*\subset\alpha$. For $\alpha\in C$, let $\mathscr{B}_\alpha=\{A': A\in\mathscr{A}_\alpha\}$; for $\alpha\notin C$, let $\mathscr{B}_\alpha=0$. We now assert

$$\forall B\subset\omega\times\omega_1(\{\alpha<\omega_1: B\cap\omega\times\alpha\in\mathscr{B}_\alpha\}\text{ is stationary}). \quad (*)$$

To prove (*), let $S = \{\alpha < \omega_1 : B^* \cap \alpha \in \mathscr{A}_\alpha\}$. S is stationary, so $S \cap C$ is stationary. If $\alpha \in S \cap C$, then $B \cap (\omega \times \alpha) = (B^* \cap \alpha)' \in \mathscr{B}_\alpha$.

Since $|\mathscr{B}_\alpha| \le \omega$, write each $\mathscr{B}_\alpha$ as $\{B_\alpha^k : k \in \omega\}$. $B_\alpha^k \subset \omega \times \alpha$. Let $B_{\alpha,n}^k = \{\xi : \langle n, \xi \rangle \in B_\alpha^k\}$. We show that for some n, $\langle B_{\alpha,n}^n : \alpha < \omega_1 \rangle$ is a $\diamondsuit$-sequence. If not, then for each n, we can find a $B_n \subset \omega_1$ such that $\{\alpha < \omega_1 : B_n \cap \alpha = B_{\alpha,n}^n\}$ is non-stationary. Let $B = \bigcup_n (\{n\} \times B_n)$; then for each n,

$$\{\alpha < \omega_1 : B \cap (\omega \times \alpha) = B_\alpha^n\}$$

is non-stationary. Since a countable union of non-stationary sets is non-stationary,

$$\{\alpha < \omega_1 : \exists n \, (B \cap (\omega \times \alpha) = B_\alpha^n)\}$$

is non-stationary, contradicting (*). □

The coding of subsets of $\omega \times \omega_1$ by subsets of ω_1 is a standard technique. It can be modified to allow $\diamondsuit$-sequences or $\diamondsuit^+$ sequences to capture subsets of $\omega_1 \times \omega_1$, functions from ω_1 into ω_1, or other objects of cardinality ω_1; see Exercise 51.

EXERCISES

(1) Prove the Δ-system lemma (1.5) directly. *Hint*. One may assume that for some n, $\forall x \in \mathscr{A} \, (|x| = n)$. Now proceed by induction on n. Observe that there is an uncountable $\mathscr{B} \subset \mathscr{A}$ such that either $\bigcap \mathscr{B} \neq 0$ or $\mathscr{B}$ is pairwise disjoint.

(2) Show that there is a family $\mathscr{A}$ of ω_ω finite sets such that no $\mathscr{B} \subset \mathscr{A}$ of cardinality ω_ω forms a Δ-system.

(3) Show that if $\kappa \le 2^\omega$ and X_α are separable spaces for $\alpha < \kappa$, then $\prod_{\alpha<\kappa} X_\alpha$ is separable. *Hint*. Consider first the space ${}^I X$, where $I \subset {}^\omega 2$ and X is separable. Let D be dense in X. Let E be the set of $\varphi \in {}^I D$, such that for some $n \in \omega$,

$$\forall f, g \in I \, (f \restriction n = g \restriction n \rightarrow \varphi(f) = \varphi(g));$$

then E is dense in X.

(4) Show that if $\kappa > 2^\omega$, then the space ${}^\kappa 2$ (where $2 = \{0, 1\}$ has the discrete topology) is not separable. *Hint*. If $D \subset {}^\kappa 2$ is countable, show that there are $\alpha < \beta$ such that $\forall f \in D \, (f(\alpha) = f(\beta))$.

(5) Show that a topological space X has the c.c.c. iff there is no sequence of open sets, $\langle U_\alpha : \alpha < \omega_1 \rangle$ such that whenever $\alpha < \beta$, $\overline{U}_\alpha$ is a proper subset of $\overline{U}_\beta$.

(6) Show that in the c.c.c. space 2^{ω_1}, there is a sequence of open sets, $\langle U_\alpha : \alpha < \omega_1 \rangle$ such that whenever $\alpha < \beta$, U_α is a proper subset of U_β.

(7) If $\mathscr{B}$ is a complete Boolean algebra, show that $\mathscr{B}$ has the c.c.c. iff there is no sequence $\langle b_\alpha : \alpha < \omega_1 \rangle$ from $\mathscr{B}$ such that $\alpha < \beta \to b_\alpha < b_\beta$.

(8) If $f, g \in \omega^\omega$, say $f <^* g$ iff $\exists n\, \forall m > n\,(f(m) < g(m))$. Let $\mathscr{F} \subset \omega^\omega$ with $|\mathscr{F}| = \kappa$. Assuming MA(κ), show that $\exists g \in \omega^\omega\, \forall f \in \mathscr{F}\,(f <^* g)$. *Hint.* $\mathbb{P}$ is the set of pairs $\langle p, F \rangle$ such that p is a finite partial function from ω to ω and F is a finite subset of $\mathscr{F}$. $\langle p, F \rangle \le \langle q, G \rangle$ iff $q \subset p$, $G \subset F$, and

$$\forall f \in G\, \forall n \in (\mathrm{dom}(p) \smallsetminus \mathrm{dom}(q))\,(p(n) > f(n)).$$

Or, the result may be deduced directly from Theorem 2.15 (see VIII Exercise A3).

(9) Let $\mathscr{B} \subset \mathscr{P}(\omega)$ be an almost disjoint family of size κ, where $\omega \le \kappa < 2^\omega$. Let $\mathscr{A} \subset \mathscr{B}$ with $|\mathscr{A}| \le \omega$. Assuming MA(κ), show that there is a $d \subset \omega$ such that $\forall x \in \mathscr{A}\,(|d \cap x| < \omega)$ and $\forall x \in \mathscr{B} \smallsetminus \mathscr{A}\,(|x \smallsetminus d| < \omega)$. *Remark.* MA($\omega$), and hence the result of this exercise when $|\mathscr{B}| = \omega$, is a theorem of ZFC. When $|\mathscr{B}| = \omega$, however, one may easily prove this result directly by a diagonal argument without using partial orders.

(10) (Hausdorff, Luzin). Show (in ZFC) that the result of Exercise 9 can be false if $|\mathscr{A}| = |\mathscr{B} \smallsetminus \mathscr{A}| = \omega_1$. *Hint.* $\mathscr{A} = \{a_\alpha : \alpha < \omega_1\}$, and $\mathscr{B} \smallsetminus \mathscr{A} = \{b_\alpha : \alpha < \omega_1\}$. Construct a_α, b_α inductively so that $a_\alpha \cap b_\alpha = 0$ but $\alpha \ne \beta \to a_\alpha \cap b_\beta \ne 0$.

(11) Give the ordinal $\omega_1 + 1$ the order topology. Show that the product $(\omega_1 + 1)^\omega$ is an example of a compact Hausdorff space which (regardless of the axioms of set theory) is the union of ω_1 closed nowhere dense sets. Show that the unit ball in a non-separable Hilbert space with the weak topology is another such example. *Hint.* For $(\omega_1 + 1)^\omega$, consider

$$\{f : \forall n\,(f(x) \ne \omega_1 \to f(n) \le \alpha)\}.$$

(12) Show that the products $[0, 1]^{\omega_1}$ and 2^{ω_1} are examples of compact c.c.c. Hausdorff spaces in which, regardless of the axioms of set theory, there is a union of ω_1 closed nowhere dense sets which is not first category. *Hint.* In $[0, 1]^{\omega_1}$, consider $\{f : \exists \alpha\,(f(\alpha) = 0)\}$. Observe that by c.c.c., if V is dense and open, then there is a dense open $W \subset V$ such that W is a countable union of basic open sets.

(13) Show that Theorem 2.20 remains true if we replace $\mathbb{R}$ by any separable metric space.

(14) Show that the following are examples of complete metric spaces which, regardless of the axioms of set theory, are the unions of ω_1 closed nowhere dense sets.

(a) D^ω, where D is an uncountable discrete space.

(b) Any non-separable Hilbert space.

(15) $\langle \mathbb{P}, \leq \rangle$ is called *separative* iff $\leq$ is a partial order in the strict sense and whenever $p \nleq q$, there is an r such that $r \leq p$ and $r \perp q$. Show that $\mathbb{P}$ is separative iff the i of Lemma 3.3 is 1–1 and satisfies

$$\forall p, q \in \mathbb{P} \left(p \leq q \leftrightarrow i(p) \leq i(q)\right).$$

(16) Show that the partial orders in Examples 5 and 6 of §2 are separative, and that the partial order used in Theorem 2.21 is not. In which cases are the orders of 2.7 and 2.22 separative?

(17) Give an example of a $\mathbb{P}$ which is not separative such that the i of 3.3 is 1–1.

(18) Prove the uniqueness of the $\mathscr{B}$ and i of Lemma 3.3. Thus, if $\mathscr{B}_1$, i_1 and $\mathscr{B}_2$, i_2 both satisfy Lemma 3.3, show that there is an isomorphism h from $\mathscr{B}_1$ to $\mathscr{B}_2$ such that $h \circ i_1 = i_2$.

(19) Prove Lemma 3.3 algebraically by representing elements of $\mathscr{B}$ by formal suprema. Thus, as a set, $\mathscr{B} = \mathscr{P}(\mathbb{P})/\sim$, where

$$S \sim T \text{ iff } \bigvee S \text{ "should ="} \bigvee T.$$

Then $[S] \leq [T]$ iff $\forall p \in S \, \neg \exists q \leq p \, \forall r \in T (q \perp r)$.

(20) Let X be any set and let Y be countable. Let $\mathbb{P}$ be the set of functions p such that $|p| < \omega$, $\text{dom}(p) \subset X$ and $\text{ran}(p) \subset Y$. Order $\mathbb{P}$ by: $p \leq q$ iff $q \subset p$. Show that $\mathbb{P}$ has the c.c.c. *Hint.* Use the Δ-system lemma (1.5).

(21) In Exercise 20, let $X = \kappa \times \omega$ and $Y = 2$. Find family $\mathscr{D}$ of κ dense sets in $\mathbb{P}$ such that if G is a filter intersecting all $D \in \mathscr{D}$, $\bigcup G$ is a function from $\kappa \times \omega$ into 2 such that the $(\bigcup G)_\alpha \in {}^\omega 2$ are all distinct, where

$$(\bigcup G)_\alpha (n) = (\bigcup G)(\alpha, n).$$

(22) If $a, b \subset \omega$, say $a \subset^* b$ iff $|a \setminus b| < \omega$ and $|b \setminus a| = \omega$. Assume MA(κ), and let $\langle X, < \rangle$ be a total order with $|X| \leq \kappa$. Show that there are $a_x \subset \omega$ for $x \in X$, such that $x < y \rightarrow a_x \subset^* a_y$. *Hint.* $\mathbb{P}$ is the set of pairs $\langle p, n \rangle$ such that $n \in \omega$, $\text{dom}(p)$ is a finite subset of X, and for each $x \in \text{dom}(p)$, $p(x) \subset n$. $\langle p, n \rangle \leq \langle q, m \rangle$ iff $m \leq n$, $\text{dom}(q) \subset \text{dom}(p)$

$\forall x \in \text{dom}(q)\,(p(x) \cap m = q(x))$, and

$$\forall x, y \in \text{dom}(q)\,(x < y \to (p(x) \setminus p(y)) \subset m).$$

Use the Δ-system lemma to prove that $\mathbb{P}$ has the c.c.c.

(23) Let $\mathscr{B}$ be the Boolean algebra $\mathscr{P}(\omega)/\text{fin}$, where fin is the ideal of finite sets. Assume MA(κ), and let $\mathscr{A}$ be any Boolean algebra of cardinality κ. Show that $\mathscr{A}$ is isomorphic to a sub-algebra of $\mathscr{B}$. *Remark*. This generalizes Exercise 22.

*(24) (Hausdorff) Show that there is an (ω_1, ω_1^*) gap in $\mathscr{P}(\omega)/\text{fin}$; i.e., find a_α, b_α in $\mathscr{P}(\omega)$ for $\alpha < \omega_1$ such that $a_\alpha \subset^* b_\alpha$, $\alpha < \beta \to (a_\alpha \subset^* a_\beta \wedge b_\beta \subset^* b_\alpha)$, and $\neg \exists c\, \forall \alpha\, (a_\alpha \subset^* c \subset^* b_\alpha)$. *Hint*. Choose a_α, b_α inductively so that for each $\alpha < \omega_1$ and $n < \omega$, $|\{\xi < \alpha: (a_\alpha \setminus b_\xi) \subset n\}| < \omega$.

(25) Assume MA(κ). Let $\mathscr{A}$ be a family of Lebesgue measurable subsets of $\mathbb{R}$, with $|\mathscr{A}| = \kappa$. Show that $\bigcup \mathscr{A}$ is Lebesgue measurable and $\mu(\bigcup \mathscr{A}) = \mu(\bigcup \mathscr{B})$ for some countable $\mathscr{B} \subset \mathscr{A}$.

(26) A partial order $\mathbb{P}$ has ω_1 as a *precaliber* iff whenever $p_\alpha \in \mathbb{P}$ for $\alpha < \omega_1$, there is an uncountable $X \subset \omega_1$ such that $\{p_\alpha : \alpha \in X\}$ has F.I.P. (for all finite $s \subset X$, $\exists q\, \forall \alpha \in s\, (q \leq p_\alpha)$). Show that MA($\omega_1$) implies that every c.c.c. $\mathbb{P}$ has ω_1 as a precaliber. *Remark*. This is like Lemma 2.23.

(27) Assume MA(ω_1). Let A_α be a Lebesgue measurable subset of $\mathbb{R}$ for $\alpha < \omega_1$, with $\mu(A_\alpha) > 0$. Show that for some uncountable $X \subset \omega_1$, $\mu(\bigcap_{\alpha \in X} A_\alpha) > 0$. *Hint*. If $\forall \alpha\, (\mu(A_\alpha) > \varepsilon)$, let

$$\mathbb{P} = \{s \subset \omega_1 : |s| < \omega \wedge \mu(\textstyle\bigcap_{\alpha \in s} A_\alpha) > \varepsilon\}.$$

Show $\mathbb{P}$ has c.c.c., and apply Exercise 26 to $\{\{\alpha\}: \alpha < \omega_1\}$.

(28) (Cantor) Let $\langle D, < \rangle$ be a countable total ordering which has no first or last element and which is dense in itself:

$$\forall x, y\, (x < y \to \exists z\, (x < z < y)).$$

Show that $\langle D, < \rangle$ is isomorphic to the rationals, $\langle \mathbb{Q}, < \rangle$. *Hint*. Well-order both D and $\mathbb{Q}$ in type ω, and construct the isomorphism inductively.

(29) Let $\langle X, < \rangle$ be a total ordering satisfying:

(a) X has no first or last element.
(b) X is connected in the order topology.
(c) X is separable in the order topology.

Show that $\langle X, <\rangle$ is isomorphic to the reals, $\langle \mathbb{R}, <\rangle$. *Hint.* Apply Exercise 28 to a countable dense $D \subset X$, and use (b) to extend the isomorphism.

(30) Assume that there is a Suslin line (Definition 4.1). Show that there is a Suslin line, $\langle X, <\rangle$ satisfying (a) and (b) of Exercise 29. *Hint.* Start with the result of Theorem 4.4, throw away the first and last element (if they exist), and take the Dedekind completion.

*(31) Let $\langle X, <\rangle$ be a total ordering which is c.c.c. in the order topology. Show that X has a dense subset of cardinality $\leq \omega_1$.

(32) Show that there is a total ordering $\langle X, <\rangle$ such that there are no increasing or decreasing sequences in X of type ω_1, but in the order topology, every separable subspace of X is nowhere dense. *Hint.* Use an Aronszajn tree to construct an Aronszajn line.

(33) Show that if κ is singular then there is a κ-Suslin tree. *Hint.* Try a disjoint union of well-orderings.

(34) Let κ be regular, and assume that there is a κ-Suslin (or Kurepa, or Aronszajn) tree. Show that there is one which is a sub-tree of ${}^{<\kappa}2$.

(35) If T, T' are κ-trees, the product, $T \times T'$ is the κ-tree whose α-th level is $\mathrm{Lev}_\alpha(T) \times \mathrm{Lev}_\alpha(T')$, with order defined by $\langle x, x'\rangle < \langle y, y'\rangle$ iff $x < y$ and $x' < y'$. Show that if T, T' are κ-Aronszajn trees, so is $T \times T'$.

(36) Show that if T is a κ-Suslin tree, then $T \times T$ is not a κ-Suslin tree.

(37) Assume $\kappa = \lambda^+$, where λ is regular and $2^{<\lambda} = \lambda$. Show that there is a κ-Aronszajn tree. *Hint.* Replace ω by λ in the proof of Theorem 5.9, but demand that $\mathrm{ran}(s_\alpha)$ is nonstationary in λ.

(38) Assume κ is regular, T is a κ-Aronszajn tree, $\lambda < \kappa$, $x \in T$, and $|\{y \in T: y > x\}| = \kappa$. Show

$$\exists \alpha > \mathrm{ht}(x, T)\big(|\{y \in \mathrm{Lev}_\alpha(T): y > x\}| \geq \lambda\big).$$

Hint. If λ is regular, each $|\mathrm{Lev}_\alpha(T)| < \lambda$, and T is well-pruned, then when $\mathrm{cf}(\alpha) = \lambda$, there is a $q(\alpha) < \alpha$ such that T does not branch between levels $q(\alpha)$ and α. The Pressing-Down Lemma will yield a contradiction.

(39) Show that any Aronszajn tree which is a sub-tree of $\{s \in {}^{<\omega_1}\omega: s$ is 1–1$\}$ cannot be a Suslin tree. *Hint.* For each $n \in \omega$, $\{s \in T: \exists \alpha(\mathrm{dom}(s) = \alpha + 1 \wedge s(\alpha) = n)\}$ is an antichain.

(40) An ω_1-Aronszajn tree T is called *special* iff T is the union of ω antichains. Show that T is special iff there is a map $f: T \to \mathbb{Q}$ such that for $x, y \in T$, $x < y \to f(x) < f(y)$. Prove that a special Aronszajn tree exists. *Hint*. Construct T, along with f, by induction. *Remark*. $\diamondsuit$ implies the existence of an Aronszajn sub-tree of $\{s \in {}^{<\omega_1}\omega : s \text{ is } 1\text{–}1\}$ which is not special (see [Baumgartner 1970]). A special Aronszajn tree cannot be Suslin.

*(41) (Baumgartner) Show that MA + $\neg$CH implies that every ω_1-Aronszajn tree is special. *Hint*. $\mathbb{P}$ is the set of finite partial order-preserving functions from T into $\mathbb{Q}$.

(42) Give ω_1 the order topology, and let $f: \omega_1 \to \mathbb{R}$ be continuous. Show that

$$\exists \alpha < \omega_1 \, \forall \beta > \alpha \, (f(\beta) = f(\alpha)).$$

Hint. Fix $\varepsilon > 0$. For each limit α, there is a $g(\alpha) < \alpha$, such that f varies by $< \varepsilon$ in $(g(\alpha), \alpha]$. Now apply the Pressing-Down Lemma (6.15).

(43) Assume $\kappa = \mathrm{cf}(\mu) > \omega$. Show that the Boolean algebras, $\mathscr{P}(\kappa)/\mathrm{Cub}^*(\kappa)$ and $\mathscr{P}(\mu)/\mathrm{Cub}^*(\mu)$ are isomorphic. *Hint*. Consider a c.u.b. $C \subset \mu$ such that $\mathrm{type}(C) = \kappa$.

(44) (Herink) Let $\kappa > \omega$ be regular, and let $\mathscr{B}$ be the Boolean algebra $\mathscr{P}(\kappa)/\mathrm{Cub}^*(\kappa)$. If $A \subset \kappa$, $[A] \in \mathscr{B}$ is its equivalence class. Show that if $A_\alpha \subset \kappa$ for $\alpha < \kappa$, then the inf, $\bigwedge_{\alpha<\kappa} [A_\alpha]$ exists and equals $[D]$, where D is the *diagonal intersection*,

$$D = \{\alpha < \kappa : \forall \beta < \alpha \, (\alpha \in A_\beta)\}.$$

(45) Let $\kappa > \omega$ be regular. Show that there are stationary $S_\alpha \subset \kappa$ for $\alpha < \kappa$ such that $\alpha < \beta \to S_\beta \subset S_\alpha$, and the diagonal intersection of the S_α is $\{0\}$. *Hint*. See Exercise 44. *Remark*. By Lemma 6.14, the S_α cannot be c.u.b.

(46) (Herink) Let $\kappa > \omega$ be regular, and $\mathscr{I} = \{X \subset \kappa : |X| < \kappa\}$. Show that the Boolean algebras $\mathscr{P}(\kappa)/\mathscr{I}$ and $\mathscr{P}(\kappa)/\mathrm{Cub}^*(\kappa)$ are not isomorphic. *Hint*. See Exercise 44.

(47) Let A be a set of infinite cardinals such that for all regular λ, $A \cap \lambda$ is not stationary in λ. Show that there is a 1–1 function g on A, such that $\forall \alpha \in A \, (g(\alpha) < \alpha)$.

(48) κ is called *strongly Mahlo* iff κ is strongly inaccessible and $\{\alpha < \kappa : \alpha$ is regular$\}$ is stationary in κ. Show that for such κ,

$$\{\alpha < \kappa : \alpha \text{ is strongly inaccessible}\}$$

is stationary in κ.

(49) Suppose that κ is strongly inaccessible and there are no κ-Aronszajn trees (such κ are called *weakly compact*). Show that whenever S is a stationary subset of κ, there is a regular $\lambda < \kappa$ such that $S \cap \lambda$ is stationary in λ. *Hint.* If not, apply Exercise 47, and find a κ-tree T such that a path through T yields a 1–1 function, g, on S such that $\forall \alpha \in S\,(g(\alpha) < \alpha)$.

(50) κ is called a strongly hyper-Mahlo iff κ is strongly Mahlo and

$$\{\alpha < \kappa : \alpha \text{ is strongly Mahlo}\}$$

is stationary in κ. Likewise, define hyper-hyper-Mahlo, etc. Show that if κ is weakly compact, then κ is strongly Mahlo, hyper-Mahlo, hyper-hyper-Mahlo, etc. *Hint.* Apply Exercise 49. *Remark.* For more on such cardinals, see [Kunen 1977].

(51) Show that the following are equivalent:

(1) $\diamondsuit$.

(2) There are $A_\alpha \subset \alpha \times \alpha$ for $\alpha < \omega_1$, such that for all $A \subset \omega_1 \times \omega_1$,

$$(\{\alpha < \omega_1 : A \cap \alpha \times \alpha = A_\alpha\} \text{ is stationary}).$$

(3) There are $f_\alpha : \alpha \to \alpha$ for $\alpha < \omega_1$ such that for each $f : \omega_1 \to \omega_1$,

$$\exists \alpha\,(f \restriction \alpha = f_\alpha \wedge \alpha > 0). \qquad (*)$$

(4) (3) with $(*)$ replaced by $\{\alpha : f \restriction \alpha = f_\alpha\}$ is stationary.

(52) Let $\kappa > \omega$ be regular. $\diamondsuit_\kappa$ is the statement that there are sets $A_\alpha \subset \alpha$ for $\alpha < \kappa$ such that $\forall A \subset \kappa(\{\alpha < \kappa : A \cap \alpha = A_\alpha\}$ is stationary). Show that $\diamondsuit_\kappa$ implies $2^{<\kappa} = \kappa$ and that there is a family of 2^κ almost disjoint stationary subsets of κ. *Remark.* $\diamondsuit_\kappa$ follows from $\mathbf{V} = \mathbf{L}$ (see VI Exercise 12).

(53) Show that for κ regular and $>\omega$, $\diamondsuit_\kappa$ is equivalent to the following statement. There are $\mathscr{A}_\alpha \subset \mathscr{P}(\alpha)$ for $\alpha < \kappa$, such that each $|\mathscr{A}_\alpha| \le \alpha$ and for each $A \subset \kappa$, $\{\alpha < \kappa : A \cap \alpha \in \mathscr{A}_\alpha\}$ is stationary.

(54) Show that $\diamondsuit$ implies that there are ω_1-Suslin trees, T, T', such that $T \times T'$ is a Suslin tree.

(55) Show that the following version of $\diamondsuit$ is inconsistent: There are $A_\alpha \subset \alpha$ for $\alpha < \omega_1$, such that for all stationary $A \subset \omega_1$, $\exists \alpha \in A\,(A \cap \alpha = A_\alpha)$.

(56) Let $S(\kappa, \lambda, \mathscr{I})$ abbreviate the statement that $\kappa > \omega$ and $\mathscr{I}$ is a κ-complete ideal on κ which contains each singleton and which is λ-*saturated*; i.e., there is no family $\{X_\alpha : \alpha < \lambda\} \subset \mathscr{P}(\kappa)$ such that each $X_\alpha \notin \mathscr{I}$ but $\alpha \neq \beta \rightarrow (X_\alpha \cap X_\beta) \in \mathscr{I}$. Show

(a) $\exists \lambda\, \exists \mathscr{I}\, S(\kappa, \lambda, \mathscr{I}) \rightarrow \kappa$ is regular.

(b) $S(\kappa, \lambda, \mathscr{I}) \wedge \lambda < \lambda' \rightarrow S(\kappa, \lambda', \mathscr{I})$.

(c) $\exists \mathscr{I}\, S(\kappa, \kappa, \mathscr{I}) \rightarrow \kappa$ is weakly inaccessible.

(see Theorem 6.11).

(57) For $\lambda \leq \kappa$, show that $S(\kappa, \lambda, \mathscr{I})$ is equivalent to the (seemingly) weaker statement that $\kappa > \omega$ and $\mathscr{I}$ is a κ-complete ideal on κ containing singletons such that there is no family $\{X_\alpha : \alpha < \lambda\} \subset \mathscr{P}(\kappa)$ such that each $X_\alpha \notin \mathscr{I}$ but $\alpha \neq \beta \rightarrow (X_\alpha \cap X_\beta) = 0$.

(58) Assume that $\kappa > \omega$ is regular and that there is no almost disjoint family of κ^+ stationary subsets of κ. Show $S(\kappa, \kappa^+, \mathrm{Cub}^*(\kappa))$. *Hint.* Use Lemma 6.14. *Remark.* It is unknown whether this situation is consistent.

(59) Assume $\exists \lambda < \kappa\, \exists \mathscr{I}\, S(\kappa, \lambda, \mathscr{I})$. Show that there are no κ-Aronszajn trees. *Hint.* Assume $\mathscr{I}$ lives on T, and find a sub-tree T' of T which contradicts Exercise 38. *Remark.* It is consistent with $\exists \mathscr{I}\, S(\kappa, \kappa, \mathscr{I})$ that there is a κ-Suslin tree; see [Kunen 1978].

(60) κ is called (2-valued) measurable iff there is an $\mathscr{I}$ such that $S(\kappa, 2, \mathscr{I})$; so, $\mathscr{I}$ is a prime ideal, or $\mathscr{I}^*$ is an ultrafilter. Show that if κ is measurable then κ is strongly inaccessible. *Hint.* If $\lambda < \kappa$, and f_α, for $\alpha < \kappa$, are distinct members of ${}^\lambda 2$, define $g\colon \lambda \rightarrow 2$ so that for $\xi < \lambda$,

$$\{\alpha : f_\alpha(\xi) = g(\xi)\} \in \mathscr{I}^*.$$

Remark. $\exists \mathscr{I}\, S(\kappa, \omega_1, \mathscr{I})$ is consistent with $\kappa = 2^\omega$, or with $\kappa < 2^\omega$; see VII Exercise H24.

(61) Show $\exists \mathscr{I}\, S(\kappa, \omega, \mathscr{I})$ implies $\exists \mathscr{J}\, S(\kappa, 2, \mathscr{J})$. *Hint.* Assume $S(\kappa, \omega, \mathscr{I})$ and find an *atom*, A; i.e., $A \subset \kappa$, $A \notin \mathscr{I}$, and $\forall X \subset A\,(X \in \mathscr{I} \vee (A \setminus X) \in \mathscr{I})$.

(62) Assume $\exists \mathscr{I}\, S(\kappa, \lambda, \mathscr{I})$ and $2^{<\lambda} < \kappa$. Show $\exists \mathscr{J}\, S(\kappa, 2, \mathscr{J})$. *Hint.* As in Exercise 61, try to find an atom.

CHAPTER III

THE WELL-FOUNDED SETS

We discuss here the Axiom of Foundation. This axiom, like the Axiom of Extensionality, has the effect of restricting the domain of discourse to those sets where mathematics actually takes place.

§1. Introduction

Some questions about sets are irrelevant to mathematics.

First irrelevant question. Is there anything which is not a set? Certainly there is in the "real world" of cows and pigs, but our axioms of set theory say nothing about this "real world", since we have declared that they talk only about sets – in fact, hereditary sets (see I §4). Furthermore, the Axiom of Extensionality has embodied in it the assertion that all things in our domain of discourse are sets. It seems likely that we have not left any interesting mathematics behind by so restricting our universe, since mathematical objects like $\mathbb{R}$ and $\mathbb{C}$ are hereditary sets and have been defined explicitly within this domain in I §11.

Second irrelevant question. Is there an x such that $x = \{x\}$? This question is independent of the existence of a physical reality, and such an x would clearly be an hereditary set. However, such an x did not occur in the construction of mathematical objects like $\mathbb{R}$ and $\mathbb{C}$.

In this chapter we define, working in ZF^-, the class **WF** of well-founded sets. Intuitively, **WF** is the class of those sets constructed from 0 by iterating the various set-theoretic operations. For example, $\mathbb{R}$ and $\mathbb{C}$ were so constructed, but if $x = \{x\}$, we shall see that $x \notin \mathbf{WF}$. We then prove some theorems indicating that all mathematics takes place within **WF**. This leads us to introduce the Axiom of Foundation, which says $\mathbf{V} = \mathbf{WF}$; equivalently, we shall henceforth study only **WF**.

We emphasize that our adopting the Axiom of Extensionality did not mean that we were asserting that there are *really* no cows and pigs – only that there are none in the domain being considered. Likewise, our adopting the Axiom of Foundation does not comment on whether there are *really* (whatever that means) any x such that $x = \{x\}$; we are simply refraining from considering such x.

§2. Properties of the well-founded sets

We shall work in ZF^- and define the class **WF** of well-founded sets by starting with 0 and iterating the power set operation. We shall then prove that **WF** is closed under the other set-theoretic operations as well.

2.1. DEFINITION. By transfinite recursion, define $R(\alpha)$ for $\alpha \in \mathbf{ON}$ by:
(*a*) $R(0) = 0$.
(*b*) $R(\alpha + 1) = \mathscr{P}(R(\alpha))$.
(*c*) $R(\alpha) = \bigcup_{\xi<\alpha} R(\xi)$ when α is a limit ordinal. □

2.2. DEFINITION. $\mathbf{WF} = \bigcup \{R(\alpha) : \alpha \in \mathbf{ON}\}$. □

So, the well-founded sets are defined to be those which are in some $R(\alpha)$.

2.3. LEMMA. *For each* α:
(*a*) $R(\alpha)$ *is transitive.*
(*b*) $\forall \xi \le \alpha \quad (R(\xi) \subset R(\alpha))$.

PROOF. Transfinite induction on α; we assume the lemma holds for all $\beta < \alpha$, and conclude it for α.

Case I. $\alpha = 0$: trivial.

Case II. α is a limit: (*b*) is immediate from the definition, and (*a*) follows from the fact that the union of transitive sets is transitive.

Case III. $\alpha = \beta + 1$. Since $R(\beta)$ is transitive, $\mathscr{P}(R(\beta)) = R(\alpha)$ is transitive and $R(\beta) \subset R(\alpha)$. This establishes (*a*) and (*b*) for α. □

So, the sets $R(\alpha)$ are increasing in α.

If $x \in \mathbf{WF}$, the least α for which $x \in R(\alpha)$ must be a successor ordinal by Definition 2.1(*c*).

2.4. DEFINITION. If $x \in \mathbf{WF}$, rank(x) is the least β such that $x \in R(\beta + 1)$. □

So, if $\beta = \text{rank}(x)$, then $x \subset R(\beta)$, $x \notin R(\beta)$, and $x \in R(\alpha)$ for all $\alpha > \beta$.

2.5. LEMMA. *For any* α, $R(\alpha) = \{x \in \mathbf{WF}: \text{rank}(x) < \alpha\}$.

PROOF. For $x \in \mathbf{WF}$, $\operatorname{rank}(x) < \alpha$ iff

$$\exists \beta < \alpha\,(x \in R(\beta + 1)) \quad \text{iff} \quad x \in R(\alpha). \quad \square$$

The following is useful when computing ranks.

2.6. LEMMA. *If $y \in \mathbf{WF}$, then*
(*a*) $\forall x \in y\,(x \in \mathbf{WF} \wedge \operatorname{rank}(x) < \operatorname{rank}(y))$, *and*
(*b*) $\operatorname{rank}(y) = \sup\{\operatorname{rank}(x) + 1 : \ x \in y\}$.

PROOF. For (*a*), let $\alpha = \operatorname{rank}(y)$, then $y \in R(\alpha + 1) = \mathscr{P}(R(\alpha))$. If $x \in y$, then $x \in R(\alpha)$, so $\operatorname{rank}(x) < \alpha$ by Lemma 2.5.

For (*b*), let $\alpha = \sup\{\operatorname{rank}(x) + 1 : x \in y\}$. By (*a*), $\alpha \leq \operatorname{rank}(y)$. Furthermore, each $x \in y$ has rank $< \alpha$, so $y \subset R(\alpha)$. Thus, $y \in R(\alpha + 1)$, so $\operatorname{rank}(y) \leq \alpha$. $\square$

Lemma 2.6(*a*) says that the class **WF** is transitive, and that we may think of the elements $y \in \mathbf{WF}$ as having been "constructed", by transfinite recursion, from well-founded sets of smaller rank. Thus, **WF** excludes sets which are built up from themselves. More formally, there is no $x \in \mathbf{WF}$ such that $x \in x$, since we would have $\operatorname{rank}(x) < \operatorname{rank}(x)$. Likewise, **WF** excludes circularities like $x \in y \wedge y \in x$, since this would yield $\operatorname{rank}(x) < \operatorname{rank}(y) < \operatorname{rank}(x)$.

Each ordinal is in **WF** and its rank is itself.

2.7. LEMMA. (a) $\forall \alpha \in \mathbf{ON}\,(\alpha \in \mathbf{WF} \wedge \operatorname{rank}(\alpha) = \alpha)$.
(b) $\forall \alpha \in \mathbf{ON}\,(R(\alpha) \cap \mathbf{ON} = \alpha)$.

PROOF. We prove (*a*) by transfinite induction on α, so assume (*a*) holds for all $\beta < \alpha$. Then, for $\beta < \alpha$, $\beta \in R(\beta + 1) \subset R(\alpha)$, so $\alpha \subset R(\alpha)$, so $\alpha \in R(\alpha + 1)$. By Lemma 2.6(*b*), $\operatorname{rank}(\alpha) = \sup\{\beta + 1 : \beta < \alpha\} = \alpha$. Thus, (*a*) holds for α.

(*b*) is immediate from (*a*) and Lemma 2.5. $\square$

WF contains not only the ordinals but also all the other sets which arise in standard mathematical construction, since **WF** is closed under these constructions.

2.8. LEMMA. (*a*) *If $x \in \mathbf{WF}$, then $\bigcup x$, $\mathscr{P}(x)$, and $\{x\} \in \mathbf{WF}$, and the rank of these sets is less than* $\operatorname{rank}(x) + \omega$.

(*b*) *If $x, y \in \mathbf{WF}$, then $x \times y$, $x \cup y$, $x \cap y$, $\{x, y\}$, $\langle x, y\rangle$, and ${}^{y}x$ are all in* **WF**, *and the rank of these sets is less than* $\max(\operatorname{rank}(x), \operatorname{rank}(y)) + \omega$.

PROOF. For (*a*), let $\alpha = \text{rank}(x)$, then $x \subset R(\alpha)$, so $\mathscr{P}(x) \subset \mathscr{P}(R(\alpha)) = R(\alpha + 1)$, so $\mathscr{P}(x) \in R(\alpha + 2)$. Similarly, $\{x\} \in R(\alpha + 2)$ and $\bigcup x \in R(\alpha + 1)$.

For (*b*), let $\alpha = \max(\text{rank}(x), \text{rank}(y))$. As in (*a*), show, e.g., $\{x,y\} \in R(\alpha + 2)$, $\langle x,y \rangle \in R(\alpha + 3)$. Any ordered pair of elements of $x \cup y$ is in $R(\alpha + 2)$, so ${}^{y}x \subset R(\alpha + 3)$, so ${}^{y}x \in R(\alpha + 4)$. We leave the rest of the details to the reader. □

It is possible to compute precisely the ranks of these set-theoretic combinations of x and y in terms of the ranks of x and y. See Exercise 4.

2.9. LEMMA. *$\mathbb{Z}$, $\mathbb{Q}$, $\mathbb{R}$, and $\mathbb{C}$ are all in $R(\omega + \omega)$.*

PROOF. By Lemma 2.8 and the definitions of these sets (I §11). □

2.10. LEMMA. $\forall x\, (x \in \mathbf{WF} \leftrightarrow x \subset \mathbf{WF})$.

PROOF. $x \in \mathbf{WF} \rightarrow x \subset \mathbf{WF}$ just restates the transitivity of **WF** (Lemma 2.6.(*a*)). If $x \subset \mathbf{WF}$, let $\alpha = \sup\{\text{rank}(y) + 1 : y \in x\}$; then $x \subset R(\alpha)$ so $x \in R(\alpha + 1)$. □

It is possible to derive the closure properties of Lemma 2.8. directly from Lemma 2.10, but Lemma 2.10 is much stronger. Any $R(\gamma)$ for limit γ satisfies the same closure properties, but any class satisfying Lemma 2.10 must contain **WF**. See Exercise 3.

We now look at the size of the $R(\alpha)$.

2.11. LEMMA. $\forall n \in \omega\, (|R(n)| < \omega)$.

PROOF. Induction on n. □

2.12. LEMMA. $|R(\omega)| = \omega$.

PROOF. Since $\omega \subset R(\omega)$, it is enough to see that $R(\omega)$ is countable. Under AC, this is immediate from Lemma 2.11. To avoid AC, note that in the proof of Lemma 2.11, we may explicitly define a well-order of $R(n)$ by induction on n. For example, given a well-order of $R(n)$, we may identify $R(n + 1)$ with ${}^{R(n)}2$ and order it lexicographically. □

A very simple 1–1 map between ω and $R(\omega)$ is given in Exercise 5.

The cardinalities of the $R(\alpha)$ increase exponentially: $|R(\omega)| = \omega$, $|R(\omega + 1)| = 2^{\omega}$, $|R(\omega + 2)| = 2^{2^{\omega}}$, etc. More generally, the following holds.

2.13. LEMMA. (AC). $|R(\omega + \alpha)| = \beth_{\alpha}$.

PROOF. Induction on α. □

We conclude this section by making a case that all reasonable mathematics takes place in **WF**.

2.14. LEMMA. (AC). (*a*) *Every group is isomorphic to a group in* **WF**.
(*b*) *Every topological space is homeomorphic to a topological space in* **WF**.

PROOF. Formally, a group is a pair $\langle G, \cdot \rangle$ where $\cdot : G \times G \to G$, but by Lemmas 2.8 and 2.10,

$$\langle G, \cdot \rangle \in \mathbf{WF} \leftrightarrow G \in \mathbf{WF} \leftrightarrow G \subset \mathbf{WF}.$$

If $\langle G, \cdot \rangle$ is any group, let $\alpha = |G|$, let f be a 1–1 map from α onto G, and define an operation $\circ$ on α by $\xi \circ \eta = f^{-1}(f(\xi) \cdot f(\eta))$. Then $\langle \alpha, \circ \rangle$ is isomorphic to $\langle G, \cdot \rangle$. We leave (*b*) to the reader. □

Lemma 2.14 cannot be proved without AC (see IV Exercise 25).

Thus, we see that **WF** contains as elements the various concrete mathematical objects like $\mathbb{R}$ and $\mathbb{C}$ (by Lemma 2.9), and identical copies of the various abstract objects like groups and spaces.

§3. Well-founded relations

This is a generalization of the notion of well-order, and will be of basic importance for later work.

Although the definition of **WF** used the Power Set Axiom in an essential way, many of the results in this section go through in $\mathrm{ZF}^- - \mathrm{P}$. This will become useful to know in IV, when we shall apply these results within interpretations in which the Power Set Axiom either fails or has not yet been shown to hold. Thus, for the rest of this chapter, we shall explicitly say which axioms of set theory our theorems presuppose. In some cases, however, when it does not matter for future applications, we shall be lazy and mark our theorems with ZF^- when in fact they go through under $\mathrm{ZF}^- - \mathrm{P}$ (see Exercise 8).

3.1. DEFINITION. ($\mathrm{ZF}^- - \mathrm{P}$). A relation R *is well-founded* on a set A *iff*

$$\forall X \subset A\, [X \neq 0 \to \exists y \in X\, (\neg \exists z \in X\, (z\, R\, y))]. \tag{1}$$

The y of (1) is called *R-minimal* in X. □

Thus, R is well-founded on A iff every non-empty subset has an R-minimal element. In particular, if R totally orders A, then R is well-founded on A iff R well-orders A. As in I §6, we do not require that $R \subset A \times A$, so that if R is well-founded on A and $B \subset A$, then R is well-founded on B.

If R is not a total order, then the X of Definition 3.1 may have more than one R-minimal element. For example, the empty relation, 0, is well-founded on any A, and any element $y \in X$ is 0-minimal in X.

The following lemma begins to show the relationship between well-founded relations and well-founded sets.

3.2. LEMMA (ZF$^-$). *If* $A \in \mathbf{WF}$, $\in$ *is well-founded on* A.

PROOF. Let X be a non-empty subset of A. Let $\alpha = \min\{\text{rank}(y) : y \in X\}$, and fix $y \in X$ with $\text{rank}(y) = \alpha$. Then y is $\in$-minimal in X by Lemma 2.6(a). □

The converse of Lemma 3.2 need not be true. For example, if $x = \{y\}$, $y = \{x\}$, and $x \neq y$, then $y \notin \mathbf{WF}$ but $\in$ is well-founded (in fact empty) on y. As a partial converse, we have the following.

3.3. LEMMA (ZF$^-$). *If A is transitive and* $\in$ *is well-founded on A, then* $A \in \mathbf{WF}$.

PROOF. By Lemma 2.10, it is sufficient to show $A \subset \mathbf{WF}$. If $A \not\subset \mathbf{WF}$, let $X = A \setminus \mathbf{WF} \neq 0$ and let y be $\in$-minimal in X. If $z \in y$, then $z \notin X$, but $z \in A$ since A is transitive; so $z \in \mathbf{WF}$. Thus, $y \subset \mathbf{WF}$, so $y \in \mathbf{WF}$ by Lemma 2.10, contradicting $y \in A \setminus \mathbf{WF}$. □

We now show that $A \in \mathbf{WF}$ iff $\in$ is well-founded on the *transitive closure* of A, which is the least transitive set containing A as a subset.

3.4. DEFINITION (ZF$^-$ − P). (a) By recursion on n define $\bigcup^0 A = A$, $\bigcup^{n+1} A = \bigcup(\bigcup^n A)$.

(b) $\text{tr cl}(A) = \bigcup\{\bigcup^n A : n \in \omega\}$. □

So $\text{tr cl}(A) = A \cup \bigcup A \cup \bigcup^2 A \cup \dots$, and has as elements the elements of A, plus elements of elements of A, plus … .

3.5. LEMMA (ZF$^-$ − P). (a) $A \subset \text{tr cl}(A)$.

(b) $\text{tr cl}(A)$ *is transitive.*

(c) *If* $A \subset T$ *and* T *is transitive, then* $\text{tr cl}(A) \subset T$.

(d) *If A is transitive, then* $\text{tr cl}(A) = A$.

(e) *If* $x \in A$, *then* $\text{tr cl}(x) \subset \text{tr cl}(A)$.

(f) $\text{tr cl}(A) = A \cup \bigcup\{\text{tr cl}(x) : x \in A\}$.

PROOF. (a) is obvious. For (b), note that $y \in \bigcup^n A \to y \subset \bigcup^{n+1} A$. For ($c$), show by induction that $\bigcup^n A \subset T$. (*d*) follows from (*a*) and (*c*) taking $A = T$. For (*e*), $x \in A \rightarrow x \in \text{tr cl}(A) \rightarrow x \subset \text{tr cl}(A)$ so apply (*c*) to x. For (*f*), let

$$T = A \cup \bigcup \{\text{tr cl}(x) : x \in A\}.$$

T is transitive, so $\text{tr cl}(A) \subset T$ by (*c*), but $T \subset \text{tr cl}(A)$ by (*a*) and (*e*). □

3.6. THEOREM (ZF$^-$). *For any set A the following are equivalent:*

(*a*) $A \in \mathbf{WF}$.

(*b*) $\text{tr cl}(A) \in \mathbf{WF}$.

(*c*) $\in$ *is well-founded on* $\text{tr cl}(A)$.

PROOF. (*a*) → (*b*): If $A \in \mathbf{WF}$, then by induction on n, $\bigcup^n A \in \mathbf{WF}$ since **WF** is closed under $\bigcup$ (Lemma 2.8). Thus, each $\bigcup^n A \subset \mathbf{WF}$, so $\text{tr cl}(A) \subset \mathbf{WF}$, so $\text{tr cl}(A) \in \mathbf{WF}$ (by Lemma 2.10).

(*b*) → (*c*): is Lemma 3.2.

(*c*) → (*a*): By (*c*) and Lemma 3.3, $\text{tr cl}(A) \in \mathbf{WF}$, so $A \subset \text{tr cl}(A) \subset \mathbf{WF}$, so $A \in \mathbf{WF}$ (by Lemma 2.10 again). □

We remark that our definition of **WF** used the Power Set Axiom in an essential way. Equivalent (*c*) is useful if one wants to give a presentation of **WF** working in ZF$^-$ − P (see Exercise 8).

§4. The Axiom of Foundation

If one is convinced by §2 that all mathematics takes place in **WF**, it is reasonable to adopt as an axiom the statement $\mathbf{V} = \mathbf{WF}$. As pointed out in §1, this does not mean that we must believe that "really" $\mathbf{V} = \mathbf{WF}$—but only that we are restricting our domain of discourse to be just **WF**. It is not hard to see informally that all the axioms of ZF$^-$ are still true under this interpretation, essentially because **WF** is closed under the various set-theoretic operations like $\bigcup$ and $\mathscr{P}$ postulated to exist by these axioms. This will be discussed more formally in IV. In this section, we content ourselves with discussing some of the consequences of adding the axiom $\mathbf{V} = \mathbf{WF}$ to ZF$^-$.

The statement $\mathbf{V} = \mathbf{WF}$ is highly non-elementary, relying on a long string of definitions. For elegance, we adopt as the official version of the axiom an equivalent to $\mathbf{V} = \mathbf{WF}$ which is very simply stated in the language of set theory.

AXIOM 2. *Foundation.*

$$\forall x (\exists y (y \in x) \rightarrow \exists y (y \in x \wedge \neg \exists z (z \in x \wedge z \in y))).$$

Equivalently, if $x \neq 0$, $\exists y \in x (x \cap y = 0)$, or every non-empty set has an $\in$-minimal element, or, if we extend the definition of well-founded to proper classes (see §5), $\in$ is well-founded on $\mathbf{V}$.

4.1. THEOREM (ZF$^-$). *The following are equivalent:*
(*a*) *the Axiom of Foundation,*
(*b*) $\forall A$ (*$\in$ is well-founded on A*),
(*c*) $\mathbf{V} = \mathbf{WF}$.

PROOF. (*a*) $\leftrightarrow$ (*b*) is immediate from the definition of well-founded. For (*b*) $\to$ (*c*), (*b*) implies that for any A, $\in$ is well-founded on tr cl(A), so $A \in \mathbf{WF}$. For (*c*) $\to$ (*b*), apply Lemma 3.2. □

Unlike the other axioms of ZFC, Foundation has no application in ordinary mathematics, since accepting it is equivalent to restricting our attention to **WF**, where all mathematics takes place anyway. Foundation does rule out certain pathologies. For example, we remarked in §2 that there is no $x \in \mathbf{WF}$ such that $x \in x$, so Foundation implies that $\neg \exists x (x \in x)$ (or, apply the axiom directly to show $\exists y \in \{x\} (y \cap \{x\} = 0)$, so $x \cap \{x\} = 0$, or $x \notin x$). Likewise, there cannot be an x, y with $x \in y \wedge y \in x$ (or, apply the axiom directly to $\{x, y\}$).

Since Foundation is equivalent to $\mathbf{V} = \mathbf{WF} = \bigcup_\alpha R(\alpha)$, it gives us a picture of all sets being created by an iterative process, starting from nothing (see Figure 4.1).

Assuming that $\in$ is well-founded on every set also simplifies certain definitions. For example, the following holds.

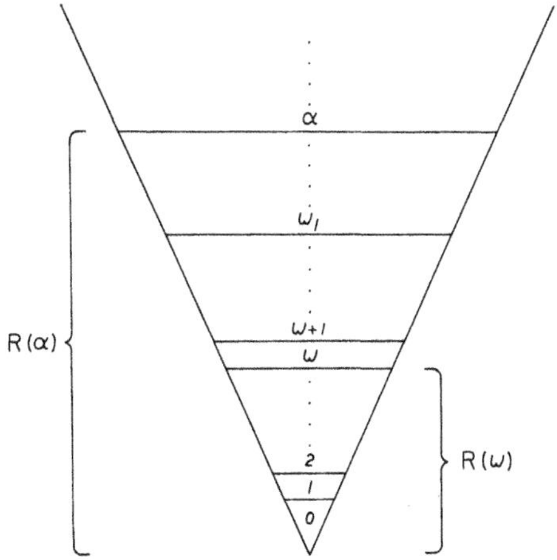

Figure 4.1. The Universe.

4.2. THEOREM (ZF $-$ P). *A is an ordinal iff A is transitive and totally ordered by* $\in$.

PROOF. To see that $\in$ well-orders A, let X be any non-empty subset of A, and apply Foundation to X to produce an $\in$-minimal element. □

This will become important in IV.

§5. Induction and recursion on well-founded relations

If R is a well-founded relation on the set A, a proof by *tranfinite induction* on R is one which establishes $\forall x \in A\, \phi(x)$ via first proving that for all $x \in A$,

$$\forall y \in A\,(y\,R\,x \rightarrow \phi(y)) \rightarrow \phi(x). \tag{1}$$

The conclusion $\forall x \in A\, \phi(x)$ is justified, since an R-minimal element of $\{x \in A: \neg\phi(x)\}$ would lead to a contradiction.

As an example, we may view the proof in Lemma 3.3 that if A is transitive and $\in$ is well-founded on A then $A \subset \mathbf{WF}$ as a transfinite induction. Here $\phi(x)$ is "x is well-founded" and (1) reduces to $x \subset \mathbf{WF} \rightarrow x \in \mathbf{WF}$.

It is often useful to consider these notions on proper classes as well.

5.1. DEFINITION. (ZF$^-$ $-$ P). $\mathbf{R}$ is well-founded on $\mathbf{A}$ iff

$$\forall X \subset \mathbf{A}\,[X \neq 0 \rightarrow \exists y \in X(\neg\exists z \in X(z\,\mathbf{R}\,y))]. \tag{2}$$ □

This is verbatim the definition of well-founded for relations on sets (Definition 3.1), but there is a formal distinction due to the way we handle classes. Definition 3.1 defines a formula in two variables R, A, whereas 5.1 is really a definition schema. Given formulas defining $\mathbf{R}$ and $\mathbf{A}$, (2) becomes an abbreviation for another formula (see I §9 for more details). For example, "$\in$ is well-founded on $\mathbf{V}$" is a sentence in the language of set theory which is easily seen to be equivalent to the Axiom of Foundation.

Note that the variable X in (2) must be considered to range over sub*sets* of A, since there is no way to quantify over classes. This can cause difficulty if we try to justify a proof by transfinite induction, since we would need the existence of an $\mathbf{R}$-minimal element of $\{x \in \mathbf{A}: \neg\phi(x)\}$, which could be a proper class (but see Exercise 17). Fortunately, the relations we shall deal with in practice will satisfy an additional condition which removes the problem.

5.2. DEFINITION (ZF$^-$ $-$ P). $\mathbf{R}$ is *set-like* on $\mathbf{A}$ iff for all $x \in \mathbf{A}$, $\{y \in A: y\,\mathbf{R}\,x\}$ is a *set*. □

For example, the $\in$ relation is set-like on any class $\mathbf{A}$, and every relation on a set is set-like.

5.3. DEFINITION ($ZF^- - P$). If $\mathbf{R}$ is set-like on $\mathbf{A}$ and $x \in \mathbf{A}$, then
(*a*) $\operatorname{pred}(\mathbf{A}, x, \mathbf{R}) = \{y \in \mathbf{A}: y \mathbf{R} x\}$.
(*b*) $\operatorname{pred}^0(\mathbf{A}, x, \mathbf{R}) = \operatorname{pred}(\mathbf{A}, x, \mathbf{R})$;
$\operatorname{pred}^{n+1}(\mathbf{A}, x, \mathbf{R}) = \bigcup \{\operatorname{pred}(\mathbf{A}, y, \mathbf{R}) : y \in \operatorname{pred}^n(\mathbf{A}, x, \mathbf{R})\}$.
(*c*) $\operatorname{cl}(\mathbf{A}, x, \mathbf{R}) = \bigcup \{\operatorname{pred}^n(\mathbf{A}, x, \mathbf{R}) : n \in \omega\}$. □

Note that these are all sets. If $\mathbf{R}$ is the $\in$ relation and $\mathbf{A}$ is transitive, then $\operatorname{pred}(\mathbf{A}, x, \mathbf{R}) = x$, $\operatorname{pred}^n(\mathbf{A}, x, \mathbf{R}) = \bigcup^n x$, and $\operatorname{cl}(\mathbf{A}, x, \mathbf{R}) = \operatorname{tr\,cl}(x)$. The fact that $\operatorname{tr\,cl}(x)$ is transitive generalizes to the following.

5.4. LEMMA ($ZF^- - P$). *If* $\mathbf{R}$ *is set-like on* $\mathbf{A}$ *and* $x \in \mathbf{A}$, *then for all* $y \in cl(\mathbf{A}, x, \mathbf{R})$, $\operatorname{pred}(\mathbf{A}, y, \mathbf{R}) \subset \operatorname{cl}(\mathbf{A}, x, \mathbf{R})$. □

5.5. THEOREM ($ZF^- - P$). *If* $\mathbf{R}$ *is well-founded and set-like on* $\mathbf{A}$, *then every non-empty subclass* $\mathbf{X}$ *of* $\mathbf{A}$ *has an* $\mathbf{R}$*-minimal element.*

PROOF. Fix $x \in \mathbf{X}$. If x is not $\mathbf{R}$-minimal in $\mathbf{A}$, then $\mathbf{X} \cap \operatorname{cl}(\mathbf{A}, x, \mathbf{R})$ is a non-empty sub*set* of $\mathbf{A}$ and thus has an $\mathbf{R}$-minimal element y. By Lemma 5.4, y is $\mathbf{R}$-minimal in $\mathbf{X}$. □

A special case of this theorem, for $\mathbf{A} = \mathbf{ON}$ and R the $\in$ relation, has been proved in I Theorem 9.2; there is a discussion there on how to state this kind of result without using classes.

By Theorem 5.5, proofs by transfinite induction on well-founded *set-like* relations are justified. We may view the proof in Theorem 4.1 that the Axiom of Foundation ($\in$ is well-founded on $\mathbf{V}$) implies $\mathbf{V} = \mathbf{WF}$ in this light; (1) becomes $x \subset \mathbf{WF} \rightarrow x \in \mathbf{WF}$.

We may also define functions by *transfinite recursion* on any well-founded set-like relation. The special case for functions on $\mathbf{ON}$ was given in I Theorem 9.3, and there is a discussion there of the precise meaning of our use of classes.

5.6. THEOREM ($ZF^- - P$). (*Transfinite Recursion*). *Assume* $\mathbf{R}$ *is well-founded and set-like on* $\mathbf{A}$. *If* $\mathbf{F}: \mathbf{A} \times \mathbf{V} \rightarrow \mathbf{V}$, *then there is a unique* $\mathbf{G}: \mathbf{A} \rightarrow \mathbf{V}$ *such that*

$$\forall x \in \mathbf{A}\, [\mathbf{G}(x) = \mathbf{F}(x, \mathbf{G} \restriction \operatorname{pred}(\mathbf{A}, x, \mathbf{R}))]. \tag{$*$}$$

PROOF. A direct generalization of the proof of I 9.3, which was the special case for $\mathbf{A} = \mathbf{ON}$ and R the $\in$-relation.

Uniqueness of $\mathbf{G}$ is easily established by transfinite induction on $\mathbf{R}$, so we turn to the proof of existence.

For brevity, write $\text{pred}(x)$ and $\text{cl}(x)$ for $\text{pred}(\mathbf{A}, x, \mathbf{R})$ and $\text{cl}(\mathbf{A}, x, \mathbf{R})$ respectively. Call a sub*set* $d \subset \mathbf{A}$ *closed* iff $\forall x \in d\,(\text{pred}(x) \subset d)$. Then each $x \in \mathbf{A}$ is in some closed set, namely $\{x\} \cup \text{cl}(x)$. If d is closed, call g a d-approximation iff g is a function with domain d and

$$\forall x \in d\,[g(x) = \mathbf{F}(x, g \restriction \text{pred}(x))].$$

As in the uniqueness proof, if g is a d-approximation and g' is a d'-approximation, then $g \restriction (d \cap d') = g' \restriction (d \cap d')$.

By transfinite induction on x show that there is a $\{x\} \cup \text{cl}(x)$-approximation. If this is true for all $y\,\mathbf{R}\,x$, let g_y be the $\{y\} \cup \text{cl}(y)$-approximation; then $h = \bigcup \{g_y : y\,\mathbf{R}\,x\}$ is a $\text{cl}(x)$-approximation and $h \cup \{\langle x, \mathbf{F}(x, h)\rangle\}$ is a $\{x\} \cup \text{cl}(x)$-approximation.

Now, define $\mathbf{G}(x)$ to be the value $g(x)$, where g is the d-approximation for some (any) closed d containing x. □

Note that if $\mathbf{R}$ were not set-like, the condition (∗) in Theorem 5.6 would not make sense.

As an application of Theorem 5.6, consider the equation

$$\text{rank}(y) = \sup\{\text{rank}(x) + 1 : x \in y\}$$

which we proved for $y \in \mathbf{WF}$ (Lemma 2.6(b)). We may now view this as defining $\text{rank}(y)$ by recursion on the $\in$ relation, which is well-founded on $\mathbf{WF}$. More generally, we may define rank as follows.

5.7. Definition ($\text{ZF}^- - \text{P}$). If $\mathbf{R}$ is well-founded and set-like on $\mathbf{A}$, define

$$\text{rank}(x, \mathbf{A}, \mathbf{R}) = \sup\{\text{rank}(y, \mathbf{A}, \mathbf{R}) + 1 : y\,\mathbf{R}\,x \wedge y \in \mathbf{A}\}, \qquad \square$$

Formally, the $\mathbf{F}$ of Theorem 5.6 is here given by

$$\mathbf{F}(x, h) = \sup\{\alpha + 1 : \alpha \in \text{ran}(h)\}.$$

5.8. Lemma (ZF^-). *If* $\mathbf{A}$ *is transitive and* $\in$ *is well-founded on* $\mathbf{A}$, *then* $\mathbf{A} \subset \mathbf{WF}$ *and* $\text{rank}(x, \mathbf{A}, \in) = \text{rank}(x)$ *for all* $x \in \mathbf{A}$.

Proof. If $\mathbf{A} \not\subset \mathbf{WF}$, let x be $\in$-minimal in $\mathbf{A} \setminus \mathbf{WF}$. Then $x \subset \mathbf{WF}$, since $\mathbf{A}$ is transitive, so $x \in \mathbf{WF}$, a contradiction (this was like the proof of Lemma 3.3). Likewise, an $\in$-minimal element of $\{x \in \mathbf{A} : \text{rank}(x, \mathbf{A}, \in) \neq \text{rank}(x)\}$ would yield a contradiction by Lemma 2.6(b). □

Definition 5.7 may be used to give a definition of rank on $\mathbf{WF}$ which does not use the Power Set Axiom (see Exercise 8).

Another example of recursion generalizes the fact that any well-ordering R of set A is isomorphic to an ordinal (I Theorem 7.6). One could view the isomorphism G as defined by $G(a) = \{G(b) : b\, R\, a\}$. More generally, the following holds.

5.9. DEFINITION ($ZF^- - P$). Let $\mathbf{R}$ be well-founded and set-like on $\mathbf{A}$. Define the *Mostowski collapsing function*, G, of $\mathbf{A}$, $\mathbf{R}$ by

$$\mathbf{G}(x) = \{\mathbf{G}(y) : y \in \mathbf{A} \wedge y\, \mathbf{R}\, x\}.$$

The *Mostowski collapse*, $\mathbf{M}$, of $\mathbf{A}$, $\mathbf{R}$ is defined to be the range of $\mathbf{G}$. □

Thus, $\mathbf{G}: \mathbf{A} \to \mathbf{M}$. $\mathbf{G}$ need not be 1–1. For example, if $\mathbf{R} = 0$, then $\mathbf{G}(x) = 0$ for all $x \in \mathbf{A}$ and $\mathbf{M} = \{0\}$ if $\mathbf{A} \neq 0$.

5.10. LEMMA ($ZF^- - P$). *With the notation of Definition* 5.9,

(*a*) $\forall x, y \in \mathbf{A}\, (x\, \mathbf{R}\, y \to \mathbf{G}(x) \in \mathbf{G}(y))$.

(*b*) $\mathbf{M}$ *is transitive.*

(*c*) (ZF^-) $\mathbf{M} \subset \mathbf{WF}$.

(*d*) (ZF^-) *If* $x \in \mathbf{A}$, *then* $\text{rank}(x, \mathbf{A}, \mathbf{R}) = \text{rank}(\mathbf{G}(x))$.

PROOF. (*a*), (*b*) are immediate from the definition of $\mathbf{G}$. For (*c*), prove that $\forall x \in \mathbf{A}\, (\mathbf{G}(x) \in \mathbf{WF})$ by induction on x. For (*d*),

$$\text{rank}(\mathbf{G}(x)) = \sup\{\text{rank}(v) + 1 : v \in \mathbf{G}(x)\} = \sup\{\text{rank}(\mathbf{G}(y)) + 1 : y\, \mathbf{R}\, x\}.$$

Then $\text{rank}(\mathbf{G}(x)) = \text{rank}(x, \mathbf{A}, \mathbf{R})$ by induction on x. □

In many cases of interest, the Mostowski collapsing function is an isomorphism. The condition for this is given by

5.11. DEFINITION ($ZF^- - P$). $\mathbf{R}$ is *extensional* on $\mathbf{A}$ iff

$$\forall x, y \in \mathbf{A}(\forall z \in \mathbf{A}(z\, \mathbf{R}\, x \leftrightarrow z\, \mathbf{R}\, y) \to x = y). \quad \square$$

This is equivalent to saying that the Axiom of Extensionality is true in $\mathbf{A}$ if $\in$ is interpreted as $\mathbf{R}$. It is sometimes convenient to note that $\mathbf{R}$ is extensional on $\mathbf{A}$ iff for all $x, y \in A$, $x \neq y \to \text{pred}(\mathbf{A}, x, \mathbf{R}) \neq \text{pred}(\mathbf{A}, y, \mathbf{R})$; for example, it is immediate from this that total orderings are extensional. Another class of examples is provided by the following.

5.12. LEMMA ($ZF^- - P$). *If* $\mathbf{N}$ *is transitive, then the* $\in$ *relation is extensional on* $\mathbf{N}$.

PROOF. $\text{pred}(\mathbf{N}, x\, \in) = x$. □

By Lemma 5.12, the Mostowski collapsing function cannot be an isomorphism unless **R** is estensional on **A**. Conversely, the following applies.

5.13. LEMMA ($ZF^- - P$). *With the notation of Definition 5.9, if* **R** *is extensional on* **A**, *then* **G** *is an isomorphism, i.e.,* **G** *is* 1–1 *and* $\forall x, y \in \mathbf{A}(x \mathbf{R} y \leftrightarrow \mathbf{G}(x) \in \mathbf{G}(y))$.

PROOF. We first show **G** is 1–1. If not, fix x to be **R**-minimal in

$$\{x \in \mathbf{A}: \exists y \in \mathbf{A}(x \neq y \wedge \mathbf{G}(x) = \mathbf{G}(y))\},$$

and fix $y \neq x$ such that $\mathbf{G}(x) = \mathbf{G}(y)$. Since **R** is extensional, one of the following two cases holds: *Case 1.* For some $z \in \mathbf{A}$, $z \mathbf{R} x$ and $\neg(z \mathbf{R} y)$. Since $\mathbf{G}(z) \in \mathbf{G}(x) = \mathbf{G}(y)$, $\mathbf{G}(z) = \mathbf{G}(w)$ for some w such that $w \mathbf{R} y$. Then $w \neq z$, and z contradicts the minimality of x. *Case 2.* For some $w \in \mathbf{A}$, $w \mathbf{R} y$ and $\neg(w \mathbf{R} x)$. Then as in Case 1, there is a z such that $z \mathbf{R} x$ and $\mathbf{G}(z) = \mathbf{G}(w)$. Again, z contradicts the minimality of x.

Since **G** is 1–1, the fact that **G** is an isomorphism is immediate from its definition. □

To summarize, we give the following.

5.14. THEOREM ($ZF^- - P$). (*Mostowski Collapsing Theorem*). *Suppose* **R** *is well-founded, set-like, and extensional on* **A**; *then there is a transitive class* **M** *and a* 1–1 *map* **G** *from* **A** *onto* **M** *such that* **G** *is an isomorphism between* **A**, **R** *and* **M**, $\in$. *Furthermore,* **M** *and* **G** *are unique.*

PROOF. Existence is proved by Lemma 5.13. If $\mathbf{G}'$, $\mathbf{M}'$ also satisfied the Theorem, then, by induction on x, $\mathbf{G}'(x) = \mathbf{G}(x)$ for all $x \in \mathbf{A}$, which implies $\mathbf{M}' = \mathbf{M}$. □

As an example of Theorem 5.14, suppose **R** well-orders **A**. If **A** is a set, then **M** is an ordinal. If **A** is a proper class, then $\mathbf{M} = \mathbf{ON}$. The fact that **R** is set-like prevents it from having "type" $>\mathbf{ON}$. For example, $2 \times \mathbf{ON}$ ordered lexicographically has "type $\mathbf{ON} + \mathbf{ON}$", but cannot be isomorphic to $\in$ on any class.

So far, the Axiom of Foundation was not needed, since well-foundedness of **R** was in the hypothesis. Foundation yields that $\in$ is well-founded, leading to the following.

5.15. COROLLARY ($ZF^- - P$). *If* $\in$ *is extensional on* **A**, *then there is a transitive* **M** *and a* 1–1 *map* **G** *from* **A** *onto* **M** *which is an isomorphism for the* $\in$ *relation—i.e.*

$$\forall x, y \in \mathbf{A}(x \in y \leftrightarrow \mathbf{G}(x) \in \mathbf{G}(y)). \quad \square$$

EXERCISES

Work in ZF^- unless otherwise indicated.

(1) Write out explicitly $R(n)$ for $n = 0, 1, 2, 3, 4, 5$. Thus, $R(3) = \{0, \{0\}, \{\{0\}\}, \{0, \{0\}\}\}$.

(2) Define by recursion $S(\alpha) = \bigcup_{\xi<\alpha} \mathscr{P}(S(\xi))$. Show that $S(\alpha) = R(\alpha)$ for all α.

(3) Let **M** be any class, such that

$$\forall x (x \subset \mathbf{M} \rightarrow x \in \mathbf{M}).$$

Show $\mathbf{WF} \subset \mathbf{M}$.

(4) Compute explicitly the ranks of $\bigcup x$, $\mathscr{P}(x)$, $\{x\}$, $x \times y$, $x \cup y$, $\{x, y\}$, $\langle x, y\rangle$, and ${}^{y}x$ in terms of $\text{rank}(x)$ and $\text{rank}(y)$. Find the ranks of $\mathbb{Z}$, $\mathbb{Q}$, $\mathbb{R}$, $\mathbb{C}$.

(5) Define $E \subset \omega \times \omega$ by: $n\, E\, m$ iff there is a 1 in the n-th place (counting from the right) in the binary representation of m. For example, $13 = 1101_2$, so $0\, E\, 13$, $2\, E\, 13$, $3\, E\, 13$, $\neg 1\, E\, 13$, and $\neg 4\, E\, 13$. Show that $\langle \omega, E\rangle \cong \langle R(\omega), \in\rangle$.

(6) Give a development of **WF** in $ZF^- - P$. Thus, define $\mathbf{WF} = \{A : \in$ is well-founded on $\text{tr cl}(A)\}$, and define, for $x \in \mathbf{WF}$, $\text{rank}(x) = \sup\{\text{rank}(y) + 1 : y \in x\}$. Verify that the results in §2 which do not mention $\mathscr{P}$, $R(\alpha)$, or $\mathbb{R}$ still hold. Prove 4.1 and 5.10(c) and (d).

(7) Work in $ZFC^- - P$. Show that a relation R on a set A is well-founded iff there is no ω-sequence, $\langle x_n : n \in \omega\rangle$, such that $\forall n (x_{n+1}\, R\, x_n)$. Show also that the Axiom of Foundation is equivalent to the non-existence of an ω-sequence, $\langle x_n : n \in \omega\rangle$, such that $\forall n (x_{n+1} \in x_n)$.

(8) Show in $ZF - P$ that the following are equivalent.

(a) x is an ordinal.

(b) x is transitive and $\forall y \in x$ (y is transitive).

(c) x is transitive and $\forall y, z \in x (y \in z \vee z = y \vee z \in y)$.

*(9) Show, in ZF, that AC is equivalent to $\forall \alpha (\mathscr{P}(\alpha)$ can be well-ordered$)$. *Hint.* Show by induction on α that $R(\alpha)$ can be well-ordered.

(10) Let $\langle T, <\rangle$ be a tree (see II §5). Show that for $x \in T$,

$$\mathrm{ht}(x, T) = \mathrm{rank}(x, T, <) = \mathbf{G}(x),$$

where $\mathbf{G}$ is the Mostowski collapsing function (5.9).

(11) Let T be a sub-tree of $A^{<\omega}$. T is called *well-founded* iff the inverse order, $\langle T, >\rangle$ is well-founded. Show that if A can be well-ordered, then T is well-founded iff T has no infinite chains.

(12) If T is a well-founded sub-tree of $A^{<\omega}$, let $\mathrm{rank}(T) = \mathrm{rank}(0, T, >)$, where 0 is the empty sequence. Show that if κ is infinite and $\alpha < \kappa^+$, there is a well-founded sub-tree of $\kappa^{<\omega}$ of rank α.

(13) Define $x \mathbf{R} y$ iff $x \in \mathrm{tr\,cl}(y)$. Show that $\mathbf{R}$ is well-founded and set-like on $\mathbf{WF}$ (*Hint.* $x \mathbf{R} y \to \mathrm{rank}(x) < \mathrm{rank}(y)$). Let $\mathbf{G}$ be the Mostowski collapsing function of $\mathbf{WF}, \mathbf{R}$. Show that $\mathbf{G}(x) = \mathrm{rank}(x)$ for each x.

(14) Define $x \mathbf{R} y$ iff $\langle x, 1\rangle \in y$. Show that $\mathbf{R}$ is well-founded and set-like on $\mathbf{WF}$. Let $\mathbf{G}$ be the Mostowski collapsing function of $\mathbf{WF}, \mathbf{R}$. Define $\check{y}$ recursively by:

$$\check{y} = \{\langle \check{x}, 1\rangle : x \in y\},$$

and show inductively that $\mathbf{G}(\check{y}) = y$. Hence, $\mathrm{ran}(\mathbf{G}) = \mathbf{WF}$.

(15) Let AR denote the Axiom of Replacement, and let AR* denote the strengthening of AR to

$$(\forall x \in A\, \exists y\, \varphi) \to (\exists B\, \forall x \in A\, \exists y \in B\, \varphi).$$

Show that every instance of AR* is provable in ZF. *Hint.* Let $\psi(x, \alpha)$ say that α is the least ordinal such that $\exists y \in R(\alpha)\, \varphi(x, y)$, and apply AR to ψ. *Remark.* AR* is not provable in ZF^-.

(16) Show, in ZFC, that for each formula $\varphi(x, y)$,

$$\forall x \in A\, \exists y\, \varphi(x, y) \to \exists f\, (f \text{ is a function} \wedge \mathrm{dom}(f) = A \wedge \forall x \in A\, \varphi(x, f(x))).$$

Hint. See Exercise 15.

(17) Show, in ZF, that if $\mathbf{R}$ is a well-founded relation on $\mathbf{A}$ (but possibly not set-like), then every non-empty subclass $\mathbf{X}$ of $\mathbf{A}$ has an $\mathbf{R}$-minimal element.
Hint. Find a γ, such that

$$\forall x \in \mathbf{A} \cap R(\gamma)\, [\exists y \in \mathbf{A}\, (y \mathbf{R} x) \to \exists y \in \mathbf{A} \cap R(\gamma)\, (y \mathbf{R} x)].$$

Remark. This and Exercise 15 may be viewed as examples of the Reflection Theorem (see IV §7).

(18) Let AF denote the Axiom of Foundation and AF* denote the axiom schema,

$$(\forall x\,(\forall y \in x\,\phi(y) \rightarrow \phi(x))) \rightarrow \forall x\,\phi(x)$$

($\phi(x)$ any formula not containing the variable y). Show, in $\mathrm{ZF}^- - \mathrm{P}$, that AF and AF* are equivalent.
Remark. AF* justifies transfinite induction on $\in$.

(19) Let ZF* denote ZF with AF* and AR* in place of AF and AR. Show, in ZF* − P − Inf:

(*a*) Transitive closure is defined; i.e., for all A, there is a least transitive T such that $A \subset T$.

(*b*) Transfinite recursion on $\in$ (5.6 where $\mathbf{A} = \mathbf{V}$ and $\mathbf{R}$ is $\in$) is justified.
Remark. One only needs AF* for this exercise. ZF* is equivalent to ZF by Exercises 15 and 18, but ZF* − P is not equivalent to ZF − P. For an application of AR*, see Exercise 20.

(20) Show, in ZFC* − P, that if A_n is countable for all $n < \omega$, then $\bigcup_n A_n$ is countable. Note that the proof in (I 10.21) of this fact used the Power Set Axiom.

CHAPTER IV

EASY CONSISTENCY PROOFS

As indicated in I §4, the consistency proofs in set theory proceed by considering interpretations other than the intended one. We shall now make a detailed study of this technique.

The consistency results of real mathematical interest involve axioms added to ZFC, such as CH or $\neg$CH. These results are taken up in Chapters VI–VIII. In this chapter we introduce the notions of relativization and absoluteness, which are basic to all our consistency results, and we present some easy applications, such as

$$\text{Con}(\text{ZF}^-) \to \text{Con}(\text{ZF}),$$

to show how these notions are used. We also prove the reflection theorem, which is used to show that ZF is not finitely axiomatizable.

§1. Three informal proofs

Intuitively, ZF is consistent because all its axioms are true *under the right interpretation*—namely, we think of the variables ranging over the well-founded sets. Briefly, we say **WF** is a *model* of ZF. Since any sentence provable from ZF is true in **WF**, ZF cannot prove both a sentence and its negation, so ZF is consistent.

Now this "proof" of consistency of ZF is unique among our consistency results in that we are also advocating **WF** as the natural place to do all mathematics, so that we take ZF as our basic theory. But given any class **M** we might consider the "unnatural" interpretation of letting all variables range over **M**. If S is any set of sentences and we have shown that S is true in **M**, then we have shown that S is *consistent*, even though we may know nothing about the *truth* of S in any "natural" interpretation.

For example, let **M** be the set $R(\omega)$. Let S be the set of all axioms of ZF except Infinity, together with the negation of the Axiom of Infinity—or, briefly,

$$S = \text{ZF} - \text{Inf} + \neg\text{Inf};$$

then $R(\omega)$ is a model of S. We shall see this in some detail in §2, but roughly, Pairing holds since if $x, y \in R(\omega)$, then $\{x, y\} \in R(\omega)$; likewise for Union and Power Set; Foundation holds since $R(\omega) \subset \mathbf{WF}$; Infinity fails since there are no infinite sets in $R(\omega)$. Thus, S is consistent, even though it is not a viable foundation for mathematics. The philosophical import of Con(S) is that, since

$$\mathrm{ZF} - \mathrm{Inf} \nvdash \mathrm{Inf},$$

we are not being redundant in listing Infinity as an axiom of ZF.

For a more trivial example (mentioned also in I §5), let $\mathbf{M} = \{0\}$. Then Extensionality and Comprehension are true in $\mathbf{M}$, but so is $\forall y\,(y = 0)$, so one cannot prove on the basis of these first two axioms that there is a non-empty set.

The reader who is not a confirmed Platonist will find our arguments above very suspect, especially regarding the consistency of ZF, since this requires believing in the existence of the object **WF**. In fact, by the Gödel Incompleteness Theorem (see I 14.3), one cannot prove the consistency of ZF by an argument formalizable within ZF.

What we actually do is to start with ZF^- as our basic theory. In ZF^- we have developed the properties of **WF**, $R(\omega)$, and $\{0\}$. We then show that if ZF^- is consistent, so are each of

$$\mathrm{ZF}, \tag{1}$$

$$\mathrm{ZF} - \mathrm{Inf} + \neg\mathrm{Inf}, \tag{2}$$

$$\text{Extensionality} + \text{Comprehension} + \forall y\,(y = 0). \tag{3}$$

So our results are actually *relative consistency results*—i.e., predicated upon the assumption that ZF^- is consistent. These relative consistency results will be accomplished by completely finitistic means, whereas the consistency of ZF^- will remain either an open question or an article of faith, depending upon one's philosophy.

Actually, once we have proved the relative consistency of ZF, we shall henceforth take ZF as our basic theory for the reasons discussed above; also, for technical reasons (see §5) it is much easier to prove relative consistency results from ZF than from ZF^-. Of course, once we have shown that

$$\mathrm{Con}(\mathrm{ZF}^-) \rightarrow \mathrm{Con}(\mathrm{ZF}),$$

any proof of $\mathrm{Con}(\mathrm{ZF}) \rightarrow \mathrm{Con}(S)$ yields a proof of $\mathrm{Con}(\mathrm{ZF}^-) \rightarrow \mathrm{Con}(S)$.

The consistency of (1), (2), and (3) above (assuming $\mathrm{Con}(\mathrm{ZF}^-)$) will be proved incidentally as Corollaries 4.4, 3.15, and 2.8 of this chapter, although we shall be more concerned with developing general methods, and shall refer to these examples mainly to indicate the kinds of methods we need to develop.

§2. Relativization

We now make the ideas in §1 precise. First, regarding the notion of truth in $\mathbf{M}$,

2.1. DEFINITION. Let $\mathbf{M}$ be any class; then for any formula ϕ we define $\phi^{\mathbf{M}}$, the *relativization* of ϕ to $\mathbf{M}$, by induction on ϕ by:

(a) $(x = y)^{\mathbf{M}}$ is $x = y$.
(b) $(x \in y)^{\mathbf{M}}$ is $x \in y$.
(c) $(\phi \wedge \psi)^{\mathbf{M}}$ is $\phi^{\mathbf{M}} \wedge \psi^{\mathbf{M}}$.
(d) $(\neg\phi)^{\mathbf{M}}$ is $\neg(\phi^{\mathbf{M}})$.
(e) $(\exists x\, \phi)^{\mathbf{M}}$ is $\exists x\,(x \in \mathbf{M} \wedge \phi^{\mathbf{M}})$. □

More formally still, $\mathbf{M}$ is really a formula $\mathbf{M}(v)$, ϕ is another formula, and we are defining, in the metatheory, a third formula $\phi^{\mathbf{M}}$. The formula in (e) should really be $\exists x\,(\mathbf{M}(x) \wedge \phi^{\mathbf{M}})$.

More briefly, $\phi^{\mathbf{M}}$ is the formula obtained from ϕ by replacing all quantifiers, $\exists x$, by $\exists x \in \mathbf{M}$. If $\phi(x_1, \ldots, x_n)$ is a formula with $x_1, \ldots, x_n$ free, then for $x_1, \ldots, x_n \in \mathbf{M}$, $\phi^{\mathbf{M}}(x_1, \ldots, x_n)$ "says" that $\phi^{\mathbf{M}}$ is true of $x_1, \ldots, x_n$ under the interpretation that the bound variables of ϕ range over $\mathbf{M}$. For $x_1, \ldots, x_n$ not in $\mathbf{M}$, the intuitive interpretation of $\phi^{\mathbf{M}}(x_1, \ldots, x_n)$ is unclear, but this turns out to be irrelevant.

In the definition of $\phi^{\mathbf{M}}$, the interpretation of the symbol $\in$ remains unchanged. It is also possible to consider reinterpreting $\in$; see §8.

We have defined $\phi^{\mathbf{M}}$ only for the official unabbreviated formulas of I §2. Note, however, that the defined *logical* abbreviations have their intended meaning. For example, $(\phi \vee \psi)^{\mathbf{M}}$ is really $\neg(\neg\phi \wedge \neg\psi)^{\mathbf{M}}$, which is by our definition $\neg(\neg(\phi^{\mathbf{M}}) \wedge \neg(\psi^{\mathbf{M}}))$, which is $\phi^{\mathbf{M}} \vee \psi^{\mathbf{M}}$. Similarly, $(\forall x\, \phi)^{\mathbf{M}}$ is $(\neg\exists x\, \neg\phi)^{\mathbf{M}}$, which is $\neg\exists x\,(x \in \mathbf{M} \wedge \neg(\phi^{\mathbf{M}}))$, which is logically equivalent to $\forall x\,(x \in \mathbf{M} \rightarrow \phi^{\mathbf{M}})$. The situation for abbreviations involving defined set-theoretic relations and operations, like $\subset$ or $\mathscr{P}$, is more complicated and will be discussed later.

2.2. DEFINITION. Let $\mathbf{M}$ be any class.

(a) For a sentence ϕ, "*ϕ is true in* $\mathbf{M}$" means $\phi^{\mathbf{M}}$.

(b) For a set of sentences S, "*S is true in* $\mathbf{M}$," or "$\mathbf{M}$ *is a model for S*," means that each sentence in S is true in $\mathbf{M}$. □

Intuitively, clauses (a) and (b) are variants on the same idea, but formally they are of different sorts. "*ϕ is true in* $\mathbf{M}$" is just an abbreviation for the sentence $\phi^{\mathbf{M}}$, whereas "S is true in $\mathbf{M}$" is really shorthand for a statement in the metatheory that for each ϕ in S, we can prove $\phi^{\mathbf{M}}$ from the axioms we are presently using.

The basic result from logic that we use to establish relative consistency results is the following.

2.3. LEMMA. *Let S and T be two sets of sentences in the language of set theory, and suppose, for some class (i.e., predicate)* $\mathbf{M}$, *we can prove from T that* $\mathbf{M} \neq 0$ *and* $\mathbf{M}$ *is a model for S. Then* $\mathrm{Con}(T) \to \mathrm{Con}(S)$.

PROOF. If S were inconsistent, we could prove $\chi \wedge \neg\chi$ from S for some (or any) sentence χ. Then, arguing from T, we can prove S is true in $\mathbf{M}$ and hence $\chi^{\mathbf{M}} \wedge \neg\chi^{\mathbf{M}}$, which would be a contradiction. Hence, T is inconsistent. □

The reader who finds this too informal will find a more formal treatment in §8.

The T of Lemma 2.3 will usually be something like ZF^- or ZF or ZFC. Examples of S are (1)–(3) of §1, or ZFC + GCH (in VI) or ZFC + $\neg$CH (in VII).

The reason we need $\mathbf{M}$ to be non-empty is that it is logically provable that $\exists x\,(x = x)$, although for definiteness we have stated it as Axiom 0. Note that $\exists x\,(x = x)^{\mathbf{M}}$ is equivalent to $\mathbf{M}$ being non-empty. In what follows, we shall always assume that the $\mathbf{M}$ we deal with is non-empty, so that Axiom 0 will be true in $\mathbf{M}$.

We next look at Axiom 1, Extensionality. Relativized to $\mathbf{M}$, it is

$$\forall x, y \in \mathbf{M}\,\big(\forall z \in \mathbf{M}\,(z \in x \leftrightarrow z \in y) \to x = y\big).$$

This is precisely the definition of $\in$ being extensional on $\mathbf{M}$ (III 5.11). Since $\in$ is extensional on any transitive set (III 5.12), we have the following.

2.4. LEMMA. *If* $\mathbf{M}$ *is transitive, the Axiom of Extensionality is true in* $\mathbf{M}$. □

Axiom 2, Comprehension, is usually not true in $\mathbf{M}$ unless $\mathbf{M}$ has been very carefully constructed. The following Lemma reduces Comprehension to a closure property of $\mathbf{M}$.

2.5. LEMMA. *Suppose that for each formula $\phi(x, z, w_1, \ldots, w_n)$ with no variable besides the displayed ones free,*

$$\forall z, w_1, \ldots, w_n \in \mathbf{M}\,\big(\{x \in z\colon \phi^{\mathbf{M}}(x, z, w_1, \ldots, w_n)\} \in \mathbf{M}\big);$$

then the Comprehension Axiom (i.e., each instance thereof) is true in $\mathbf{M}$.

PROOF. We must check that for each ϕ as above,

$$\forall z, w_1, \ldots, w_n \in \mathbf{M}\; \exists y \in \mathbf{M}\big(x \in y \leftrightarrow x \in z \wedge \phi^{\mathbf{M}}(x, z, w_1, \ldots, w_n)\big),$$

since this is a relativized instance of Comprehension. Given $z, w_1, \ldots, w_n$, take $y = \{x \in z : \phi^{\mathbf{M}}(x, z, w_1, \ldots, w_n)\} \in \mathbf{M}$; then for all x, and hence for all $x \in \mathbf{M}$,

$$x \in y \leftrightarrow \phi^{\mathbf{M}}(x, z, w_1, \ldots, w_n). \quad \square$$

It is easy to see that the condition of the Lemma is also necessary for Comprehension to hold in **M** if **M** is transitive. This condition will often be difficult to check in our work in later chapters, since one must consider the meaning of all possible formulas relativized to **M**. However, in many of the simple models considered in this chapter, Comprehension will hold trivially, since the following will apply.

2.6. Corollary. *If* $\forall z \in \mathbf{M}\,(\mathscr{P}(z) \subset \mathbf{M})$, *then the Comprehension Axiom is true in* **M**. $\square$

We now can prove the easiest of the three consistency results mentioned in §1.

2.7. Theorem. ZF$^-$. *If* $\mathbf{M} = \{0\}$, *then Axioms* 0–2, *together with* $\forall y\,(y = 0)$, *are true in* **M**.

Proof. Here, we must consider $\forall y\,(y = 0)$ an abbreviation of $\forall y\,\forall x\,(x \notin y)$, which is true in **M** since $0 \notin 0$. Axioms 0 and 1 are true in **M** since **M** is transitive and non-empty. Axiom 2 is true in **M** by Corollary 2.6, since every subset of every element of **M** (i.e., 0) is in **M**. $\square$

Thus, by Lemma 2.3, the following holds.

2.8. Corollary. Con(ZF$^-$) $\rightarrow$ Con(Extensionality + Comprehension + $\forall y\,(y = 0)$). $\square$

The proof of Theorem 2.7 raises an important point. We have only defined relativization for basic formulas in the language of set theory. If we wish to relativize a formula containing a defined notion, such as $y = 0$, we must express it in our original language. Now this should cause us no trouble in principle, since we have taken the position that the only real formulas are those with just $\in$ and $=$, and anything else is just an abbreviation for a real formula which we were too lazy to write down (see I §8). But many interesting mathematical statements, such as CH, and even some basic axioms, such as AC, are expressed using quite a number of defined notions, and we would like to be able to check their truth or falsity in a model without actually writing out the unabbreviated statement.

If we only had defined relations, there would be little problem, since in unabbreviating an expression we merely replace the relation with the formula it abbreviates. For example, $z \subset x$ abbreviates

$$\forall v (v \in z \to v \in x),$$

so $(z \subset x)^{\mathbf{M}}$ abbreviates

$$\forall v \in \mathbf{M} (v \in z \to v \in x),$$

which is equivalent to $z \cap \mathbf{M} \subset x$. Now, if we wish to check whether a statement involving $\subset$, like, e.g., the Power Set Axiom,

$$\forall x \exists y \forall z (z \subset x \to z \in y),$$

holds in **M**, it is not necessary to write out the unabbreviated statement. The Power Set Axiom relativized to **M** is equivalent to

$$\forall x \in \mathbf{M} \exists y \in \mathbf{M} \forall z \in \mathbf{M} (z \cap \mathbf{M} \subset x \to z \in y).$$

In the special case that **M** is transitive, which will be the case in most of the examples we study, this becomes even simpler. Then, $z \cap \mathbf{M} = z$ for $z \in \mathbf{M}$, so, for $z, y \in \mathbf{M}$,

$$(z \subset y)^{\mathbf{M}} \leftrightarrow z \subset y$$

(or $\subset$ is *absolute* for **M** in the terminology of §3). Thus, for transitive **M**, the Power Set Axiom holds in **M** iff

$$\forall x \in \mathbf{M} \exists y \in \mathbf{M} \forall z \in \mathbf{M} (z \subset x \to z \in y).$$

Equivalently, we may state the following.

2.9. LEMMA. *If* **M** *is transitive, the Power Set Axiom holds in* **M** *iff*

$$\forall x \in \mathbf{M} \exists y \in \mathbf{M} (\mathscr{P}(x) \cap \mathbf{M} \subset y). \quad \square$$

In handling defined operations and constants, we must be more careful. As pointed out in I §8, if S is a set of axioms and

$$S \vdash \forall x_1, \ldots, x_n \exists! y \, \phi(x_1, \ldots, x_n, y),$$

we may "define" $F(x_1, \ldots, x_n)$ to be the y such that $\phi(x_1, \ldots, x_n, y)$; formally, expressions using F are abbreviations for expressions not using F. We have not been specific as to which of a large number of possible unabbreviations one should take, since they are all equivalent *on the basis of* S (there is a certain amount of logic being swept under the rug here; see §8). But they need not be equivalent in a class in which S fails.

For example, let $\phi(y)$ be $\forall v (v \notin y)$. Then as long as S contains Comprehension and Extensionality $S \vdash \exists! y \, \phi(y)$, and we "define" 0 to be that y. Now $0 \in z$ could abbreviate either

$$\psi(z)\colon \exists y (\phi(y) \wedge y \in z), \quad \text{or} \quad \chi(z)\colon \forall y (\phi(y) \to y \in z);$$

these are equivalent if $\exists! y\, \phi(y)$. Let M be $\{a, b, c\}$, where

$$a = 0, \qquad b = \{0\}, \quad \text{and} \quad c = \{\{\{0\}\}\};$$

then $\phi^M(a)$ and $\phi^M(c)$ are both true, so $\psi^M(b)$ is true, while $\chi^M(b)$ is false.

To avoid this problem we only consider relativization of expressions containing F to a class $\mathbf{M}$ if we have already checked that

$$\forall x_1, \ldots, x_n\, \exists! y\, \phi(x_1, \ldots, x_n, y) \tag{$*$}$$

is true in $\mathbf{M}$. Usually, $\mathbf{M}$ will have been shown to satisfy the axioms of set theory from which ($*$) was proved in our discussions in Chapter I. If ($*$) holds in $\mathbf{M}$, we also use $F^{\mathbf{M}}(x_1, \ldots, x_n)$ for the unique $y \in \mathbf{M}$ such that $\phi^{\mathbf{M}}(x_1, \ldots, x_n, y)$. If we have achieved some insight into what $F^{\mathbf{M}}$ is, we can check the truth in $\mathbf{M}$ of expressions involving F without unravelling all the definitions. For example, if $M = \{1, 2\}$, then $\exists! y\, \forall v\, (v \notin y)$ is true in M, and $0^M = 1$. Since $1 \in 2$, we see that (the sentence abbreviated by) $\exists x\, (0 \in x)$ is true in M also. If $M = \{0\}$, then $0^M = 0$, and $\exists x\, (0 \in x)$ is false in M.

We can now apply these ideas to make precise some of our informal statements about $R(\omega)$ and $\mathbf{WF}$ in §1. Let $\mathbf{N}$ be either one of these. $\mathbf{N}$ is transitive, so it satisfies Extensionality. If $z \in \mathbf{N}$, then $\mathscr{P}(z) \in \mathbf{N}$ (see III, 2.8), so $\mathscr{P}(z) \subset \mathbf{N}$. Thus, $\mathbf{N}$ satisfies Comprehension (by Corollary 2.6), and Power Set (by Lemma 2.9). That $\mathbf{N}$ satisfies Pairing and Union follows from the fact that $\mathbf{N}$ is closed under the pairing and union operators (see III, 2.8), plus the following general fact.

2.10. Lemma. *If*

$$\forall x, y \in \mathbf{M}\, \exists z \in \mathbf{M}\, (x \in z \wedge y \in z), \quad \text{and} \quad \forall x \in \mathbf{M}\, \exists z \in \mathbf{M}\, \left(\bigcup x \subset z\right),$$

then the Pairing and Union Axioms are true in $\mathbf{M}$. □

The Replacement Axiom, like Comprehension, is often difficult to check since it involves considering an arbitrary formula, but, also like Comprehension, it is easy in $R(\omega)$ and $\mathbf{WF}$. First, it is convenient to translate the relativization of the axiom to obtain the following.

2.11. Lemma. *Suppose we can show, for each formula* $\phi(x, y, A, w_1, \ldots, w_n)$ *and each* $A, w_1, \ldots, w_n \in \mathbf{M}$: *if*

$$\forall x \in A\, \exists! y \in \mathbf{M}\, \phi^{\mathbf{M}}(x, y, A, w_1, \ldots, w_n),$$

then

$$\exists Y \in \mathbf{M}\left(\{y\colon \exists x \in A\, \phi^{\mathbf{M}}(x, y, A, w_1, \ldots, w_n)\} \subset Y\right).$$

Then the Replacement Scheme is true in $\mathbf{M}$. □

Applying this to $\mathbf{N} = R(\omega)$ or $\mathbf{WF}$, let

$$Y = \{y \in \mathbf{N}: \exists x \in A\ \phi^{\mathbf{N}}(x, y, A. w_1, \ldots, w_n)\};$$

then $Y \subset \mathbf{N}$, so, if $\mathbf{N} = \mathbf{WF}$, $Y \in \mathbf{N}$ (see III 2.10). If $\mathbf{N} = R(\omega)$, then $|Y| \leq |A| < \omega$, so for some n, $Y \subset R(n)$, and $Y \in R(n+1) \subset \mathbf{N}$. Thus, in either $\mathbf{N}$, Replacement is true..

The Axiom of Foundation relativized to a class $\mathbf{M}$ is

$$\forall x \in \mathbf{M}(\exists y \in \mathbf{M}(y \in x) \to \exists y \in \mathbf{M}(y \in x \wedge \neg \exists z \in \mathbf{M}(z \in x \wedge z \in y))).$$

If $\mathbf{M} \subset \mathbf{WF}$, then given $x \in \mathbf{M}$, we may take $y \in \mathbf{M} \cap x$ to be of least rank to see that the axiom holds (this is a minor modification of the proof that $\in$ is well-founded on any element of $\mathbf{WF}$ (see III, 3.2)). In particular, we see just in ZF$^-$ that Foundation is true in $R(\omega)$ and $\mathbf{WF}$. More generally:

2.12. LEMMA (ZF$^-$). *The Axiom of Foundation is true in any* $\mathbf{M} \subset \mathbf{WF}$. □

We have now shown:

2.13. LEMMA (ZF$^-$). $\mathbf{WF}$ *and* $R(\omega)$ *are models of* ZF − Inf. □

The Axiom of Infinity,

$$\exists x (0 \in x \wedge \forall y \in x (S(y) \in x)),$$

involves the defined notions 0 and S. Intuitively, the axiom is true in $\mathbf{WF}$ (take $x = \omega$) and false in $R(\omega)$, but a rigorous proof of this involves checking that 0 and S mean the same in $R(\omega)$ and $\mathbf{WF}$ as they do in $\mathbf{V}$; that is, 0 and S are *absolute* for $R(\omega)$ and $\mathbf{WF}$. This is quite easy, but rather than presenting a specific argument for 0 and S, we turn to a study of absoluteness in general.

3. Absoluteness

3.1. DEFINITION. Let ϕ be a formula with at most $x_1, \ldots, x_n$ free.

(1) If $\mathbf{M} \subset \mathbf{N}$, ϕ is *absolute* for $\mathbf{M}$, $\mathbf{N}$ iff

$$\forall x_1, \ldots, x_n \in \mathbf{M}\,(\phi^{\mathbf{M}}(x_1, \ldots, x_n) \leftrightarrow \phi^{\mathbf{N}}(x_1, \ldots, x_n)).$$

(2) ϕ is *absolute* for $\mathbf{M}$ iff ϕ is absolute for $\mathbf{M}$, $\mathbf{V}$; equivalently

$$\forall x_1, \ldots, x_n \in \mathbf{M}\,(\phi^{\mathbf{M}}(x_1, \ldots, x_n) \leftrightarrow \phi(x_1, \ldots, x_n)). \quad \square$$

Note that if ϕ is absolute for $\mathbf{M}$ and absolute for $\mathbf{N}$ and $\mathbf{M} \subset \mathbf{N}$, then ϕ is absolute for $\mathbf{M}$, $\mathbf{N}$.

In this section, and in §5, we shall develop methods for showing easily that some (*but not all*) ϕ are absolute for many of the models we shall be discussing. This will involve pointing out certain rules for inductively building complex absolute formulas from simple ones, so that we can eventually recognize as absolute a large number of mathematical concepts. A trivial example of such an induction principle is given by the following.

3.2. LEMMA. *If* $\mathbf{M} \subset \mathbf{N}$ *and* ϕ *and* ψ *are both absolute for* $\mathbf{M}$, $\mathbf{N}$, *then so are* $\neg\phi$ *and* $\phi \wedge \psi$. □

Since $x = y$ and $x \in y$ are absolute for all $\mathbf{M}$, and any formula without quantifiers is built up from such atomic formulas using $\neg$ and $\wedge$, we have the following.

3.3. COROLLARY. *If* ϕ *is quantifier-free, then* ϕ *is absolute for any* $\mathbf{M}$. □

Unfortunately, even very simple formulas, such as $x \subset y$ (i.e., $\forall z\,(z \in x \rightarrow z \in y)$) do involve quantifiers and can fail to be absolute. For example, if $\mathbf{M} = \{0, a\}$, where $a = \{\{0\}\}$, then $(a \subset 0)^{\mathbf{M}}$ but $a \not\subset 0$. Fortunately, if $\mathbf{M}$ is transitive, as will be most models we study, then $x \subset y$ will be absolute for $\mathbf{M}$ (see the discussion before Lemma 2.9). The next few lemmas point out the general principles operating here.

3.4. LEMMA. *If* $\mathbf{M} \subset \mathbf{N}$ *are both transitive and* ϕ *is absolute for* $\mathbf{M}$, $\mathbf{N}$, *then so is* $\exists x \in y\, \phi$ (*i.e.*, $\exists x (x \in y \wedge \phi)$).

PROOF. Write ϕ as $\phi(x, y, z_1, \ldots, z_n)$, displaying its other free variables, then for any $y, z_1, \ldots, z_n \in \mathbf{M}$,

$$[\exists x (x \in y \wedge \phi(y, z_1, \ldots, x_n))]^{\mathbf{M}} \leftrightarrow \exists x (x \in y \wedge \phi^{\mathbf{M}}(y, z_1, \ldots, z_n)) \leftrightarrow$$

$$\exists x\,(x \in y \wedge \phi^{\mathbf{N}}(y, z_1, \ldots, z_n)) \leftrightarrow [\exists x\,(x \in y \wedge \phi \quad (y, z_1, \ldots, z_n))]^{\mathbf{N}}$$

For the middle $\leftrightarrow$, we just applied the assumed absoluteness of ϕ. The first $\leftrightarrow$ used the transitivity of $\mathbf{M}$ to write $\exists x\,(x \in y \wedge \cdots)$ instead of $\exists x \in \mathbf{M}\,(x \in y \wedge \cdots)$. Likewise, the third $\leftrightarrow$ used the transitivity of $\mathbf{N}$. □

We call $\exists x \in y$ a *bounded quantifier* and a formula in which all quantifiers are bounded is called Δ_0. More formally:

3.5. DEFINITION. The Δ_0 formulas are those built up inductively by the following rules:

(1) $x \in y$ and $x = y$ are Δ_0.

(2) If ϕ, ψ are Δ_0, so are $\neg\phi$ and $\phi \wedge \psi$.

(3) If ϕ is Δ_0, so is $\exists x\,(x \in y \wedge \phi)$. □

3.6. COROLLARY. *If* **M** *is transitive and* ϕ *is* Δ_0, *then* ϕ *is absolute for* **M**. □

The usefulness of this result is limited by the fact that one rarely sees a Δ_0 formula. Even $x \subset y$ is really $\forall z\,(z \in x \rightarrow z \in y)$, which abbreviates

$$\neg \exists z \neg (z \in x \rightarrow z \in y),$$

which is not Δ_0, although it is logically equivalent to

$$\neg \exists z \in x\,(\neg z \in y),$$

which is Δ_0. In practice, Corollary 3.6 is used in conjunction with the following remark.

3.7. LEMMA. *Suppose* $\mathbf{M} \subset \mathbf{N}$, *and both* **M** *and* **N** *are models for a set of sentences S, such that*

$$S \vdash \forall x_1, \ldots, x_n\,(\phi(x_1, \ldots, x_n) \leftrightarrow \psi(x_1, \ldots, x_n));$$

then ϕ *is absolute for* **M**, **N** *iff* ψ *is.* □

In particular, $\forall x \in y$ is essentially a bounded quantifier, since $\forall x \in y\ \chi$ is logically equivalent to $\neg \exists x \in y \neg \chi$.

Applying Lemma 3.7 with **M** transitive, $\mathbf{N} = \mathbf{V}$, and S the empty set of sentences, we see that $x \subset y$ is absolute for **M**. This was a very long-winded way of proving a fact we already knew from §2, except that we shall now proceed to establish many more absoluteness results by this method. First, since many defined notions are functions rather than relations, we make the following definition.

3.8. DEFINITION. If $\mathbf{M} \subset \mathbf{N}$, and $F(x_1, \ldots, x_n)$ is a defined function, we say F is *absolute* for **M**, **N** if the formula $F(x_1, \ldots, x_n) = y$. □

More formally, suppose $F(x_1, \ldots, x_n)$ was "defined" as the unique y such that $\phi(x_1, \ldots, x_n, y)$. In accordance with our earlier conventions, we shall only discuss absoluteness of F for **M**, **N** if we know that the statement

$$\forall x_1, \ldots, x_n\, \exists! y\, \phi(x_1, \ldots, x_n, y)$$

is true in both **M** and **N**. Assuming this, F is absolute for **M**, **N** iff ϕ is iff for all $x_1, \ldots, x_n \in \mathbf{M}$, $F^{\mathbf{M}}(x_1, \ldots, x_n) = F^{\mathbf{N}}(x_1, \ldots, x_n)$. See §8 for a still more formal treatment.

3.9. THEOREM. *The following relations and functions were defined in* $\mathrm{ZF}^- - \mathrm{P} - \mathrm{Inf}$ *by formulas provably equivalent in* $\mathrm{ZF}^- - \mathrm{P} - \mathrm{Inf}$ *to* Δ_0 *formulas. They are thus absolute for any transitive* **M** *which is a model for* $\mathrm{ZF}^- - \mathrm{P} - \mathrm{Inf}$.

(a) $x \in y$,
(b) $x = y$,
(c) $x \subset y$,
(d) $\{x, y\}$,
(e) $\{x\}$,
(f) $\langle x, y \rangle$,
(g) 0,
(h) $x \cup y$,
(i) $x \cap y$,
(j) $x \setminus y$,
(k) $S(x)$ (*i.e.*, $x \cup \{x\}$),
(l) x *is transitive*,
(m) $\bigcup x$,
(n) $\bigcap x$ (*where* $\bigcap 0 = 0$).

Proof. That they were indeed defined in $\mathrm{ZF}^- - \mathrm{P} - \mathrm{Inf}$ may be verified by referring to Chapter I, but we made no effort there to use Δ_0 formulas in the definition. We now simply go down the list and, using only the axioms $\mathrm{ZF}^- - \mathrm{P} - \mathrm{Inf}$, check that the defining formulas are equivalent to Δ_0 formulas.

(a), (b), and (c) were discussed above. For (d),

$$z = \{x, y\} \leftrightarrow [x \in z \wedge y \in z \wedge \forall w \in z (w = x \vee w = y)],$$

and the formula on the right is logically equivalent to a Δ_0 formula. $z = \{x\}$ is similar. Since $\langle x, y \rangle = \{\{x\}, \{x, y\}\}$,

$$z = \langle x, y \rangle \leftrightarrow [\exists w \in z (w = \{x\}) \wedge \exists w \in z (w = \{x, y\}) \wedge \forall w \in z (w = \{x\} \vee w = \{x, y\})],$$

and the formula on the right is equivalent to a Δ_0 formula obtained by replacing the occurrences of $w = \{x\}$ and $w = \{x, y\}$ by the Δ_0 formulas to which they are equivalent. For (g), (h), (i), and (k),

$$z = 0 \leftrightarrow [\forall w \in z (w \neq w)].$$

$$z = x \cup y \leftrightarrow [\forall w \in z (w \in x \vee w \in y) \wedge x \subset z \wedge y \subset z].$$

$$z = x \cap y \leftrightarrow [\forall w \in x (w \in y \rightarrow w \in z) \wedge z \subset x \wedge z \subset y].$$

$$z = S(x) \leftrightarrow [x \in z \wedge x \subset z \wedge \forall w \in z (w = x \vee w \in x)].$$

(j) is similar to (i).

Finally, for (l)–(n),

$$x \text{ is transitive} \leftrightarrow [\forall v \in x \, \forall z \in v (z \in x)].$$

$$y = \bigcup x \leftrightarrow [\forall v \in x (v \subset y) \wedge \forall z \in y \, \exists v \in x (z \in v)].$$

$$y = \bigcap x \leftrightarrow [\forall v \in x (y \subset v) \wedge \forall v \in x \, \forall z \in v (\forall w \in x (z \in w) \rightarrow z \in y) \wedge (x = 0 \rightarrow y = 0)].$$

In (n), $\bigcap 0$ "should be" $\mathbf{V}$, but we redefine it in an ad hoc way to be the set 0. □

We remark that the main purpose of this book is to discuss models of full ZFC, not fragments thereof. Nevertheless, the fact that Theorem 3.9 is valid for models of $\mathrm{ZF}^- - \mathrm{P} - \mathrm{Inf}$ will be useful since we shall be able to establish basic absoluteness results before having proved a given model satisfies ZFC, and in fact we shall use these absoluteness results to aid in verifying some of the axioms of ZFC.

The reader may have noticed a much quicker proof of absoluteness of ordered pairing (3.9(f)). Once one knows that unordered pairing means the same in $\mathbf{M}$ and in $\mathbf{V}$, the same must be true of any composition of these operations—in particular of ordered pairing. Thus, the following holds.

3.10. LEMMA. *Absolute notions are closed under composition. That is, suppose* $\mathbf{M} \subset \mathbf{N}$, $\phi(x_1, \ldots, x_n)$, $F(x_1, \ldots, x_n)$, *and* $G_i(y_1, \ldots, y_m)$ $(i = 1, \ldots, n)$ *are all absolute for* $\mathbf{M}$, $\mathbf{N}$; *then so are the formula*

$$\phi(G_1(y_1, \ldots, y_n), \ldots, G_n(y_1, \ldots, y_m)),$$

and the function

$$F(G_1(y_1, \ldots, y_m), \ldots, G_n(y_1, \ldots, y_m)).$$

PROOF. For the case $n = m = 1$. If $y \in \mathbf{M}$, then

$$(\phi(G(y)))^{\mathbf{M}} \leftrightarrow \phi^{\mathbf{M}}(G^{\mathbf{M}}(y)) \leftrightarrow \phi^{\mathbf{N}}(G^{\mathbf{N}}(y)) \leftrightarrow (\phi(G(y)))^{\mathbf{N}},$$

since $G^{\mathbf{M}}(y) = G^{\mathbf{N}}(y)$ and ϕ is absolute for $\mathbf{M}$, $\mathbf{N}$. Similarly,

$$F(G(y))^{\mathbf{M}} = F^{\mathbf{M}}(G^{\mathbf{M}}(y)) = F^{\mathbf{N}}(G^{\mathbf{N}}(y)) = F(G(y))^{\mathbf{N}}. \quad \square$$

Now, a simpler way to prove $\langle x, y \rangle$ is absolute is to write

$$\langle x, y \rangle = F(G_1(x, y), G_2(x, y)),$$

where $G_1(x, y)$ is $\{x\}$ and $F(x, y) = G_2(x, y) = \{x, y\}$, and use the fact that G_1, G_2, and F are absolute by Theorem 3.9 (d) and (e). The longer proof of 3.9(f) established the stronger result that ordered pairing is a Δ_0 function. It is *not* true in general that a composition of Δ_0 functions is Δ_0 (see Exercise 13), although it is true that it requires the Axiom of Foundation to provide an explicit counterexample (by Exercise 35). In particular, the functions and relations proved absolute in the next theorem are in fact provably Δ_0 (see Exercise 12).

3.11. THEOREM. *The following relations and functions are absolute for any transitive model for* $\mathrm{ZF}^- - \mathrm{P} - \mathrm{Inf}$.

(a) *z is an ordered pair.*
(b) $A \times B$.
(c) *R is a relation.*
(d) $\mathrm{dom}(R)$.
(e) $\mathrm{ran}(R)$.

(f) *R is a function.*
(g) $R(x)$.
(h) *R is a* 1–1 *function.*

PROOF. z is an ordered pair $\leftrightarrow [\exists x \in \bigcup z\, \exists y \in \bigcup z(z = \langle x, y\rangle)]$, and the formula on the right is obtained by substituting absolute functions in an absolute relation, so is absolute by Lemma 3.10. To see formally why Lemma 3.10 applies here, we write

$$z \text{ is an ordered pair} \leftrightarrow \phi(G_1(z), G_2(z), G_3(z)),$$

where $G_1(z) = G_2(z) = \bigcup z$, which is absolute by Theorem 3.9, $G_3(z) = z$, and $\phi(a, b, c)$ is

$$\exists x \in a\, \exists y \in b\,(c = \langle x, y\rangle),$$

which is absolute since it is obtained by bounded quantification of the absolute formula $c = \langle x, y\rangle$. In the future, we shall replace this formality with the words "by substitution".

For (b)–(h):

$$C = A \times B \leftrightarrow [\forall x \in \forall y \in B(\langle x, y\rangle \in C) \wedge$$
$$\forall z \in C\, \exists x \in A\, \exists y \in B(z = \langle x, y\rangle)].$$

$$R \text{ is a relation} \leftrightarrow [\forall z \in R\,(z \text{ is an ordered pair})].$$

$$A = \text{dom}(R) \leftrightarrow [\forall x \in A\, \exists y \in \bigcup\bigcup R\,(\langle x, y\rangle \in R) \wedge$$
$$\forall x \in \bigcup\bigcup R\, \forall y \in \bigcup\bigcup R(\langle x, y\rangle \in R \rightarrow x \in A)].$$

$$R \text{ is a function} \leftrightarrow [R \text{ is a relation} \wedge \forall x \in \bigcup\bigcup R\, \forall y \in \bigcup\bigcup R\, \forall y' \in \bigcup\bigcup R$$
$$(\langle x, y\rangle \in R \wedge \langle x, y'\rangle \in R \rightarrow y = y')].$$

$$y = R(x) \leftrightarrow [(\phi(x) \wedge \langle x, y\rangle \in R) \vee (\neg\phi(x) \wedge y = 0)],$$

where $\phi(x)$ is

$$\exists v \in \bigcup\bigcup R\,(\langle x, v\rangle \in R \wedge \forall w \in \bigcup\bigcup R\,(\langle x, w\rangle \in R \rightarrow v = w)).$$

$$R \text{ is a 1–1 function} \leftrightarrow [R \text{ is a function} \wedge$$
$$\forall x \in \text{dom}(R)\, \forall x' \in \text{dom}(R)\,(R(x) = R(x') \rightarrow x = x')].$$

Thus, the stated notions are obtained from absolute notions by substitution, bounded quantification, and propositional connectives, and are thus absolute as claimed. We have omitted (e) since it is just like (d). For (g), $R(x)$ is really a defined function of *two* variables R and x, which, if our notation were more consistent we would call, say, appl(R, x). appl(R, x) is the unique y such that $\langle x, y\rangle \in R$ if $\exists!(\langle x, y\rangle \in R)$; if $\neg\exists! y(\langle x, y\rangle \in R)$, we set $\mathrm{appl}(R, x) = 0$. □

There are, of course, infinitely many more properties of functions that one might check are absolute–such as, e.g., "f maps A onto A and has no fixed points." Rather than listing in Theorem 3.11 anything that could possibly occur, we shall simply say "this is absolute by the methods of §3."

Our absoluteness results make it very easy to verify the Axiom of Infinity in a model.

3.12. LEMMA. *Let* $\mathbf{M}$ *be a transitive model for* $\mathrm{ZF}^- - \mathrm{P} - \mathrm{Inf}$. *If* $\omega \in \mathbf{M}$, *then the Axiom of Infinity is true in* $\mathbf{M}$.

PROOF. By the absoluteness of 0 and S, Infinity relativized to $\mathbf{M}$ is equivalent to

$$\exists x \in \mathbf{M}\,(0 \in x \;\wedge\; \forall y \in x\,(S(y) \in x)),$$

which is seen to be true by taking $x = \omega$. □

The same argument shows that Infinity is false in $R(\omega)$, since any $x \in \mathbf{WF}$ containing 0 and closed under S has infinite rank. The following theorem concludes our discussion of $R(\omega)$.

3.13. THEOREM (ZF^-). $R(\omega)$ *is a model of* $\mathrm{ZFC} - \mathrm{Inf} + (\neg\,\mathrm{Inf})$.

PROOF. By the above and Lemma 2.13, we need only check that AC is true in $R(\omega)$. To prove this, we must check that

$$\forall A \in R(\omega)\, \exists R \in R(\omega)\, [(R \text{ well-orders } A)^{R(\omega)}].$$

Fix $A \in R(\omega)$. We know, even without assuming AC, that A is finite and thus can be well-ordered. Let $R \subset A \times A$ well-order A; then $R \in R(\omega)$. The fact that $(R \text{ well-orders } A)^{R(\omega)}$ follows from the following Lemma, which completes the proof of the Theorem. □

3.14. LEMMA (ZF^-). *Suppose* $\mathbf{M}$ *is a transitive model of* $\mathrm{ZF}^- - \mathrm{P} - \mathrm{Inf}$. *Let* $A, R \in \mathbf{M}$ *and suppose that* R *well-orders* A. *Then* $(R \text{ well-orders } A)^{\mathbf{M}}$.

PROOF. That $(R \text{ totally orders } A)^{\mathbf{M}}$ follows by the methods of Theorem 3.11, since this is expressed using basic properties of pairs and quantification over A. For well-ordering, we must check that $(\forall X\, \phi(X, A, R))^{\mathbf{M}}$, where $\phi(X, A, R)$ is

$$X \subset A \;\wedge\; X \neq 0 \rightarrow \exists y \in X\, \forall z \in X\, (\langle z, y\rangle \notin R).$$

Now ϕ is absolute for $\mathbf{M}$ by the methods of Theorem 3.11. It is thus sufficient to check that $\forall X \in \mathbf{M}\, \phi(X, A, R)$, which follows since R well-orders A, so in fact $\forall X\ \phi(X, A, R)$. □

This Lemma illustrates the general fact that a universal quantification of an absolute formula relativizes *down* from $\mathbf{V}$ to $\mathbf{M}$, but it may not relativize *up*. It is conceivable that $(R \text{ well-orders } A)^{\mathbf{M}}$, but R fails to well-order A, since there may be an $X \subset A$ which is not in $\mathbf{M}$ that fails to have an R-least member (see Exercise 21). Well-ordering is absolute if we assume Foundation (see Theorem 5.4), but other essentially universal notions, like the property of being a cardinal, fail to be (see the remarks after Corollary 7.11).

By Lemma 2.3, Theorem 3.13 implies the following relative consistency result.

3.15. COROLLARY. $\mathrm{Con}(\mathrm{ZF}^-) \rightarrow \mathrm{Con}(\mathrm{ZFC} - \mathrm{Inf} + \neg\mathrm{Inf})$. □

§4. The last word on Foundation

In this section we finish our discussion of $\mathbf{WF}$. Henceforth the Axiom of Foundation will be assumed without comment, and our basic system will be ZF (or ZFC in VII and VIII).

4.1. THEOREM (ZF^-). *All axioms of* ZF *are true in* $\mathbf{WF}$.

PROOF. By Lemmas 2.13 and 3.12. □

We consider AC. The following absoluteness result is very particular to $\mathbf{WF}$ and will have no further application.

4.2. LEMMA. *Let* $A \in \mathbf{WF}$. *Then* A *can be well-ordered iff* $(A \textit{ can be well-ordered})^{\mathbf{WF}}$.

PROOF. If A can be well-ordered, let $R \subset A \times A$ well-order A. $A \times A \in \mathbf{WF}$ by III 2.8, so $R \in \mathbf{WF}$ by III 2.10. By Lemma 3.14, $(R \text{ well-orders } A)^{\mathbf{WF}}$, so $(A \text{ can be well-ordered})^{\mathbf{WF}}$. Conversely, if $(A \text{ can be well-ordered})^{\mathbf{WF}}$, fix $R \in \mathbf{WF}$ such that $(R \text{ well-orders } A)^{\mathbf{WF}}$; then, as in the proof of Lemma 3.14, R totally orders A and every non-0 subset of A *in* $\mathbf{WF}$ has an R-least member. But every subset of A *is* in $\mathbf{WF}$ (by III 2.10), so R well-orders A. □

4.3. COROLLARY (ZF^-). $\mathrm{AC} \rightarrow (\mathrm{AC})^{\mathbf{WF}}$. □

The converse of 4.3 need not hold since it is consistent that every well-founded set can be well-ordered but some non-well-founded set cannot (see Exercise 25).

Since ZF^- proves that **WF** is a model for ZF and ZFC^- proves that **WF** is a model for ZFC, we have the following.

4.4. Corollary. $\mathrm{Con}(\mathrm{ZF}^-) \rightarrow \mathrm{Con}(\mathrm{ZF})$ *and* $\mathrm{Con}(\mathrm{ZFC}^-) \rightarrow \mathrm{Con}(\mathrm{ZFC})$. □

We shall, in V and VI, give proofs of $\mathrm{Con}(\mathrm{ZF}) \rightarrow \mathrm{Con}(\mathrm{ZFC})$.

Corollary 4.4 provides a formal justification for using Foundation as a basic axiom when proving relative consistency results. This will be a great technical convenience, since it enables us to establish the absoluteness of many more properties (see §5). However, our reason for using Foundation transcends mere convenience. We have seen that **WF** contains all reasonable mathematical objects (see III §2), and furthermore, reasonable mathematical notions such as well-ordering are absolute for **WF** (for more examples see Exercise 3), so we might as well live in **WF** while doing our mathematics.

Models constructed later for other set-theoretic statements, e.g., GCH or ¬GCH, will have a more ad hoc character, and will not be accompanied by a plausibility argument that the statement should be a basic axiom.

Intuitively, ZF axiomatizes the well-founded sets whereas ZF^- axiomatizes the hereditary sets (see I §4). It might have been more satisfying philosophically to have begun with an axiomatization, ZF^{--}, of "all there is," including cows and pigs. We did not do that because it would require a more extensive revision of ZF than merely dropping one axiom as in ZF^-. To formulate ZF^{--}, we work in a language which has a one-place predicate symbol $\mathrm{Set}(x)$ in addition to the binary $\in$. We then write down all the axioms of ZF^- suitably modified using Set; e.g., extensionality becomes

$$\forall x\, \forall y\, (\mathrm{Set}(x) \wedge \mathrm{Set}(y) \wedge \forall z\, (z \in x \leftrightarrow z \in y) \rightarrow x = y).$$

Note that the variable z is not relativized to Set since members of sets may fail to be sets. ZF^- is then essentially the theory $\mathrm{ZF}^{--} + \forall x\, \mathrm{Set}(x)$. One can, in ZF^{--}, define the classes **WF** and **HS**, where

$$\mathbf{HS} = \{x\colon \mathrm{Set}(x) \wedge \forall n \in \omega\, \forall y \in \textstyle\bigcup^n x\, (\mathrm{Set}(y))\}.$$

Arguing in ZF^{--}, one can show that $\mathbf{WF} \subset \mathbf{HS} \subset \mathbf{V}$, **HS** satisfies ZF^-, and **WF** satisfies ZF. Thus,

$$\mathrm{Con}(\mathrm{ZF}^{--}) \rightarrow \mathrm{Con}(\mathrm{ZF}).$$

§5. More absoluteness

Now that Foundation is a basic axiom, the well-foundedness of the $\in$ relation can be used to establish a large number of new absoluteness results. The basic result is the absoluteness of being an ordinal, which we state in the next theorem together with some related results.

5.1. THEOREM. *The following relations and functions were defined in* ZF − P *by formulas provably equivalent in* ZF − P *to* Δ_0 *formulas. They are thus absolute for transitive models of* ZF − P.

(a) *x is an ordinal.*	(f) 0.
(b) *x is a limit ordinal.*	(g) 1.
(c) *x is a successor ordinal.*	(h) 2.
(d) *x is a finite ordinal.*	...
(e) ω.	(z) 20.

PROOF. On the basis of ZF − P, x is an ordinal iff x is transitive and totally ordered by $\in$ (see III 4.2). Now "x is transitive" is equivalent to a Δ_0 formula by Theorem 3.9, and "x is totally ordered by $\in$" is expressed by quantifying over x:

$$\forall y \in x \, \forall z \in x \, (y \in z \ \vee \ y = z \ \vee \ z \in y) \ \wedge \ \text{etc.},$$

so is also Δ_0. This proves (a).

For (b), x is a limit ordinal iff x is an ordinal, $\forall y \in x \, \exists z \in x (y \in z)$, and $x \neq 0$, and this is all Δ_0 ($x = 0$ is Δ_0 by Theorem 3.9). For (c), x is a successor ordinal iff x is an ordinal and is neither a limit ordinal nor 0. For (d), x is a finite ordinal iff x and all $y \in x$ are either 0 or successor ordinals.

(d) is equivalent to saying that the predicate $x \in \omega$ is expressible by a Δ_0 formula, whereas (e) makes the same assertion about $x = \omega$. To verify (e), note that $x = \omega$ iff x is a limit ordinal and $\forall y \in x \, (y$ is not a limit ordinal).

(f) was done in Theorem 3.9. For (g)–(z), recall that $x = S(y)$ is Δ_0 by 3.9, and observe:

$$x = 1 \leftrightarrow \exists y \in x \big(y = 0 \ \wedge \ x = S(y)\big).$$
$$x = 2 \leftrightarrow \exists y \in x \big(y = 1 \ \wedge \ x = S(y)\big).$$
$$\dots$$
$$x = 20 \leftrightarrow \exists y \in x \big(y = 19 \ \wedge \ x = S(y)\big). \quad \Box$$

We remark that actually the fact that **M** satisfied Infinity is needed only in (e), to show that the constant ω exists, and Foundation can be avoided in (f)–(z), but we shall not bother to keep track of this sort of thing now.

The fact that the notions of ordinal and the set of finite ordinals are absolute implies the absoluteness of associated notions, such as well-ordering and finite set.

5.2. LEMMA. *If* **M** *is a transitive model of* ZF − P, *then every finite subset of* **M** *is in* **M**.

PROOF. By induction on n, show

$$\forall x \subset \mathbf{M}(|x| = n \rightarrow x \in \mathbf{M}).$$

For $n = 0$, this is the absoluteness of 0. If we know it for n and $x \subset \mathbf{M}$ has $n + 1$ elements, let $y \in x$; then $y \in \mathbf{M}$, $(x \setminus \{y\}) \in \mathbf{M}$, and $x = \{y\} \cup (x \setminus \{y\})$, so by absoluteness of pairing, union, and $\setminus$ (see Theorem 3.9), $x \in \mathbf{M}$. □

5.3. THEOREM. *The following are absolute for any transitive* $\mathbf{M}$ *satisfying* ZF − P.

(a) *x is finite.*
(b) A^n.
(c) $A^{<\omega} (= \bigcup \{A^n : n \in \omega\})$.

PROOF. On the basis of ZF − P, x is finite iff $\exists f\, \phi(x, f)$, where $\phi(x, f)$ says:

$$f \text{ is a function } \wedge \operatorname{dom}(f) = x \wedge \operatorname{ran}(f) \in \omega \wedge f \text{ is } 1\text{–}1,$$

which is absolute for $\mathbf{M}$ by Theorems 3.11 and 5.1. It is thus sufficient to show that for $x \in \mathbf{M}$,

$$\exists f \in \mathbf{M}\, \phi(x, f) \leftrightarrow \exists f\, \phi(x, f).$$

The direction from left to right is obvious. The implication from right to left will follow from the fact that

$$\phi(x, f) \rightarrow f \in \mathbf{M}.$$

To see this, note that $\phi(x, f)$ implies that f is a finite set of ordered pairs of elements of $\mathbf{M}$. $\mathbf{M}$ is closed under pairing by absoluteness of pairing, so $f \in \mathbf{M}$ by Lemma 5.2.

For (b) and (c), note that we must consider A^n to be a defined function of two variables $F(A, n)$, where we set $F(A, x) = 0$ if $x \notin \omega$. $A^{<\omega}$ is a defined function $G(A)$ of one variable. Both F and G are defined on the basis of ZF − P (see I 7.21 and discussion following).

Now, for (b) we must check that for A, $x \in \mathbf{M}$, $F(A, x) = F^{\mathbf{M}}(A, x)$. By absoluteness of ω, $F^{\mathbf{M}}(A, x) = 0$ unless $x \in \omega$, and by absoluteness of notions involving functions, $n \in \omega$ implies

$$F^{\mathbf{M}}(A, n) = \{f \in \mathbf{M} : f \text{ is a function } \wedge \operatorname{dom}(f) = n \wedge \operatorname{ran}(f) \subset A\},$$

which equals $F(A, n)$ as in part (a). (c) is similar. □

5.4. THEOREM. *The following are absolute for any transitive model satisfying* ZF − P.

(a) *R well-orders A.*
(b) type(A, R).

PROOF. For (a), it is sufficient to prove that if A, $R \in \mathbf{M}$, then

$$(R \text{ well-orders } A)^{\mathbf{M}} \rightarrow (R \text{ well-orders } A),$$

since the other direction was given by Lemma 3.14. Now it is a theorem of ZF − P (I 7.6) that every well-ordering is isomorphic to an ordinal. Thus, if $(R \text{ well-orders } A)^{\mathbf{M}}$, there are f, $\alpha \in \mathbf{M}$, such that

$$(\alpha \text{ is an ordinal and } f \text{ is an isomorphism from } \langle A, R \rangle \text{ to } \alpha)^{\mathbf{M}}.$$

But this formula is absolute for $\mathbf{M}$ by Theorem 5.1 and the methods of §3, so α is really an ordinal and f an isomorphism, so R in fact well-orders A in type α. This argument also establishes the absoluteness of type(A, R). □

Most ordinal arithmetic is absolute. For example:

5.5. THEOREM. *The following are absolute for any transitive model satisfying* ZF − P:

(a) $\alpha + 1$.
(b) $\alpha - 1$ (*Def. I* 9.4).
(c) $\alpha + \beta$.
(d) $\alpha \cdot \beta$.

PROOF. $\alpha + 1$ is $S(\alpha)$. For (b),

$$x = \alpha - 1 \leftrightarrow (\alpha \text{ is a successor ordinal} \wedge \alpha = x + 1) \vee (\alpha \text{ is not a successor ordinal} \wedge x = \alpha).$$

For (d),

$$\alpha \cdot \beta = \text{type}(\beta \times \alpha, \text{lex}(\beta, \alpha)),$$

where lex(β, α), the lexicographic order on $\beta \times \alpha$, is easily seen to be absolute by the methods of §3. (c) is similar. □

If one thinks of + and · as defined by transfinite recursion, then their absoluteness can also be proved via a general result on the absoluteness of such definitions, which we discuss next.

Since our theorem on recursive definition (III 5.6) was stated in terms of classes, we remark on what relativization and absoluteness mean for classes. Formally, a class, $\mathbf{A}$, is just a formula, $\mathbf{A}(x)$ (see I §9), but we are thinking intuitively of $\mathbf{A} = \{x: \mathbf{A}(x)\}$. By $\mathbf{A}^{\mathbf{M}}$ we mean $\{x \in \mathbf{M}: \mathbf{A}(x)^{\mathbf{M}}\}$, so that $\mathbf{A}$ is absolute for $\mathbf{M}$ iff $\mathbf{A}^{\mathbf{M}} = \mathbf{A} \cap \mathbf{M}$. For example, $\mathbf{V}(x)$ is $x = x$, which is always absolute, and $\mathbf{V}^{\mathbf{M}} = \mathbf{M}$; $\mathbf{ON}^{\mathbf{M}} = \mathbf{ON} \cap \mathbf{M}$ if $\mathbf{M}$ is a transitive model of ZF − P. Classes which are relations in more than one variable are treated similarly. Thus, if $\mathbf{R} \subset \mathbf{V} \times \mathbf{V}$ (i.e., $\mathbf{R}(x, y)$ is a formula and we are thinking of $\mathbf{R} = \{\langle x, y\rangle : \mathbf{R}(x, y)\}$), then $\mathbf{R}^{\mathbf{M}} = \{\langle x, y\rangle \in \mathbf{M} \times \mathbf{M}: \mathbf{R}(x, y)^{\mathbf{M}}\}$, and $\mathbf{R}$ is absolute for $\mathbf{M}$ iff $\mathbf{R}^{\mathbf{M}} = \mathbf{R} \cap (\mathbf{M} \times \mathbf{M})$.

We also wish to relativize classes which are functions. Say $\mathbf{G}: \mathbf{V} \to \mathbf{V}$ (i.e., $\mathbf{G}(x, y)$ is a formula and $\forall x\, \exists! y\, \mathbf{G}(x, y)$). We are thinking of $\mathbf{G}$ as the class of ordered pairs, $\{\langle x, y\rangle : \mathbf{G}(x, y)\}$, but, logically, our use of the functional notation $\mathbf{G}(x)$ is equivalent to using the formula $\mathbf{G}(x, y)$ to introduce a defined operation, so we follow the same conventions discussed in §2 regarding such definitions. Thus, we do not use the function $\mathbf{G}^{\mathbf{M}}$ unless we know that $(\forall x\, \exists! y\, \mathbf{G}(x, y))^{\mathbf{M}}$, in which case $\mathbf{G}^{\mathbf{M}}: \mathbf{M} \to \mathbf{M}$, and $\mathbf{G}$ is absolute for $\mathbf{M}$ iff $\mathbf{G}^{\mathbf{M}} = \mathbf{G} \restriction \mathbf{M}$. Observe that it must be made clear from context that we are indeed regarding $\mathbf{G}$ as a function, since absoluteness of $\mathbf{G}$ as a relation would require only that $\mathbf{G}^{\mathbf{M}} = \mathbf{G} \cap (\mathbf{M} \times \mathbf{M})$, and would not require $\mathrm{dom}(\mathbf{G}^{\mathbf{M}})$ to be equal to $\mathbf{M}$.

5.6. THEOREM. *Let* $\mathbf{R}$ *be a relation which is well-founded and set-like on* $\mathbf{A}$ *and* $\mathbf{F}: \mathbf{A} \times \mathbf{V} \to \mathbf{V}$. *Let* $\mathbf{G}: \mathbf{A} \to \mathbf{V}$ *be defined (as in III 5.6) so that*

$$\forall x \in \mathbf{A}\, [\mathbf{G}(x) = \mathbf{F}(x, \mathbf{G} \restriction \mathrm{pred}(\mathbf{A}, x, \mathbf{R}))].$$

Let $\mathbf{M}$ *be a transitive model of* ZF − P *and assume*

(1) $\mathbf{F}$ *is absolute for* $\mathbf{M}$.

(2) $\mathbf{R}$ *and* $\mathbf{A}$ *are absolute for* $\mathbf{M}$, ($\mathbf{R}$ *is set-like on* $\mathbf{A}$)$^{\mathbf{M}}$, *and*

$$\forall x \in \mathbf{M}\, (\mathrm{pred}(\mathbf{A}, x, \mathbf{R}) \subset \mathbf{M}).$$

Then $\mathbf{G}$ *is absolute for* $\mathbf{M}$.

PROOF. Note first that ($\mathbf{R}$ is well-founded on $\mathbf{A}$)$^{\mathbf{M}}$ since $\mathbf{R}^{\mathbf{M}} = \mathbf{R} \cap \mathbf{M} \times \mathbf{M}$ is well-founded on $\mathbf{A}^{\mathbf{M}} = \mathbf{A} \cap \mathbf{M}$, so any non-empty subset of $\mathbf{A}^{\mathbf{M}}$ in $\mathbf{M}$ has an $\mathbf{R}^{\mathbf{M}}$-minimal element. Thus, we may apply transfinite recursion within $\mathbf{M}$ to define $\mathbf{G}^{\mathbf{M}}: \mathbf{A}^{\mathbf{M}} \to \mathbf{M}$ such that

$$\forall x \in \mathbf{A}^{\mathbf{M}}\, [\mathbf{G}^{\mathbf{M}}(x) = \mathbf{F}^{\mathbf{M}}(x, \mathbf{G}^{\mathbf{M}} \restriction \mathrm{pred}^{\mathbf{M}}(\mathbf{A}^{\mathbf{M}}, x, \mathbf{R}^{\mathbf{M}}))].$$

But then $\mathbf{G}^{\mathbf{M}} = \mathbf{G} \restriction \mathbf{A}^{\mathbf{M}}$ by transfinite induction; that is, an $\mathbf{R}$-minimal element of $\{x \in \mathbf{A}^{\mathbf{M}}: \mathbf{G}^{\mathbf{M}}(x) \neq \mathbf{G}(x)\}$ would, by our assumed absoluteness statements, lead to a contradiction. □

We remark that in most important applications, such as when $\mathbf{R}$ is $\in$ and $\mathbf{A}$ is $\mathbf{V}$ or $\mathbf{ON}$, assumption (2) of Theorem 5.6 is trivial to verify.

5.7. THEOREM. *The following are absolute for any transitive model of* ZF − P:

(a) α^{β} (*ordinal exponentiation*).

(b) $\mathrm{rank}(x)$ ($= \mathrm{rank}(x, \mathbf{V}, \in)$; *see III* 5.8).

(c) $\mathrm{tr\,cl}(x)$.

PROOF. α^β is defined by recursion on β (see I 9.5). rank(x) is defined by recursion on x. For tr cl(x), first, define $\bigcup^n x$ by recursion on n:

$$\bigcup^y(x) = \begin{cases} 0 & \text{if} \quad y \notin \omega, \\ x & \text{if} \quad y = 0, \\ \bigcup(\bigcup^{y-1} x) & \text{if} \quad 0 \in y \in \omega. \end{cases}$$

Then $\bigcup^y(x)$ is an absolute function of variables y, x, so

$$\text{tr cl}(x) = \bigcup\{\bigcup^n(x) \colon n \in \omega\}$$

is absolute. □

It is important for Theorem 5.7(b) that we think of rank(x) as being officially defined recursively. Our original definition was in terms of the $R(\alpha)$ (see III 2.4), but if $\mathbf{M}$ does not satisfy the Power Set Axiom, then $R(\alpha)^{\mathbf{M}}$ is not defined. Under the Power Set Axiom, the two definitions are equivalent by III 5.8.

If $\mathbf{M}$ does satisfy the Power Set Axiom, then $\mathscr{P}^{\mathbf{M}}$ and $R(\alpha)^{\mathbf{M}}$ are defined, but are not usually absolute (even though rank is by Theorem 5.7(b)).

5.8. LEMMA. *Let* $\mathbf{M}$ *be a transitive model for* ZF; *then*

(a) $\mathscr{P}(x)^{\mathbf{M}} = \mathscr{P}(x) \cap \mathbf{M}$ *if* $x \in \mathbf{M}$.

(b) $R(\alpha)^{\mathbf{M}} = R(\alpha) \cap \mathbf{M}$ *if* $\alpha \in \mathbf{M}$.

PROOF. (a) follows from the absoluteness of $\subset$. (b) follows from the absoluteness of rank and the fact that $R(\alpha) = \{x \colon \text{rank}(x) < \alpha\}$. □

In VII and VIII, our $\mathbf{M}$ will be a countable set (this is possible by §7), in which case $\mathscr{P}(x)^{\mathbf{M}} \neq \mathscr{P}(x)$ unless x is finite, and $R(\alpha)^{\mathbf{M}} \neq R(\alpha)$ unless α is finite or ω.

§6. The $H(\kappa)$

We develop here a very important way of producing, in ZFC, set models of ZFC − P.

6.1. DEFINITION. For any infinite cardinal κ, $H(\kappa) = \{x \colon |\text{tr cl}(x)| < \kappa\}$. □

AC is not needed for the definition, since we have taken $|y| < \kappa$ to mean that y is well-orderable and $|y| < \kappa$, but AC will be needed later in developing the properties of $H(\kappa)$.

The elements of $H(\kappa)$ are said to be *hereditarily* of cardinality $<\kappa$. $H(\omega)$ is the set of *hereditarily finite* sets; $H(\omega_1)$ is the set of *hereditarily countable* sets. That each $H(\kappa)$ is a set and not a proper class follows from the following.

6.2. LEMMA. *For any infinite* κ, $H(\kappa) \subset R(\kappa)$.

PROOF. Fix $x \in H(\kappa)$. We show $\text{rank}(x) < \kappa$. Let $t = \text{tr cl}(x)$ and $S = \{\text{rank}(y) \colon y \in t\}$; so $S \subset \mathbf{ON}$. We check first that S is an ordinal. To see this, let α be the least ordinal not in S; then $\alpha \subset S$. If $\alpha \neq S$, let β be the least element of S larger than α, and fix $y \in t$ with $\text{rank}(y) = \beta$. Then, since t is transitive $\forall z \in y\,(\text{rank}(z) < \alpha)$, so

$$\text{rank}(y) = \sup\{\text{rank}(z) + 1 \colon z \in y\} \leq \alpha,$$

a contradiction. Thus, $\alpha = S$. (*Remark.* $\alpha = \text{rank}(x) = \{\text{rank}(y) \colon y \in \text{tr cl}(x)\}$. See also III Exercise 13).

Since $|t| < \kappa$, $\alpha < \kappa$. Since, $x \subset t \subset R(\alpha)$, $\text{rank}(x) \leq \alpha < \kappa$. $\square$

In most cases, $H(\kappa)$ is a proper subset of $R(\kappa)$. For example,

$$\mathscr{P}(\omega) \in R(\omega_1) \setminus H(\omega_1).$$

More generally, the following holds.

6.3. LEMMA. (AC). *If* κ *is regular*, $H(\kappa) = R(\kappa)$ *iff* $\kappa = \omega$ *or* κ *is strongly inaccessible.*

PROOF. If $\kappa = \omega$ or is strongly inaccessible then, by an easy induction on $\alpha < \kappa$,

$$\forall \alpha < \kappa\,(|R(\alpha)| < \kappa).$$

Now, if $\text{rank}(x) = \alpha < \kappa$, $\text{tr cl}(x) \subset R(\alpha)$, so $|\text{tr cl}(x)| < \kappa$. Thus, $R(\kappa) \subset H(\kappa)$, so $R(\kappa) = H(\kappa)$ by Lemma 6.2.

If $\kappa > \omega$ is regular and not strongly inaccessible, fix $\lambda < \kappa$ with $2^\lambda \geq \kappa$. Then, $\mathscr{P}(\lambda) \in R(\kappa) \setminus H(\kappa)$. $\square$

For singular κ, see Exercise 5. The next Lemma lists the basic properties of $H(\kappa)$.

6.4. LEMMA. *For any infinite* κ,

(a) $H(\kappa)$ *is transitive.*

(b) $H(\kappa) \cap \mathbf{ON} = \kappa$.

(c) *If* $x \in H(\kappa)$, *then* $\bigcup x \in H(\kappa)$.

(d) *If* $x, y \in H(\kappa)$, *then* $\{x, y\} \in H(\kappa)$.

(e) *If* $x \in H(\kappa)$ *and* $y \subset x$, *then* $y \in H(\kappa)$.
(f) (AC) *If* κ *is regular, then* $\forall x\,(x \in H(\kappa) \leftrightarrow x \subset H(\kappa) \wedge |x| < \kappa)$.

PROOF. (a) follows from the fact that $y \in x \to \mathrm{tr\,cl}(y) \subset \mathrm{tr\,cl}(x)$. (b) follows from the fact that $\mathrm{tr\,cl}(\alpha) = \alpha$. (c)–(e) are similar. For (f), if $x \subset H(\kappa) \wedge |x| < \kappa$, then, since

$$\mathrm{tr\,cl}(x) = x \cup \bigcup \{\mathrm{tr\,cl}(y) \colon y \in x\}$$

(see III 3.5), $\mathrm{tr\,cl}(x)$ is the union of $<\kappa$ sets of cardinality $<\kappa$, which has cardinality $<\kappa$ (under AC) since κ is regular. □

6.5. THEOREM (AC). *If* κ *is regular and* $>\omega$, *then* $H(\kappa)$ *is a model of* ZFC − P.

PROOF. Extensionality holds because $H(\kappa)$ is transitive, and Foundation holds in any model (see Lemma 2.12). The rest of the axioms of ZF − P − Inf are done exactly as for $R(\omega)\,(=H(\omega))$ and **WF** (see Lemma 2.13), using Lemma 6.4. Thus, 6.4(c) gives us the Union Axiom, 6.4(d) gives us Pairing, 6.4(e) yields Comprehension, and 6.4(f) yields Replacement.

Next, since $H(\kappa)$ is a model for ZF − P − Inf and $\omega \in H(\kappa)$ by 6.4(b), the Axiom of Infinity is true in $H(\kappa)$ (see Lemma 3.12).

Finally, to see that AC is true in $H(\kappa)$, it is sufficient to check that

$$\forall A \in H(\kappa)\, \exists R \in H(\kappa)\,(R \text{ well-orders } A),$$

since well-ordering is absolute for $H(\kappa)$ (see Theorem 5.4). Fix $A \in H(\kappa)$ and let $R \subset A \times A$ well-order A (by AC in **V**). Then $R \subset H(\kappa)$ by 6.4(d) so $R \in H(\kappa)$ by 6.4(f). □

6.6. THEOREM (AC). *If* κ *is regular and* $>\omega$, *the following are equivalent*
(a) $H(\kappa)$ *satisfies* ZFC.
(b) $H(\kappa) = R(\kappa)$.
(c) κ *is strongly inaccessible.*

PROOF. (b)$\leftrightarrow$(c) is Lemma 6.3. For (a), $H(\kappa)$ will satisfy the Power Set Axiom iff

$$\forall x \in H(\kappa)\, \exists y \in H(\kappa)\, \forall z \in H(\kappa)\,(z \subset x \to z \in y).$$

Since $z \subset x \in H(\kappa) \to z \in H(\kappa)$, and $H(\kappa)$ satisfies Comprehension, the Power Set Axiom holds in $H(\kappa)$ iff

$$\forall x \in H(\kappa)\,(\mathscr{P}(x) \in H(\kappa)).$$

This is true if $H(\kappa) = R(\kappa)$, but false if for some $\lambda < \kappa$, $2^\lambda \geq \kappa$, since then $\lambda \in H(\kappa)$ but $\mathscr{P}(\lambda) \notin H(\kappa)$. □

Taking κ not inaccessible, we have

$$\mathrm{Con}(\mathrm{ZFC}) \rightarrow \mathrm{Con}(\mathrm{ZFC} - \mathrm{P} + \neg(\text{Power Set Axiom})).$$

Thus, the Power Set Axiom is not provable from the other axioms of ZFC. Actually, taking $\kappa = \omega_1$, we have a stronger statement.

6.7. Corollary. $\mathrm{Con}(\mathrm{ZFC}) \rightarrow \mathrm{Con}(\mathrm{ZFC} - \mathrm{P} + \forall x\,(x \textit{ is countable}))$.

Proof. In ZFC, we define the model $H(\omega_1)$ for ZFC − P. If $x \in H(\omega_1)$, then x is countable, and any function from ω onto x is in $H(\omega_1)$, so $(x \text{ is countable})^{H(\omega_1)}$. □

If κ is strongly inaccessible, one can easily check that basic cardinal arithmetic is absolute for $H(\kappa)$ (see Exercise 2). We note here only the following.

6.8. Lemma (AC). *If κ is strongly inaccessible, then "α is strongly inaccessible" is absolute for $H(\kappa)$.* □

In particular, if κ is the first strong inaccessible, then $H(\kappa)$ is a model for ZFC plus the non-existence of strong inaccessibles.

6.9. Corollary. $\mathrm{Con}(\mathrm{ZFC}) \rightarrow \mathrm{Con}(\mathrm{ZFC} + \neg\exists\alpha\,(\alpha \textit{ is strongly inaccessible}))$.

Proof. Let $\mathrm{Si}(\kappa)$ abbreviate "κ is strongly inaccessible". Formally, working in ZFC, one cannot prove $\exists\kappa\,\mathrm{Si}(\kappa)$ (by this Corollary), and so one cannot define the least such κ. Instead, define

$$\mathbf{M} = \{x\colon \forall\kappa(\mathrm{Si}(\kappa) \rightarrow x \in H(\kappa))\}.$$

So, in ZFC one cannot decide whether $\mathbf{M} = \mathbf{V}$ or $\mathbf{M}$ is $H(\kappa)$ for the least strong inaccessible κ, but in either case, $\mathbf{M}$ satisfies $\mathrm{ZFC} + \neg\exists\alpha\,\mathrm{Si}(\alpha)$. □

An obvious question is whether one can, working in ZFC, produce a model for $\mathrm{ZFC} + \exists\kappa\,\mathrm{Si}(\kappa)$. The answer is no; see §10 for more discussion.

§7. Reflection theorems

Is there a *set*, M, which is a model for all of ZFC? In the last section we saw that $H(\kappa)$, for regular $\kappa > \omega$, was a model for ZFC − P, but the Power Set Axiom fails in $H(\kappa)$ unless κ is strongly inaccessible, and one cannot prove in ZFC that a strongly inaccessible cardinal exists. Can one, arguing just from ZFC, produce a set model for all of ZFC?

This is almost the case. Given any finite list of axioms, $\phi_1, \dots, \phi_n$ of ZFC, one can prove in ZFC that there is a transitive set M in which $\phi_1 \wedge \cdots \wedge \phi_n$ is true. These M will become very important in our discussion of forcing in VII. Our construction of M will indicate that it is "true" platonistically that one can make M satisfy all of ZFC, but by results related to the Gödel Incompleteness Theorem, this platonistic argument cannot be formalized within ZFC (see §10).

We first comment on some technical points which were not too important in earlier sections, but which are worth noting now. We do not actually have a formula of set theory which says of the set, M, that M is a model for ZFC or ZFC − P (but see also §10). Thus, some of our statements regarding such models involved a certain abuse of notation. When we said, in Theorem 6.5, that $H(\kappa)$ was a model for ZFC − P, we really meant that for each axiom ϕ of ZFC − P,

$$\text{ZFC} \vdash \forall \kappa (\kappa > \omega \wedge \kappa \text{ regular} \rightarrow \phi^{H(\kappa)}).$$

We shall see in this section that for any finite list, $\phi_1, \dots, \phi_n$ of axioms of ZFC,

$$\text{ZFC} \vdash \exists M (\phi_1^M \wedge \cdots \wedge \phi_n^M).$$

M can be taken to be either a suitable $R(\beta)$ or a suitable countable transitive set.

At this stage, it is not even clear how, within ZFC, one may phrase the question of whether there is a set model for all of ZFC. Intuitively, we could let ϕ_i ($i \in \omega$) enumerate the axioms of ZFC and let χ be the sentence

$$\exists M (\phi_0^M \wedge \phi_1^M \wedge \phi_2^M \wedge \cdots),$$

but χ is not even a legitimate finite sentence in the language of set theory. Platonistically, χ makes sense and will (informally) be "seen" to be "true" by the arguments of this section.

This difficulty in formally handling all of ZFC simultaneously turns out to be not much of a handicap, since we only use a finite amount of ZFC at a time. Thus, for example, if ZFC $\vdash \psi$, then there is a finite list of axioms of ZFC, $\phi_1, \dots, \phi_n$ such that $\phi_1, \dots, \phi_n \vdash \psi$ (see I 2.1). Likewise, our results of §§3, 5 showing that certain formulas were absolute for transitive models of ZF − P actually established absoluteness for transitive models of sufficiently large finite fragments of ZF − P.

7.1. LEMMA. *For each of the formulas χ proved in §§3, 5 to be absolute for transitive models of* ZF − P (*see* 3.9, 3.11, 5.1, 5.3, 5.4, 5.5, *and* 5.7), *we* *can find axioms* $\phi_1, \ldots, \phi_n$ *of* ZF − P *such that*

$$\mathrm{ZF} - \mathrm{P} \vdash \forall M\,(M \textit{ transitive} \wedge \phi_1^M \wedge \cdots \wedge \phi_n^M \rightarrow \chi \textit{ is absolute for } M).$$

Likewise for the defined functions proved to be absolute.

Proof. Absoluteness for defined functions reduces to absoluteness of their defining formulas. We agreed in § 3 to discuss relativizing a function $F(x_1, \ldots, x_n)$ to M only when its defining formula $\chi(x_1, \ldots, x_n, y)$ satisfies the uniqueness criterion in M,

$$\forall x_1, \ldots, x_n \, \exists ! y \, \chi(x_1, \ldots, x_n, y).$$

In all cases, the uniqueness criterion was a theorem of ZF − P, and thus of some finite collection of axioms of ZF − P.

In Theorems 3.9 and 5.1, we proved formulas absolute because they were provably equivalent to Δ_0 formulas in ZF − P, and thus in some finite sub-theory. In other theorems in §§ 3, 5, we obtained new absoluteness results from old ones either by composition (using Lemma 3.10), which put no additional requirement on M, or by transfinite recursion (using Theorem 5.6), which requires only that M satisfy enough of ZF − P to justify the particular recursion under consideration. □

One can also state a version of Lemma 7.1 for proper class models (see Exercise 26).

We now discuss the general procedure for trying to produce set models of all of ZFC. The reader familiar with model theory will recognize this as an application of the downward Löwenheim–Skolem–Tarski theorem to **V**. The main idea is to try to find a set M such that *every* formula is absolute for M. In particular, if ϕ is a sentence, then $\phi^M \leftrightarrow \phi$, so if ϕ is an axiom of ZFC, ϕ^M will be true. Actually, because of problems discussed above, if we work within ZFC we can only prove the existence of an M for which an arbitrarily prescribed finite list of formulas are absolute.

The results of §§ 3, 5 and Lemma 7.1 are irrelevant for the rest of this section, as they involved absoluteness of certain formulas of a very special form, whereas here we are trying to construct models for which all formulas are absolute. Lemma 7.1 justifies the usefulness of models for finite fragments of ZFC, and will be used in VI and VII in conjunction with the other results we prove here.

We begin by giving a criterion (Lemma 7.3), due to Tarski and Vaught, for when a list of formulas is absolute for M.

7.2. Definition. A list of formulas, $\phi_1, \ldots, \phi_n$, is *subformula closed* iff every subformula of a formula in the list is listed also. □

We consider here our $\phi_1, \ldots, \phi_n$ to be in the original official language of set theory, as in I §2. If $\phi_1, \ldots, \phi_n$ is subformula closed, then each ϕ_i is either atomic (ie., of form $x \in y$ or $x = y$), or a negation, $\neg\phi_j$ for some $j = 1, \ldots, n$, or a conjunction, $\phi_j \wedge \phi_k$ for some j, k, or an existential quantification $\exists x\, \phi_j$ for some j. Since every formula has only finitely many subformulas, any finite list of formulas may be expanded to a bigger finite list which is subformula closed.

7.3. Lemma. *Let* $\mathbf{M}$ *and* $\mathbf{N}$ *be classes with* $\mathbf{M} \subset \mathbf{N}$. *Let* $\phi_1, \ldots, \phi_n$ *be a subformula closed list of formulas; then the following are equivalent:*

(a) $\phi_1, \ldots, \phi_n$ *are absolute for* $\mathbf{M}$, $\mathbf{N}$.

(b) *Whenever* ϕ_i *is of the form* $\exists x\, \phi_j(x, y_1, \ldots, y_l)$ *(with free variables of* ϕ_j *displayed),*

$$\forall y_1, \ldots, y_l \in \mathbf{M}\, [\exists x \in \mathbf{N}\, \phi_j^{\mathbf{N}}(x, y_1, \ldots, y_l) \rightarrow \exists x \in \mathbf{M}\, \phi_j^{\mathbf{N}}(x, y_1, \ldots, y_l)].$$

Proof. For (a) → (b), fix $y_1, \ldots, y_l \in \mathbf{M}$ and assume $\exists x \in \mathbf{N}\, \phi_j^{\mathbf{N}}(x, y_1, \ldots, y_l)$. Then $\phi_i^{\mathbf{N}}(y_1, \ldots, y_l)$, so, by absoluteness of ϕ_i, $\phi_i^{\mathbf{M}}(y_1, \ldots, y_l)$, or $\exists x \in \mathbf{M}\, \phi_j^{\mathbf{M}}(x, y_1, \ldots, y_l)$, so by absoluteness of ϕ_j, $\exists x \in \mathbf{M}\, \phi_j^{\mathbf{N}}(x, y_1, \ldots, y_l)$.

For (b) → (a), we check, by induction on the length of ϕ_i, that ϕ_i is absolute for $\mathbf{M}$, $\mathbf{N}$. Thus, assume we have checked absoluteness for all formulas on the list shorter than ϕ_i. If ϕ_i is atomic it is obviously absolute. If ϕ_i is $\phi_j \wedge \phi_k$, the absoluteness of (the shorter) ϕ_j and ϕ_k implies the absoluteness of ϕ_i. Likewise if ϕ_i is $\neg\phi_j$. Now suppose ϕ_i is $\exists x\, \phi_j(x, y_1, \ldots, y_l)$, and fix $y_1, \ldots, y_l \in \mathbf{M}$; then

$$\phi_i^{\mathbf{M}}(y_1, \ldots, y_l) \leftrightarrow \exists x \in \mathbf{M}\, \phi_j^{\mathbf{M}}(x, y_1, \ldots, y_l) \leftrightarrow \exists x \in \mathbf{M}\, \phi_j^{\mathbf{N}}(x, y_1, \ldots, y_l) \leftrightarrow$$
$$\exists x \in \mathbf{N}\, \phi_j^{\mathbf{N}}(x, y_1, \ldots, y_l) \leftrightarrow \phi_i^{\mathbf{N}}(y_1, \ldots, y_l).$$

The first and last ↔ just applied the definition of relativization, the second ↔ applied absoluteness of ϕ_j, and the third ↔ applied condition (b). □

The usefulness of Lemma 7.3 is that condition (b) involves only truth of formulas in the larger class, $\mathbf{N}$. In our first application, we take $\mathbf{N} = \mathbf{V}$ and try to find a set, $M = R(\beta)$ such that $\phi_1, \ldots, \phi_n$ are absolute for M. (b) may be viewed as a closure condition on M.

7.4. Theorem. *The Reflection Theorem. Given any formulas* $\phi_1, \ldots, \phi_n$,

$$\mathrm{ZF} \vdash \forall\alpha\, \exists\beta > \alpha\, (\phi_1, \ldots, \phi_n \textit{ are absolute for } R(\beta)).$$

The proof of Theorem 7.4 uses very little about the structure of the $R(\alpha)$. To emphasize this, we shall prove a more general form of 7.4 which will be needed anyway in VI. Theorem 7.4 is the special case of the following when $\mathbf{Z} = \mathbf{V}$ and $Z(\alpha) = R(\alpha)$.

7.5. THEOREM. *Suppose* **Z** *is a class and, for each* α, $Z(\alpha)$ *is a set, and assume*
(1) $\alpha < \beta \to Z(\alpha) \subset Z(\beta)$.
(2) *If* γ *is a limit ordinal*, $Z(\gamma) = \bigcup_{\alpha<\gamma} Z(\alpha)$.
(3) $\mathbf{Z} = \bigcup_{\alpha\in\mathbf{ON}} Z(\alpha)$.
Then, for any formulas $\phi_1, \ldots, \phi_n$,

$$\forall\alpha\,\exists\beta > \alpha\big(\phi_1, \ldots, \phi_n \textit{ are absolute for } Z(\beta), \mathbf{Z}\big). \tag{$*$}$$

The reader should note that although Theorem 7.5 is a direct generalization of Theorem 7.4, we have stated 7.5 more informally. A more technically correct, but less readable, statement of Theorem 7.5 would be that given $\phi_1, \ldots, \phi_n$ and a definition of **Z** and $Z(\alpha)$, the sentence which asserts that conditions (1)–(3) imply $(*)$ is provable from ZF.

PROOF OF THEOREM 7.5. We apply Lemma 7.3 with $\mathbf{N} = \mathbf{Z}$, and try to find an $\mathbf{M} = Z(\beta)$ such that (b) of 7.3 applies. We may assume the list $\phi_1, \ldots, \phi_n$ is subformula closed; if the original list is not, we simply expand it to one that is.

For each $i = 1, \ldots, n$, define a function $\mathbf{F}_i : \mathbf{ON} \to \mathbf{ON}$ as follows. If ϕ_i is of the form $\exists x\, \phi_j(x, y_1, \ldots, y_l)$, let $\mathbf{G}_i(y_1, \ldots, y_l)$ be 0 if $\neg\exists x \in \mathbf{Z}\, \phi_j^{\mathbf{Z}}(x, y_1, \ldots, y_l)$, and the least η such that

$$\exists x \in Z(\eta)\, \phi_j^{\mathbf{Z}}(x, y_1, \ldots, y_l)$$

if $\exists x \in \mathbf{Z}\phi_j^{\mathbf{Z}}(x, y_1, \ldots, y_l)$. Let

$$\mathbf{F}_i(\xi) = \sup\{\mathbf{G}_i(y_1, \ldots, y_l) \colon y_1, \ldots, y_l \in Z(\xi)\};$$

this sup exists by the Replacement Axiom. If ϕ_i is not an existential quantification, set $\mathbf{F}_i(\xi) = 0$.

By Lemma 7.3, if β is a limit ordinal and for each i, $\forall\xi < \beta\big(\mathbf{F}_i(\xi) < \beta\big)$, then $\phi_1, \ldots, \phi_n$ will be absolute for $Z(\beta)$, **Z**. Fix α; we show how to find such a $\beta > \alpha$. Let $\beta_0 = \alpha$, and let β_{p+1} be the largest of

$$\beta_p + 1, \quad \mathbf{F}_1(\beta_p), \ldots, \mathbf{F}_n(\beta_p);$$

this defines, by recursion, β_p for $p \in \omega$. Let $\beta = \sup\{\beta_p \colon p \in \omega\}$. Since $\alpha = \beta_0 < \beta_1 < \beta_2 < \cdots$, β is a limit ordinal $>\alpha$. Note that

$$\xi < \xi' \to \mathbf{F}_i(\xi) \le \mathbf{F}_i(\xi').$$

If $\xi < \beta$, then $\xi < \xi_p$ for some p, so $\mathbf{F}_i(\xi) \le \mathbf{F}_i(\beta_p) \le \beta_{p+1} < \beta$. □

In theorem 7.4, we may take each ϕ_i to be a sentence, in which case

$$\mathrm{ZF} \vdash \forall\alpha\,\exists\beta > \alpha\left(\bigwedge_{i=1}^{n} (\phi_i^{R(\beta)} \leftrightarrow \phi_i)\right), \tag{1}$$

where we use $\bigwedge_{i=1}^{n} \psi_i$ to abbreviate $\psi_1 \wedge \psi_2 \wedge \cdots \wedge \psi_n$. In particular, if ϕ_i is an axiom, then $\mathrm{ZF} \vdash \phi_i$, so $\mathrm{ZF} \vdash \forall\alpha\, \exists\beta > \alpha\, (\bigwedge_{i=1}^{n} \phi_i^{R(\beta)})$. If we want $R(\beta)$ to satisfy AC, we must argue from ZFC to conclude AC is true in **V**, even though just in ZF we produce a β such that $\mathrm{AC}^{R(\beta)} \leftrightarrow \mathrm{AC}$. More generally, the following holds.

7.6. COROLLARY. *Let S be any set of axioms extending* ZF, *and $\phi_1, \dots, \phi_n$ any axioms of S; then*

$$S \vdash \forall\alpha\, \exists\beta > \alpha \left(\bigwedge_{i=1}^{n} \phi_i^{R(\beta)} \right).$$

PROOF. Apply (1) (since S extends ZF) plus the fact that $S \vdash \phi_i$ for each i. □

Corollary 7.6 is a pure existence theorem; we do not have a simple description of the β that works. It is easy to find $\phi_1, \dots, \phi_n$ in ZF which force β to equal ω_β (see Exercise 10).

A consequence of Corollary 7.6 is that theories like ZF or ZFC are not finitely axiomatizable:

7.7. COROLLARY. *Let S be any set of axioms extending* ZF, *and $\phi_1, \dots, \phi_n$ any axioms of S. If from $\phi_1, \dots, \phi_n$ one can prove all axioms of S, then S is inconsistent.*

PROOF. Assume that from $\phi_1, \dots, \phi_n$ one can prove all of S. We argue from S and produce a contradiction.

Fix β to be the least ordinal such that

$$\bigwedge_{i=1}^{n} \phi_i^{R(\beta)};$$

then all axioms of S holds in $R(\beta)$. Since S extends ZF, all the basic absoluteness results of §5 hold for $R(\beta)$. In particular, by Lemma 5.8, if $\alpha \in R(\beta)$, then $R(\alpha)^{R(\beta)} = R(\alpha) \cap R(\beta)$, which is $R(\alpha)$, so the function $R(\alpha)$ is absolute for $R(\beta)$. Since S proves

$$\exists\alpha \bigwedge_{i=1}^{n} \phi_i^{R(\alpha)},$$

this must be true in $R(\beta)$, so

$$\exists\alpha < \beta \bigwedge_{i=1}^{n} \phi_i^{R(\alpha)},$$

contradicting the definition of β as the least such ordinal. □

Platonistically, we could carry out the arguments of 7.3–7.6 with all formulas simultaneously and show that $\forall\alpha\,\exists\beta > \alpha\,(\bigwedge_{i\in\omega}\phi_i^{R(\beta)})$, where the ϕ_i list all axioms of ZFC. But this argument cannot be formalized within ZFC. Corollary 7.7 shows that no finite list of axioms of ZFC is equivalent to ZFC. In fact, given axioms $\phi_1, \ldots, \phi_n$ of ZFC, the first $R(\beta)$ which is a model of $\bigwedge_{i=1}^{n}\phi_i$ is not a model of ZFC, and the proof of Corollary 7.7 explicitly produces a theorem of ZFC, namely $\exists\alpha\,(\bigwedge_{i=1}^{n}\phi_i^{R(\alpha)})$, which is false in this $R(\beta)$.

By slightly modifying the proof of Theorem 7.4 we can obtain *countable* sets A for which a given list of formulas is absolute. Of course, now A cannot be an $R(\beta)$, nor can A be transitive, since $\mathscr{P}$ cannot be absolute for a countable transitive model. Non-transitive models are not of much use in themselves, but we may apply the Mostowski collapse (III 5.15) to them to obtain transitive models. We again state our results in the more general framework of Theorem 7.5.

7.8. THEOREM (AC). *Let* $\mathbf{Z}$ *be any class and let* $\phi_1, \ldots, \phi_n$ *be any formulas; then*

$$\forall X \subset \mathbf{Z}\,\exists A\,[X \subset A \subset \mathbf{Z} \wedge (\phi_1, \ldots, \phi_n \text{ are absolute for } A, \mathbf{Z}) \wedge \wedge |A| \leq \max(\omega, |X|)].$$

PROOF. As before, assume the list $\phi_1, \ldots, \phi_n$ is subformula closed. Let $Z(\alpha) = \mathbf{Z} \cap R(\alpha)$, and note that $\mathbf{Z}$ and the $Z(\alpha)$ satisfy the hypothesis of Theorem 7.5. Now fix α such that $X \subset Z(\alpha)$, and, by Theorem 7.5, fix $\beta > \alpha$ such that $\phi_1, \ldots, \phi_n$ are absolute for $Z(\beta), \mathbf{Z}$. We shall find $A \subset Z(\beta)$. By AC, fix a well-order $\lhd$ of $Z(\beta)$.

If ϕ_i has l_i free variables, $y_1, \ldots, y_{l_i}$, define a function $H_i : Z(\beta)^{l_i} \to Z(\beta)$ as follows. If ϕ_i is $\exists x\,\phi_j(x, y_1, \ldots, y_{l_i})$, and

$$\exists x \in Z(\beta)\,\phi_j^{Z(\beta)}(x, y_1, \ldots, y_{l_i}),$$

let $H_i(y_1, \ldots, y_{l_i})$ be the $\lhd$-first such x. If

$$\neg\exists x \in Z(\beta)\,\phi_j^{Z(\beta)}(x, y_1, \ldots, y_{l_i}),$$

or if ϕ_i is not an existential quantification, let $H_i(y_1, \ldots, y_{l_i})$ be the $\lhd$-first element of $Z(\beta)$. If $l_i = 0$, identify H_i with an element of $Z(\beta)$.

By Lemma 7.3, if A is closed under each H_i, then each ϕ_i will be absolute for $A, Z(\beta)$, and hence for $A, \mathbf{Z}$. Thus, we simply take A to be the closure of X under $H_1, \ldots, H_i$. The fact that $|A| \leq \max(\omega, |X|)$ follows from I 10.23. □

The functions H_i are called *Skolem functions* for the ϕ_i.

The proof of Theorem 7.8 seems somewhat inelegant, since we applied the same sort of argument twice—once to get $Z(\beta)$ and again to get A. Unfortunately, the approach of starting out with a well-order $\lhd$ of $\mathbf{Z}$ and picking A directly cannot be carried out in ZFC, since $\mathbf{Z}$ may be a proper class. Even in set theories which allow quantification over proper classes (see I §12), it is not provable from AC (which says every *set* can be well-ordered), that proper classes can be well-ordered.

We now wish to apply the Mostowski collapsing isomorphism to the A of Theorem 7.8 to obtain a transitive model. As would be expected, an isomorphism preserves all properties. Thus, the following applies.

7.9. Lemma. *Let G be a* 1–1 *map from A onto M which is an isomorphism for the $\in$ relation; then for each formula $\phi(x_1, \dots, x_n)$,*

$$\forall x_1, \dots, x_n \in A\, [\phi(x_1, \dots, x_n)^A \leftrightarrow \phi(G(x_1), \dots, G(x_n))^M].$$

Proof. Induction on ϕ. □

In particular, for ϕ a sentence we have $\phi^A \leftrightarrow \phi^M$.

7.10. Corollary (AC). *Let $\mathbf{Z}$ be any transitive class. Let $\phi_1, \dots, \phi_n$ be sentences. Then*

$$\forall X \subset Z[X \text{ is transitive} \to \exists M[x \subset M \wedge \bigwedge_{i=1}^{n} (\phi_i^M \leftrightarrow \phi_i^{\mathbf{Z}}) \wedge \wedge\, M \text{ is transitive} \wedge |M| \le \max(\omega, |X|)]]$$

Proof. We may assume that ϕ_n is the Axiom of Extensionality; if not, just add it to the list. Let A be as in the conclusion of Theorem 7.8; then $\phi_i^A \leftrightarrow \phi_i^{\mathbf{Z}}$. Since $\mathbf{Z}$ is transitive, Extensionality is true in $\mathbf{Z}$, and hence in A. Thus, by III 5.15, there is an $\in$-isomorphism, G, from A onto some transitive set, M. To see that $X \subset M$, note that, for $x \in X$

$$G(x) = \{G(y)\colon y \in A \wedge y \in x\} = \{G(y)\colon y \in x\},$$

since X is transitive. Hence, $G(x) = x$ for all $x \in X$ by $\in$-induction on x. □

As a special case, we may take $\mathbf{Z}$ to be $\mathbf{V}$ and $X = \omega$. Then, the following holds.

7.11. Corollary. *Let S be any set of axioms extending* ZFC *and $\phi_1, \dots, \phi_n$ any axioms of S; then*

$$S \vdash \exists M\left(|M| = \omega \wedge M \text{ is transitive} \wedge \bigwedge_{i=1}^{n} \phi_i^M\right). \quad \square$$

In particular, in ZFC we can prove the existence of a countable transitive model M for any desired finite fragment of ZFC. By listing enough axioms, we can ensure that all the specific absoluteness results of §3 and §5 hold for M, and that $\mathscr{P}^M(x)$ and ω_α^M are defined. But these concepts are not absolute. ω_1^M is a countable ordinal which M "thinks" is uncountable; but that just means that there is no function *in* M from ω onto ω_1^M. Likewise, $\mathscr{P}^M(\omega) = \mathscr{P}(\omega) \cap M$ is really countable, but not by someone living in M. The fact that sets in M can be really countable but uncountable from the point of view of M (i.e., "countable" is not absolute) is known as the Skolem Paradox.

As with earlier results, a Platonist would believe that there is a countable transitive model for all of ZFC, since one could go through the proof of Corollary 7.11 using all formulas at once; but this argument cannot be formalized within ZFC.

8. Appendix 1: More on relativization

We sketch here a more formal treatment than that in §2. There is a general notion of relativization in logic, but we shall discuss only the special case of interest for set theory.

A *relative interpretation* of set theory into itself consists of two formulas, $\mathbf{M}(x, v)$ and $\mathbf{E}(x, y, v)$, with no free variables other than the ones shown. We think of v as a parameter in defining the class $\{x : \mathbf{M}(x, v)\}$ with binary relation $\{\langle x, y\rangle : \mathbf{E}(x, y, v)\}$. If ϕ is a formula, we define $\phi^{\mathbf{M},\mathbf{E}}$ by replacing $x \in y$ by $\mathbf{E}(x, y, v)$ and restricting the bound variables to range over $\mathbf{M}$. In §§1–7, $\mathbf{E}(x, y, v)$ was always $x \in y$. In the case of **WF**, where we discuss a fixed model, the parameter v does not appear in $\mathbf{M}$. However, when discussing set models, $\mathbf{M}(x, v)$ is formally $x \in v$, so that results in §7 of the form $\exists M\,(\phi^M)$ would be written $\exists v\,(\phi^{x\in v, x\in y})$ in our present notation.

To see that relativization yields relative consistency proofs, one must first check the following.

8.1. Lemma. *If $\phi_1, \ldots, \phi_n, \psi$ are sentences, $\phi_1, \ldots, \phi_n \vdash \psi$, and $\mathbf{M}, \mathbf{E}$ are as above, then*

$$\vdash \forall v\big[[\exists x\, \mathbf{M}(x, v) \wedge \phi_1^{\mathbf{M},\mathbf{E}} \wedge \cdots \wedge \phi_n^{\mathbf{M},\mathbf{E}}] \rightarrow \psi^{\mathbf{M},\mathbf{E}}\big].$$

Proof. Similar to the easy direction of the Gödel Completeness Theorem. □

A more formal statement of Lemma 2.3 is as follows.

8.2. Lemma. *Let S, T be sets of sentences in the language of set theory, and suppose that for some $\mathbf{M}, \mathbf{E}$ we know that for each finite list $\phi_1, \ldots, \phi_n$ of* *axioms for S,*

$$T \vdash \exists v\,[\exists x\,\mathbf{M}(x,v) \wedge \phi_1^{\mathbf{M},\mathbf{E}} \wedge \cdots \wedge \phi_n^{\mathbf{M},\mathbf{E}}];$$

then $\mathrm{Con}(T) \to \mathrm{Con}(S)$.

PROOF. Assume S is inconsistent. Then there are $\phi_1, \dots, \phi_n$ in S such that $\phi_1, \dots, \phi_n \vdash \chi \wedge \neg\chi$ for some sentence χ. By Lemma 8.1,

$$T \vdash \exists v\,[\chi^{\mathbf{M},\mathbf{E}} \wedge \neg(\chi^{\mathbf{M},\mathbf{E}})],$$

since $(\chi \wedge \neg\chi)^{\mathbf{M},\mathbf{E}}$ is $\chi^{\mathbf{M},\mathbf{E}} \wedge \neg(\chi^{\mathbf{M},\mathbf{E}})$. Thus, T is inconsistent. □

Note that Lemma 8.2 yields a completely finitistic relative consistency proof when S and T are effectively described, even though S and T may formalize concepts involving infinite sets.

We now discuss how to handle formally relativization of defined concepts. Let $\mathscr{L} = \{\in\}$ be the language of set theory, S a set of axioms in $\mathscr{L}$, and S' an extension of S by definitions in some bigger language $\mathscr{L}'$ (see I §13). If $\phi'(x_1, \dots, x_n)$ is a formula of $\mathscr{L}'$, there is a formula $\phi(x_1, \dots, x_n)$ of $\mathscr{L}$ such that

$$S' \vdash \forall x_1, \dots, x_n (\phi(x_1, \dots, x_n) \leftrightarrow \phi'(x_1, \dots, x_n))$$

(see I 13.1), and we *define* $\phi'^{\mathbf{M},\mathbf{E}}$ to be $\phi^{\mathbf{M},\mathbf{E}}$. Now this definition presupposes that we have fixed some specific way of obtaining ϕ from ϕ'. Suppose ψ is another formula of $\mathscr{L}$ such that

$$S' \vdash \forall x_1, \dots, x_n (\psi(x_1, \dots, x_n) \leftrightarrow \phi'(x_1, \dots, x_n)).$$

Then $\psi^{\mathbf{M},\mathbf{E}}$ may not be equivalent to $\phi^{\mathbf{M},\mathbf{E}}$. However, note that ϕ and ψ are provably equivalent in S', and thus in S, since S' is a conservative extension of S (see I 13.2). Thus, if $\mathbf{M}$, $\mathbf{E}$ is a model of S (or of any finite subtheory of S which proves the equivalence of ϕ and ψ), then $\phi^{\mathbf{M},\mathbf{E}}$ and $\psi^{\mathbf{M},\mathbf{E}}$ are equivalent. In general, we shall only discuss $\phi'^{\mathbf{M},\mathbf{E}}$ in those cases where we have checked that $\mathbf{M}$, $\mathbf{E}$ is a model for S (or for the finite subtheory of S on which the definitions are based), so that we know that the meaning of $\phi'^{\mathbf{M},\mathbf{E}}$ does not depend on the particular procedure for eliminating the defined symbols.

§9. Appendix 2: Model theory in the metatheory

If we are willing to accept infinitistic methods in the metatheory, it is instructive to apply results from model theory to ZF and related theories.

A structure $\mathfrak{A} = \langle A, E \rangle$ for the language of set theory is a set A with a relation $E \subset A \times A$. If S is a set of sentences in this language, the Gödel

Completeness Theorem says

$$\mathrm{Con}(S) \leftrightarrow \exists \mathfrak{A} (\mathfrak{A} \models S),$$

where $\mathfrak{A} \models S$ means that all the sentences of S are true in $\mathfrak{A}$.

This point of view replaces some of the proof theory in our consistency results by model theory. For example, the proof in §4 of $\mathrm{Con}(\mathrm{ZF}^-) \rightarrow \mathrm{Con}(\mathrm{ZF})$ may be viewed as follows: Assume $\mathrm{Con}(\mathrm{ZF}^-)$, and let $\mathfrak{A} \models \mathrm{ZF}^-$. Let

$$B = \{a \in A : \mathfrak{A} \models a \text{ is well-founded}\};$$

then, by the arguments of §4,

$$\langle B, E \cap B \times B \rangle \models \mathrm{ZF},$$

so ZF is consistent.

In the metatheory, we may now carry out the same analysis of structures that we previously considered to be formalized within ZF. For example, call $\mathfrak{A}$ *well-founded* iff every non-empty subset of A has an E-minimal element. If $\mathfrak{A}$ is well-founded and extensional, then $\mathfrak{A}$ is isomorphic to some transitive set M with the real $\in$ relation.

Not every consistent $S \supset \mathrm{ZF}$ has a well-founded model (see §10). However, presumably ZFC has a well-founded model, since ZFC is true in $\mathbf{V}$, so the arguments of §7 may be applied in the metatheory (using all formulas simultaneously) to produce an $R(\beta)$ which is a model for ZFC.

Some readers may find the preceding argument too platonistic, but anyone who believes in infinite sets will believe in the set $R(\omega)$, which is a well-founded model for ZFC – Inf (see Theorem 3.13). Actually, even Intuitionists (but not Finitists) believe that Con(ZFC – Inf), since ZFC – Inf is interpretable within Peano Arithmetic (see Exercise 30), which is intuitionistically consistent (see [Kleene 1952]). The tenets of Intuitionism are sufficiently vaguely stated to make it unclear whether there can be an intuitionistically acceptable proof of Con(ZF).

Even a Finitist believes that every theory with a finite model is consistent. As an example of such a theory, let S be Extensionality + Comprehension + $\forall y (y = 0)$, which has $\{0\}$ as a model. Thus, $\mathrm{Con}(S)$ is true finitistically; it is unnecessary to hypothesize $\mathrm{Con}(\mathrm{ZF}^-)$ to prove $\mathrm{Con}(S)$ as we did in Corollary 2.8.

§10. Appendix 3: Model theory in the formal theory

On a different tack from §9, we may consider basic model theory to be formalized within ZF (altough many deeper results require ZFC). In this case, we may consider the metatheory to be finitistic.

A step toward formalizing model theory was made in I §14, where we formalized basic syntactical notions. Carrying this a bit further, we may write a formula CON(X), with free variable X, which "says" that X is a set of sentences in the language of set theory from which one cannot derive a contradiction.

We may likewise formalize semantic notions and write a formula

$$\mathfrak{M} \models x[s]$$

in three free variables which "says" that $\mathfrak{M}$ is a pair $\langle M, E \rangle$, x is a formula, $s \in M^{<\omega}$, and x is true in $\mathfrak{M}$ using s to assign values in M to the free variables of x. When x is a sentence, we delete the $[s]$. $\mathfrak{M} \models X$ is $\forall x \in X\, (\mathfrak{M} \models x)$. We may then state the Gödel Completeness Theorem as

$$\forall X\, (\mathrm{CON}(X) \leftrightarrow \exists \mathfrak{M}\, (\mathfrak{M} \models X)),$$

and prove it within ZF (AC can be avoided here since the language is countable).

The above is merely an elaboration of the fact that model theory, like any other branch of mathematics, may be formalized within ZFC (or ZF if Choice is not used). The discussion becomes more interesting, and also more confusing, when we try to compare metatheoretic objects with formal ones. We may lessen this confusion slightly by using the Quine corner convention, as in I §14.

Note that we now have two distinct notions of truth in $\mathfrak{M}$. Let $\phi(x_1, \dots, x_n)$ be a formula. On the one hand, we may write the relativization $\phi^{M,E}(x_1, \dots, x_n)$ as in §8; this does not require formalizing logic within ZF. On the other hand, we may form the constant $\ulcorner \phi \urcorner$ as in I §14 and write $\langle M, E \rangle \models \ulcorner \phi \urcorner [\langle x_1, \dots, x_n \rangle]$. That these two notions are equivalent is proved in the metatheory by a straightforward induction on ϕ. Formally, we state the equivalence as follows.

10.1. Lemma. *For each* $\phi(x_1, \dots, x_n)$,

$$\mathrm{ZF} \vdash \forall \langle M, E \rangle\, \forall x_1, \dots, x_n\, (\phi^{M,E}(x_1, \dots, x_n) \leftrightarrow$$
$$\leftrightarrow \langle M, E \rangle \models \ulcorner \phi \urcorner [\langle x_1, \dots, x_n \rangle]). \quad \Box$$

If S is a recursive set of sentences in the metatheory, we may use a representing formula χ_S for S (see I 14.1) to add a constant $\ulcorner S \urcorner = \{x \colon \chi_S(x)\}$ which denotes S in the formal theory. This notation is actually not quite precise, since $\ulcorner S \urcorner$ really depends on the particular formula we have chosen to represent S.

We now have for each such S, a sentence CON($\ulcorner S \urcorner$) in the language of set theory asserting that S is consistent. CON($\ulcorner S \urcorner$) is equivalent to the CON^S of I §14. The Gödel Incompleteness Theorem (I 14.3) shows that if

S is consistent and extends ZF (or actually $\text{ZF}^- - \text{P} - \text{Inf}$; see I Exercise 23) then $S \nvdash \text{CON}(\ulcorner S \urcorner)$. (*Caution*: this presupposes that we used a "reasonable" χ_S to represent S—see Exercise 36).

It may be possible, however, to prove from S the consistency of weaker theories. For example, $\text{ZFC} \vdash \text{CON}(\ulcorner \text{ZFC} - \text{P} \urcorner)$. To see this, we may modify the proof of Theorem 6.5 to formalize within ZFC an argument that $H(\omega_1)$ is a model for ZFC − P. Note that this says more than what we proved in §6, which was that for each axiom ϕ of ZFC − P, $\text{ZFC} \vdash \phi^{H(\omega_1)}$. Likewise, let I denote the axiom $\exists \kappa$ (κ is strongly inaccessible); then we may modify Theorem 6.6 to show

$$\text{ZFC} + \text{I} \vdash \text{CON}(\ulcorner \text{ZFC} \urcorner). \tag{1}$$

These facts place limits on the possible kinds of finitistic relative consistency proofs. We have a finitistic proof of

$$\text{Con}(\text{ZFC}) \to \text{Con}(\text{ZFC} + \neg \text{I})$$

(See Corollary 6.9), but there is no finitistic proof of

$$\text{Con}(\text{ZFC}) \to \text{Con}(\text{ZFC} + \text{I}) \tag{2}$$

unless ZFC is inconsistent. To see this, suppose we had such a proof, then it could be formalized within ZFC to yield

$$\text{ZFC} \vdash \text{CON}(\ulcorner \text{ZFC} \urcorner) \to \text{CON}(\ulcorner \text{ZFC} + \text{I} \urcorner),$$

which, by (1), implies

$$\text{ZFC} + \text{I} \vdash \text{CON}(\ulcorner \text{ZFC} + \text{I} \urcorner),$$

so ZFC + I is inconsistent by the Incompleteness Theorem. Thus, by (2), ZFC is inconsistent.

If we now allow infinitistic methods in the metatheory, we may view the preceding discussion using model theory in the metatheory as we did in §9.

Let $\mathfrak{A} = \langle A, E \rangle \models \text{ZF}$. For each natural number n, let $n^{\mathfrak{A}} \in A$ be the interpretation of $\ulcorner n \urcorner$ in $\mathfrak{A}$. We call $\mathfrak{A}$ an *ω-model* iff these are the only natural numbers of $\mathfrak{A}$; that is, iff there is no $a \in A$ such that $\mathfrak{A} \models$ "$a \in \omega$" but $a \neq n^{\mathfrak{A}}$ for each n. If there is such an a, we call it a *non-standard*, or *infinitely large* natural number. Every well-founded model is an ω-model, but not conversely. By the compactness theorem, any $\mathfrak{A}$ has an elementary extension which is not an ω-model. By the omitting types theorem, any countable ω-model has an elementary extension which is an ω-model but not well-founded.

If $\mathfrak{A} \models \text{ZF}$, then for each formula ϕ in the metatheory, there is a corresponding $\phi^{\mathfrak{A}} \in A$, where $\phi^{\mathfrak{A}}$ is the interpretation of $\ulcorner \phi \urcorner$ in $\mathfrak{A}$. If $\mathfrak{A}$ is an ω-model, then these are the only formulas of $\mathfrak{A}$, but if $\mathfrak{A}$ is not an ω-model, then $\mathfrak{A}$ has non-standard formulas whose lengths are infinitely large natural numbers.

Presumably, ZF is consistent and has a well-founded model (see §9), but it is easy to produce consistent $S \supset \text{ZF}$ with no ω-model. For example, let S be ZF + $\neg\text{CON}(\ulcorner\text{ZF}\urcorner)$. S is consistent by the Gödel Incompleteness Theorem. If $\mathfrak{A} \models S$, then $\mathfrak{A}$ contains a proof of contradiction from ZF. However, since ZF is really consistent this must be a non-standard proof, so $\mathfrak{A}$ is not an ω-model.

This S provides another curious example. In the metatheory, let ϕ_n $(n \in \omega)$ be a recursive listing of S. We may now formalize this listing within S. In S, we may define c to be the largest $x \in \omega$ such that

$$\exists \beta \, (R(\beta) \models \{\phi_n \colon n < x\}).$$

This is a meaningful definition within S since $S \vdash \neg(\text{CON}(\ulcorner S \urcorner)$ (since $S \vdash (\neg \text{Con}(\ulcorner \text{ZF} \urcorner) \wedge \ulcorner \text{ZF} \urcorner \subset \ulcorner S \urcorner))$. By the reflection theorem, $S \vdash (c > \ulcorner n \urcorner)$ for all standard n, so whenever $\mathfrak{A} \models S$, $c^{\mathfrak{A}}$ is a non-standard natural number. Now, in S define b to be the least ordinal β such that $R(\beta) \models \{\phi_n \colon n < c\}$. Then $R(b)^{\mathfrak{A}}$ is a model for each standard axiom ϕ_n of S, but fails to satisfy some non-standard axioms. This example shows that a consistent theory S can define a set model for itself in a weak sense; i.e., S defines a transitive set, $R(b)$, such that for each axiom ϕ of S, $S \vdash (R(b) \models \phi)$, even though S cannot prove (and in fact refutes) $\exists \mathfrak{M} (\mathfrak{M} \models \ulcorner S \urcorner)$.

In the preceding paragraph one may replace S by any recursive extension of ZF which is ω-inconsistent (see Exercise 32).

For more on the model theory of set theory, see [Keisler–Morley 1968] and [Keisler–Silver 1971].

EXERCISES

Work in ZF unless otherwise indicated.

(1) Which axioms of ZF are true in **ON**?

(2) (AC) Let κ be strongly inaccessible. Check that the following are absolute for $R(\kappa)$:

(a) $\mathscr{P}(x)$.
(b) ω_α.
(c) $\beth_\alpha$.
(d) $R(\alpha)$.
(e) $\text{cf}(\alpha)$.
(f) α is strongly inaccessible.

(3) Check, in ZFC^-, that (a)–(f) of the preceding exercise are absolute for **WF**, along with

(g) $\exists\alpha$ (α is strongly inaccessible).

(4) (AC) For $\kappa > \omega$, show that $|H(\kappa)| = 2^{<\kappa}$.

(5) (AC) For $\kappa > \omega$, show that $H(\kappa) = R(\kappa)$ iff $\kappa = \beth_\kappa$.

(6) Zermelo set theory, Z, is ZF without Replacement. ZC is Z plus AC. Show that for any limit $\gamma > \omega$, $R(\gamma)$ is a model for Z; assuming AC in **V**, show that such $R(\gamma)$ are models for ZC.

(7) (AC) Show that for all $\kappa > \omega$, $H(\kappa)$ is a model for Z $-$ P. Show that the Power Set Axiom is true in $H(\kappa)$ iff $\kappa = \beth_\gamma$ for some limit γ. Show that Replacement fails in $H(\beth_\omega)$.

(8) Show that in $R(\omega + \omega)$, it is not true that every well-ordering is isomorphic to an ordinal. *Hint.* Consider $2 \times \omega$, ordered lexicographically. Track down the specific instance of Replacement which fails in $R(\omega + \omega)$.

(9) Argue within Z, and prove that $A \times B$ exists for any sets A and B. *Hint.* $A \times B \subset \mathcal{P}\mathcal{P}(A \cup B)$. Show that within Z one may develop the basic properties of functions and well-ordering (see I §6), and the basic properties of ω, $\mathbb{R}$, and $\mathbb{C}$ (see I §§7, 11). Furthermore verify that within ZC one may develop at least 99% of modern mathematics.

(10) Find a sentence ϕ such that for any β, if ϕ is absolute for $R(\beta)$, then $\beta = \omega_\beta$. Then, find a formula $\psi(x)$, such that for any non-0 transitive M, if $\psi(x)$ is absolute for M, then $M = R(\beta)$ for some β such that $\beta = \omega_\beta$. *Hint.* ϕ will be enough of ZF to guarantee that $\forall\alpha$ (ω_α exists). $\psi(x)$ can be "$\phi \wedge (x \text{ is an } R(\alpha))$."

(11) Show that the relation "R is well-founded on A" is absolute for transitive models of ZF $-$ P.

(12) Verify that all the relations and functions listed in Theorem 3.11 were defined in $ZF^- - P - \text{Inf}$ by formulas provably equivalent to Δ_0 formulas.

*(13) Let $F(n) = n + n$ for $n \in \omega$, and $F(x) = 0$ for $x \notin \omega$. Show that F is not a Δ_0 function (i.e., $F(x) = y$ is not equivalent to a Δ_0 formula), but F is the composition of two Δ_0 functions. *Hint.* Showing F is not Δ_0 requires some model theory. Show first that $\{m \in \omega : m \text{ is even}\}$ is not first-order definable in $\langle \omega, < \rangle$.

(14) A formula in the language of set theory is called Σ_i iff it is of the form $\exists y_1 \cdots \exists y_n\,\theta$, where θ is Δ_0. Π_1 formulas are of the form $\forall y_1 \cdots \forall y_n\,\theta$ for Δ_0 θ. Let $\phi(x_1, \ldots, x_m)$ be Σ_1 and $\psi(x_1, \ldots, x_m)$ be Π_1, and suppose $\mathbf{M} \subset \mathbf{N}$ and both $\mathbf{M}$ and $\mathbf{N}$ are transitive. Show that for all $x_1, \ldots, x_m \in \mathbf{M}$,

$$\phi^{\mathbf{M}}(x_1, \ldots, x_m) \rightarrow \phi^{\mathbf{N}}(x_1, \ldots, x_n) \quad \text{and} \quad \psi^{\mathbf{N}}(x_1, \ldots, x_m) \rightarrow \psi^{\mathbf{M}}(x_1, \ldots, x_n).$$

Thus, Σ_1-s relativize up and Π_1-s relativize down.

(15) Let S be a set of sentences. A formula $\phi(x_1, \ldots, x_n)$ is provably Δ_1 from S ($\vdash_S \Delta_1$) iff there is a Σ_1 $\psi_\Sigma(x_1, \ldots, x_n)$ and a Π_1 $\psi_\Pi(x_1, \ldots, x_n)$ such that $S \vdash [\forall x_1, \ldots, x_n (\phi(x_1, \ldots, x_n) \leftrightarrow \psi_\Sigma(x_1, \ldots, x_n) \leftrightarrow \psi_\Pi(x_1, \ldots, x_n))]$. Assume $\mathbf{M} \subset \mathbf{N}$ and both are transitive models for S. Show that if ϕ is $\vdash_S \Delta_1$, then ϕ is absolute for $\mathbf{M}$, $\mathbf{N}$.

(16) Check that the formulas shown in §§3, 5 to be absolute for transitive models of ZF − P are in fact $\vdash_{\text{ZF}-\text{P}} \Delta_1$.

(17) Show that for any formula ϕ, the following are equivalent:

(a) ϕ is $\vdash_{\text{ZF}} \Delta_1$.

(b) For some finite set of axioms S of ZF, ZF $\vdash \forall M\,[(M$ transitive $\wedge$ $(\bigwedge S)^M) \rightarrow \phi$ is absolute for $M]$, where $\bigwedge S$ denotes the conjunction of the sentences in S.

Hint. For $b \rightarrow a$, use reflection.

(18) Let $\mathbf{F}$ be a 1–1 function from $\mathbf{V}$ onto $\mathbf{V}$. Define $\mathbf{E} \subset \mathbf{V} \times \mathbf{V}$ by $x\,\mathbf{E}\,y$ iff $x \in \mathbf{F}(y)$. Show (in ZFC) that $\mathbf{V}, \mathbf{E}$ is a model of ZFC$^-$ (see §8 for the definition of relativization to $\mathbf{V}, \mathbf{E}$).

(19) Use the preceding exercise to show the consistency of ZFC$^-$ + $\exists x\,(x = \{x\})$, assuming Con(ZFC). *Hint*. Let $\mathbf{F}(0) = 1$, $\mathbf{F}(1) = 0$. Likewise, show the consistency of ZFC$^-$ + $\exists x\,\exists y\,(x = \{y\} \wedge y = \{x\} \wedge x \neq y)$.

(20) Assume Con(ZFC), and show the consistency of ZFC$^-$ plus the following modified Mostowski Collapsing Theorem: If R is an extensional relation on the set A, then $\langle A, R\rangle$ is isomorphic to $\langle M, \in\rangle$ for some transitive set M. Note R is *not* assumed to be well-founded.

*(21) Show that there is no finite $S \subset$ ZFC$^-$ such that one can prove in ZFC$^-$ that "R well-orders A" is absolute for transitive models of S. *Hint*. Use the previous exercise, plus some model theory.

(22) For any set A, define $R(\alpha, A)$ by: $R(0, A) = \{A\} \cup \text{tr cl}(A)$, $R(\alpha + 1, A) = \mathscr{P}(R(\alpha, A))$, and $R(\alpha, A) = \bigcup_{\xi<\alpha} R(\xi, A)$ when α is a limit ordinal. Let

$\mathbf{WF}(A) = \bigcup_{\alpha \in \mathbf{ON}} R(\alpha, A)$. Show in ZF$^-$ that $\mathbf{WF}(A)$ is a transitive model of ZF$^-$, and that AC implies $\mathrm{AC}^{\mathbf{WF}(A)}$.

(23) Assuming Con(ZF$^-$), show that it is consistent with ZF$^-$ to have $\mathbf{V} = \mathbf{WF}(U)$, where U is an infinite set and $\forall x \in U\,(x = \{x\})$.

*(24) (Fraenkel–Mostowski) Show $\mathrm{Con}(\mathrm{ZF}^-) \to \mathrm{Con}(\mathrm{ZF}^- + \neg\mathrm{AC})$. *Hint.* Assume $\mathbf{V} = \mathbf{WF}(U)$ as in Exercise 23. Let G be the group of all permutations of U, and, for $B \subset U$, let

$$G_B = \{\pi \in G\colon \forall x \in B(\pi(x) = x)\}.$$

For $\pi \in G$, define an automorphism π_* of $\mathbf{V}$, so that $\pi_*(x) = \pi(x)$ for $x \in U$, and $\pi_*(y) = \{\pi_*(z)\colon z \in y\}$ for all y. Let

$$\mathbf{A} = \{y\colon \exists B \subset U\,(|B| < \omega \wedge \forall \pi \in G_B(\pi_*(y) = y))\},$$

and

$$\mathbf{HA} = \{y \in \mathbf{A}\colon \mathrm{tr\,cl}(y) \subset \mathbf{A}\}.$$

Show that $\mathbf{HA}$ is a transitive model of ZF$^-$ + "U cannot be well-ordered".

(25) Assuming Con(ZFC$^-$), show that it is consistent with ZF$^-$ to have $\neg$AC, $\mathrm{AC}^{\mathbf{WF}}$, and a group $\langle H, \cdot \rangle$ which is not isomorphic to a well-founded group. *Hint.* See Exercise 24. H can be any group in $\mathbf{HA}$ which cannot be well-ordered in $\mathbf{HA}$.

(26) Prove a version of Lemma 7.1 for proper class models.

(27) Prove 7.8 and 7.10 in ZFC* $-$ P (see III Exercise 19).

(28) Show that AF* (see III Exercise 18) is true relativized to any set.

(29) (Barwise) Show that it is consistent to have ZF $-$ P $-$ Inf, together with set x whose transitive closure does not exist. *Remark.* By Exercise 24 and III Exercise 19, one cannot, in ZF, produce a model for this with the real $\in$ relation. *Hint.* Start with distinct x_n such that $\forall n\,(x_n = \{x_{n+1}\})$ (see Exercises 18–20). Let $M_0 = \{x_n\colon n \in \omega\}$, $M_{k+1} = \{y \subset M_k\colon |y| < \omega\}$, and $M = \bigcup_k M_k$.

(30) Let PA be first order Peano Arithmetic. Show that PA and ZF $-$ Inf are relatively interpretable in each other, so that

$$\mathrm{Con}(\mathrm{PA}) \leftrightarrow \mathrm{Con}(\mathrm{ZF} - \mathrm{Inf})$$

is provable finitistically. *Hint.* To interpret ZF $-$ Inf in PA, see III Exercise 5.

*(31) Show that within ZF one may define a countable set M and an $E \subset M \times M$ such that for each axiom ϕ of ZF,

$$\mathrm{ZF} \vdash \phi^{M,E}.$$

Hint. This involves formalizing the proof of the Gödel Completeness Theorem within ZF.

(32) A set of axioms, S, extending $\mathrm{ZF}^- - \mathrm{P} - \mathrm{Inf}$ is called *ω-inconsistent* iff there is a formula $\phi(x)$, such that $S \vdash \phi(\ulcorner n \urcorner)$ for all n but $S \vdash \neg \forall x \in \omega\, \phi(x)$. Show that if S is a recursive extension of ZF which is ω-inconsistent, then in S one can define an ordinal b such that for each axiom ϕ of S,

$$S \vdash \phi^{R(b)}.$$

Hint. See the end of § 10.

(33) Assume that $S \cup \{\chi\}$ and $S \cup \{\neg\chi\}$ are both ω-inconsistent. Show that S is ω-inconsistent. *Hint.* Consider

$$(\chi \to \varphi(x)) \wedge (\neg\chi \to \psi(x)),$$

where $\varphi(x)$ demonstrates the ω-inconsistency of $S \cup \{\chi\}$ and $\psi(x)$ demonstrates the ω-inconsistency of $S \cup \{\neg\chi\}$.

*(34) Let S be a recursive extension of ZF, and assume that for some formula $\psi(x)$, $S \vdash \exists x\, \psi(x)$ and $S \vdash \forall x\, (\psi(x) \to \varphi^x)$ for each axiom φ of S. Show that S is ω-inconsistent. *Hint.* Let χ be the statement

$$\forall x\, (\psi(x) \to x \models \ulcorner S \urcorner).$$

$S \cup \{\neg\chi\}$ is ω-inconsistent. But also $S \cup \{\chi\}$ is ω-inconsistent; to prove this, argue within $S \cup \{\chi\}$, and let

$$\alpha = \min\{\mathrm{rank}(M)\colon\ M \models \ulcorner S \urcorner \text{ and } M \text{ is transitive}\};$$

then

$$\forall M\, ((\mathrm{rank}(M) = \alpha \wedge M \models \ulcorner S \urcorner \wedge M \text{ is transitive}) \to \neg\chi^M),$$

which produces an ω-inconsistency as with $S \cup \{\neg\chi\}$.

*(35) Assume ZF^- is ω-consistent. Show that for each formula φ, there is a Δ_0 formula ψ such that the following statement is consistent with ZF^-: If φ is absolute for all transitive models of ZF^-, then φ and ψ are equivalent. *Hint.* Let T be the theory which is described in Exercise 20; T is ω-consistent if ZF^- is. If there is no such ψ, then

$$T \vdash \varphi \text{ is absolute for transitive models of } \mathrm{ZF}^- \tag{1}$$

and T is consistent with the statement:

$$\text{For no } \Delta_0 \text{ formula } \psi \text{ does } \mathrm{ZF}^- \vdash \text{“}\varphi \text{ and } \psi \text{ are equivalent”}. \qquad (2)$$

Say φ is $\varphi(x_1, \ldots, x_n)$. Applying (2), plus some model theory, plus the fact that every model is isomorphic to a transitive model, prove within T that (2) implies

$$\exists \text{ transitive } M, N\,(M \models \mathrm{ZF}^- \wedge N \models ZF^- \wedge \\ \wedge\ \exists x_1, \ldots, x_n \in M \cap N\,(\varphi(x_1, \ldots, x_n)^M \wedge \neg\varphi(x_1, \ldots, x_n)^N)).$$

But this contradicts (1).

(36) Show that there is a formula $\chi(x)$, such that

(a) χ represents ZF; i.e.,

$\phi \in \mathrm{ZF} \rightarrow \mathrm{ZF} \vdash \chi(\ulcorner\phi\urcorner)$ and $\phi \notin \mathrm{ZF} \rightarrow \mathrm{ZF} \vdash \neg\chi(\ulcorner\phi\urcorner)$, and

(b) If $\ulcorner \mathrm{ZF} \urcorner$ is added via the definition

$\ulcorner \mathrm{ZF} \urcorner = \{x\colon \chi(x)\}$, then $\mathrm{ZF} \vdash \mathrm{CON}(\ulcorner \mathrm{ZF} \urcorner)$.

CHAPTER V

DEFINING DEFINABILITY

Call a set, a, *definable* iff there is some property $P(x)$, expressed in English (augmented by mathematical symbols), such that a is the unique object satisfying P:

$$\forall x\,(P(x) \leftrightarrow x = a).$$

For example, $\mathbb{R}$, ω_8, and 34 are definable. Since there are only countably many English expressions, there are only countably many definable sets. In particular, not all ordinals are definable. Let δ be the first non-definable ordinal; but we have just defined it; $P(x)$ here is:

$$x \text{ is an ordinal and } x \text{ is not definable and } \forall y < x\ (y \text{ is definable}).$$

This "paradox", like the one about the first number not definable in forty words or less (see I §2), arises from a careless use of the imprecise notion of "property". When we make the notion precise, using formulas, then the "paradox" disappears, as there is no way to write a formula to express "x is defined by a formula", so the discussion of δ above simply cannot take place within ZF. Similar problems involved with discussing all formulas simultaneously occurred in IV §7.

In §1 of this chapter we shall show how to write a formula of two free variables, x and A, which says that $x \in A$ and x is definable by a formula of set theory relativized to A. These considerations will be important in §2, where we shall discuss the ordinal definable sets and prove Con(ZF) $\rightarrow$ Con(ZFC), and in VI, where we shall discuss the constructible sets and prove Con(ZF) $\rightarrow$ Con(ZFC + GCH).

Since the results of this chapter will culminate in the construction, within ZF, of a model for ZFC, it is important to note that all of our discussion in §§1 and 2 takes place within ZF; that is, AC is not used. In fact, in §1 we may work within ZF $-$ P.

§1. Formalizing definability

Our basic notion will be a defined function of two variables, $\mathrm{Df}(A, n)$. Intuitively, $\mathrm{Df}(A, n)$ will be the set of n-place relations on A which are definable by a formula with n free variables relativized to A. We may then express that an element, x, of A, is definable by a formula relativized to A by saying $\{x\} \in \mathrm{Df}(A, 1)$.

We would like to define $\mathrm{Df}(A, n)$ as the set of subsets of A^n of the form

$$\{\langle x_1, \ldots, x_n \rangle : \phi^A(x_1, \ldots, x_n)\}$$

for some formula ϕ with n free variables. However, since there are infinitely many such ϕ, it is not immediately clear how to transcribe our informal concept of Df into a formal definition within ZF. One way to accomplish this is to first formalize within ZF the syntax of first-order logic (as outlined in I §14), and then formalize the model-theoretic notion of satisfaction (as outlined in IV §10), but it is rather tedious to carry out such a treatment in detail. Instead, we shall simply define $\bigcup_n \mathrm{Df}(A, n)$ as the least set of relations on A containing basic relations such as $\{\langle x, y \rangle \in A^2 : x \in y\}$ and closed under the operations of intersection, complementation, and projection (corresponding to the logical $\wedge$, $\neg$, and $\exists$). This approach brings out the fact that the notion of definability does not depend on the specific syntactical details of a development of first-order logic. If one takes the trouble to formalize logic within ZF, it is easily seen that the two approaches are equivalent (see Exercise 10).

1.1. Definition. If $n \in \omega$ and $i, j < n$,

(a) $\mathrm{Proj}(A, R, n) = \{s \in A^n : \exists t \in R\,(t \restriction n = s)\}$.

(b) $\mathrm{Diag}_\in(A, n, i, j) = \{s \in A^n : s(i) \in s(j)\}$.

(c) $\mathrm{Diag}_=(A, n, i, j) = \{s \in A^n : s(i) = s(j)\}$.

(d) By recursion on $k \in \omega$, define $\mathrm{Df}'(k, A, n)$ (for all n simultaneously) by:

(1) $\mathrm{Df}'(0, A, n) = \{\mathrm{Diag}_\in(A, n, i, j) : i, j < n\} \cup \{\mathrm{Diag}_=(A, n, i, j) : i, j < n\}$.

(2) $\mathrm{Df}'(k+1, A, n) = \mathrm{Df}'(k, A, n) \cup \{A^n \setminus R : R \in \mathrm{Df}'(k, A, n)\} \cup$

$\{R \cap S : R, S \in \mathrm{Df}'(k, A, n)\} \cup \{\mathrm{Proj}(A, R, n) : R \in \mathrm{Df}'(k, A, n+1)\}$.

(e) $\mathrm{Df}(A, n) = \bigcup \{\mathrm{Df}'(k, A, n) : k \in \omega\}$. □

1.2. Lemma. *If $R, S \in \mathrm{Df}(A, n)$, then $A^n \setminus R \in \mathrm{Df}(A, n)$ and $R \cap S \in \mathrm{Df}(A, n)$. If $R \in \mathrm{Df}(A, n+1)$, then* $\mathrm{Proj}(A, R, n) \in \mathrm{Df}(A, n)$. □

The next Lemma shows that, as we intended, $\mathrm{Df}(A, n)$ contains every relation on A which is definable by a formula relativized to A.

1.3. LEMMA. *Let $\phi(x_0, \ldots, x_{n-1})$ be any formula whose free variables are among $x_0, \ldots, x_{n-1}$; then*

$$\forall A\left[\{s \in A^n: \phi^A(s(0), \ldots, s(n-1))\} \in \mathrm{Df}(A, n)\right]. \qquad (*)$$

PROOF. By induction on the length of ϕ. If ϕ is $x_i \in x_j$, then $(*)$ follows from the fact that $\mathrm{Diag}_\in(A, n, i, j) \in \mathrm{Df}(A, n)$. Similarly if ϕ is $x_i = x_j$. If ϕ is $\psi \wedge \chi$, and we know $(*)$ for ψ and χ, then $(*)$ for ϕ follows from the fact that $\mathrm{Df}(A, n)$ is closed under finite intersections. If ϕ is $\neg\psi$, use closure of $\mathrm{Df}(A, n)$ under complements. Finally, assume ϕ is $\exists y\, \psi$. If y is not one of the variables $x_0, \ldots, x_{n-1}$, then applying $(*)$ for ψ,

$$\{s \in A^n: \phi^A(s(0), \ldots, x(n-1))\} =$$
$$= \mathrm{Proj}\left(A, \{t \in A^{n+1}: \psi^A(t(0), \ldots, t(n))\}, n\right) \in \mathrm{Df}(A, n).$$

If y is x_j, then $\phi(x_0, \ldots, x_{n-1})$ is $\exists x_j\, \psi(x_0, \ldots, x_{n-1})$, so x_j is not free in ϕ. Let z be a variable not occurring anywhere in ϕ, let $\psi'(x_0, \ldots, x_{n-1}, z)$ be $\psi(x_0, \ldots, x_{j-1}, z, x_{j+1}, \ldots, x_{n-1})$, and let ϕ' be $\exists z\, \psi'$; then ϕ' and ϕ are logically equivalent and the preceding argument shows that $(*)$ holds for ϕ', and hence for ϕ. □

We remark that whereas Lemma 1.2 is (an abbreviation of) one statement, which is provable in ZF, 1.3 is really a lemma schema; namely, 1.3 asserts that for each ϕ, the sentence $(*)$ is provable in ZF.

Intuitively, the converse of Lemma 1.3 says that any element of $\mathrm{Df}(A, n)$ is defined by a formula, but in trying to state this we run into the usual problems involved in mentioning all formulas at once. Let $\phi_0, \phi_1, \phi_2, \ldots$ list all formulas with free variables among $x_0, \ldots, x_{n-1}$; then platonistically,

$$\forall R\, \forall Y \in \mathrm{Df}(A, n) \bigvee_{i \in \omega} \left(Y = \{s \in A^n: \phi_i^A(s(0), \ldots, s(n-1))\}\right) \qquad (**)$$

is true. The "proof" is: if $Y \in \mathrm{Df}(A, n)$, then $Y \in \mathrm{Df}'(k, A, n)$ for some k, so we can prove $(**)$ by induction on k. The induction step uses the fact that the ways new relations are put into $\mathrm{Df}'(k+1, A, n)$ correspond to the ways formulas are built up using $\neg$, $\wedge$, $\exists$. Formally, of course, $(**)$ is not a sentence in the language of set theory.

Nevertheless, the intuitive idea that $\mathrm{Df}(A, n)$ is precisely the set of n-place relations on A definable by a formula relativized to A is useful heuristically. It suggests facts which "should" be true, and which may then be proved rigorously by other means. For example, $\mathrm{Df}(A, n)$ "should" be countable, since there are only countably many formulas. Formally, we may prove in ZF that $\mathrm{Df}(A, n)$ is countable using the fact that it was defined as the closure of a countable set under finitary operations (or see Corollary 1.6 below).

Or, $\mathrm{Df}(A, n)$ "should" be absolute for transitive models of ZF − P, since if $A \in \mathbf{M}$ and $\mathbf{M}$ is transitive, then all elements of A are in $\mathbf{M}$, so relativizing a formula to A means the same thing within $\mathbf{M}$ as within $\mathbf{V}$. Formally we shall prove absoluteness using the way Df was defined in 1.1 (see Lemma 1.7 below). For technical reasons, it will be important in §2 and in VI that not only Df, but also a function from ω which enumerates it, is absolute, so we shall begin by proving countability of $\mathrm{Df}(A, n)$ somewhat more explicitly than would otherwise be necessary.

1.4. DEFINITION. By recursion on $m \in \omega$, $\mathrm{En}(m, A, n)$ is defined (for all n simultaneously) by the following clauses:

(a) If $m = 2^i \cdot 3^j$ and $i, j < n$, then $\mathrm{En}(m, A, n) = \mathrm{Diag}_{\in}(A, n, i, j)$.
(b) If $m = 2^i \cdot 3^j \cdot 5$ and $i, j < n$, then $\mathrm{En}(m, A, n) = \mathrm{Diag}_{=}(A, n, i, j)$.
(c) If $m = 2^i \cdot 3^j \cdot 5^2$, then $\mathrm{En}(m, A, n) = A^n \setminus \mathrm{En}(i, A, n)$.
(d) If $m = 2^i \cdot 3^j \cdot 5^3$, then $\mathrm{En}(m, A, n) = \mathrm{En}(i, A, n) \cap \mathrm{En}(j, A, n)$.
(e) If $m = 2^i \cdot 3^j \cdot 5^4$, then $\mathrm{En}(m, A, n) = \mathrm{Proj}(A, \mathrm{En}(i, A, n+1), n)$.
(f) If m is not of the form specified in one of (a)–(e), then $\mathrm{En}(m, A, n) = 0$. □

1.5. LEMMA. *For any n and A,* $\mathrm{Df}(A, n) = \{\mathrm{En}(m, A, n) \colon m \in \omega\}$.

PROOF. First, by induction on m, $\forall n\,(\mathrm{En}(m, A, n) \in \mathrm{Df}(A, n))$; observe (for clause (f)) that $0 \in \mathrm{Df}(A, n)$ since $\mathrm{Df}(A, n)$ is closed under intersections and complements. Next, by induction on k,

$$\forall n\,(\mathrm{Df}'(k, A, n) \subset \{\mathrm{En}(m, A, n) \colon m \in \omega\}). \quad \square$$

1.6. COROLLARY. $|\mathrm{Df}(A, n)| \leq \omega$. □

1.7. LEMMA. *The defined functions* Df *and* En *are absolute for transitive models of* ZF − P.

PROOF. The absoluteness of Proj, $\mathrm{Diag}_{\in}$, $\mathrm{Diag}_{=}$, Df′, Df, and En are easily checked successively by the methods of IV §5. The fact that functions defined recursively using absolute notions are absolute (IV 5.6) is applied several times. For En, we also use the absoluteness of ordinal exponentiation (IV 5.7). □

The logician will realize that Df is a shortcut for formalizing logical syntax and satisfaction within ZF, whereas En is a shortcut for Gödel numbering. Further model-theoretic notions may likewise be expressed using Df and En, as we point out in the rest of this section. This material is not needed for most of this book, but it is important for advanced work on $\mathbf{L}$, such as the proof of $\diamondsuit^+$ in $\mathbf{L}$ (see VI 4.10).

Using En, we may state the following improvement of Lemma 1.3; the proofs are almost identical.

1.8. LEMMA. *Let $\phi(x_0, \ldots, x_{n-1})$ be any formula whose free variables are among $x_0, \ldots, x_{n-1}$; then for some m,*

$$\forall A\, [\{s \in A^n : \phi^A(s(0), \ldots, s(n-1))\} = \mathrm{En}(m, A, n)]. \quad \square \qquad (**)$$

Formally, Lemma 1.8 is an assertion in the metatheory that for each ϕ we can find an m such that (**) is provable in ZF. Informally, we think of ϕ as the m-th formula in n variables under the enumeration En.

Using the predicate En, we may write a predicate $A \prec B$ which "says" that $A \subset B$ and every formula ϕ is absolute for A, B (see IV, Definition 3.1). The relation $A \prec B$ for sets is a special case of the model-theoretic notion of *elementary submodel.*

1.9. DEFINITION. $A \prec B$ (A is *elementarily included* in B) iff $A \subset B$ and

$$\forall n, m\, [\mathrm{En}(m, A, n) = \mathrm{En}(m, B, n) \cap A^n]. \quad \square$$

1.10. LEMMA. *For each $\phi(x_0, \ldots, x_{n-1})$,*

$$\forall A, B\, [A \prec B \rightarrow \phi \text{ is absolute for } A, B].$$

PROOF. By Lemma 1.8. $\square$

We now prove the downward Löwenheim–Skolem–Tarski theorem (1.11), which says that given a B, there is a countable $A \prec B$. This is almost the same as IV 7.8. Theorem 1.11 is stronger than IV 7.8, which only produced an A such that a given finite list of formulas was absolute for A, B. But also IV 7.8 was stronger, in that it allowed B to be a proper class. Note that Lemma 1.10 would not make sense when B is a proper class, as $\mathrm{En}(m, B, n)$ would not be defined.

1.11. THEOREM (AC).

$$\forall B\, \forall X \subset B\, \exists A\, [X \subset A \subset B \wedge A \prec B \wedge |A| \leq \max(\omega, |X|)].$$

PROOF. Fix a well-ordering, $\lhd$, of B. For each $m, n \in \omega$, we define $H_{mn}: B^n \rightarrow B$. Informally, H_{mn} is the Skolem function for the m-th formula in n variables. Formally, if m is of the form $2^i \cdot 3^j \cdot 5^4$ and $s \in \mathrm{En}(m, B, n)$ (i.e., "the m-th formula in n variables is existential and is true of the n-tuple s"), then $H_{mn}(s)$ is the $\lhd$-first $x \in B$ such that $s^\frown\langle x\rangle \in \mathrm{En}(i, B, n+1)$. Otherwise, let $H_{mn}(s)$ be the $\lhd$-first element of B.

Let A be the closure of X under the H_{mn}; then $|A| \leq \max(\omega, |X|)$ (see I

10.23). It is a straightforward induction on m to show that

$$\forall n\,[\mathrm{En}(m, A, n) = \mathrm{En}(m, B, n) \cap A^n]. \quad \Box$$

It is worth noting what the above means when $n = 0$. $A^0 = \{0\}$ (0 is the unique function with empty domain), so the only subsets of A^0 are 0 and $1 = \{0\}$. For $n = 0$, $\phi(s(0), \ldots, s(n-1))$ is ϕ, so that Lemma 1.8 says that for each sentence ϕ, there is an m such that for all A, $\mathrm{En}(m, A, 0)$ is 1 iff ϕ^A is true and 0 iff ϕ^A is false. Then, Lemma 1.10 says that when $A \prec B$, $\phi^A \leftrightarrow \phi^B$ for each sentence ϕ.

1.12. DEFINITION. $A \equiv B$ (A and B are *elementarily equivalent*) iff

$$\forall m\,(\mathrm{En}(m, A, 0) = \mathrm{En}(m, B, 0)). \quad \Box$$

Informally, $A \equiv B$ means that A and B satisfy the same sentences. It might thus be expected that if A and B are isomorphic, then $A \equiv B$.

1.13. LEMMA. *Let $G: A \to B$ be a 1–1 map which is an isomorphism for the $\in$ relation; then*

(1) $\forall n, m\, \forall x \in A^n\,(s \in \mathrm{En}(m, A, n) \leftrightarrow G \circ s \in \mathrm{En}(m, B, n))$.

(2) $A \equiv B$.

PROOF. (1) is by induction on n, and (2) is the special case of (1) with $m = 0$. $\Box$

$A \prec B$ implies that $A \equiv B$, but $A \equiv B$ does not imply that $A \subset B$ or $B \subset A$. For example, if $A = \{a\}$ and $B = \{b\}$, then $A \equiv B$ by Lemma 1.13. A less trivial application of 1.13, where G is a Mostowski collapsing isomorphism, occurs in the proof of $\diamondsuit^+$ in **L** (see VI 4.10).

§2. Ordinal definable sets

Informally, a set a is ordinal definable iff it is definable from some finite sequence of ordinals; i.e., iff there is a property $P(y_1, \ldots, y_n, x)$ and ordinals $\alpha_1, \ldots, \alpha_n$ such that

$$\forall x\,(P(\alpha_1, \ldots, \alpha_n, x) \leftrightarrow x = a).$$

Let **OD** be the class of ordinal definable sets. Observe that each ordinal is in **OD** ($P(y_1, x)$ is $y_1 = x$), so **ON** $\subset$ **OD**.

At first sight, the concept of "ordinal definable" is just as suspect as that of "definable", although, since **ON** $\subset$ **OD**, **OD** seems to avoid the specific paradox of the first non-definable ordinal mentioned at the beginning of this chapter. However, there is a trick, due to Gödel, which enables us to make a legitimate definition of **OD** within ZF. The technical details of this treatment are worked out in [Myhill–Scott 1971].

The key idea is that (informally still) a is ordinal definable iff a is ordinal definable within some $R(\beta)$—i.e., iff there is a $P(y_1, \ldots, y_n, x)$, $\alpha_1, \ldots, \alpha_n$, and $\beta > \max(\alpha_1, \ldots, \alpha_n, \operatorname{rank}(a))$ such that

$$\forall x \in R(\beta)\,(P(\alpha_1, \ldots, \alpha_n, x) \leftrightarrow x = a)^{R(\beta)}. \qquad (*)$$

To see this, if $a \in \mathbf{OD}$, then we have

$$\forall x\,(P(\alpha_1, \ldots, \alpha_n, x) \leftrightarrow x = a)$$

for some $\alpha_1, \ldots, \alpha_n, P$. By the Reflection Theorem (IV 7.4), we may find a $\beta > \max(\alpha_1, \ldots, \alpha_n, \operatorname{rank}(a))$ such that P is absolute for $R(\beta)$, whence $(*)$ will hold. Conversely, if $(*)$ holds, then $(*)$ shows that a is definable in $\mathbf{V}$ from the ordinals $\alpha_1, \ldots, \alpha_n$, together with β; it is the addition of β at this point which would make our argument break down if we tried to replace "ordinal definable" by "definable".

Now we may, using our Df of §1, express within ZF the assertion "a is ordinal definable within some $R(\beta)$", so we shall take this as our official definition of **OD**.

2.1. Definition. **OD** is the class of all sets a such that

$$\exists \beta > \operatorname{rank}(a)\,\exists n\,\exists s \in \beta^n\,\exists R \in \operatorname{Df}(R(\beta), n+1)$$
$$\forall x \in R(\beta)\,(s^\frown\langle x\rangle \in R \leftrightarrow x = a). \quad \square$$

2.2. Theorem. *For each formula $\phi(y_1, \ldots, y_n, x)$ with at most the displayed variables free,*

$$\forall \alpha_1 \cdots \forall \alpha_n\,\forall a\,[\forall x\,(\phi(\alpha_1, \ldots, \alpha_n, x) \leftrightarrow x = a) \rightarrow a \in \mathbf{OD}].$$

Proof. Fix $\alpha_1, \ldots, \alpha_n, a$ and assume $\forall x\,(\phi(\alpha_1, \ldots, \alpha_n, x) \leftrightarrow x = a)$. By the Reflection Theorem (IV 7.4), fix $\beta > \max(\alpha_1, \ldots, \alpha_n, \operatorname{rank}(a))$ so that ϕ is absolute for $R(\beta)$. Let

$$R = \{\langle y_1, \ldots, y_n, x\rangle \in R(\beta)^{n+1} : \phi(y_1, \ldots, y_n, x)\}.$$

If we let $s = \langle \alpha_1, \ldots, \alpha_n\rangle \in \beta^n$, then

$$\forall x \in R(\beta)\,(s^\frown\langle x\rangle \in R \leftrightarrow x = a).$$

But also, by absoluteness of ϕ,

$$R = \{\langle y_1, \ldots, y_n, x\rangle \in R(\beta)^{n+1} : \phi(y_1, \ldots, y_n, x)^{R(\beta)}\},$$

so $R \in \operatorname{Df}(R(\beta), n+1)$ by Lemma 1.3. Thus, $a \in \mathbf{OD}$. $\square$

Formally, Theorem 2.2, like Lemma 1.3 and the Reflection Theorem, is a schema, asserting that for each ϕ, the displayed formula is provable.

The converse to Theorem 2.2 should state that if $a \in \mathbf{OD}$, then there is a ϕ which defines a from ordinals, and we run into the same problem formalizing this within ZF as we had with a converse to Lemma 1.3. However, we shall in fact produce one specific $\phi(y, x)$ and prove that

$$\forall a \left(a \in \mathbf{OD} \rightarrow \exists \alpha \, \forall x \left(\phi(\alpha, a) \leftrightarrow x = a\right)\right).$$

In fact, ϕ will define a function from $\mathbf{ON}$ onto $\mathbf{OD}$. The idea for this is that if $a \in \mathbf{OD}$, then a is put into $\mathbf{OD}$ via Definition 2.1 by a certain finite number of ordinals: the s, n, β of 2.1, together with an m such that the R of 2.1 is $\mathrm{En}(m, R(\beta), n)$, where En enumerates Df (see Definition 1.4). This gives us a way of mapping $\mathbf{ON}^{<\omega}$ onto $\mathbf{OD}$. Furthermore, $\mathbf{ON}$ may be mapped onto $\mathbf{ON}^{<\omega}$ (without using AC) by a generalization of the proof that $|\kappa \times \kappa| = \kappa$ (I 10.12).

2.3. Definition. If $s, t \in \mathbf{ON}^{<\omega}$, define $s \lhd t$ iff

(a) $\max(\mathrm{ran}(s)) < \max(\mathrm{ran}(t))$ or

(b) $\max(\mathrm{ran}(s)) = \max(\mathrm{ran}(t)) \wedge \mathrm{dom}(s) < \mathrm{dom}(t)$ or

(c) $\max(\mathrm{ran}(s)) = \max(\mathrm{ran}(t)) \wedge \mathrm{dom}(s) = \mathrm{dom}(t) \wedge \exists k \in \mathrm{dom}(z) \left(s \restriction k = t \restriction k \wedge s(k) < t(k)\right)$. □

So, in clause (c), if s and t have the same length (= domain) and the same maximum value, we compare them lexicographically.

It is easily seen that $\lhd$ well-orders $\mathbf{ON}^{<\omega}$. Furthermore, the predecessors of any $t \in \mathbf{ON}^{<\omega}$ form a set. In fact, if $\alpha = \max(\mathrm{ran}(t))$, then any $s \lhd t$ must be in $(\alpha + 1)^{<\omega}$ by virtue of clause (a) of Definition 2.3. Since $|(\alpha + 1)^{<\omega}| = |\alpha|$ for any infinite α (see I 10.13), we see that if κ is an uncountable cardinal, then $\kappa^{<\omega}$ is ordered in type κ and is an initial segment of $\mathbf{ON}^{<\omega}$. $\omega^{<\omega}$ is ordered in type ω^2.

2.4. Definition. $\mathrm{Enon}(\gamma)$ is the γ-th element of $\mathbf{ON}^{<\omega}$ in the order $\lhd$. □

2.5. Lemma. Enon *is a* 1–1 *map from* $\mathbf{ON}$ *onto* $\mathbf{ON}^{<\omega}$. □

Formally, we may think of Enon as defined by transfinite recursion. Or, we may think of Enon^{-1} as the Mostowski collapse of the relation $\lhd$ on $\mathbf{ON}^{<\omega}$ (see III 5.14).

Using our enumeration Enon of ordinal sequences, we may define an enumeration Enod of $\mathbf{OD}$.

2.6. Definition. For $\gamma \in \mathbf{ON}$, $\mathrm{Enod}(\gamma)$ is defined as follows:

(a) If $\mathrm{Enon}(\gamma) = s^\frown\langle \beta, n, m\rangle$, where $n, m \in \omega$, $s \in \beta^{<\omega}$, $\mathrm{dom}(s) = n$, and for some (unique) $a \in R(\beta)$,

$$\forall x \in R(\beta)\left(s^\frown\langle x\rangle \in \mathrm{En}(m, R(\beta), n+1) \leftrightarrow x = a\right),$$

then $\mathrm{Enod}(\gamma) = a$.

(b) If the hypotheses of (a) fail, then $\mathrm{Enod}(\gamma) = 0$. □

2.7. LEMMA. $\mathbf{OD} = \{\mathrm{Enod}(\gamma)\colon \gamma \in \mathbf{ON}\}$.

PROOF. This is immediate from the fact that En enumerates Df (see Lemma 1.5), together with the fact that $0 \in \mathbf{OD}$. To see that $0 \in \mathbf{OD}$, apply Theorem 2.2 with $\phi(y, x)$ the formula $y = x$ and α the ordinal 0. □

Observe that Lemma 2.7 yields the promised converse of Theorem 2.2. If $\phi(y, x)$ is $(y \in \mathbf{ON} \wedge x = \mathrm{Enod}(y))$, then

$$\forall a \in \mathbf{OD}\, \exists \alpha\, \forall x\, (\phi(\alpha, a) \leftrightarrow x = a).$$

2.2 and 2.7 can be used to show that **OD** contains all ordinals and is closed under every definable set-theoretic operation. For example, the following holds.

2.8. LEMMA. (a) $\mathbf{ON} \subset \mathbf{OD}$.

(b) *If* $w, z \in \mathbf{OD}$, *then* $\{w, z\} \in \mathbf{OD}$.

(c) *If* $w \in \mathbf{OD}$, *then* $\bigcup w \in \mathbf{OD}$ *and* $\mathscr{P}(w) \in \mathbf{OD}$.

PROOF. For (a), apply Theorem 2.2 with $\phi(y, x)$ the formula $y = x$ to conclude $\alpha \in \mathbf{OD}$ for all α. For (b), say $w = \mathrm{Enod}(\alpha_1)$ and $z = \mathrm{Enod}(\alpha_2)$. Let $a = \{w, z\}$, let $\phi(y_1, y_2, x)$ be the formula

$$y_1 \in \mathbf{ON} \wedge y_2 \in \mathbf{ON} \wedge x = \{\mathrm{Enod}(y_1), \mathrm{Enod}(y_2)\};$$

then an application of Theorem 2.2 yields $a \in \mathbf{OD}$. The two parts of (c) are likewise proved using Theorem 2.2; here $\phi(y, x)$ is, respectively, $y \in \mathbf{ON} \wedge x = \bigcup \mathrm{Enod}(y)$ and $y \in \mathbf{ON} \wedge x = \mathscr{P}(\mathrm{Enod}(y))$. □

It might now be conjectured that **OD** is a model of ZF, with Lemma 2.8 being used to establish the Pairing, Union, and Power Set axioms in **OD**. Unfortunately, **OD** need not be transitive, so we have problems with the Axiom of Extensionality. In fact, Extensionality is false in **OD** unless $\mathbf{V} = \mathbf{OD}$ (see Exercise 5), a situation which, although consistent, is unlikely; we shall return to the possibility that $\mathbf{V} = \mathbf{OD}$ at the end of this section.

We obtain a model of ZF by using, instead of **OD**, the class of hereditarily ordinal definable sets—i.e., those $x \in \mathbf{OD}$ such that all members of x, members of members of x, and etc. are in **OD**.

2.9. DEFINITION. $\mathbf{HOD} = \{x \in \mathbf{OD}: \text{tr cl}(x) \subset \mathbf{OD}\}$. □

2.10. LEMMA. $\mathbf{ON} \subset \mathbf{HOD} \subset \mathbf{OD}$ *and* $\mathbf{HOD}$ *is transitive.* □

2.11. LEMMA. *For any set* a, *if* $a \in \mathbf{OD}$ *and* $a \subset \mathbf{HOD}$, *then* $a \in \mathbf{HOD}$. □

2.12. LEMMA. *For each* α, $(R(\alpha) \cap \mathbf{HOD}) \in \mathbf{HOD}$.

PROOF. Clearly, $(R(\alpha) \cap \mathbf{HOD}) \subset \mathbf{HOD}$, so by Lemma 2.11 it is sufficient to show $(R(\alpha) \cap \mathbf{HOD}) \in \mathbf{OD}$, which is true since it is definable from α. More formally, we apply Theorem 2.2, where $\phi(y, x)$ is the formula $y \in \mathbf{ON} \wedge x = R(y) \cap \mathbf{HOD}$. □

2.13. THEOREM (ZF). *All axioms of* ZFC *hold in* $\mathbf{HOD}$.

PROOF. Extensionality holds since $\mathbf{HOD}$ is transitive and Foundation holds in any class (see IV 2.4 and IV 2.12, respectively).

For Comprehension, it is sufficient (by IV 2.5) to show that for each formula $\psi(v, z, w_1, \ldots, w_n)$,

$$\forall z, w_1, \ldots, w_n \in \mathbf{HOD}\left(\{v \in z: \psi^{\mathbf{HOD}}(v, z, w_1, \ldots, w_n)\} \in \mathbf{HOD}\right).$$

Fix $z, w_1, \ldots, w_n \in \mathbf{HOD}$, with $z = \text{Enod}(\alpha_0)$ and each $w_i = \text{Enod}(\alpha_i)$. Let

$$a = \{v \in z: \psi^{\mathbf{HOD}}(v, z, w_1, \ldots, w_n)\};$$

then a is the unique set x satisfying

$$\phi(\alpha_0, \alpha_1, \ldots, \alpha_n, x), \quad \text{where} \quad \phi(y_0, y_1, \ldots, y_n, x)$$

is

$$y_0, y_1, \ldots, y_n \in \mathbf{ON} \wedge x = \{v \in \text{Enod}(y_0):$$

$$\psi^{\mathbf{HOD}}(v, \text{Enod}(y_0), \text{Enod}(y_1), \ldots, \text{Enod}(y_n))\}.$$

Thus, by Theorem 2.2, $a \in \mathbf{OD}$. Since $a \subset z \in \mathbf{HOD}$ and $\mathbf{HOD}$ is transitive, $a \subset \mathbf{HOD}$, so $a \in \mathbf{HOD}$ by Lemma 2.11.

The Axioms of Pairing, Union, Replacement, and Power Set all assert that $\mathbf{HOD}$ contains large enough sets, and are all proved in a similar way using Lemma 2.12. For example, for Pairing, it is sufficient to show that for each $x, y \in \mathbf{HOD}$ there is a $w \in \mathbf{HOD}$ such that $x \in w \wedge y \in w$; so take $w = R(\alpha) \cap \mathbf{HOD}$, where $\alpha > \max(\text{rank}(x), \text{rank}(y))$. Of course, since $\mathbf{HOD}$ satisfies Comprehension, we now know $\{x, y\} \in \mathbf{HOD}$, a fact which could also have been deduced from Lemma 2.8(b), but this is irrelevant here. The discussion of Union, Replacement, and Power Set are similar.

The Axiom of Infinity holds since $\omega \in \mathbf{HOD}$ (see IV 3.2), so **HOD** satisfies ZF.

Finally, to verify AC in **HOD**, it is sufficient, by the absoluteness of well-ordering (IV 5.4), to show that for all $A \in \mathbf{HOD}$ there is an $R \in \mathbf{HOD}$ such that R well-orders A. Fix $A = \mathrm{Enod}(\alpha) \in \mathbf{HOD}$. Since $A \subset \mathbf{OD}$, we may well-order the elements of A in the order in which they first appear under the enumeration Enod; thus, let

$$R = \{\langle x, y\rangle \in A \times A : \exists \xi (x = \mathrm{Enod}(\xi) \wedge \forall \eta \leq \xi (y \neq \mathrm{Enod}(\eta))\}.$$

$R \in \mathbf{OD}$ since R is definable from α; the rigorous proof quotes, as usual, Theorem 2.2. Since $R \subset A \times A \subset \mathbf{HOD}$, $R \in \mathbf{HOD}$ by Lemma 2.11. □

All axioms of ZFC except Choice and Comprehension were verified in **HOD** using just very general properties of **HOD**, rather than the specific definition of **HOD**. For an abstract statement of what is going on here, see Exercise 6.

Some mathematicians might find the definitions of **OD** and **HOD** somewhat fishy because of their extremely non-constructive nature. For example, it is consistent that there is an $x \subset \omega$, such that x is ordinal definable but not definable from ordinals within any $R(\beta)$ with β less than the first strongly inaccessible cardinal (see VIII Exercise J13). Thus, the predicate $x \in \mathbf{OD}$ is, unlike most "reasonable" predicates (see IV Exercise 2), not necessarily absolute for $R(\kappa)$ when κ is strongly inaccessible. One must really know about all sets in **V** to determine whether $x \in \mathbf{OD}$.

Aside from philosophical difficulties, the non-constructive nature of **OD** makes it very difficult to deal with. Most interesting questions about **HOD** have answers which are independent of ZFC. For example, it cannot be decided within ZFC + GCH whether **HOD** is a model for CH (see VIII Exercise J14).

In our discussion of **HOD**, observe that we never decided whether **OD** was a proper sub-class of **V**. In fact, the statement $\mathbf{V} = \mathbf{OD}$ is consistent with (see VI) and independent from (see VII Exercise E5) ZFC + GCH.

It is tempting to suspect that **HOD** is a model for the statement $\mathbf{V} = \mathbf{OD}$, but this need not be true. Clearly, if $x \in \mathbf{HOD}$, then $x \in \mathbf{OD}$, but this need not imply $(x \in \mathbf{OD})^{\mathbf{HOD}}$, since the definition of **OD** (Definition 2.1) need not be absolute. The statement $(\mathbf{V} = \mathbf{OD})^{\mathbf{HOD}}$, (i.e., $\forall x \in \mathbf{HOD}\,[(x \in \mathbf{OD})^{\mathbf{HOD}}]$) is itself not decidable within ZFC + GCH (see VIII Exercise J14).

Since, by Theorem 2.13, we can work in ZF and prove that **HOD** is a model of ZFC, we have the following.

2.14. COROLLARY. $\mathrm{Con(ZF)} \rightarrow \mathrm{Con(ZFC)}$. □

However, since nothing further can be proved about **HOD** within ZF, or even within ZFC + GCH, **HOD** does not provide us with any more relative consistency results. In VI, we shall define the model **L**. **L** is also obtained using the notion of definability, but in a much more constructive way, which will enable us to check that not only ZFC, but also GCH and $\diamondsuit^+$, hold relativized to **L**.

EXERCISES

Work in ZF unless otherwise indicated.

(1) Let $\mathrm{Df}^*(A, n) = \mathrm{Df}(A, n) \cup \omega$. Show that there is an En* such that 1.3 and 1.5–1.7 hold for Df*. *Remark.* $\mathrm{Df}^*(A, n)$ is not always a subset of $\mathscr{P}(A^n)$, and thus need not equal $\mathrm{Df}(A, n)$.

(2) Show that there are Df and En* satisfying 1.3 and 1.5–1.7, such that

$$\forall A\, \forall n \big(\mathrm{Df}(A, n) \subset \mathrm{Df}^*(A, n) \subset \mathscr{P}(A^n)\big), \quad \text{and}$$

$$\exists A\, \exists n \big(\mathrm{Df}(A, n) \neq \mathrm{Df}^*(A, n)\big).$$

(3) Let S be ZF + $\neg\mathrm{CON}(\ulcorner \mathrm{ZF} \urcorner)$. Show that in S one can define a Df and En* satisfying 1.3 and 1.5–1.7 such that (provably from S),

$$\forall A\, \forall n \big(\mathrm{Df}^*(A, n) \subset \mathrm{Df}(A, n)\big), \quad \text{and} \quad \exists A\, \exists n \big(\mathrm{Df}^*(A, n) \neq \mathrm{Df}(A, n)\big).$$

(4) Suppose Df* and En* are functions for which Lemmas 1.3 and 1.5 hold, and suppose we define **OD*** by replacing Df by Df* in the definition of **OD**. Show that **OD*** = **OD**. *Hint.* 2.7 plus 2.2* yield **OD** $\subset$ **OD***.

(5) Show that the following are equivalent:
  (a) **V** = **OD**.
  (b) **V** = **HOD**.
  (c) **OD** is transitive.
  (d) Extensionality is true in **OD**.
Hint. For each α, $R(\alpha) \in$ **OD** and $\big(R(\alpha) \cap$ **OD**$\big) \in$ **OD**.

(6) Suppose **M** is a transitive class satisfying the Comprehension Axiom, and suppose that $\forall x \subset \mathbf{M}\, \exists y \in \mathbf{M} (x \subset y)$. Show that **M** is a model of ZF. Conversely, show that if **M** is a transitive proper class and satisfies ZF (or a suitable finite sub-theory of ZF), then $\forall x \subset \mathbf{M}\, \exists y \in \mathbf{M} (x \subset y)$.

(7) Show that there is a finite conjunction ϕ of axioms of ZF, such that whenever **M** is a transitive proper class satisfying ϕ, **M** satisfies all axioms

of ZF. *Hint.* Apply IV 7.5 to the $\mathbf{M} \cap R(\alpha)$. *Remark.* Compare this result with the fact that ZF is not finitely axiomatizable (IV 7.7).

(8) Use the predicate DfR to write a predicate $\text{Mod}_{\text{ZF}}(A)$ which "says" that A is a transitive model of ZF.

(9) If T is transitive, $\mathbf{OD}(T)$ is, intuitively, the class of those sets definable from a finite number of elements of $\mathbf{ON} \cup T \cup \{T\}$. Show how to make this definition rigorous, define $\mathbf{HOD}(T)$, and show that $\mathbf{HOD}(T)$ satisfies ZF. Show that $\mathbf{HOD}(T)$ satisfies AC iff T has a well-ordering in $\mathbf{HOD}(T)$. *Remark.* $\mathbf{HOD}(\mathscr{P}(\omega))$ need not satisfy AC; see VII Exercise E4.

(10) Formalize model theory with ZF as indicated in IV §10. Prove within ZF that $\text{Df}(A. n)$ is the set of n-place relations on A which are first-order definable in the structure $\langle A, \in \rangle$. Likewise, show that our notions of $A \prec B$ and $A \equiv B$ defined in §1 are equivalent to the usual model-theoretic definitions of $\langle A, \in \rangle \prec \langle B, \in \rangle$ and $\langle A, \in \rangle \equiv \langle B, \in \rangle$, respectively.

CHAPTER VI

THE CONSTRUCTIBLE SETS

In this chapter we work in ZF and define the class **L** of constructible sets. **L** is a transitive proper class model of ZFC + GCH, and satisfies a large number of useful combinatorial principles, such as $\diamondsuit$ and $\diamondsuit^+$.

We shall define $\mathbf{L} = \bigcup_{\alpha \in \mathbf{ON}} L(\alpha)$, where the sets $L(\alpha)$ are constructed by recursion as follows: $L(0) = 0$, and $L(\alpha) = \bigcup_{\xi < \alpha} L(\xi)$ when α is a limit. Given $L(\alpha)$, $L(\alpha + 1)$ will be the set of subsets of $L(\alpha)$ which are definable from a finite number of elements of $L(\alpha)$ by a formula relativized to $L(\alpha)$. In the precise definition (1.4), we shall make the "definable by a formula" rigorous by using the Df of V §1.

There is a close analogy between the $L(\alpha)$ and the $R(\alpha)$. The only difference in their definitions is that $R(\alpha + 1)$ contains all subsets of $R(\alpha)$, whereas $L(\alpha + 1)$ contains just definable subsets of $L(\alpha)$. This difference will become very important, but because of the similarities, many of the simpler properties of the $L(\alpha)$ and $R(\alpha)$ are the same.

§1. Basic properties of L

In this section we work within ZF, give a rigorous definition of **L**, and develop some of its basic properties. We begin by defining the *definable power set* operation, $\mathscr{D}$. Intuitively, $\mathscr{D}(A)$ is the set of subsets of A which are definable from a finite number of elements of A by a formula relativized to A.

1.1. DEFINITION. $\mathscr{D}(A) = \{X \subset A : \exists n \in \omega\, \exists s \in A^n\, \exists R \in \mathrm{Df}(A, n+1)$
$(X = \{x \in A : s^\frown\langle x \rangle \in R\})\}$. □

$\mathscr{D}$ is related to formulas ϕ in the metatheory by the following.

1.2. LEMMA. *Let $\phi(v_0, v_1, \ldots, v_{n-1}, x)$ be any formula, with all free variables shown; then*

$$\forall A\, \forall v_0, \ldots, v_{n-1} \in A\, [\{x \in A : \phi^A(v_0, v_1, \ldots, v_{n-1}, x)\} \in \mathscr{D}(A)].$$

PROOF. By the corresponding Lemma V 1.3 for Df. □

Formally, Lemma 1.2 is, like V 1.3, really a schema in the metatheory, asserting that for each ϕ, the displayed formula is provable in ZF.

1.3. LEMMA. *For any A,*

(*a*) $\mathscr{D}(A) \subset \mathscr{P}(A)$.

(*b*) *If A is transitive, then $A \subset \mathscr{D}(A)$.*

(*c*) $\forall X \subset A(|X| < \omega \rightarrow X \in \mathscr{D}(A))$.

(*d*) (AC) $|A| \geq \omega \rightarrow |\mathscr{D}(A)| = |A|$.

PROOF. (*a*) is immediate from the definition of $\mathscr{D}$. For (*b*), Lemma 1.2, applied with the formula $x \in v$, yields

$$\forall v \in A\,[\{x \in A\colon x \in v\} \in \mathscr{D}(A)],$$

which, if A is transitive, reduces to $\forall v \in A\,[v \in \mathscr{D}(A)]$.

For (*c*), first recall that if $R, S \in \mathrm{Df}(A, n+1)$, then

$$(A^{n+1} \smallsetminus R) \in \mathrm{Df}(A, n+1) \quad \text{and} \quad R \cap S \in \mathrm{Df}(A, n+1)$$

(see V 1.2); hence also

$$R \cup S = A^{n+1} \smallsetminus ((A^{n+1} \smallsetminus R) \cap (A^{n+1} \smallsetminus S)) \in \mathrm{Df}(A, n+1),$$

and $0 = R \cap (A^{n+1} \smallsetminus R) \in \mathrm{Df}(A, n+1)$. Next, by induction on $m \leq n$, show that

$$E_n^m = \{t \in A^{n+1}\colon \exists i < m\,(t(n) = t(i))\} \in \mathrm{Df}(A, n+1);$$

the induction step uses the fact that E_n^{m+1} is the union of E_n^m with $\{t \in A^{n+1}\colon t(n) = t(m)\}$, which is in $\mathrm{Df}(A, n+1)$ by definition (see V 1.1 (*c*)). Thus, for any $s \in A^n$,

$$\mathrm{ran}(s) = \{x \in A\colon s^\frown\langle x\rangle \in E_n^n\} \in \mathscr{D}(A),$$

so $\forall n < \omega\, \forall X \subset A\,(|X| \leq n \rightarrow X \in \mathscr{D}(A))$.

For (*d*), if we assume AC and $|A| \geq \omega$, then $|A^n| = |A|$ for all n. Since $|\mathrm{Df}(A, n+1)| \leq \omega$ (see V 1.6), we have $|\mathscr{D}(A)| \leq |A|$. $|\mathscr{D}(A)| \geq |A|$ follows from the fact that $\forall x \in A\,(\{x\} \in \mathscr{D}(A))$, which is a special case of (*c*).

Those readers who think that (*c*) is a trivial consequence of Lemma 1.2 should refer to Exercises 19 and 20.

1.4. DEFINITION. By transfinite recursion define $L(\alpha)$ for $\alpha \in \mathbf{ON}$ by:

(*a*) $L(0) = 0$.

(*b*) $L(\alpha + 1) = \mathscr{D}(L(\alpha))$.

(*c*) $L(\alpha) = \bigcup_{\xi < \alpha} L(\xi)$ when α is a limit ordinal. □

1.5. DEFINITION. $\mathbf{L} = \bigcup\{L(\alpha): \alpha \in \mathbf{ON}\}$. □

Many (but not all) of the results proved in III §2 about the $R(\alpha)$ just used that some specific definable subset of $R(\alpha)$ is in $R(\alpha + 1)$; these results hold for the $L(\alpha)$ as well. We proceed in analogy with III §2.

1.6. LEMMA. *For each* α:

(*a*) $L(\alpha)$ *is transitive.*

(*b*) $\forall \xi \leq \alpha\,(L(\xi) \subset L(\alpha))$.

PROOF. By induction α, almost verbatim as in III 2.3. We assume the lemma holds for all $\beta < \alpha$, and we attempt to prove it for α. This is trivial when α is 0 or a limit, so assume $\alpha = \beta + 1$. Then, $L(\beta)$ is transitive and $L(\alpha) = \mathscr{D}(L(\beta))$, so $L(\beta) \subset L(\alpha) \subset \mathscr{P}(L(\beta))$ (by 1.3 (*b*)), which implies both (*a*) and (*b*) for α. □

If $x \in \mathbf{L}$, the least α for which $x \in L(\alpha)$ must be a successor ordinal by Definition 1.4 (*c*).

1.7. DEFINITION. If $x \in \mathbf{L}$, $\rho(x)$, the **L**-*rank* of x, is the least β such that $x \in L(\beta + 1)$. □

Exactly as in III 2.5, we have the following.

1.8. LEMMA. *For any* α,

$$L(\alpha) = \{x \in \mathbf{L}: \rho(x) < \alpha\}. \quad \square$$

As with the $R(\alpha)$,if $x \in \mathbf{L}$ and $\beta = \rho(x)$, then $x \subset L(\beta)$, $x \notin L(\beta)$, but $x \in L(\beta + 1)$. Unlike the $R(\alpha)$, there are often subsets of $L(\beta)$ which are in $\mathbf{L}$ but not in $L(\beta + 1)$. The ordinals, however, appear at the same place in both constructions.

1.9. LEMMA. (*a*) $\forall \alpha \in \mathbf{ON}\,(\alpha \in \mathbf{L} \wedge \rho(\alpha) = \alpha)$.

(*b*) $\forall \alpha \in \mathbf{ON}\,(L(\alpha) \cap \mathbf{ON} = \alpha)$.

PROOF. (*a*) is immediate from (*b*), so we prove (*b*) by induction on α. The induction for α a limit or 0 is trivial, so we assume $\alpha = \beta + 1$ and $L(\beta) \cap \mathbf{ON} = \beta$, and try to conclude $L(\alpha) \cap \mathbf{ON} = \alpha$.

Now, since $L(\beta) \subset L(\alpha) \subset \mathscr{P}(L(\beta))$, we have $\beta \subset L(\alpha) \cap \mathbf{ON} \subset \alpha$, so we shall be done if we show that $\beta \in L(\alpha)$. Recall (IV 5.1) that there is a Δ_0 formula $\phi(x)$ such that

$$\forall x\,(x \text{ is an ordinal} \leftrightarrow \phi(x)).$$

Since Δ_0 formulas are absolute for all transitive sets,

$$\beta = L(\beta) \cap \mathbf{ON} = \{x \in L(\beta) : \phi^{L(\beta)}(x)\},$$

so $\beta \in \mathscr{D}(L(\beta)) = L(\alpha)$ by Lemma 1.2. □

Another application of Lemma 1.2 is the following.

1.10. Lemma. *$L(\alpha) \in L(\alpha + 1)$.*

Proof. $L(\alpha) = \{x \in L(\alpha) : (x = x)^{L(\alpha)}\}$, which is in $\mathscr{D}(L(\alpha)) = L(\alpha + 1)$ by Lemma 1.2. □

We may also prove analogues of parts of III 2.8 and III 2.9 asserting that, for example, if $x \in \mathbf{L}$, then $\bigcup x \in \mathbf{L}$, but this will follow anyway by absoluteness of $\bigcup$ once we know that $\mathbf{L}$ is a model for ZF. Going through the proof of III 2.8 would, however, also give us an estimate on $\rho(\bigcup x)$ in terms of $\rho(x)$; see Exercise 5. Unlike **WF**, we cannot assert that **L** contains $\mathbb{R}$ or $\mathbb{C}$, or is closed under non-absolute functions such as $\mathscr{P}$.

We now compare the $R(\alpha)$ and $L(\alpha)$ construction more closely.

1.11. Lemma. *$L(\alpha) \subset R(\alpha)$ for all α.*

Proof. Transfinite induction on α. □

1.12. Lemma. *Every finite subset of $L(\alpha)$ is in $L(\alpha + 1)$.*

Proof. By Lemma 1.3(*c*). □

1.13. Lemma. (*a*) $\forall n \in \omega\,(L(n) = R(n))$.
(*b*) $L(\omega) = R(\omega)$.

Proof. (*a*) is by induction on *n*, using Lemma 1.12. (*b*) follows from (*a*). □

1.14. Lemma (AC). *For all $\alpha \geq \omega$, $|L(\alpha)| = |\alpha|$.*

Proof. Since $\alpha \subset L(\alpha)$, $|\alpha| \leq |L(\alpha)|$. We prove $|\alpha| = |L(\alpha)|$ by transfinite induction on α; assume $\alpha \geq \omega$ and $\forall \beta < \alpha\,(\beta \geq \omega \to |L(\beta)| = |\beta|)$; then $\forall \beta < \alpha\,(|L(\beta)| \leq |\alpha|)$ (since $|L(n)| = |R(n) < \omega$ for $n < \omega$). If α is a limit, then $L(\alpha) = \bigcup_{\beta<\alpha} L(\beta)$ is a union of $|\alpha|$ sets of cardinality $\leq |\alpha|$, so (by AC), has cardinality $\leq |\alpha|$. If $\alpha = \beta + 1$, then $|L(\beta)| = |\beta| = |\alpha|$, and $L(\alpha) = \mathscr{D}(L(\beta))$, so $|L(\alpha)| = |\alpha|$ by Lemma 1.3(*d*). □

One can in fact prove Lemma 1.14 without AC (see Exercise 1). Lemma 1.14 shows that just by cardinality considerations, $L(\alpha) \neq R(\alpha)$ for "most" $\alpha > \omega$, since, for example, if $\omega < \alpha < \omega_1$, $L(\alpha)$ is countable but $R(\alpha)$ is uncountable. In fact, if $\alpha > \omega$, then $|L(\alpha)| = |R(\alpha)|$ iff $\alpha = \beth_\alpha$ (Exercise 2). Furthermore, $\mathscr{P}(\omega) \subset R(\omega + 1)$, and there is no reason to think that every subset of ω is in **L** at all; if $\mathscr{P}(\omega) \not\subset \mathbf{L}$ (which is consistent by VII), then $L(\alpha) \neq R(\alpha)$ for any $\alpha > \omega$. However, it is also consistent that $\mathbf{V} = \mathbf{L}$ (by Corollary 3.4), and this implies $L(\alpha) = R(\alpha)$ whenever $\alpha = \beth_\alpha$ (see Exercise 3).

§2. ZF in L

We work in ZF and prove that all axioms of ZF are true in **L**.

2.1. THEOREM (ZF). **L** *is a model of* ZF.

PROOF. Extensionality holds in **L** because **L** is transitive, and Foundation holds in any class (see IV 2.4 and 2.12).

To verify Comprehension, it is sufficient, by IV 2.5, to verify, for each formula $\psi(x, z, v_1, \ldots, v_n)$ with all free variables shown, that

$$\forall z, v_1, \ldots, v_n \in \mathbf{L}\left(\{x \in z\colon \psi^{\mathbf{L}}(x, z, v_1, \ldots, v_n)\} \in \mathbf{L}\right).$$

Now the definition of $L(\alpha + 1) = \mathscr{D}(L(\alpha))$ was cooked up to make this work, but $\mathscr{D}(L(\alpha))$ involves relativizations to $L(\alpha)$, not to **L**. To remedy this, we apply Reflection. Fix $z, v_1, \ldots, v_n \in \mathbf{L}$, and fix α so that $z, v_1, \ldots, v_n \in L(\alpha)$. Now let $\beta > \alpha$ be such that ψ is absolute for $L(\beta)$, **L** (by IV 7.5); then

$$\{x \in z\colon \psi^{\mathbf{L}}(x, z, v_1, \ldots, v_n)\} = \{x \in L(\beta)\colon \phi^{L(\beta)}(x, z, v_1, \ldots, v_n)\},$$

where ϕ is $x \in z \wedge \psi$, so this set is in $\mathscr{D}(L(\beta)) = L(\beta + 1)$ by Lemma 1.2.

The Axioms of Pairing, Union, Replacement, and Power Set all involve only the existence of large enough sets, and are easily verified in **L**. For example, to verify Replacement, it is sufficient to show (by IV 2.11) that for each formula $\phi(x, y, A, w_1 \ldots, w_n)$ and each $A, w_1, \ldots, w_n \in \mathbf{L}$, if we assume

$$\forall x \in A\ \exists! y \in \mathbf{L}\ \phi^{\mathbf{L}}(x, y, A, w_1, \ldots, w_n), \tag{1}$$

then we can conclude

$$\exists Y \in \mathbf{L}\left(\{y\colon \exists x \in A\ \phi^{\mathbf{L}}(x, y, A, w_1, \ldots, w_n)\} \subset Y\right). \tag{2}$$

Now assuming (1), let

$$\alpha = \sup\{\rho(y) + 1\colon \exists x \in A\ \phi^{\mathbf{L}}(x, y, A, w_1, \ldots, w_n)\};$$

then taking $Y = \mathrm{L}(\alpha)$, which is in **L** by Lemma 1.10, (2) is satisfied.

Finally, Infinity holds in $\mathbf{L}$ because $\omega \in \mathbf{L}$. □

The fact that only the Comprehension Axiom required any work is formalized in V Exercise 6.

§3. The Axiom of Constructibility

Observe that in our discussion of $\mathbf{L}$, we have not proved that $\mathbf{L}$ is a proper sub-class of $\mathbf{V}$. In fact, although the statement $\mathbf{V} = \mathbf{L}$ seems unlikely, it is consistent with ZF, as we shall see in this section. In VII we shall show that $\mathbf{V} \neq \mathbf{L}$ is also consistent with ZF.

3.1. Definition. The *Axiom of Constructibility* is the statement $\mathbf{V} = \mathbf{L}$; i.e., $\forall x\, \exists \alpha\, (x \in L(\alpha))$. □

We shall prove the consistency of ZF + $\mathbf{V} = \mathbf{L}$ by showing that $\mathbf{L}$, which we have just seen is a model for ZF, satisfies $\mathbf{V} = \mathbf{L}$ as well. Now the fact that $\mathbf{V} = \mathbf{L}$ holds in $\mathbf{L}$ is not quite as trivial as $1 = 1$ (or $\mathbf{L} = \mathbf{L}$). It is obvious that $\forall x \in \mathbf{L}\, \exists \alpha\, (x \in L(\alpha))$, but $(\mathbf{V} = \mathbf{L})^{\mathbf{L}}$ says $\forall x \in \mathbf{L}\, \exists \alpha \in \mathbf{L}\, ((x \in L(\alpha))^{\mathbf{L}})$, and to prove $(\mathbf{V} = \mathbf{L})^{\mathbf{L}}$, we must first verify the absoluteness of the formula $x \in L(\alpha)$.

The situation for $\mathbf{L}$ may be contrasted with that for **HOD** (see V §2). It is not provable in ZF that $(\mathbf{V} = \mathbf{HOD})^{\mathbf{HOD}}$; the obvious "proof" fails because of the non-absoluteness of **HOD**.

3.2. Lemma. *The function $L(\alpha)$ is absolute for transitive models of* ZF − P.

Proof. We saw in V 1.7 that Df was absolute. It then follows easily from the methods of IV §5 that $\mathscr{D}$ is absolute. Since $L(\alpha)$ was defined by transfinite recursion from $\mathscr{D}$, it is absolute as well (see IV 5.6). □

3.3. Theorem (ZF). $\mathbf{L}$ *is a model of* ZF + $\mathbf{V} = \mathbf{L}$.

Proof. For ZF in $\mathbf{L}$, see Theorem 2.1. To show that $(\mathbf{V} = \mathbf{L})^{\mathbf{L}}$, we must show

$$\forall x \in \mathbf{L}\, \exists \alpha \in \mathbf{L}\, ((x \in L(\alpha))^{\mathbf{L}}).$$

Fix $x \in \mathbf{L}$. Now fix α such that $x \in L(\alpha)$. $\alpha \in \mathbf{L}$ since $\mathbf{ON} \subset \mathbf{L}$, and $(x \in L(\alpha))^{\mathbf{L}}$ by Lemma 3.2. □

3.4. Corollary. $\mathrm{Con(ZF)} \rightarrow \mathrm{Con(ZF} + \mathbf{V} = \mathbf{L})$. □

There is a strong analogy between our consistency proofs of the Axiom of Constructibility, $\mathbf{V} = \mathbf{L}$, and of the Axiom of Foundation, $\mathbf{V} = \mathbf{WF}$ (see IV §4), but an important distinction must be drawn. Working in ZF^-, we defined **WF**, and showed it was a model for ZF^- plus Foundation; this provided a formal proof of

$$\text{Con}(\text{ZF}^-) \rightarrow \text{Con}(\text{ZF}^- + \mathbf{V} = \mathbf{WF}).$$

But we also presented an argument that all mathematics takes place in **WF** anyway, so that we might as well add $\mathbf{V} = \mathbf{WF}$ as an axiom, making our basic system ZF. However, there is no reason to believe that all mathematical objects, or even all subsets of ω, lie in **L**. Thus, we do not consider $\mathbf{V} = \mathbf{L}$ to be a plausible basic axiom to add to ZF, but we merely use it as a tool to obtain relative consistency results. Its usefulness lies in the observation that any sentence provable from $\text{ZF} + \mathbf{V} = \mathbf{L}$ is consistent with ZF, and also true in **L**. If ϕ is a sentence of some mathematical complexity, it is usually easier to prove $\phi^{\mathbf{L}}$ by assuming $\mathbf{V} = \mathbf{L}$ and proving ϕ, so that during the argument we do not have to worry about relativizing everything to **L**. We shall carry out this procedure with AC + GCH in §4, and with $\diamondsuit$ and $\diamondsuit^+$ in §5.

We conclude this section with a number of other consequences of the absoluteness of $L(\alpha)$. These involve minimality properties of the model **L**.

3.5. THEOREM. *If* $\mathbf{M}$ *is any transitive proper class model of* $\text{ZF} - \text{P}$, *then* $\mathbf{L} = \mathbf{L}^{\mathbf{M}} \subset \mathbf{M}$.

PROOF. We note first that $\mathbf{ON} \subset \mathbf{M}$. To see this, fix $\alpha \in \mathbf{ON}$. Since $\mathbf{M} \not\subset R(\alpha)$, there is an $x \in \mathbf{M}$ with $\text{rank}(x) \geq \alpha$. But the rank function is absolute for $\mathbf{M}$ (IV 5.7), so $\text{rank}(x) \in \mathbf{M}$, and hence $\alpha \in \mathbf{M}$ since $\mathbf{M}$ is transitive.

Now, by absoluteness of $L(\alpha)$ (and **ON**),

$$\mathbf{L}^{\mathbf{M}} = \{x \in \mathbf{M}\colon (\exists \alpha\,(x \in L(\alpha)))^{\mathbf{M}}\} = \bigcup \{L(\alpha) : \alpha \in \mathbf{ON}\} = \mathbf{L},$$

so $\mathbf{L} = \mathbf{L}^{\mathbf{M}} \subset \mathbf{M}$. $\square$

As usual with proper classes, some care must be taken to see that Theorem 3.5 really makes sense. To be more formal, we note that there is a finite conjunction ϕ of axioms of $\text{ZF} - \text{P}$ so that the notions of ordinal, rank, and $L(\alpha)$ are absolute for transitive models of ϕ (see IV 7.1). Then we may express Theorem 3.5 by saying that for each class (i.e., predicate) **M**, the statement.

$$\mathbf{M} \text{ is a transitive proper class } \wedge \phi^{\mathbf{M}} \rightarrow \mathbf{L} \subset \mathbf{M}$$

is provable.

We now look at analogous results for set models.

3.6. Definition. $o(M) = M \cap \mathbf{ON}$. □

3.7. Lemma. *If M is any transitive set, $o(M) \in \mathbf{ON}$, and is the first ordinal not in M.* □

3.8. Theorem. *There is a finite conjunction ψ of axioms of* ZF − P *such that*

$$\forall M\,(M \text{ transitive} \wedge \psi^M \rightarrow (L(o(M)) = \mathbf{L}^M \subset M)).$$

Proof. ψ is a conjunction of the ϕ discussed above with enough axioms to prove that there is no largest ordinal. Now, if M is transitive and ψ^M, then $o(M)$ is a limit ordinal, so $L(o(M)) = \bigcup_{\alpha \in M} L(\alpha)$. But

$$\mathbf{L}^M = \{x \in M : (\exists \alpha\,(x \in L(\alpha)))^M\} = \bigcup_{\alpha \in M} L(\alpha)$$

by absoluteness of $L(\alpha)$, so $L(o(M)) = \mathbf{L}^M \subset M$. □

Similar to minimality statements, there is a uniqueness theorem for transitive models of ZF + $\mathbf{V} = \mathbf{L}$; such a model is either $\mathbf{L}$ if it is a proper class, or some $L(\delta)$ if it is a set. As usual, one needs only a finite number of axioms.

3.9. Theorem *There is a finite conjunction χ of axioms of* ZF − P + $\mathbf{V} = \mathbf{L}$ *such that*

(*a*) *If $\mathbf{M}$ is a transitive proper class and $\chi^{\mathbf{M}}$, then $\mathbf{M} = \mathbf{L}$.*
(*b*) $\forall M\,(M \text{ transitive} \wedge \chi^M \rightarrow M = L(o(M)))$.

Proof. χ is just the ψ of Theorem 3.8, conjuncted with $\mathbf{V} = \mathbf{L}$. Thus, if $\mathbf{M}$ is transitive and $\chi^{\mathbf{M}}$, then $(\forall x\,(x \in \mathbf{L}))^{\mathbf{M}}$, so $\mathbf{M} = \mathbf{L}^{\mathbf{M}}$. (*a*) and (*b*) now follow from Theorems 3.5 and 3.8, respectively. □

An example of a transitive set model of ZF − P + $\mathbf{V} = \mathbf{L}$ is $L(\kappa)$ for any regular $\kappa > \omega$ (see Corollary 4.12).

The minimality of $\mathbf{L}$ enables us to establish the consistency of $\mathbf{V} = \mathbf{HOD}$.

3.10. Theorem (ZF). $\mathbf{V} = \mathbf{L} \rightarrow \mathbf{V} = \mathbf{HOD}$.

Proof. Applying Theorem 3.5 to $\mathbf{HOD}$ yields $\mathbf{L} \subset \mathbf{HOD}$, so

$$\mathbf{V} = \mathbf{L} \rightarrow \mathbf{V} \subset \mathbf{HOD}. \quad \square$$

3.11. Corollary. Con(ZF) → Con(ZF + $\mathbf{V} = \mathbf{HOD}$). □

3.12. Corollary (ZF). $\mathbf{V} = \mathbf{L} \rightarrow$ AC.

PROOF. This is immediate from Theorem 3.10 plus the fact that AC always holds in **HOD**. □

In the next section we give a direct proof of $\mathbf{V} = \mathbf{L} \to$ AC which does not rely on the development of **HOD**.

§4. AC and GCH in L

We shall work in ZF and show that $\mathbf{V} = \mathbf{L}$ implies AC and GCH. By the results of the last section, this will yield that, as a theorem of ZF, $\mathrm{AC}^{\mathbf{L}}$ and $\mathrm{GCH}^{\mathbf{L}}$ hold, so that $\mathbf{L}$ is a model of ZFC + GCH.

We consider AC first. Although we have already seen (Corollary 3.12) that $\mathbf{V} = \mathbf{L} \to$ AC, we shall give a much more direct proof here by explicitly defining well-orders of the $L(\alpha)$. The definition will proceed inductively. The key idea is that once we have a well-order of $L(\alpha)$, this induces a lexicographic order on $L(\alpha)^{<\omega}$, and we can then well-order $L(\alpha + 1) = \mathscr{D}(L(\alpha))$ by enumerating the countable number of ways subsets of $L(\alpha)$ can be defined from elements of $L(\alpha)^{<\omega}$. A rigorous presentation of this uses the enumeration En (Definition V 1.4) of Df.

4.1. DEFINITION. By recursion on α, define well-orders $\lhd_\alpha = \lhd(\alpha)$ of $L(\alpha)$ as follows. $\lhd = 0$. If α is a limit,

$$\lhd_\alpha = \{\langle x, y\rangle \in L(\alpha) \times L(\alpha) : \rho(x) < \rho(y) \vee (\rho(x) = \rho(y) \wedge \langle x, y\rangle \in \lhd(\rho(x) + 1))\}.$$

Given $\lhd_\alpha$, let $\lhd_\alpha^n$ be the induced lexicographic order on $L(\alpha)^n$:

$$s \lhd_\alpha^n t \leftrightarrow \exists k < n\, (s \restriction k = t \restriction k \wedge s(k) \lhd_\alpha t(k)).$$

If $X \in L(\alpha + 1) = \mathscr{D}(L(\alpha))$, let n_X be the least n such that

$$\exists s \in L(\alpha)^n\, \exists R \in \mathrm{Df}(L(\alpha), n + 1)\, (X = \{x \in L(\alpha) : s^\frown\langle x\rangle \in R\}).$$

Let s_X be the $\lhd_\alpha^{n_X}$ – least $s \in L(\alpha)^{n_X}$ such that

$$\exists R \in \mathrm{Df}(L(\alpha), n_X + 1)\, (X = \{x \in L(\alpha) : s^\frown\langle x\rangle \in R\}),$$

and let m_X be the least $m \in \omega$ such that

$$X = \{x \in L(\alpha) : s_X{}^\frown\langle x\rangle \in \mathrm{En}(m, L(\alpha), n_X)\}.$$

For $X, Y \in L(\alpha + 1)$, define $X \lhd_{\alpha+1} Y$ iff
(*a*) $X, Y \in L(\alpha) \wedge X \lhd_\alpha Y$, *or*
(*b*) $X \in L(\alpha) \wedge Y \notin L(\alpha)$, *or*

(c) $X, Y \notin L(\alpha) \wedge [(n_X < n_Y) \vee (n_X = n_Y \wedge s_X \lhd_\alpha^{n_X} s_Y) \vee$
$(n_X = n_Y \wedge s_X = s_Y \wedge m_X < m_Y)]$. □

Observe that in Definition 4.1, we must check inductively as we go along that $\lhd_\alpha$ is indeed a well-order of $L(\alpha)$, since the definition of s_X presupposes that $\lhd_\alpha$ is a well-order. A very formal presentation would first define $\lhd_{\alpha+1}$ to be 0 if $\lhd_\alpha$ is not a well-order, and then prove as a lemma that $\lhd_\alpha$ is a well-order for all α; so Definition 4.1 is really a definition and a lemma combined.

4.2. LEMMA. $\mathbf{V} = \mathbf{L} \rightarrow$ AC.

PROOF. If $x \in \mathbf{L}$, then $x \subset L(\alpha)$ for some α, and the well-order $\lhd_\alpha$ well-orders x. □

We may also use the $\lhd_\alpha$ to well-order all of $\mathbf{L}$.

4.3. DEFINITION. $x <_L y$ iff $x, y, \in \mathbf{L}$ and

$$\rho(x) < \rho(y) \vee (\rho(x) = \rho(y) \wedge \langle x, y \rangle \in \lhd(\rho(x) + 1)). \quad \square$$

Thus, we order elements of $\mathbf{L}$ first by their $\mathbf{L}$-rank; the elements of the same $\mathbf{L}$-rank α are ordered by $\lhd_{\alpha+1}$. This is the same procedure used in the inductive construction of the $\lhd_\alpha$, so the following is obtained.

4.4. LEMMA. *$<_L$ well-orders $\mathbf{L}$. Each $L(\alpha)$ is an initial segment of $\mathbf{L}$ under $<_L$, and $<_L$ restricted to $L(\alpha)$ is $\lhd_\alpha$.* □

Unlike the analogous well-order obtained for **HOD**, the well-order of **L** is absolute in the following sense.

4.5. LEMMA. (*a*) *The function $\lhd(\alpha)$ is absolute for transitive models of* ZF − P.

(*b*) *If $\mathbf{M}$ is a transitive proper class model for* ZF − P, *then $<_L$ is absolute for $\mathbf{M}$.*

(*c*) *If M is a transitive set model of* ZF − P, *$x, y \in M$, and $x, y \in \mathbf{L}^M$, then $x <_L y$ iff $(x <_L y)^M$.*

PROOF. (*a*) follows from the absoluteness of functions defined by recursion. (*b*) follows from (*a*), since $\mathbf{L} \subset \mathbf{M}$ (by Theorem 3.5). (*c*) also follows from (*a*), since if $x, y \in \mathbf{L}^M$, then $x, y \in L(\alpha)$ for some $\alpha < o(M)$ (by Theorem 3.8). □

We cannot actually assert in (*c*) that $<_L$ is absolute for M. For example, we shall see in VII how to construct a transitive $M \in \mathbf{L}$ such that M is a model for an arbitrary finite fragment of ZF + $\mathbf{V} \neq \mathbf{L}$. Then, if x, y are in $M \smallsetminus \mathbf{L}^M$, x and y will be compared by $<_L$ but not by $(<_L)^{\mathbf{M}}$. Likewise the predicate $x \in \mathbf{L}$ is not absolute, although the function $L(\alpha)$ is. If $x \in \mathbf{L}$ but $(x \notin \mathbf{L})^M$, then $x \in L(\alpha)$ for some $\alpha > \mathrm{o}(M)$.

We turn now to GCH. Assuming $\mathbf{V} = \mathbf{L}$, CH will follow from the fact that every subset of ω is constructed at some countable stage, i.e., $\mathscr{P}(\omega) \subset L(\omega_1)$. Since $|L(\omega_1)| = \omega_1$ (by Lemma 1.14), $2^\omega \leq \omega_1$. More generally, the following holds.

4.6. THEOREM. *If* $\mathbf{V} = \mathbf{L}$, *then for all infinite ordinals* α, $\mathscr{P}(L(\alpha)) \subset L(\alpha^+)$.

PROOF. Let χ be a finite conjunction of axioms of ZF + $\mathbf{V} = \mathbf{L}$ such that

$$\forall M\,(M \text{ transitive} \wedge \chi^M \rightarrow M = L(\mathrm{o}(M))).$$

This is possible by Theorem 3.9.

Assume $\mathbf{V} = \mathbf{L}$, and fix $A \in \mathscr{P}(L(\alpha))$. Let $X = L(\alpha) \cup \{A\}$. Then $|X| = |\alpha|$ by Lemma 1.14 (1.14 uses AC, but we have just seen that $\mathbf{V} = \mathbf{L} \rightarrow$ AC). By a Löwenheim–Skolem argument followed by the Mostowski Collapsing Theorem there is a transitive M such that $|M| = |\alpha|$, $X \subset M$, and $\chi^M \leftrightarrow \chi^{\mathbf{V}}$ (we have applied IV 7.10, with $\mathbf{Z} = \mathbf{V}$). But $\chi^{\mathbf{V}}$ is true by $\mathbf{V} = \mathbf{L}$, so χ^M holds, whence $M = L(\mathrm{o}(M))$. Since $|M| = |\alpha|$, $|\mathrm{o}(M)| < \alpha^+$. Thus,

$$A \in L(\mathrm{o}(M)) \subset L(\alpha^+). \quad \square$$

4.7. COROLLARY (ZF). $\mathbf{V} = \mathrm{L} \rightarrow$ AC + GCH.

PROOF. AC is Lemma 4.2. For GCH, Theorem 4.6 yields, for each cardinal $\kappa \geq \omega$, $\mathscr{P}(\kappa) \subset \mathscr{P}(L(\kappa)) \subset L(\kappa^+)$, whence $2^\kappa \leq |L(\kappa^+)| = \kappa^+$ by Lemma 1.14. $\square$

4.8. COROLLARY (ZF). $(\mathrm{AC} + \mathrm{GCH})^{\mathbf{L}}$. $\square$

4.9. COROLLARY. Con(ZF) $\rightarrow$ Con(ZFC + GCH). $\square$

Many proofs from $\mathbf{V} = \mathbf{L}$, including the proofs of GCH, $\diamondsuit$, and $\diamondsuit^+$, involve an argument which a Platonist would like to describe as taking an elementary submodel of $\mathbf{V}$—i.e., constructing a set $W \subset \mathbf{V}$ such that all formulas are absolute for W, $\mathbf{V}$ (and then applying the Mostowski Collapsing Theorem to obtain a transitive M as in Theorem 4.6). Formally, however, we must modify these platonistic arguments somewhat to avoid talking about all formulas simultaneously.

The procedure we followed above was to obtain a W for which a certain finite list of formulas is absolute. A different approach, exemplified by our proof of $\diamondsuit^+$ in §5, is to formalize the notion of elementary submodel, using the relation $W \prec N$ for *sets* N (see V 1.9), and apply it with N equal to a suitable $L(\kappa)$ rather than $\mathbf{L}$.

There is also a much more combinatorial approach to $\mathbf{L}$, developed by [Gödel 1940] in an attempt to explain his work to non-logicians. There, rather than iterating applications of $\mathcal{D}$, he defines eight very simple functions, $F_1, \dots, F_8$ (for example, $F_1(x, y) = \{x, y\}$ and $F_3(x, y) = x \setminus y$), and iterates applications of these to construct $\mathbf{L}$. The Löwenheim–Skolem argument reduces to closing under $F_1, \dots, F_8$. This approach has the merit of removing all vestiges of logic from the treatment of $\mathbf{L}$.

We conclude this section with some results on the existence of set models of $\mathbf{V} = \mathbf{L}$. We saw in IV 7.11 that in ZFC that one can prove the existence of countable transitive models for any desired finite fragment of ZFC. This can now be improved to the following.

4.10. THEOREM (ZF). *Let $\phi_1, \dots, \phi_n$ be any axioms of* ZF + $\mathbf{V} = \mathbf{L}$; *then*

$$\exists M \left(|M| = \omega \wedge M \textit{ is transitive} \wedge \bigwedge_{i=1}^{n} \phi_i^M \right). \qquad (*)$$

PROOF. The fact that $(*)$ follows from ZF + $\mathbf{V} = \mathbf{L}$ is a direct application of IV 7.11. Thus, working in ZF, we know that $(*)$ holds relativized to $\mathbf{L}$. Fix $M \in \mathbf{L}$ satisfying the conditions of $(*)$ in $\mathbf{L}$. We now check, by absoluteness that M satisfies $(*)$ in $\mathbf{V}$.

$(|M| = \omega)^{\mathbf{L}}$, so there is an $f \in \mathbf{L}$ such that (f maps ω 1–1 onto $M)^{\mathbf{L}}$. Then f maps ω 1–1 onto M, so $|M| = \omega$. (M is transitive)$^{\mathbf{L}}$ implies M is transitive. Finally, since $\mathbf{L}$ is transitive, each $(\phi_i^M)^{\mathbf{L}}$ is equivalent to ϕ_i^M, so M satisfies $(*)$ in $\mathbf{V}$. □

If enough axioms of ZF + $\mathbf{V} = \mathbf{L}$ are listed in Theorem 4.10, then M must be of the form $L(\delta)$ for some δ with $\omega < \delta < \omega_1$ (see Theorem 3.9), but, as usual with such reflection theorems, we have no simple combinatorial description of what such a δ looks like. We now show (Corollary 4.12) that $L(\kappa)$, for regular $\kappa > \omega$, satisfies all axioms of ZF + $\mathbf{V} = \mathbf{L}$ *except* possibly the Power Set Axiom.

4.11. LEMMA. *If* $\mathbf{V} = \mathbf{L}$, *then* $L(\kappa) = H(\kappa)$ *whenever* $\kappa > \omega$ *and* κ *is regular.*

PROOF. $L(\omega) = R(\omega) = H(\omega)$, so assume $\kappa > \omega$. If $x \in L(\kappa)$, then $x \in L(\alpha)$ for some $\alpha < \kappa$, whence $\mathrm{tr\,cl}(x) \subset L(\alpha)$, so $|\mathrm{tr\,cl}(x)| \leq |L(\alpha)| < \kappa$, so $x \in H(\kappa)$. Thus, $L(\kappa) \subset H(\kappa)$. If $L(\kappa) \neq H(\kappa)$, fix $A \in H(\kappa) \setminus L(\kappa)$ with $A \cap (H(\kappa) \setminus L(\kappa)) = 0$ (by Foundation). $H(\kappa)$ is transitive, so

$$A \subset H(\kappa) \cap L(\kappa) = L(\kappa),$$

so $A \subset L(\alpha)$ for some α with $\omega < \alpha < \kappa$ since $|A| < \kappa$ and κ is regular. But then $A \in L(\alpha^+) \subset L(\kappa)$ by Theorem 4.6, which is a contradiction. $\square$

Lemma 4.11 also holds when κ is singular (Exercise 4).

4.12. COROLLARY (ZF). *If $\kappa > \omega$ and κ is regular, then all axioms of* ZF $-$ P + $\mathbf{V} = \mathbf{L}$ *hold in $L(\kappa)$. If κ is weakly inaccessible, then the Power Set Axiom holds also in $L(\kappa)$.*

PROOF. If $\mathbf{V} = \mathbf{L}$, then this follows immediately from Lemma 4.11 and the corresponding facts about the $H(\kappa)$ (see IV 6.5 and 6.6; observe that by GCH, all weak inaccessibles are strong inaccessibles). In ZF, proceed as in Theorem 4.10; observe that if κ is regular then (κ is regular)$^{\mathbf{L}}$ and if κ is weakly inaccessible, then (κ is weakly inaccessible)$^{\mathbf{L}}$. $\square$

4.13. COROLLARY. Con(ZF) $\rightarrow$ Con(ZFC + GCH + $\neg\exists\alpha$ (α *is weakly inaccessible*)).

PROOF. This is an improvement of IV 6.9, and the proof is very similar. Let Wi(κ) abbreviate "κ is weakly inaccessible". Let

$$\mathbf{M} = \{x \in \mathbf{L}\colon \forall\kappa\,(\mathrm{Wi}(\kappa)^{\mathbf{L}} \rightarrow x \in L(\kappa))\}.$$

In ZF, one cannot decide whether $\mathbf{M} = \mathbf{L}$ or $\mathbf{M} = L(\kappa)$ for the least κ which is weakly inaccessible in $\mathbf{L}$, but in either case $\mathbf{M}$ satisfies ZFC + GCH + $\neg\exists\alpha\,(\mathrm{Wi}(\alpha))$. $\square$

§5. $\diamondsuit$ and $\diamondsuit^+$ in L

We show here that $\mathbf{V} = \mathbf{L}$ implies the combinatorial principles $\diamondsuit$ and $\diamondsuit^+$ of II §7. Of course, we need only prove $\diamondsuit^+$, since $\diamondsuit^+ \rightarrow \diamondsuit$ by II 7.14. The proof of $\diamondsuit^+$ is a modification of the proof of GCH. If $A \subset \omega_1$, we applied (in Theorem 4.6) a Skolem closure argument to $L(\omega_1) \cup \{A\}$ to show $A \in L(\omega_2)$. Here, we apply a Skolem closure argument to $\{\omega_1, A\}$ to show that A gets "captured" on a c.u.b. set of countable ordinals. We phrase our argument using the $\prec$ defined in V 1.9.

5.1. LEMMA. *$\{\rho < \omega_1\colon L(\rho) \prec L(\omega_1)\}$ is unbounded in ω_1.*

PROOF. Let $H_{nm}\colon L(\omega_1)^n \rightarrow L(\omega_1)$ be Skolem functions for $L(\omega_1)$ defined verbatim as in the proof of V 1.11 (so the B there is $L(\omega_1)$, and $\lhd$ is any well-order of $L(\omega_1)$). Then, $L(\rho) \prec L(\omega_1)$ whenever $L(\rho)$ is closed under all the H_{nm}. But the set of all such ρ is c.u.b. in ω_1 (see II 6.13). $\square$

Observe that by Corollary 4.12, $L(\rho) \prec L(\omega_1)$ implies that $L(\rho)$ is a model of ZF − P + V = L.

5.2. THEOREM. V = L *implies* $\diamondsuit^+$ *(and hence also* $\diamondsuit$*).*

PROOF. Assume V = L. For $\alpha < \omega_1$, let q(α) be the least $\rho > \alpha$ such that $L(\rho) \prec L(\omega_1)$. Let $\mathscr{A}_\alpha = \mathscr{P}(\alpha) \cap L(q(\alpha))$. By Lemma 5.1, $q(\alpha) < \omega_1$, so $\mathscr{A}_\alpha$ is countable. We shall show that $\langle \mathscr{A}_\alpha : \alpha < \omega_1 \rangle$ is a $\diamondsuit^+$-sequence.

We begin with some remarks on the Löwenheim–Skolem method applied to $L(\omega_2)$. Use the order $<_L$ on $L(\omega_2)$ to define Skolem functions K_{nm}: $L(\omega_2)^n \to L(\omega_2)$. Thus, $K_{nm}(s) = 0$ unless m is of the form $2^i \cdot 3^j \cdot 5^4$ and $s \in \mathrm{En}(m, L(\omega_2), n)$, in which case $K_{nm}(s)$ is the $<_L$-first $x \in L(\omega_2)$ such that $s^\frown\langle x \rangle \in \mathrm{En}(i, L(\omega_2), n+1)$.

Fix an $X \subset L(\omega_2)$, and let $Y = \mathrm{cl}(X)$, the closure of X under the K_{nm}. Then $Y \prec L(\omega_2)$, so in particular all axioms of ZF − P + V = L hold relativized to Y. Since one of these axioms is Extensionality, $\in$ is extensional on Y, so the Mostowski collapsing function on Y is an isomorphism.

If $\phi(x)$ is "x is the first uncountable ordinal", then $(\exists x\, \phi(x))^{L(\omega_2)}$ so $(\exists x\, \phi(x))^Y$. If $\gamma \in Y$ satisfies $\phi(\gamma)^Y$, then, since $Y \prec L(\omega_2)$, $\phi(\gamma)^{L(\omega_2)}$, so $\gamma = \omega_1$; thus, $\omega_1 \in Y$. Likewise, $0 \in Y$, $\omega \in Y$, and Y is closed under the ordinal successor function S, so $\omega + \omega \subset Y$. Furthermore, if $\zeta \in Y$ and $\zeta < \omega_1$, then (ζ is countable)^Y, so there is an $f \in Y$ with $(f''\omega = \zeta)^Y$, whence $(f''\omega = \zeta)^{L(\omega_2)}$, so $f''\omega = \zeta$. Since $\omega \subset Y$, we have each $f(n) \in Y$, so $\zeta \subset Y$. Thus, either $\omega_1 \subset Y$ or $Y \cap \omega_1 \in \omega_1$.

From now on, assume X is countable. Then Y is countable; so let $\alpha = Y \cap \omega_1$, the first countable ordinal not in Y. Let G be the Mostowski isomorphism from Y onto some transitive set M. Since all axioms of ZF − P + V = L are true in M, $M = L(\delta)$ for some δ. M is countable, so $\delta < \omega_1$. Since G is an isomorphism on the ordinals and $Y \cap (\omega_1 + 1) = \alpha \cup \{\omega_1\}$, $G(\zeta) = \zeta$ for all $\zeta < \alpha$ and $G(\omega_1) = \alpha$, so $\alpha < \delta < \omega_1$. Since $\phi(\omega_1)^Y$, we have $\phi(\alpha)^M$, so (α is uncountable)^M, whereas (α is countable)$^{L(\omega_1)}$, so (α is countable)$^{L(q(\alpha))}$. Thus $\alpha < \delta < q(\alpha) < \omega_1$, since a map from ω onto α gets constructed by stage $q(\alpha)$ but not by stage α. Next, suppose $A \in Y$ and $A \subset \omega_1$. By definition of G (III 5.9), $G(A) = \{G(x) : x \in Y \wedge x \in A\}$, so $G(A) = \{G(\zeta) : \zeta \in A \cap \alpha\} = A \cap \alpha$. Thus, $A \cap \alpha = G(A) \in L(\delta)$, so $A \cap \alpha \in \mathscr{A}_\alpha$.

We now return to showing that $\langle \mathscr{A}_\alpha : \alpha < \omega_1 \rangle$ is a $\diamondsuit^+$-sequence. Fix $A \subset \omega_1$. Let

$$C = \{\alpha < \omega_1 : \alpha = \mathrm{cl}(\alpha \cup \{A\}) \cap \omega_1\}.$$

C is easily seen to be closed in ω_1, and C is unbounded since for any $\beta < \omega_1$, $\mathrm{cl}(\beta \cup \{A\}) \cap \omega_1$ is an ordinal $\geq \beta$ in C. Now fix $\alpha \in C$; we shall show that $A \cap \alpha \in \mathscr{A}_\alpha$ and $C \cap \alpha \in \mathscr{A}_\alpha$. We apply the above discussion with $Y =$

$\mathrm{cl}(\alpha \cup \{A\})$. Since $A \in Y$, we have already shown that $A \cap \alpha \in \mathscr{A}_\alpha$. However, $C \notin Y$, and we cannot show that $C \cap \alpha \in L(\delta)$ (see Exercise 11), but we are only asserting that $C \cap \alpha \in L(q(\alpha))$.

There are two ways one may attempt to define analogues of the Skolem functions K_{nm} on $L(\delta)$. Let $K^1_{nm}\colon L(\delta)^n \to L(\delta)$ be the function induced via the isomorphism G—namely

$$K^1_{nm}(G(y_1), \ldots, G(y_n)) = G(K_{nm}(y_1, \ldots, y_n)).$$

Let cl^1 denote closure with respect to the K^1_{nm}; then since $G(A) = A \cap \alpha$,

$$C \cap \alpha = \{\beta < \alpha\colon \beta = \mathrm{cl}^1(\beta \cup \{A \cap \alpha\}) \cap \alpha\}.$$

Define $K^2_{nm} : L(\delta)^n \to L(\delta)$ using $\mathrm{En}(m, L(\delta), n)$ directly. Thus $K^2_{nm}(s) = 0$ unless m is of the form $2^i \cdot 3^j \cdot 5^4$ and $s \in \mathrm{En}(m, L(\delta), n)$, in which case $K^2_{nm}(s)$ is the $<_L$-first $x \in L(\delta)$ such that $s^\frown\langle x\rangle \in \mathrm{En}(i, L(\delta), n+1)$. Let cl^2 denote closure with respect to the K^2_{nm} and let

$$C^2 = \{\beta < \alpha\colon \beta = \mathrm{cl}^2(\beta \cup \{A \cap \alpha\}) \cap \alpha\}.$$

$L(\delta) \in L(q(\alpha))$ and $L(q(\alpha)) \prec L(\omega_1)$ is a transitive model for ZF $-$ P $+$ **V** $=$ **L**, so absoluteness of the various concepts involved implies that $C^2 \in L(q(\alpha))$. We shall thus be done if we can show that the K^1_{nm} and K^2_{nm} are the same, so that $C \cap \alpha = C^2$. Equivalently, we must show that for $y_1, \ldots, y_n \in Y$,

$$K^2_{nm}(G(y_1), \ldots, G(y_n)) = G(K_{nm}(y_1, \ldots, y_n)).$$

Assume that $n = 2^i \cdot 3^j \cdot 5^4$ and that $s = \langle y_1, \ldots, y_n\rangle \in \mathrm{En}(m, L(\omega_2), n)$, since the discussion if this does not hold is similar (but easier). Then $s \in \mathrm{En}(m, Y, n)$ (since $Y \prec L(\omega_2)$), and $K_{nm}(s)$ is the $<_L$-first $x \in Y$ such that $s^\frown\langle x\rangle \in \mathrm{En}(i, Y, n+1)$. Also, by V 1.13,

$$\langle G(y_1), \ldots, G(y_n)\rangle \in \mathrm{En}(m, L(\delta), n),$$

and $K^2_{nm}(G(y_1), \ldots, G(y_n))$ is the $<_L$-first element $G(x) \in L(\delta)$ such that $s^\frown\langle x\rangle \in \mathrm{En}(i, Y, n+1)$ (applying V 1.13 to $\mathrm{En}(i, Y, n+1)$). That this is the same as $G(K_{nm}(s))$ follows from the fact that for $x, z \in Y$, $x <_L z$ iff $G(x) <_L G(z)$, since $x <_L z$ iff $(x <_L z)^{L(\omega_2)}$ iff $(x <_L z)^Y$ iff $(G(x) <_L G(z))^{L(\delta)}$ iff $G(x) <_L G(z)$; the first and last "iff" used absoluteness, the second "iff" used $Y \prec L(\omega_2)$, and the third "iff" used IV 7.9. □

There is also a direct proof of the weaker axiom, $\diamondsuit$, in **L**, which is of interest because it generalizes to all regular cardinals $\geq \omega_1$, whereas $\diamondsuit^+$ need not hold on such cardinals; see Exercises 13–17.

EXERCISES

(1) Show, without using AC, that $|L(\alpha)| = |\alpha|$ for all $\alpha \geq \omega$. *Hint.* Consider the proofs of 4.10 and 4.12.

(2) Show, using AC, that for $\alpha > \omega$, $|L(\alpha)| = |R(\alpha)|$ iff $\alpha = \beth_\alpha$.

(3) Show that if $\mathbf{V} = \mathbf{L}$, then for $\alpha > \omega$, $L(\alpha) = R(\alpha)$ iff $\alpha = \beth_\alpha$.

(4) Show that if $\mathbf{V} = \mathbf{L}$, then $L(\kappa) = H(\kappa)$ for all infinite cardinals κ.

(5) Assuming $x, y \in \mathbf{L}$, compute explicitly ρ of $\bigcup x$, $\{x\}$, $x \times y$, $x \cup y$, $\{x, y\}$, and $\langle x, y\rangle$ in terms of $\rho(x)$ and $\rho(y)$. Show $\mathbb{Z}$, $\mathbb{Q} \in L(\omega + \omega)$. Assuming $\mathbb{R}$ to be defined by Dedekind cuts (as in I § 11), show $\mathbb{R}^{\mathbf{L}} = \mathbb{R} \cap \mathbf{L}$ and $\rho(\mathbb{R}^{\mathbf{L}}) = \omega_1^{\mathbf{L}}$.

(6) For any set A, let $L(0, A) = \{A\} \cup \mathrm{tr\,cl}(A)$, $L(\alpha + 1, A) = \mathscr{D}(L(\alpha, A))$, and $L(\alpha, A) = \bigcup_{\xi<\alpha} L(\xi, A)$ when α is a limit. $\mathbf{L}(A) = \bigcup\{L(\alpha, A) \colon \alpha \in \mathbf{ON}\}$. Show that $\mathbf{L}(A)$ is a transitive model for ZF, and that AC holds in $\mathbf{L}(A)$ iff $\mathrm{tr\,cl}(A)$ has a well-ordering in $\mathbf{L}(A)$. Furthermore, show that $\mathbf{L}(A)$ is the least transitive proper class model for ZF containing the element A. *Remark.* $\mathbf{L}(\mathscr{P}(\omega))$ need not satisfy AC (see VII Exercise E4).

(7) Assume $\exists S \subset \omega_1\, (\mathbf{V} = \mathbf{L}(S))$. Show that AC and GCH hold. *Hint.* For CH, show that if $A \in \mathscr{P}(\omega)$, then

$$\exists \beta, \gamma < \omega_1\, (A \in L(\beta, S \cap \gamma)).$$

(8) Assume $\exists S \subset \omega_1\, (\mathbf{V} = \mathbf{L}(S))$. Prove $\diamondsuit^+$. *Hint.* In the notation of the proof of Theorem 5.2, $q(\alpha)$ is now the least $\rho > \alpha$ such that $L(\rho, S \cap \alpha) \prec L(\omega_1 S \cap \alpha)$. The Löwenheim–Skolem method is applied in $L(\omega_2, S)$. To prove $\delta < q(\alpha)$: *Case I.* For some $\sigma < \omega_1$, $\omega_1 = \omega_1^{\mathbf{L}(S \cap \sigma)}$, then the argument in Theorem 5.2 works. *Case II.* For all α, ω_1 is inaccessible in $\mathbf{L}(S \cap \alpha)$; then $L(q(\alpha), S \cap \alpha)$ contains cardinals of $\mathbf{L}(S \cap \alpha)$ above α, whereas $L(\delta, S \cap \alpha)$ does not.

(9) Show that the following are equivalent:

(i) $\diamondsuit^+$.

(ii) There is a Kurepa tree $T \subset 2^{<\omega_1}$, such that

$$\forall f \in 2^{\omega_1}\, \exists g \in 2^{\omega_1}\, (f \in \mathbf{L}(g) \wedge g \text{ is a path through } T).$$

(iii) There is an ω_1-tree $T \subset 2^{<\omega_1}$ and an $h \in 2^{\omega_1}$, such that

$$\forall f \in 2^{\omega_1}\ \exists g \in 2^{\omega_1}\ (f \in \mathbf{L}(g, h) \wedge g \text{ is a path through } T).$$

Here, $\mathbf{L}(g, h)$ means $\mathbf{L}(\{g, h\})$. *Hint.* (iii) $\to$ (i) is like Exercise 8. Now, for $g \in T \cap 2^{\alpha}$, $q(\alpha, g)$ is the least $\rho > \alpha$ such that

$$L(\rho, \{g, h \restriction \alpha, T \cap 2^{<\alpha}\}) \prec L(\omega_1, \{g, h \restriction \alpha, T \cap 2^{<\alpha}\}),$$

and $q(\alpha) = \sup\{q(\alpha, g)\colon g \in T \cap 2^{\alpha}\}$. The Löwenheim–Skolem method is applied in $L(\omega_2, h)$; note that we may assume that $T \in \mathbf{L}(h)$.

(10) Show that $\{\rho\colon L(\rho) \prec L(\omega_1)\}$ is c.u.b. in ω_1. Note that this is not quite proved in Lemma 5.1.

(11) In the proof of $\diamondsuit^+$ in Theorem 5.2, show that $C \notin \mathrm{cl}(\omega_1 \cup \{A\})$ and that if α is a limit point of C then $C \cap \alpha \notin L(\delta)$. *Hint.* If $D \subset \omega_1$ is c.u.b. and $D \in \mathrm{cl}(\omega_1 \cup \{A\})$, then $D \cap \beta$ is unbounded in β for every $\beta \in C$.

(12) For any set P, show that there is a least transitive proper class model $\mathbf{M}$ for ZF such that $P \cap \mathbf{M} \in \mathbf{M}$. This model is called $\mathbf{L}^P$. Show that $\mathbf{L}^P$ always satisfies AC. *Hint.* $L^P(0) = 0$. $L^P(\alpha + 1) = \mathscr{D}^P(L^P(\alpha))$, where $\mathscr{D}^P(A)$ is the set of subsets of A first-order definable in the structure $(A; \in, P \cap A)$ from a finite number of elements of A. *Remark.* If $P = \mathscr{P}(\omega)$, $\mathbf{L}^P = \mathbf{L}$, whereas $L(P)$ need not equal $\mathbf{L}$.

(13) Assume $\mathbf{V} = \mathbf{L}$ and prove $\diamondsuit$ directly. *Hint.* Let $\langle A_\alpha, C_\alpha \rangle$ be the $<_L$-first pair of subsets of α such that C_α is c.u.b. in α and $\neg \exists \xi \in C_\alpha (A_\alpha \cap \xi = A_\xi)$. $\langle A_\alpha, C_\alpha \rangle = \langle 0, 0 \rangle$ if there is no such pair. Then $\langle A_\alpha\colon \alpha < \omega_1 \rangle$ is a $\diamondsuit$-sequence.

(14) Assume $\mathbf{V} = \mathbf{L}$ and prove that $\diamondsuit(\kappa, E)$ holds for all regular $\kappa > \omega$ and stationary $E \subset \kappa$, where $\diamondsuit(\kappa, E)$ says that there are $A_\alpha \subset \alpha$ for $\alpha \in E$ such that for all $A \subset \kappa$, $\{\alpha\colon A \cap \alpha = A_\alpha\}$ is stationary.

(15) Let $\kappa > \omega$ be regular and $E \subset \kappa$ be stationary. $\diamondsuit^+(\kappa, E)$ says that there are $\mathscr{A}_\alpha \subset \mathscr{P}(\alpha)$ for $\alpha \in E$ such that $|\mathscr{A}_\alpha| \leq \alpha$ and for all $A \subset \kappa$, there is a c.u.b. $C \subset \kappa$ such that $\forall \alpha \in E \cap C (A \cap \alpha \in \mathscr{A}_\alpha \wedge C \cap \alpha \in \mathscr{A}_\alpha)$. Assume $\mathbf{V} = \mathbf{L}$, and let $E = \{\alpha < \kappa\colon \mathrm{cf}(\alpha) = \omega\}$. Show $\diamondsuit^+(\kappa, E)$. *Remark.* $\diamondsuit^+(\kappa) \to \diamondsuit^+(\kappa, E)$, whereas $\diamondsuit(\kappa, E) \to \diamondsuit(\kappa)$, where $\diamondsuit^+(\kappa)$ is $\diamondsuit^+(\kappa, \kappa)$ and $\diamondsuit(\kappa)$ is $\diamondsuit(\kappa, \kappa)$.

(16) $\diamondsuit^*(\kappa, E)$ is the weakening of $\diamondsuit^+(\kappa, E)$ obtaining by removing the condition, $C \cap \alpha \in \mathscr{A}_\alpha$. Show, in ZFC, that $\diamondsuit^*(\kappa, E) \to \diamondsuit(\kappa, E)$. *Hint.* See II 7.14 and II Exercise 53.

(17) κ is *ineffable* iff $\kappa > \omega$, κ is regular, and whenever $A_\alpha \subset \alpha$ for $\alpha < \kappa$, there is a stationary $S \subset \kappa$ such that

$$\forall \alpha, \beta \in S\,(\alpha < \beta \rightarrow A_\alpha = A_\beta \cap \alpha).$$

Show in ZFC, that if κ is ineffable, then κ is strongly inaccessible, $\diamondsuit(\kappa)$, and $\neg\diamondsuit^*(\kappa)$. *Remark.* See [Baumgartner 1973] for more on such cardinals. If κ is measurable, then κ is ineffable in $\mathbf{V}$ and in $\mathbf{L}$.

(18) Suppose Df* and En* are functions for which V 1.3, V 1.5, and V 1.7 hold. Use these to define $\mathscr{D}^*$ and $\mathbf{L}^*$. Show $\mathbf{L}^* = \mathbf{L}$. *Hint.* Show $\mathbf{L} \subset \mathbf{L}^*$ and $\mathbf{L}^* \subset \mathbf{L}$.

(19) Show that one can define, in ZF, Df* and En* satisfying V 1.3, V 1.5, V 1.7 such that 1.3(c) of this chapter,

$$\forall X \subset A\,(|X| < \omega \rightarrow X \in \mathscr{D}^*(A))$$

is not provable in ZFC (assuming Con(ZF)). *Hint.* Define $\mathrm{Df}^*(A, n) = \bigcup_{k<\alpha} \mathrm{Df}'(k, A, n)$, where $\alpha = \omega$ if CON(⌜ZF⌝), and the least Gödel number of a contradiction if ¬CON(⌜ZF⌝). *Remark.* Thus, in showing $\mathbf{L} \subset \mathbf{L}^*$ in Exercise 18, one must verify that one can develop basic properties of $\mathbf{L}^*$ without using 1.3(c).

(20) What is wrong with the following "proof" of 1.3(c)? Let
$X = \{a_0, \ldots, a_{n-1}\}$. Then, by 1.2,
$X = \{x \in A\colon \phi^A(a_0, \ldots, a_{n-1}x)\} \in \mathscr{D}(A)$, where ϕ is
$x = a_0 \vee \ldots \vee x = a_{n-1}$.

(21) Consider logic to be formalized within ZF (as in IV §10), and let SM be the sentence.

$$\exists M(M \text{ is transitive} \wedge M \models \ulcorner \mathrm{ZF} \urcorner).$$

Show that one can prove, in ZF + SM, that there is a δ such that

$$(L(\delta) \models \ulcorner \mathrm{ZF} \urcorner) \wedge \forall M\,((M \text{ transitive} \wedge M \models \ulcorner \mathrm{ZF} \urcorner) \rightarrow L(\delta) \subset M).$$

Further, show that this $L(\delta)$ is a model for ZFC + $\mathbf{V} = \mathbf{L}$ + $\neg SM$. Use this to obtain a proof of

$$\mathrm{Con(ZF)} \rightarrow \mathrm{Con}(\mathrm{ZF} + \neg SM). \qquad (*)$$

Remarks. $L(\delta)$ is called the *minimal model.* (*) also follows from the Gödel Incompleteness Theorem (I 14.3).

*(22) Second-order logic allows quantification over all subsets of a structure. Define $\mathscr{D}^2(A)$ by using second-order definability over A, and define

$\mathbf{L}^2$ using $\mathscr{D}^2$ rather than $\mathscr{D}$. Show $\mathbf{L}^2 = \mathbf{HOD}$. Likewise, define $\mathbf{L}^n$ using n-th order logic, and show $\mathbf{L}^n = \mathbf{HOD}$ if $n \geq 2$.

*(23) Let

$$\mathscr{D}^-(A) = \{X \subset A\colon \exists R \in \mathrm{Df}(A, 1)(X = \{x \in A : \langle x \rangle \in R\})\},$$

so here we do not allow elements of A as parameters. Define $\mathbf{L}^-$ by replacing $\mathscr{D}$ by $\mathscr{D}^-$ in the definition of $\mathbf{L}$. Show $\mathbf{L}^- = \mathbf{L}$. *Hint*. By induction on α, show $\exists \beta\, (\langle L(\xi) : \xi < \alpha \rangle \in L^-(\beta))$.

CHAPTER VII

FORCING

The method of constructibility discussed in VI produced one model—**L**—and hence established only the consistency of statements true in **L**, such as GCH or $\diamondsuit^+$. Forcing, on the other hand, is a general technique for producing a wide variety of models satisfying diverse mathematical properties.

§1. General remarks

There are two obstacles to understanding forcing—one mathematical and the other metamathematical.

The mathematical difficulty is that one must become proficient in handling partial orders, dense sets, and filters. The reader who is familiar with Martin's Axiom (see II §2) has already come a long way towards mastering this difficulty, although it will now become necessary to consider the relativization of these concepts to various models of set theory.

The metamathematical difficulty is that to prove the consistency of ZF + $\mathbf{V} \neq \mathbf{L}$ (or of any stronger theory, such as ZFC + $\neg$CH), we cannot, as we are used to, simply work in ZF or ZFC and define a transitive model for the desired axioms.

To appreciate the difficulty, suppose we were able to work within ZFC, define a transitive proper class **N**, and prove that each axiom of ZF + $\mathbf{V} \neq \mathbf{L}$ is true **N**. Then, by minimality of **L**, we would have $\mathbf{L} \subset \mathbf{N}$; but also $\mathbf{L} \neq \mathbf{N}$, since $\mathbf{V} = \mathbf{L}$ is true in **L** and false in **N**. Thus, arguing in ZFC, we could prove that there is a proper extension of **L**, so ZFC $\vdash \mathbf{V} \neq \mathbf{L}$, which is impossible (assuming Con(ZFC)), since ZFC + $\mathbf{V} = \mathbf{L}$ is consistent (by VI 3.4).

The naive way to sidestep this difficulty is simply to produce a transitive *set* model N for ZF + $\mathbf{V} \neq \mathbf{L}$. The above argument applied to N, using minimality of **L** for set models (VI 3.8) yields only $L(\mathrm{o}(N)) \subset N$ and $L(\mathrm{o}(N)) \neq N$, but that does not contradict $\mathbf{V} = \mathbf{L}$; if $x \in N \setminus L(\mathrm{o}(N))$, then x can still be in **L**, in which case $\rho(x) > \mathrm{o}(N)$.

Naively still, our general procedure will be as follows. We start with any countable transitive model M for ZFC. M is called the *ground model*. We shall describe a general procedure for finding countable transitive models N for ZFC such that $M \subset N$ and $o(M) = o(N)$. An N obtained by our procedure will be called a *generic extension* of M. As long as we succeed in making $M \neq N$, N will satisfy $\mathbf{V} \neq \mathbf{L}$, since, by VI 3.8

$$\mathbf{L}^N = L(o(N)) = L(o(M)) = \mathbf{L}^M \subset M.$$

However, we shall in fact be able to make N satisfy $\neg$CH, or CH + $2^{\omega_1} = \omega_5$, or a wide variety of other statements by varying certain details in our construction.

Unfortunately, this naive approach is not quite correct. By results related to the Gödel Incompleteness Theorem, one cannot argue within ZFC and produce any set models at all for ZFC; see IV §§7 and 10, as well as §9 of this chapter for a further discussion of this point. Fortunately, however, we can produce in ZFC countable transitive models M for any desired finite list of axioms of ZFC, or even of ZFC + $\mathbf{V} = \mathbf{L}$ (see IV 7.11 and VI 4.10).

The method of forcing will then be used to show how to produce models N for any given finite list of axioms of ZFC + $\mathbf{V} \neq \mathbf{L}$ (or ZFC + $\neg$CH, etc.); such N will be generic extensions of models M for suitably many axioms of ZFC.

The formal structure of our proof of

$$\mathrm{Con(ZFC)} \rightarrow \mathrm{Con(ZFC} + \mathbf{V} \neq \mathbf{L})$$

will be as follows. Assume we can derive a contradiction from ZFC + $\mathbf{V} \neq \mathbf{L}$. Then there is a finite list of axioms, $\phi_1, \ldots, \phi_n$ of ZFC + $\mathbf{V} \neq \mathbf{L}$ such that

$$\phi_1 \cdots \phi_n \vdash \psi \wedge \neg\psi$$

for some (or any) ψ. But, by the method of forcing, we shall show that

$$\mathrm{ZFC} \vdash \exists N\,(\phi_1^N \wedge \cdots \wedge \phi_n^N),$$

so

$$\mathrm{ZFC} \vdash \exists N\,(\psi^N \wedge \neg\psi^N),$$

whence ZFC is inconsistent. This method in fact produces a completely finitistic relative consistency proof, since we define explicitly how to construct an inconsistency in ZFC given one in ZFC + $\mathbf{V} \neq \mathbf{L}$.

The advantage of this approach is that when studying forcing we may temporarily forget the metamathematical niceties in the previous paragraph, and just assume naively that we have a countable transitive M satisfying all of ZFC (or even ZFC + $\mathbf{V} = \mathbf{L}$). We may then concentrate on the mathematical problems involved in constructing the generic extension N of M, satisfying, e.g., ZFC + $\neg$CH. Once the construction is understood, we may check that the previous paragraph applies to yield a finistic proof of

$$\mathrm{Con(ZFC)} \rightarrow (\mathrm{ZFC} + \neg\mathrm{CH}).$$

In §9 we shall return to the metamathematics of forcing, and shall discuss several other approaches for doing the same thing. In the meantime, when we say "let M be a countable transitive model for ZFC", the reader may consider this to be an abbreviation for "let M be a countable transitive model for enough axioms of ZFC to carry out the argument at hand."

§2. Generic extensions

Let M be a countable transitive model for ZFC. If $\langle \mathbb{P}, \leq \rangle$ is a partial order (in the sense of our discussion of MA in II §2) and $\langle \mathbb{P}, \leq \rangle \in M$, then $\langle \mathbb{P}, \leq \rangle$ will yield a method of obtaining a generic extension, N, of M, which is also a model of ZFC. By varying $\langle \mathbb{P}, \leq \rangle$, we shall be able to produce a wide variety of relative consistency results.

For technical reasons, it will be convenient to restrict our attention to partial orders with a largest element. One can do forcing without this restriction (Exercise B1), but it is slightly more cumbersome, and it produces no more consistency results (Exercise B2). Most partial orders occurring "naturally", such as in applications of MA in II §2, have a largest element anyway. To avoid excess verbiage, we define the following.

2.1. Definition. A p.o. is a triple, $\langle \mathbb{P}, \leq, \mathbb{1} \rangle$ such that $\leq$ partially orders $\mathbb{P}$ and $\mathbb{1}$ is a largest element of $\mathbb{P}$ (i.e., $\forall p \in \mathbb{P}\,(p \leq \mathbb{1})$). c.t.m. abbreviates "countable transitive model." □

Following standard abuses of notation, we shall often write $\mathbb{P}$ when we mean $\langle \mathbb{P}, \leq, \mathbb{1} \rangle$. Thus, $\mathbb{P} \in M$ means $\mathbb{P} \in M$, $\leq \in M$, and $\mathbb{1} \in M$ (although $\mathbb{1} \in M$ follows from $\mathbb{P} \in M$ by transitivity of M). If two p.o.'s are under discussion, we may refer to them as, e.g., $\langle \mathbb{P}, \leq_{\mathbb{P}}, \mathbb{1}_{\mathbb{P}} \rangle$ and $\langle \mathbb{Q}, \leq_{\mathbb{Q}}, \mathbb{1}_{\mathbb{Q}} \rangle$. Formally, of course, the set $\mathbb{P}$ does not determine its ordering, $\leq_{\mathbb{P}}$; and $\langle \mathbb{P}, \leq_{\mathbb{P}} \rangle$ may not determine $\mathbb{1}_{\mathbb{P}}$, since the fact that we do not require $\leq_{\mathbb{P}}$ to be a partial order in the strict sense means that there could be many largest elements (see II 2.1 and following discussion).

2.2. Definition. Let $\mathbb{P}$ be a p.o. G is $\mathbb{P}$-*generic* (i.e., $\langle \mathbb{P}, \leq, \mathbb{1} \rangle$-generic) over M iff G is a filter on $\mathbb{P}$ and for all dense $D \subset \mathbb{P}$, $D \in M \rightarrow G \cap D \neq 0$. □

2.3. Lemma. *If M is countable and $p \in \mathbb{P}$, then there is a G which is $\mathbb{P}$-generic over M such that $p \in G$.*

PROOF. Exactly like the proof of MA(ω) (see II 2.6(c)). Let $D_n (n \in \omega)$ enumerate all dense subsets of $\mathbb{P}$ which are in M. Inductively choose a sequence $q_n (n \in \omega)$ so that

$$p = q_0 \geq q_1 \geq \cdots$$

and $q_{n+1} \in D_n$. Let G be the filter generated by $\{q_n : n \in \omega\}$. □

It is important to keep track of what is absolute for M and what is not. In our intended applications, M will be a c.t.m. for ZFC and $\langle \mathbb{P}, \leq, \mathbb{1} \rangle$ will be in M. It is then easily seen by the methods of IV §3 that notions like "p.o." or "dense" are absolute for M. However, the enumeration of the D_n takes place outside of M. By absoluteness,

$$\{D \in M : D \text{ is dense in } \mathbb{P}\} = \{D : D \text{ is dense in } \mathbb{P}\}^M,$$

but this set will not usually be countable in M (countable is *not* absolute).

2.2 and 2.3 did not require that M is a model for anything. But it will become important as we go along that M satisfy at least some of ZFC, to ensure that various dense sets we construct actually lie in M. This occurs, for example, in the proof of the next lemma, which says that in most cases $G \notin M$.

2.4. LEMMA. *If M is a transitive model of* ZF − P, $\mathbb{P} \in M$ *is a* p.o. *such that*

$$\forall p \in \mathbb{P}\, \exists q, r \in \mathbb{P}\, (q \leq p \wedge r \leq p \wedge q \perp r), \tag{1}$$

and G is $\mathbb{P}$-generic over M, then $G \notin M$.

PROOF. If $G \in M$, then $D = \mathbb{P} \setminus G \in M$, since set-theoretic difference is absolute. Also, D is dense: if $p \in \mathbb{P}$ and q, r are as in (1), then q, r cannot both be in G (since G is a filter); thus, p has an extension in D.

However, $G \cap D = 0$, contradicting the definition of generic. □

The proof of Lemma 2.4 only required M to satisfy a very weak fragment of ZF − P, but there is no reason to try to keep track of precisely which finite set of axioms of ZFC are required for M at each step.

We remark that if condition (1) fails for $\mathbb{P}$, then there is a filter G on $\mathbb{P}$ which intersects *all* dense subsets of $\mathbb{P}$, and if $\mathbb{P} \in M$, then $G \in M$ (see Exercise A1). Any application of MA or forcing to such a $\mathbb{P}$ will be trivial. Thus, almost all partial orders considered in II, VII, or VIII satisfy (1), although (1) is never needed in the abstract treatment of MA or forcing.

We may now unveil slightly more about generic extensions. Let M be a c.t.m. for ZFC, with $\mathbb{P}$ a p.o. in M and G $\mathbb{P}$-generic over M. We shall show how to construct another c.t.m. for ZFC, called $M[G]$, which will satisfy

$M \subset M[G]$, $o(M) = o(M[G])$, and $G \in M[G]$. $M[G]$ will be the least extension of M to a c.t.m. for ZFC containing G. The fact that $G \in M[G]$ will imply, by Lemma 2.4, that in most cases $M \neq M[G]$.

The particular axioms of set theory that $M[G]$ satisfies beyond ZFC will be very sensitive to the combinatorial properties satisfied by $\mathbb{P}$ *in* M; most of these properties are *not* absolute. For example, consider the c.c.c. (Def. II 2.3). If M is a c.t.m. and $\mathbb{P} \in M$, then in $\mathbf{V}$, $\mathbb{P}$ is countable and thus trivially has the c.c.c. But $\mathbb{P}$ may well fail to have the c.c.c. in M.

Working within M, one may construct the various c.c.c. p.o.'s considered in II, plus many more which are not c.c.c. (in M). These all can be used for generic extensions.

We now return to the basic theory, which works equally well with any p.o. in M. The whole procedure of constructing $M[G]$ may seem rather complicated at first, but once over this hurdle, the techniques of cooking up a $\mathbb{P}$ to produce a desired consistency result will be reduced to (sometimes very difficult) problems in the combinatorics of partial orders.

The first step is to define $M[G]$. Roughly, this will be the set of all sets which can be constructed from G by applying set-theoretic processes definable in M. Each element of $M[G]$ will have a *name* in M, which tells how it has been constructed from G. We use letters τ, σ, π to range over names.

People living within M will be able to comprehend a name, τ, for an object in $M[G]$, but they will not in general be able to decide the object, τ_G, that τ names, since this will require a knowledge of G.

2.5. Definition. τ is a $\mathbb{P}$-name iff τ is a relation and

$$\forall \langle \sigma, p \rangle \in \tau\,[\sigma \text{ is a } \mathbb{P}\text{-name} \wedge p \in \mathbb{P}]. \quad \square$$

This definition does not mention models or any order on $\mathbb{P}$. The collection of $\mathbb{P}$-names will be a proper class if $\mathbb{P} \neq 0$.

Definition 2.5 must be understood as a definition by transfinite recursion. Formally, one defines the characteristic function of the $\mathbb{P}$-names, $\mathbf{H}(\mathbb{P}, \tau)$, by

$$\mathbf{H}(\mathbb{P}, \tau) = 1 \text{ iff } \tau \text{ is a relation} \wedge \forall \langle \sigma, p \rangle \in \tau\,[\mathbf{H}(\mathbb{P}, \sigma) = 1 \wedge p \in \mathbb{P}].$$

$$\mathbf{H}(\mathbb{P}, \tau) = 0 \text{ otherwise.}$$

Then, τ is defined to be a $\mathbb{P}$-name iff $\mathbf{H}(\mathbb{P}, \tau) = 1$. For a fixed $\mathbb{P}$, $\mathbf{H}(\mathbb{P}, \tau)$ is defined from $\mathbf{H} \restriction \text{tr cl}(\tau)$ using concepts absolute for transitive models of ZF $-$ P, so $\mathbf{H}$ is absolute for transitive models of ZF $-$ P. (We are using IV 5.6, where $x\,\mathbf{R}\,y$ iff $x \in \text{tr cl}(y)$). Thus also, the concept "τ is a $\mathbb{P}$-name" is absolute for transitive models of ZF $-$ P. For more practice in such recursions, see III, Exercises 13 and 14.

2.6. DEFINITION. $\mathbf{V}^{\mathbb{P}}$ is the class of $\mathbb{P}$-names. If M is a transitive model of ZFC and $\mathbb{P} \in M$, $M^{\mathbb{P}} = \mathbf{V}^{\mathbb{P}} \cap M$. Or, by absoluteness,

$$M^{\mathbb{P}} = \{\tau \in M: (\tau \text{ is a } \mathbb{P}\text{-name})^M\}. \quad \square$$

When forcing over M, use is made only of the $\mathbb{P}$-names in $M^{\mathbb{P}}$, which we may think of as being defined within M.

2.7. DEFINITION. $\text{val}(\tau, G) = \{\text{val}(\sigma, G): \exists p \in G\,(\langle \sigma, p\rangle \in \tau)\}$. We also write τ_G for $\text{val}(\tau, G)$. $\square$

$\text{val}(\tau, G)$ is defined by transfinite recursion on τ, as was "τ is a $\mathbb{P}$-name".

2.8. DEFINITION. If M is a transitive model of ZFC, $\mathbb{P} \in M$, and $G \subset \mathbb{P}$, then

$$M[G] = \{\tau_G: \tau \in M^{\mathbb{P}}\}. \quad \square$$

$\text{dom}(\tau) = \{\sigma: \exists p\,(\langle \sigma, p\rangle \in \tau)\}$, the usual definition of domain (although τ is usually not a function). By absoluteness, the M-people know $\text{dom}(\tau)$, and they may think of $\text{dom}(\tau)$ as a set of names for objects which may possibly be in τ_G.

$\text{val}(\tau, G)$ was defined by transfinite recursion using absolute concepts, and is thus absolute for transitive models of ZF $-$ P for the same reason "τ is a $\mathbb{P}$-name" was. Of course, the absoluteness of $\text{val}(\tau, G)$ says nothing for M unless $G \in M$, which will usually be false. It does yield the following.

2.9. LEMMA. *Under the notation of Definition 2.8, if N is a transitive model of ZFC with $M \subset N$ and $G \in N$, then $M[G] \subset N$.*

PROOF. For each $\tau \in M^{\mathbb{P}}$, $\tau \in N$ and $G \in N$, so $\text{val}(\tau, G) = (\text{val}(\tau, G))^N \in N$. $\square$

Thus, once we check that $M[G]$ is indeed a transitive extension of M containing G and satisfying ZFC, it will be the least such extension.

We pause for some examples in our intended framework where M is a c.t.m. for ZFC and $\mathbb{P}$ is a p.o. in M. 0 is a $\mathbb{P}$-name, since it trivially satisfies Definition 2.5, and $0_G = 0$ for any G by Definition 2.7. If $p \in \mathbb{P}$, then $\{\langle 0, p\rangle\} \in M^{\mathbb{P}}$, and

$$\text{val}(\{\langle 0, p\rangle\}, G) = \begin{cases} \{0\} & \text{if } p \in G, \\ 0 & \text{if } p \notin G. \end{cases}$$

There will always be generic G with $p \in G$ (by Lemma 2.3), and, assuming $\exists q \in \mathbb{P}\,(q \perp p)$, there will be generic G with $p \notin G$. Thus, τ_G can depend on

G. However, in some special cases, τ_G is independent of G. We saw $0_G = 0$, and, $\mathrm{val}(\{\langle 0, \mathbb{1}\rangle\}, G) = \{0\}$ for all generic G since any non-empty filter contains $\mathbb{1}$. More generally,

$$\mathrm{val}(\{\langle \sigma_i, \mathbb{1}\rangle : i \in I\}, G) = \{\mathrm{val}(\sigma_i, G) : i \in I\}.$$

This observation enables us to see that any element $x \in M$ is represented in a canonical way by a name, called $\check{x}$.

2.10. Definition. If $\mathbb{P}$ is a p.o., define the $\mathbb{P}$-name $\check{x}$ recursively by: $\check{x} = \{\langle \check{y}, \mathbb{1}_{\mathbb{P}}\rangle : y \in x\}$. □

Formally, $\check{x}$ depends on $\mathbb{1}_{\mathbb{P}}$ as well as x but the p.o. $\langle \mathbb{P}, \leq_{\mathbb{P}}, \mathbb{1}_{\mathbb{P}}\rangle$ will always be clear from context. Definition 2.10 is another definition by recursion and is easily seen to be absolute for transitive models of ZFC, so if $x \in M$, then $\check{x} \in M$. As examples $\check{0} = 0$, $\check{1} = \{0\}\check{} = \{\langle 0, \mathbb{1}\rangle\}$, $\check{2} = \{\langle \check{0}, \mathbb{1}\rangle, \langle \check{1}, \mathbb{1}\rangle\}$, etc. We just saw that $\mathrm{val}(\check{0}, G) = 0$ and $\mathrm{val}(\check{1}, G) = 1$.

2.11. Lemma. *If M is a transitive model of* ZFC, $\mathbb{P}$ *is a* p.o. *in M, and G is a non-empty filter on* $\mathbb{P}$, *then*

(a) $\forall x \in M\,[\check{x} \in M^{\mathbb{P}} \wedge \mathrm{val}(\check{x}, G) = x]$.

(b) $M \subset M[G]$.

Proof. For (a), absoluteness of $\check{}$ implies $\check{x} \in M^{\mathbb{P}}$. $\mathrm{val}(\check{x}, G) = x$ is proved by induction on x, using

$$\mathrm{val}(\check{x}, G) = \{\mathrm{val}(\check{y}, G) : y \in x\}.$$

(b) is immediate from (a). □

We may now see that $G \in M[G]$ by cooking up a name that represents it.

2.12. Definition. If $\mathbb{P}$ is a p.o., $\Gamma = \{\langle \check{p}, p\rangle : p \in \mathbb{P}\}$. □

Γ of course depends on $\mathbb{P}$, which will always be clear from context. Unlike names of the form $\check{x}$, the object named by Γ depends on G. By absoluteness, Γ is in M if $\mathbb{P}$ is.

2.13. Lemma. *Under the hypotheses of Lemma* 2.11, $\Gamma_G = G$. *Hence,* $G \in M[G]$.

Proof. $\Gamma_G = \{(\check{p})_G : p \in G\} = \{p : p \in G\} = G$. □

Two more easy facts about $M[G]$ are the following.

2.14. LEMMA. *Under the hypotheses of Lemma* 2.11, $M[G]$ *is transitive.*

PROOF. Immediate from Definitions 2.7 and 2.8. □

2.15. LEMMA. *Under the hypotheses of Lemma* 2.11,
(a) $\forall \tau \in M^{\mathbb{P}} (\text{rank}(\tau_G) \leq \text{rank}(\tau))$.
(b) $o(M[G]) = o(M)$.

PROOF. (a) is by induction on τ. For (b), we have $M[G] \cap \mathbf{ON} \subset M \cap \mathbf{ON}$ by (a) and the fact that $\text{rank}(\tau) \in M$ for all $\tau \in M$. Thus, $M[G] \cap \mathbf{ON} = M \cap \mathbf{ON}$ since $M \subset M[G]$. □

As a further example, of building names, we check that $M[G]$ satisfies some of the easier axioms of ZFC. Thus, pairing holds because given $\sigma, \tau \in M^{\mathbb{P}}$, we can define a name $\text{up}(\sigma, \tau) \in M^{\mathbb{P}}$ which always names $\{\sigma_G, \tau_G\}$.

2.16. DEFINITION. (a) $\text{up}(\sigma, \tau) = \{\langle \sigma, \mathbb{1} \rangle, \langle \tau, \mathbb{1} \rangle\}$.
(b) $\text{op}(\sigma, \tau) = \text{up}(\text{up}(\sigma, \sigma), \text{up}(\sigma, \tau))$. □

2.17. LEMMA. *Under the hypotheses of Lemma* 2.11, *if* $\sigma, \tau \in M^{\mathbb{P}}$, *then*
(a) $\text{up}(\sigma, \tau) \in M^{\mathbb{P}}$ *and* $\text{val}(\text{up}(\sigma, \tau), G) = \{\sigma_G, \tau_G\}$.
(b) $\text{op}(\sigma, \tau) \in M^{\mathbb{P}}$ *and* $\text{val}(\text{op}(\sigma, \tau), G) = \langle \sigma_G, \tau_G \rangle$. □

2.18. LEMMA. *Under the hypotheses of Lemma* 2.11, *the Axioms of Extensionality, Foundation, Pairing, and Union are true in* $M[G]$.

PROOF. Extensionality holds because $M[G]$ is transitive, Foundation is true relativized to any class, and Pairing is immediate from Lemma 2.17(a). For Union, it is sufficient to show that if $a \in M[G]$, then there is a $b \in M[G]$ such that $\bigcup a \subset b$ (see IV 2.10). Fix $\tau \in M^{\mathbb{P}}$ such that $a = \tau_G$; let $\pi = \bigcup \text{dom}(\tau)$; then $\pi \in M^{\mathbb{P}}$, so $b = \pi_G \in M[G]$. If c is any element of a, $c = \sigma_G$ for some $\sigma \in \text{dom}(\tau)$. Since $\sigma \subset \pi$, $c = \sigma_G \subset \pi_G = b$ (by applying Definition 2.7 to σ and π). Thus, $\bigcup a \subset b$. □

Observe that in proving Lemma 2.18, we did not show that $\bigcup a \in M[G]$, although this will follow once we show that $M[G]$ satisfies the Comprehension Axiom. For a direct proof, see Exercise A6. Observe also that we have not yet used the notion of generic in a non-trivial way; our last six lemmas are true for any $G \subset \mathbb{P}$, such that $\mathbb{1} \in G$. The fact that G intersects the dense sets of M becomes important in the development of the concept of forcing in §3, which is then used to show $M[G]$ satisfies ZFC in §4.

We conclude this section with some additional technical facts which will be useful later when we wish to show that a generic filter intersects some sets which are not dense.

2.19. Definition. If $E \subset \mathbb{P}$ and $p \in \mathbb{P}$, then E is *dense below* p iff

$$\forall q \leq p \, \exists r \leq q \, (r \in E). \quad \square$$

2.20. Lemma. *Assume that M is a transitive model of* ZFC, $\mathbb{P} \in M$, $E \subset \mathbb{P}$, *and* $E \in M$. *Let G be $\mathbb{P}$-generic over M; then*

(a) *Either $G \cap E \neq 0$, or $\exists q \in G \;\; \forall r \in E \, (r \perp q)$.*

(b) *If $p \in G$ and E is dense below p, then $G \cap E \neq 0$.*

Proof. For (a), let

$$D = \{p\colon \exists r \in E \, (p \leq r)\} \cup \{q\colon \forall r \in E \, (r \perp q)\}.$$

D is dense, since if $q \in \mathbb{P}$, and $q \notin D$, then fix $r \in E$ with r and q compatible; if $p \leq r$ and $p \leq q$, then p is an extension of q in D. Thus, $G \cap D \neq 0$, which implies (a).

For (b), if $G \cap E = 0$, then, by (a), fix $q \in G$ with $\forall r \in E \, (r \perp q)$. Let $q' \in G$ with $q' \leq q$ and $q' \leq p$, and then, since E is dense below p, let $r \in E$ with $r \leq q'$; then $r \leq q$, contradicting $r \perp q$. $\square$

§3. Forcing

Let us consider first a specific example. Fix a c.t.m. M for ZFC, and let $\mathbb{P}$ be the set of finite partial functions from ω to 2 ordered by reverse inclusion (as in II §2, Example 5). $\mathbb{1}_{\mathbb{P}}$ is the empty function. $\langle \mathbb{P}, \leq, \mathbb{1} \rangle \in M$, since its definition is absolute for transitive models of ZFC (or ZF − P).

If G is a filter on $\mathbb{P}$, $f_G = \bigcup G$ is a function with $\mathrm{dom}(f_G) \subset \omega$. For each n, we let, as in II §2, $D_n = \{p \in \mathbb{P}\colon n \in \mathrm{dom}(p)\}$; then D_n is dense, and $D_n \in M$ (again by absoluteness of its definition). Thus, if G is $\mathbb{P}$-generic over M, $G \cap D_n \neq 0$ for all n, so $\mathrm{dom}(f_G) = \omega$.

We now show $f_G \in M[G]$. Since $G \in M[G]$ and $f_G = \bigcup G$, $f_G \in M[G]$ will follow immediately from the absoluteness of $\bigcup$ for transitive models of ZF, once we have shown $M[G]$ satisfies ZFC (or enough of ZF − P to obtain absoluteness of $\bigcup$). However, we may check $f_G \in M[G]$ directly. Let

$$\Phi = \{\langle (\langle n, m \rangle)\check{\ }, p \rangle\colon p \in \mathbb{P} \wedge n \in \mathrm{dom}(p) \wedge p(n) = m\}.$$

Since $\mathrm{val}((\langle n, m \rangle)\check{\ }, G) = \langle n, m \rangle$ (see Lemma 2.11),

$$\Phi_G = \{\langle n, m \rangle\colon \exists p \in G \, (n \in \mathrm{dom}(p) \wedge p(n) = m)\} = f_G.$$

Thus, $f_G \in M[G]$.

If G is $\mathbb{P}$-generic over M, then $G \notin M$ by Lemma 2.4. Also, $f_G \notin M$, for let $E = \{p\colon p \not\subset f\}$; then E is dense and $G \cap E = 0$. If $f_G \in M$, then also $E \in M$, contradicting the definition of generic. Note the similarity of this argument with the one for $\neg\text{MA}(2^\omega)$ (II 2.6).

We now bring in the idea of *forcing*. In II §2, we had the intuitive idea that elements $p \in \mathbb{P}$ were conditions which say something about G or some object (such as f_G above), which we plan to construct from G. We continue with this motivation, but now in the context of models, rather than of MA.

People living in M cannot construct a G which is $\mathbb{P}$-generic over M. They may believe on faith that there exists a being to whom their universe, M, is countable. Such a being will have a generic G and an $f_G = \bigcup G$. The people in M do not know what G and f_G are but they have names for them, Γ and Φ. They may also read the preceding few paragraphs and thus figure out certain properties of G and f_G; for example, f_G is a function from ω to 2. They do not know what $f_G(0)$ is, since that depends on the particular G chosen. But they can see that $f_G(0)$ will be 0 if $\{\langle 0,0\rangle\} \in G$ and 1 if $\{\langle 0,1\rangle\} \in G$. More generally, they can construct a *forcing language*, where a sentence ψ of the forcing language uses the names in $M^{\mathbb{P}}$ to assert something about $M[G]$; an example of such a ψ is $\Phi(\check{0}) = \check{1}$. The person in M may not know whether a given ψ is true in $M[G]$. The truth or falsity of ψ in $M[G]$ will in general depend on G. We write $p \Vdash \psi$ (*p forces* ψ) to mean that for all G which are $\mathbb{P}$-generic over M, if $p \in G$, then ψ is true in $M[G]$. For example,

$$\{\langle 0,0\rangle\} \Vdash \Phi(\check{0}) = \check{0}, \quad \text{and} \quad \{\langle 0,1\rangle\} \Vdash \Phi(\check{0}) = \check{1}.$$

Also,

$$\mathbb{1} \Vdash \Phi \text{ is a function from } \check{\omega} \text{ into } \check{2}, \quad \text{and} \quad \mathbb{1} \Vdash \Phi = \bigcup \Gamma;$$

i.e., these last two sentences are true for all generic G. Now, people living in M can figure out all the above forcing facts without even seeing a generic G. This illustrates the following.

Fact 1. It may be decided within M whether or not $p \Vdash \psi$.

This will be very important not only for proving that $M[G]$ satisfies ZFC, but for applying forcing later, since the people of M will have to be able to apply their combinatorial techniques to construct various complicated $\mathbb{P}$ for which the desired axioms of set theory (beyond ZFC) are forced to be true in $M[G]$.

Fact 1 is at first surprising, since the notion $p \Vdash \psi$ seems to require a knowledge of all generic G, but a person in M may always decide whether $p \Vdash \psi$ by going through the kind of analysis we have used in our examples.

It is immediate from the definition of $\Vdash$ that if G is $\mathbb{P}$-generic over M

and $p \Vdash \psi$ for some $p \in G$, then ψ is true in $M[G]$. As a converse of this observation, we shall show the following.

Fact 2. If G is $\mathbb{P}$-generic over M and ψ is true in $M[G]$, then for some $p \in G$, $p \Vdash \psi$.

For example, if ψ is $\Phi(\check{0}) = \check{0}$ and ψ is true (i.e., $f_G(0) = 0$), then $p(0) = 0$ for some $p \in G$. If this $p \in H$, where H is another generic filter, then $f_H(0) = 0$ also; i.e., ψ will be true in $M[H]$. Thus, $p \Vdash \psi$.

We now leave our specific example and turn to a more rigorous discussion of forcing with an arbitrary $\mathbb{P}$. The actual theorem expressing Facts 1 and 2 (Theorem 3.6) will form the backbone of our forcing technique.

3.1. Definition. Let $\phi(x_1, \ldots, x_n)$ be a formula with all free variables shown; let M be a c.t.m. for ZFC, $\mathbb{P}$ a p.o. in M, $\tau_1, \ldots, \tau_n \in M^{\mathbb{P}}$, and $p \in \mathbb{P}$; then $p \Vdash_{\mathbb{P},M} \phi(\tau_1, \ldots, \tau_n)$ iff

$$\forall G\,[(G \text{ is } \mathbb{P}\text{-generic over } M \ \wedge\ p \in G) \rightarrow \phi^{M[G]}(\text{val}(\tau_1, G), \ldots, \text{val}(\tau_n, G))]. \qquad \Box \quad (1)$$

The subscript $\mathbb{P}$ on $\Vdash_{\mathbb{P},M}$ should really be $\langle \mathbb{P}, \leq, \mathbb{1} \rangle$. We shall usually just write $\Vdash$ when there is only one partial order and one ground model M under discussion.

Intuitively, the $\phi(\tau_1, \ldots, \tau_n)$ in Definition 3.1 is a sentence of the *forcing language*; this idea could be made rigorous by formalizing logic within set theory and defining the forcing language to be the first-order language whose one binary relation symbol is $\in$, and whose constant symbols are the elements of $M^{\mathbb{P}}$. Instead, our approach is not actually to define a forcing language. Formally, 3.1 is a definition schema in the metatheory. For each formula $\phi(x_1, \ldots, x_n)$, with free variables among $x_1, \ldots, x_n$, we can define another formula $\text{Force}_\phi(\tau_1, \ldots, \tau_n, \mathbb{P}, \leq, \mathbb{1}, M, p)$, which asserts (1), along with $\langle \mathbb{P}, \leq, \mathbb{1} \rangle \in M$, $p \in \mathbb{P}$, and $\tau_1, \ldots, \tau_n \in M^{\mathbb{P}}$.

As an exercise in understanding Definition 3.1, one may verify the following.

3.2. Lemma. *In the notation of Definition* 3.1,

(a) $(p \Vdash \phi(\tau_1, \ldots, \tau_n) \wedge q \leq p) \rightarrow q \Vdash \phi(\tau_1, \ldots, \tau_n)$.

(b) $(p \Vdash \phi(\tau_1, \ldots, \tau_n)) \wedge (p \Vdash \psi(\tau_1, \ldots, \tau_n))$ *iff* $p \Vdash (\phi(\tau_1, \ldots, \tau_n) \wedge \psi(\tau_1, \ldots, \tau_n))$. $\Box$

Now, the notion "$p \Vdash \phi(\tau_1, \ldots, \tau_n)$" has been defined in $\mathbf{V}$, not M, and involves a knowledge of all possible generic G. By Fact 1, we should be able to decide within M whether $p \Vdash \phi(\tau_1, \ldots, \tau_n)$; we translate this rigorously by defining another relation, $p \Vdash^* \phi(\tau_1, \ldots, \tau_n)$ and showing that for all ϕ,

$$p \Vdash \phi(\tau_1, \ldots, \tau_n) \leftrightarrow (p \Vdash^* \phi(\tau_1, \ldots, \tau_n))^M.$$

Thus, $p \Vdash \phi(\tau_1, \ldots, \tau_n)$ will be equivalent to some statement relativized to M.

After this section, we shall rarely refer back to the details of the definition of $\Vdash^*$, although we shall frequently use Facts 1 and 2 (Theorem 3.6) and their Corollary, 3.7. Thus, the reader who is bored by these details may simply skip directly to Theorem 3.6. There are as many different (equivalent) definitions of $\Vdash^*$ as there are texts in set theory; see, e.g., Exercises B3 and B4 for a somewhat slicker approach.

The most difficult part of our definition of $\Vdash^*$ will be when $\phi(\tau_1, \tau_2)$ is $\tau_1 = \tau_2$. As a simple example of what to expect, suppose $\tau_1 = \{\langle \pi_1, s \rangle\}$ and $\tau_2 = \{\langle \pi_2, s \rangle\}$, and we are trying to tell a person in M which p force $\tau_1 = \tau_2$. If $p \perp s$, then $p \Vdash \tau_1 = \tau_2$, since whenever $p \in G$, $s \notin G$, so $\tau_{1_G} = 0 = \tau_{2_G}$. If $p \leq s$, then whenever $p \in G$, $\tau_{1_G} = \{\pi_{1_G}\}$ and $\tau_{2_G} = \{\pi_{2_G}\}$, so $p \Vdash \tau_1 = \tau_2$ iff $p \Vdash \pi_1 = \pi_2$. It is instructive to check that for any p, $p \Vdash \tau_1 = \tau_2$ iff

$$\forall q (q \leq p \wedge q \leq s \rightarrow q \Vdash \pi_1 = \pi_2),$$

but we mainly wish to emphasize that in the definition of $\Vdash^*$, the question of whether $p \Vdash^* \tau_1 = \tau_2$ must depend on whether $q \Vdash^* \pi_1 = \pi_2$ for various $q \in \mathbb{P}$, $\pi_1 \in \mathrm{dom}(\tau_1)$, and $\pi_2 \in \mathrm{dom}(\tau_2)$.

We begin by defining, in $\mathbf{V}$, the notion $p \Vdash^* \phi(\tau_1, \ldots, \tau_n)$; so this definition does not mention any model. However, in our intended application, we shall consider only the relativized notion $(p \Vdash^* \phi(\tau_1, \ldots, \tau_n))^M$, where M is the ground model.

3.3. Definition. Fix a p.o. $\mathbb{P}$. The following clauses define the notion $p \Vdash^* \phi(\tau_1, \ldots, \tau_n)$ where $\phi(x_1, \ldots, x_n)$ is a formula with all free variables shown, $p \in \mathbb{P}$, and $\tau_1, \ldots, \tau_n \in \mathbf{V}^{\mathbb{P}}$.

(a) $p \Vdash^* \tau_1 = \tau_2$ iff

(α) for all $\langle \pi_1, s_1 \rangle \in \tau_1$,

$$\{q \leq p \colon q \leq s_1 \rightarrow \exists \langle \pi_2, s_2 \rangle \in \tau_2 (q \leq s_2 \wedge q \Vdash^* \pi_1 = \pi_2)\}$$

is dense below p, and

(β) for all $\langle \pi_2, s_2 \rangle \in \tau_2$,

$$\{q \leq p \colon q \leq s_2 \rightarrow \exists \langle \pi_1, s_1 \rangle \in \tau_1 (q \leq s_1 \wedge q \Vdash^* \pi_1 = \pi_2)\}$$

is dense below p.

(b) $p \Vdash^* \tau_1 \in \tau_2$ iff

$$\{q \colon \exists \langle \pi, s \rangle \in \tau_2 (q \leq s \wedge q \Vdash^* \pi = \tau_1)\}$$

is dense below p.

(c) $p \Vdash^* (\phi(\tau_1, \ldots, \tau_n) \wedge \psi(\tau_1, \ldots, \tau_n))$ iff

$$p \Vdash^* \phi(\tau_1, \ldots, \tau_n) \quad \text{and} \quad p \Vdash^* \psi(\tau_1, \ldots, \tau_n).$$

(d) $p \Vdash^* \neg\phi(\tau_1, \ldots, \tau_n)$ iff there is no $q \leq p$ such that $q \Vdash^* \phi(\tau_1, \ldots, \tau_n)$.

(e) $p \Vdash^* \exists x\, \phi(x, \tau_1, \ldots, \tau_n)$ iff

$$\{r\colon \exists\sigma \in V^{\mathbb{P}}\, (r \Vdash^* \phi(\sigma, \tau_1, \ldots, \tau_n))\}$$

is dense below p. □

A casual inspection of Definition 3.3 will reveal that the definition is circular, and must thus be a recursion, but the exact nature of this recursion deserves some additional comment. It is intended that clause (a) be applied first to define the notion $p \Vdash^* \tau_1 = \tau_2$. Formally, we are defining a function $\mathbf{F}: V^{\mathbb{P}} \times V^{\mathbb{P}} \to \mathscr{P}(\mathbb{P})$, where $\mathbf{F}(\langle\tau_1, \tau_2\rangle)$ is intended to be

$$\{p \in \mathbb{P}\colon p \Vdash^* \tau_1 = \tau_2\}.$$

$\mathbf{F}$ is defined by transfinite recursion on the relation $\mathbf{R}$, where

$$\langle\pi_1, \pi_2\rangle\, \mathbf{R} \,\langle\tau_1, \tau_2\rangle$$

iff $\pi_1 \in \mathrm{dom}(\tau_1)$ and $\pi_2 \in \mathrm{dom}(\tau_2)$. $\mathbf{R}$ is clearly set-like, and $\mathbf{R}$ is well-founded because $\langle\pi_1, \pi_2\rangle\, \mathbf{R}\, \langle\tau_1, \tau_2\rangle$ implies $\mathrm{rank}(\pi_1) < \mathrm{rank}(\tau_1)$.

Once the notion $p \Vdash^* \tau_1 = \tau_2$ is defined, clause (b) defines the notion $p \Vdash^* \tau_1 \in \tau_2$ explicitly. Now that $\Vdash^*$ is defined for atomic formulas, clauses (c)–(e) define $\Vdash^*$ for all formulas by a straightforward induction on length. Formally, the induction takes place in the metatheory. As with $\Vdash$, for each formula $\phi(x_1, \ldots, x_n)$ we are defining a formula

$$\mathrm{Force}^*_\phi(\tau_1, \ldots, \tau_n, \mathbb{P}, \leq, p).$$

For atomic formulas, the recursion used in defining $\Vdash^*$ involves only absolute concepts and is thus absolute for transitive models of ZF − P. More precisely, we are using the absoluteness of $\mathbf{R}$ (see above), plus the absoluteness of $\{\langle\pi_1, \pi_2\rangle\colon \langle\pi_1, \pi_2\rangle\, \mathbf{R}\, \langle\tau_1, \tau_2\rangle\}$ to conclude the absoluteness of $\mathbf{F}$ (see IV 5.6). However, $\Vdash^*$ for arbitrary ϕ is not absolute; the $\exists\sigma \in \mathbf{V}^{\mathbb{P}}$ in clause (e) becomes $\exists\sigma \in M^{\mathbb{P}}$ when relativized to a model M. In any case, in checking Fact 1, we are only interested in looking at $\Vdash^*$ relativized to M.

As motivation for the specific details of clauses (a)–(e) of the definition of $\Vdash^*$, we may think of $(p \Vdash^* \phi)^M$ as an attempt by a person living in M to decide $\Vdash$. We shall eventually prove Fact 1, that $\Vdash$ is definable in M, by showing that $p \Vdash \phi$ iff $(p \Vdash^* \phi)^M$. Thus, we use, as the inductive clauses in the definition of $\Vdash^*$, relations which $\Vdash$ itself satisfies. We can then try to prove Fact 1 by induction on ϕ.

To see that $\Vdash$ indeed satisfies (a)–(e) sometimes requires some argument. For (c), it is immediate from the definition of $\Vdash$ that $p \Vdash (\phi \wedge \psi)$ iff $p \Vdash \phi$ and $p \Vdash \psi$ (see Lemma 3.2). Regarding (d), assume that $\neg \exists q \leq p (q \Vdash \phi)$; to show $p \Vdash \neg \phi$, assume not. Then, there is a generic G with $p \in G$ and ϕ true in M. By Fact 2, there is an $r \in G$ such that $r \Vdash \phi$. Let $q \in G$ with $q \leq r$ and $q \leq p$; then $q \Vdash \phi$ (by Lemma 3.2), contradicting $\neg \exists q \leq p (q \Vdash \phi)$.

Clause (e), relativized to M, says $(p \Vdash^* \exists x\, \phi(x))^M$ iff

$$\{r \leq p\colon \exists \sigma \in M^{\mathbb{P}} \left(r \Vdash^* \phi(\sigma)\right)^M\}$$

is dense below p. To check this (in one direction) for $\Vdash$, suppose $D = \{r \leq p\colon \exists \sigma \in M^{\mathbb{P}} \left(r \Vdash \phi(\sigma)\right)\}$ is dense below p. By Fact 1, $D \in M$. Thus, whenever G is generic over M and $p \in G$, $G \cap D \neq 0$, so there is a $\sigma \in M^{\mathbb{P}}$ and $r \in G$ with $r \Vdash \phi(\sigma)$; the $(\phi(\sigma_G))^{M[G]}$, so $(\exists x\, \phi(x))^{M[G]}$. Thus, $p \Vdash \exists x\, \phi(x)$.

Of course, the arguments in the preceding two paragraphs are useful only for motivation, since to verify that $\Vdash$ satisfies clauses (d) and (e), we are appealing to Facts 1 and 2, which have not yet been proved. The reader may find it a useful exercise to complete our (circular) justification of Definition 3.3.

We now proceed on a somewhat different tack to obtain a rigorous proof of Facts 1 and 2. As a preliminary lemma, we prove the following.

3.4. Lemma. *For p and $\phi(\tau_1, \ldots, \tau_n)$ as in Definition 3.3, the following are equivalent*:

(1) $p \Vdash^* \phi(\tau_1, \ldots, \tau_n)$.

(2) $\forall r \leq p \left(r \Vdash^* \phi(\tau_1, \ldots, \tau_n)\right)$.

(3) $\{r\colon r \Vdash^* \phi(\tau_1, \ldots, \tau_n)\}$ *is dense below* p.

Proof. Observe first that (2) $\to$ (1) and (2) $\to$ (3) are trivial. Next, assume $\phi(\tau_1, \tau_2)$ is either $\tau_1 = \tau_2$ or $\tau_1 \in \tau_2$. (1) $\to$ (2) follows from the fact that if D is dense below p and $r \leq p$, then D is dense below r. (3) $\to$ (1) follows from the fact that if $\{r\colon D$ is dense below $r\}$ is dense below p, then D is dense below p. Note that in both cases, we do not refer to the precise details of the definition of $p \Vdash^* \phi(\tau_1, \tau_2)$; rather, we need only that the definition involves certain sets being dense below p.

Now that the equivalence of (1)–(3) has been checked for atomic ϕ, it is easily checked for all ϕ by induction, using clauses (c)–(e) of Definition 3.3. The only place the inductive hypothesis is used in this argument is in the step for $\wedge$. □

We now express the relationship between $\Vdash^*$ and truth in $M[G]$. This is the key to relating $\Vdash^*$ to $\Vdash$.

3.5. Theorem. *Let $\phi(x_1, \ldots, x_n)$ be a formula with all free variables shown.*

Let M be a transitive model for ZFC, $\mathbb{P}$ *a* p.o. *in M, and $\tau_1, \dots, \tau_n \in M^{\mathbb{P}}$. Let G be $\mathbb{P}$-generic over M; then*

(1) *If $p \in G$ and $(p \Vdash^* \phi(\tau_1, \dots, \tau_n))^M$, then $(\phi(\mathrm{val}(\tau_1, G), \dots, \mathrm{val}(\tau_n, G)))^{M[G]}$.*

(2) *If $\phi(\mathrm{val}(\tau_1, G), \dots, \mathrm{val}(\tau_n, G))^{M[G]}$, then $\exists p \in G((p \Vdash^* \phi(\tau_1, \dots, \tau_n))^M)$.*

PROOF. When $\phi(\tau_1, \tau_2)$ is $\tau_1 = \tau_2$, the proof proceeds by transfinite induction, using clause (a) of Definition 3.3. The fact that this is indeed an induction on a well-founded relation is seen in precisely the same way that we justified the definition of $\Vdash^*$ for such ϕ. Since $\Vdash^*$ for atomic formulas is absolute for M, we may drop the relativizations to M.

To check (1), we assume $p \in G$ and $p \Vdash^* \tau_1 = \tau_2$. We must show $\tau_{1_G} = \tau_{2_G}$. We shall show $\tau_{1_G} \subset \tau_{2_G}$ using (α) of Definition 3.3(a); the proof of $\tau_{2_G} \subset \tau_{1_G}$ using (β) is the same. Every element of τ_{1_G} is of the form π_{1_G}, where $\langle \pi_1, s_1 \rangle \in \tau_1$ for some $s_1 \in G$. We must show that $\pi_{1_G} \in \tau_{2_G}$. Fix $r \in G$ with $r \le p$ and $r \le s_1$. Then $r \Vdash^* \tau_1 = \tau_2$ (by Lemma 3.4), so (by Lemma 2.20(b)), there is $q \in G$ such that $q \le r$ and such that $q \le s_1$ implies

$$\exists \langle \pi_2, s_2 \rangle \in \tau_2 (q \le s_2 \wedge q \Vdash^* \pi_1 = \pi_2). \qquad (*)$$

But $q \le s_1$, so fix $\langle \pi_2, s_2 \rangle$ as in $(*)$; then $s_2 \in G$, so $\pi_{2_G} \in \tau_{2_G}$. Also, by (1) for $\pi_1 = \pi_2$, $q \Vdash^* \pi_1 = \pi_2$ implies $\pi_{1_G} = \pi_{2_G}$, so $\pi_{1_G} \in \tau_{2_G}$.

To check (2), assume $\tau_{1_G} = \tau_{2_G}$. Let D be the set of all $r \in \mathbb{P}$ such that either $r \Vdash^* \tau_1 = \tau_2$, or

$$(\alpha')\ \exists \langle \pi_1, s_1 \rangle \in \tau_1 (r \le s_1 \wedge \forall \langle \pi_2, s_2 \rangle \in \tau_2\, \forall q \in \mathbb{P}$$
$$((q \le s_2 \wedge q \Vdash^* \pi_1 = \pi_2) \rightarrow q \perp r)),$$

or

$$(\beta')\ \exists \langle \pi_2, s_2 \rangle \in \tau_2 (r \le s_2 \wedge \forall \langle \pi_1, s_1 \rangle \in \tau_1\, \forall q \in \mathbb{P}$$
$$((q \le s_1 \wedge q \Vdash^* \pi_1 = \pi_2) \rightarrow q \perp r)).$$

First note that no $r \in G$ can satisfy (α') or (β'), for suppose $r \in G$ and $\langle \pi_1, s_1 \rangle \in \tau_1$ as in (α'); then $s_1 \in G$ so $\pi_{1_G} \in \tau_{1_G} = \tau_{2_G}$, so fix $\langle \pi_2, s_2 \rangle \in \tau_2$ with $s_2 \in G$ and $\pi_{1_G} = \pi_{2_G}$; then, by (2) for $\pi_1 = \pi_2$, fix $q_0 \in G$ with $q_0 \Vdash^* \pi_1 = \pi_2$; now fix $q \in G$ with $q \le q_0$ and $q \le s_2$; since $q \Vdash^* \pi_1 = \pi_2$ (see Lemma 3.4), we have $q \perp r$ (by (α')), $q \in G$, and $r \in G$, a contradiction. If $\neg \exists r \in G (r \Vdash^* \tau_1 = \tau_2)$, then $D \cap G = 0$. Since $D \in M$ by absoluteness, we shall be done if we can check that D is dense. Fix $p \in \mathbb{P}$. Either $p \Vdash^* \tau_1 = \tau_2$ or (α) or (β) of Definition 3.3 (a) fails. If (α) fails, then, applying the definition of "dense below p," fix $\langle \pi_1, s_1 \rangle \in \tau_1$ and $r \le p$ such that

$$\forall q \le r (q \le s_1 \wedge \forall \langle \pi_2, s_2 \rangle \in \tau_2 (\neg (q \Vdash^* \pi_1 = \pi_2 \wedge q \le s_2))). \qquad (\dagger)$$

In particular, $r \le s_1$. If $\langle \pi_2, s_2 \rangle \in \tau_2$, $q \le s_2$, and $q \Vdash^* \pi_1 = \pi_2$ then $q \perp r$, since a common extension q' of q and r would contradict ($\dagger$). Thus, $r \le p$

and r satisfies (α'). Likewise, if (β) fails, there is an $r \leq p$ satisfying (β').

Now, assume $\phi(\tau_1, \tau_2)$ is $\tau_1 \in \tau_2$. To check (1), assume $p \in G$ and $p \Vdash^* \tau_1 \in \tau_2$; then

$$D = \{q\colon \exists \langle \pi, s \rangle \in \tau_2 (q \leq s \wedge q \Vdash^* \pi = \tau_1)\}$$

is dense below p, so fix $q \in G \cap D$, and fix $\langle \pi, s \rangle \in \tau_2$ so that $q \leq s$ and $q \Vdash^* \pi = \tau_1$. Since $s \in G$ and $\langle \pi, s \rangle \in \tau_2$, $\pi_G \in \tau_{2_G}$ by definition of τ_{2_G}. Since $q \in G$ and $q \Vdash^* \pi = \tau_1$, $\pi_G = \tau_{1_G}$ by (1) applied to $\pi = \tau_1$. Thus, $\tau_{1_G} \in \tau_{2_G}$.

To check (2) for $\tau_1 \in \tau_2$, assume $\tau_{1_G} \in \tau_{2_G}$. By definition of τ_{2_G}, there is a $\langle \pi, s \rangle \in \tau_2$ such that $s \in G$ and $\pi_G = \tau_{1_G}$. By (2) for $\pi = \tau_1$, there is an $r \in G$ such that $r \Vdash^* \pi = \tau_1$. Let $p \in G$ be such that $p \leq s$ and $p \leq r$. Then $\forall q \leq p (q \leq s \wedge q \Vdash^* \pi = \tau_1)$, so $p \Vdash^* \tau_1 \in \tau_2$ (we have verified a statement stronger than that required by Definition 3.3(b)).

This concludes the proof of (1) and (2) for atomic ϕ. We now prove (1) and (2) simultaneously for all ϕ by induction on ϕ; formally, this induction takes place in the metatheory. There are six parts to this, since the induction steps must be done for $\neg$, $\wedge$, and $\exists$, and (1) and (2) must be checked. Since $\Vdash^*$ is not absolute when ϕ has quantifiers, it is now important that we relativize $\Vdash^*$ to M.

In the following, we shall, for brevity, drop explicit mention of the $\tau_1, \ldots, \tau_n$, since they may easily be filled in.

(1) $\neg$: We assume (1) and (2) for ϕ, and we conclude (1) for $\neg\phi$. Assume $p \in G$ and $(p \Vdash^* \neg\phi)^M$; we must show $\neg\phi^{M[G]}$. But if $\phi^{M[G]}$, then by (2) for ϕ, there is a $q \in G$ with $(q \Vdash^* \phi)^M$. Let $r \in G$ with $r \leq p$ and $r \leq q$; then $(r \Vdash^* \phi)^M$, contradicting the definition of $p \Vdash^* \neg\phi$.

(2) $\neg$: Assume $(\neg\phi)^{M[G]}$, and let

$$D = \{p\colon (p \Vdash^* \phi)^M \vee (p \Vdash^* \neg\phi)^M\}.$$

$D \in M$ and D is dense by the definition of $\Vdash^*$ applied within M, so fix $p \in D \cap G$. If $(p \Vdash^* \neg\phi)^M$, we are done. If $(p \Vdash^* \phi)^M$, then by (1) for ϕ, we have $\phi^{M[G]}$, a contradiction.

(1) $\wedge$: We assume (1) and (2) for ϕ and ψ, and we conclude (1) for $\phi \wedge \psi$. Assume $p \in G$ and $(p \Vdash^* (\phi \wedge \psi))^M$; then $(p \Vdash^* \phi)^M$ and $(p \Vdash^* \psi)^M$, so $\phi^{M[G]}$ and $\psi^{M[G]}$, so $(\phi \wedge \psi)^{M[G]}$.

(2) $\wedge$: Assume $(\phi \wedge \psi)^{M[G]}$. By (2) for ϕ and ψ, there are $p, q \in G$ such that $(p \Vdash^* \phi)^M$ and $(q \Vdash^* \psi)^M$. Let $r \in G$ be such that $r \leq p$ and $r \leq q$; then $(r \Vdash^* \phi)^M$ and $(r \Vdash^* \psi)^M$, so $(r \Vdash^* \phi \wedge \psi)^M$.

(1) $\exists$: Assume $p \in G$ and $(p \Vdash^* \exists x\, \phi(x))^M$; then

$$\{r\colon \exists \sigma \in M^{\mathbb{P}} \left(r \Vdash^* \phi(\sigma)\right)^M\}$$

is dense below p and in M, so fix $r \in G$ and $\sigma \in M^{\mathbb{P}}$ with $\left(r \Vdash^* \phi(\sigma)\right)^M$. By (1) for ϕ, $\left(\phi(\sigma_G)\right)^{M[G]}$, so $\left(\exists x\, \phi(x)\right)^{M[G]}$.

(2) $\exists$: Assume $(\exists x\, \phi(x))^{M[G]}$ and fix $\sigma \in M^{\mathbb{P}}$ with $(\phi(\sigma_G))^{M[G]}$. By (2) for ϕ, fix $p \in G$ so that $(p \Vdash^* \phi(\sigma))^M$; then $\forall r \le p\,((r \Vdash^* \phi(\sigma))^M)$, so $(p \Vdash^* \exists x\, \phi(x))^M$ (we have verified a statement stronger than that required by Definition 3.3(e)). $\square$

Finally, we may state and prove Facts 1 and 2 formally.

3.6. THEOREM. *Let M be a* c.t.m. *for* ZFC *and* $\mathbb{P}$ *a* p.o. *in M; let $\phi(x_1, \dots, x_n)$ be a formula with all free variables shown; let $\tau_1, \dots, \tau_n \in M^{\mathbb{P}}$.*

(1) *For all $p \in \mathbb{P}$,*

$$p \Vdash \phi(\tau_1, \dots, \tau_n) \leftrightarrow (p \Vdash^* \phi(\tau_1, \dots, \tau_n))^M.$$

(2) *For all G which are $\mathbb{P}$-generic over M,*

$$\phi(\tau_{1G}, \dots, \tau_{nG})^{M[G]} \leftrightarrow \exists p \in G\,(p \Vdash \phi(\tau_1, \dots, \tau_n)).$$

PROOF. In (1), the implication from right to left is immediate from Theorem 3.5(1) and the definition of $\Vdash$. For the implication from left to right, assume $p \Vdash \phi(\tau_1, \dots, \tau_n)$. To show $(p \Vdash^* \phi(\tau_1, \dots, \tau_n))^M$, it is sufficient (by Lemma 3.4) to show that $D = \{r\colon (r \Vdash^* \phi(\tau_1, \dots, \tau_n))^M$ is dense below p. If not, let $q \le p$ be such that $\neg \exists r \le q\,(r \in D)$. Then, by definition of $\Vdash^*$,

$$(q \Vdash^* \neg\phi(\tau_1, \dots, \tau_n))^M,$$

whence, by (1) from right to left, $q \Vdash \neg\phi(\tau_1, \dots, \tau_n)$. Let G be $\mathbb{P}$-generic over M with $q \in G$; then $(\neg\phi(\mathrm{val}(\tau_1, G), \dots, \mathrm{val}(\tau_n, G)))^{M[G]}$, but also $p \in G$, since $p \ge q$, so $(\phi(\mathrm{val}(\tau_1, G), \dots, \mathrm{val}(\tau_n, G)))^{M[G]}$, a contradiction.

For (2), the implication from left to right follows from (1) and from Theorem 3.5(2), which asserts the same thing about $\Vdash^*$. The implication from right to left is immediate from the definition of $\Vdash$. $\square$

In practice, Theorem 3.6(1) will be used to show that various sets defined using $\Vdash$ actually lie in M. For example for fixed $\tau_1, \dots, \tau_n \in M^{\mathbb{P}}$

$$\{p \in \mathbb{P}\colon p \Vdash \phi(\tau_1, \dots, \tau_n)\}$$

in in M, since this set is equal to

$$\{p \in \mathbb{P}\colon (p \Vdash^* \phi(\tau_1, \dots, \tau_n))^M\},$$

which lies in M by Comprehension in M. Likewise, e.g., for fixed σ, $\tau_2, \dots, \tau_n \in M^{\mathbb{P}}$,

$$\{\langle p, \tau_1\rangle \in \mathbb{P} \times \mathrm{dom}(\sigma)\colon p \Vdash \phi(\tau_1, \dots, \tau_n)\} \in M.$$

Theorem 3.6(2) will be important because it relates truth in $M[G]$ to $\Vdash$. The following additional facts about $\Vdash$ will also be useful.

3.7. COROLLARY. *Let M be a* c.t.m. *for* ZFC, $\mathbb{P}$ *a* p.o. *in* M, *and* $\sigma, \tau_1, \ldots, \tau_n \in M^{\mathbb{P}}$; *then*

(a) $\{p \in \mathbb{P}: (p \Vdash \phi(\tau_1, \ldots, \tau_n)) \vee (p \Vdash \neg\phi(\tau_1, \ldots, \tau_n))\}$ *is dense.*

(b) $p \Vdash \neg\phi(\tau_1, \ldots, \tau_n)$ *iff* $\neg\exists q \leq p\, (q \Vdash \phi(\tau_1, \ldots, \tau_n))$.

(c) $p \Vdash \exists x\, \phi(x, \tau_1, \ldots, \tau_n)$ *iff*

$$\{r \leq p: \exists\sigma \in M^{\mathbb{P}}\, (r \Vdash \phi(\sigma, \tau_1, \ldots, \tau_n))\}$$

is dense below p.

(d) *If* $p \Vdash \exists x\, (x \in \sigma \wedge \phi(x, \tau_1, \ldots, \tau_n))$, *then*

$$\exists q \leq p\, \exists\pi \in \mathrm{dom}(\sigma)\, (q \Vdash \phi(\pi, \tau_1, \ldots, \tau_n)).$$

PROOF. (a)–(c) are true of $\Vdash^*$ by definition, and thus hold for $\Vdash$ by Theorem 3.6(1). For (d), fix a generic G with $p \in G$. By definition of $\Vdash$, there is an $a \in \sigma_G$ such that $(\phi(a, \tau_1, \ldots, \tau_n))^{M[G]}$. $a = \pi_G$ for some $\pi \in \mathrm{dom}(\sigma)$. By Theorem 3.6(2), there is an $r \in G$ such that $r \Vdash \phi(\pi, \tau_1, \ldots, \tau_n)$. If q is a common extension of p and r, then $q \leq p$ and $q \Vdash \phi(\pi, \tau_1, \ldots, \tau_n)$. □

§4. ZFC in $M[G]$

We now apply the results of §3 to show that our generic extension is a model of ZFC. It will be convenient to verify AC in a form slightly different from the usual one.

4.1. LEMMA (ZF). AC *holds iff*

$$\forall x\, \exists\alpha \in \mathbf{ON}\, \exists f\, (f \textit{ is a function} \wedge \mathrm{dom}(f) = \alpha \wedge x \subset \mathrm{ran}(f)). \quad (*)$$

PROOF. If x, α, and f are as in $(*)$, we may define a well-order of x as follows. Let $g(z) = \min(f^{-1}\{z\})$ for $z \in x$; then g maps x 1–1 into α. If we let $y\, R\, z \leftrightarrow g(y) < g(z)$, then R well-orders x. □

4.2. THEOREM. *Let M be a* c.t.m. *for* ZFC, $\langle\mathbb{P}, \leq, \mathbb{1}\rangle$ *a* p.o. *in* M, *and G $\mathbb{P}$-generic over M; then $M[G]$ satisfies* ZFC.

PROOF. We have already verified Extensionality, Foundation, Pairing, and Union (see Lemma 2.18). Let us check Comprehension. To do this we must see that whenever $\sigma, \tau_1, \ldots, \tau_n \in M^{\mathbb{P}}$ and $\phi(x, v, y_1, \ldots, y_n)$ is any formula,

$$\{a \in \sigma_G: (\phi(a, \sigma_G, \tau_{1G}, \ldots, \tau_{nG}))^{M[G]}\} \in M[G].$$

Let

$$\rho = \{\langle\pi, p\rangle \in \mathrm{dom}(\sigma) \times \mathbb{P}: p \Vdash (\pi \in \sigma \wedge \phi(\pi, \sigma, \tau_1, \ldots, \tau_n))\}.$$

$\rho \in M^{\mathbb{P}}$ by definability of forcing (Theorem 3.6(1)). We now verify that $\rho_G = \{a \in \sigma_G \colon \phi(a)^{M[G]}\}$; for brevity, we suppress mention of $\tau_1, \ldots, \tau_n$ in the rest of this argument. First, any element of ρ_G is of the form π_G where $\langle \pi, p \rangle \in \rho$ for some $p \in G$. By definition of ρ, $p \Vdash (\pi \in \sigma \wedge \phi(\pi))$ so, by the definition of $\Vdash$, $\pi_G \in \sigma_G$ and $\phi(\pi_G)^{M[G]}$. Thus $\rho_G \subset \{a \in \sigma_G \colon \phi(a)^{M[G]}\}$. To show equality, assume $a \in \sigma_G$ and $\phi(a)^{M[G]}$. $a = \pi_G$ for some $\pi \in \mathrm{dom}(\sigma)$. Then $(\pi_G \in \sigma_G \wedge \phi(\pi_G))^{M[G]}$. Since any statement true in $M[G]$ is forced (Theorem 3.6(2)), there is a $p \in G$ such that $p \Vdash (\pi \in \sigma \wedge \phi(\pi))$; then $\langle \pi, p \rangle \in \rho$, so $\pi_G \in \rho_G$.

Next, we verify Replacement. For this we must check that for each formula $\phi(x, v, r, z_1, \ldots, z_n)$ and each $\sigma_G, \tau_{1_G}, \ldots, \tau_{n_G} \in M[G]$, if

$$(\forall x \in \sigma_G\, \exists! y\, \phi(x, y, \sigma_G, \tau_{1_G}, \ldots, \tau_{n_G}))^{M[G]},$$

then there is a $\rho \in M^{\mathbb{P}}$ such that

$$\forall x \in \sigma_G\, \exists y \in \rho_G\, (\phi(x, y, \sigma_G, \tau_{1_G}, \ldots, \tau_{n_G}))^{M[G]}.$$

Again, suppress mention of $\tau_1, \ldots, \tau_n$. Let $S \in M$ be such that $S \subset M^{\mathbb{P}}$ and

$$\forall \pi \in \mathrm{dom}(\sigma)\, \forall p \in \mathbb{P}\, [\exists \mu \in M^{\mathbb{P}}\, (p \Vdash \phi(\pi, \mu)) \rightarrow \exists \mu \in S\, (p \Vdash \phi(\pi, \mu))];$$

S exists because by Theorem 3.6(1), $p \Vdash \phi(\pi, \mu)$ is defined by a formula relativized to M, so by reflection in M we may take $S = R(\alpha)^{(M)} \cap M^{\mathbb{P}}$ for a suitable α (see IV 7.4). Let $\rho = S \times \{\mathbb{1}\}$; then $\rho_G = \{\mu_G \colon \mu \in S\}$. Fix $x \in \sigma_G$. We show $\exists y \in \rho_G\, (\phi(x, y))^{M[G]}$. $x = \pi_G$ for some $\pi \in \mathrm{dom}(\sigma)$. By assumption, $(\exists y\, \phi(\pi_G, y))^{M[G]}$, so for some $v \in M^{\mathbb{P}}$, $\phi(\pi_G, v_G)^{M[G]}$, and by Theorem 3.6(2), there is a $p \in G$ such that $p \Vdash \phi(\pi, v)$. There is then a $\mu \in S$ such that $p \Vdash \phi(\pi, \mu)$, so we have $\mu_G \in \rho_G$ and $(\phi(\pi_G, \mu_G))^{M[G]}$.

We remark that it looks like we have proved a stronger form of Replacement which weakens the $\exists! y$ in the hypothesis to $\exists y$. But this "stronger" axiom is in fact a version of reflection and is derivable in ZF (see III Exercise 15).

We have now checked all axioms of ZF $-$ P in $M[G]$ except Infinity. But now Infinity holds also, since $\omega (= (\check{\omega})_G)$ is in $M[G]$. Thus, $M[G]$ satisfies ZF $-$ P.

For the Power Set Axiom, fix $\sigma_G \in M[G]$. We shall produce a $\rho \in M^{\mathbb{P}}$ such that $\forall x \in M[G]\, (x \subset \sigma_G \rightarrow x \in \rho_G)$. Let $\rho = S \times \{\mathbb{1}\}$, where

$$S = \{\tau \in M^{\mathbb{P}} \colon \mathrm{dom}(\tau) \subset \mathrm{dom}(\sigma)\} = (\mathscr{P}(\mathrm{dom}(\sigma) \times \mathbb{P}))^M.$$

Fix any $\mu \in M^{\mathbb{P}}$ such that $\mu_G \subset \sigma_G$. We show $\mu_G \in \rho_G$. Let

$$\tau = \{\langle \pi, p \rangle \colon \pi \in \mathrm{dom}(\sigma) \wedge p \Vdash \pi \in \mu\};$$

then $\tau \in S$ so $\tau_G \in \rho_G$, so we shall be done if we can show $\mu_G = \tau_G$. To see that $\mu_G \subset \tau_G$, note that since $\mu_G \subset \sigma_G$, any element of μ_G is of the form π_G

for some $\pi \in \text{dom}(\sigma)$; since $\pi_G \in \mu_G$, there is a $p \in G$ such that $p \Vdash \pi \in \mu$, whence $\langle \pi, p \rangle \in \tau$, so $\pi_G \in \tau_G$. To see that $\tau_G \subset \mu_G$, note that any element of τ_G is of the form π_G where $\langle \pi, p \rangle \in \tau$ for some $p \in G$; then $p \Vdash \pi \in \mu$, so $\pi_G \in \mu_G$.

The key to the proof of the Power Set Axiom in $M[G]$ is that in M there is a set of names which contains representatives for any possible subset of σ_G, even though the collection of all μ such that $\mu_G \subset \sigma_G$ (or even $\mu_G = 0$) is usually not contained in a set of M (see Exercise A9).

We now know that ZF holds in $M[G]$. To check that AC holds in $M[G]$, we shall verify the equivalent of AC presented in Lemma 4.1. Fix $x = \sigma_G \in M[G]$. By AC^M, let $\text{dom}(\sigma) = \{\pi_\gamma : \gamma < \alpha\}$, where the function which takes γ to π_γ is in M. Let

$$\tau = \{\text{op}(\check{\gamma}, \pi_\gamma) : \gamma < \alpha\} \times \{\mathbb{1}\}$$

(see Definition 2.16). Then $\tau \in M$ and $\tau_G = \{\langle \gamma, \pi_{\gamma G} \rangle : \gamma < \alpha\}$, so τ_G is a function with $\text{dom}(\tau_G) = \alpha$ and $\sigma_G \subset \text{ran}(\tau_G)$.

Thus, all axioms of ZFC hold in $M[G]$. □

Our next task is to show how to design $\mathbb{P}$ to produce $M[G]$ satisfying desired additional axioms, but we mention now one immediate consequence of our results so far.

4.3. Corollary. *Let M be* c.t.m. *for* ZFC, *then there is a* c.t.m. $N \supset M$ *such that N satisfies* ZFC + $\mathbf{V} \neq \mathbf{L}$.

Proof. With the notation of Theorem 4.2, just choose $\mathbb{P}$ such that $G \notin M$. This is true whenever $\mathbb{P}$ satisfies the condition of Lemma 2.4; for example, let $\mathbb{P}$ be finite partial functions from ω to 2. Let $N = M[G]$, then, since $o(N) = o(M)$ (see Lemma 2.15), $\mathbf{L}^N = \mathbf{L}^M \subset M$, so N satisfies $\mathbf{V} \neq \mathbf{L}$. □

As pointed out in §1, Corollary 4.3 yields the following.

4.4. Corollary. Con(ZFC) $\rightarrow$ Con(ZFC + $\mathbf{V} \neq \mathbf{L}$). □

As we now proceed with the development of forcing, we shall often be discussing the relation $p \Vdash \phi$ where ϕ is a statement of some mathematical complexity. Then, as usual, ϕ will not be explicitly displayed as a formula in the official language of set theory; rather, we shall express ϕ using standard mathematical notation, which we consider to be an abbreviation for a formula of set theory. It is then worth noting that we do not have to worry about the exact way we write the unabbreviated formula, since two formulas which are equivalent in ZFC are forced by the same conditions. More precisely, the following holds.

4.5. LEMMA. (a) *Let* $\phi(x_1, \dots, x_n)$ *and* $\psi(x_1, \dots, x_n)$ *be formulas, and assume*

$$\mathrm{ZFC} \vdash \forall x_1, \dots, x_n\,(\phi(x_1, \dots, x_n) \to \psi(x_1, \dots, x_n));$$

then for any c.t.m. M *for* ZFC, p.o. $\mathbb{P} \in M$, $p \in \mathbb{P}$, *and* $\tau_1, \dots, \tau_n \in M^{\mathbb{P}}$,

$$(p \Vdash \phi(\tau_1, \dots, \tau_n)) \to (p \Vdash \psi(\tau_1, \dots, \tau_n)).$$

(b) *If we assume also that*

$$\mathrm{ZFC} \vdash \forall x_1, \dots, x_n\,(\phi(x_1, \dots, x_n) \leftrightarrow \psi(x_1, \dots, x_n)),$$

then we may conclude

$$(p \Vdash \phi(\tau_1, \dots, \tau_n)) \leftrightarrow (p \Vdash \psi(\tau_1, \dots, \tau_n)).$$

PROOF. In (a), for any G which is $\mathbb{P}$-generic over M, $M[G]$ satisfies ZFC, so

$$\phi(\tau_{1G}, \dots, \tau_{nG})^{M[G]} \to \psi(\tau_{1G}, \dots, \tau_{nG})^{M[G]}.$$

(a) thus follows from the definition of $\Vdash$. (b) follows from (a). □

§5. Forcing with finite partial functions

The most famous relative consistency proof produced by forcing is that of Con(ZFC + $\neg$CH). The methods of this section allow us to construct models in which 2^ω is ω_2, ω_5, ω_{ω_1} or anything else not obviously contradictory.

Throughout this section, M is a fixed c.t.m. for ZFC. We consider forcing over M with *finite partial functions* from one set I into another set J.

5.1. DEFINITION.

$$\mathrm{Fn}(I, J) = \{p \colon |p| < \omega \wedge p \text{ is a function} \wedge \mathrm{dom}(p) \subset I \wedge \mathrm{ran}(p) \subsetneq J\}.$$

Order $\mathrm{Fn}(I, J)$ by: $p \leq q \leftrightarrow p \supset q$. □

$\mathrm{Fn}(I, J)$ is a p.o., with largest element $\mathbb{1} = 0$ (the empty function). Since "finite" is absolute, so is $\mathrm{Fn}(I, J)$, so if $I, J \in M$, then $\mathrm{Fn}(I, J) = \mathrm{Fn}(I, J)^M \in M$. $\mathrm{Fn}(\omega, 2)$ was discussed briefly at the beginning of §3, and the elementary discussion of $\mathrm{Fn}(I, J)$ in general is similar.

If G is a filter in $\mathrm{Fn}(I, J)$, $\bigcup G$ is a function with $\mathrm{dom}(\bigcup G) \subset I$ and $\mathrm{ran}(\bigcup G) \subset J$. If $J \neq 0$, $D_i = \{p \in \mathrm{Fn}(I, J) \colon i \in \mathrm{dom}(p)\}$ is dense for all $i \in I$; furthermore, by absoluteness, $D_i \in M$ if $I, J \in M$; thus, if G is generic over M, $G \cap D_i \neq 0$ for each $i \in I$, whence $\mathrm{dom}(\bigcup G) = I$. Likewise, if I is infinite, $\{p \in \mathrm{Fn}(I, J) \colon j \in \mathrm{ran}(p)\}$ is dense and in M, so $\mathrm{ran}(\bigcup G) = J$. We have thus proved the following.

5.2. LEMMA. *If $I, J \in M$, I is infinite, $J \neq 0$, and G is* $\mathrm{Fn}(I, J)$*-generic over* M, *then* $\bigcup G$ *is a function from* I *onto* J. □

A simple application of this kind of partial order is that the notion of cardinal need not be absolute for M, $M[G]$. Thus, let κ be an uncountable cardinal of M; i.e., $\kappa \in M$ and (κ is an uncountable cardinal)^M. Let $\mathbb{P} = \mathrm{Fn}(\omega, \kappa)$, and let G be $\mathbb{P}$-generic over M. Then $\bigcup G \in M[G]$ by absoluteness of $\bigcup$), and G is a function from ω onto κ, so in $M[G]$, κ is a countable ordinal. We say that $\mathbb{P}$ *collapses* κ.

With a different I, J, we can use $\mathrm{Fn}(I, J)$ to obtain a model in which CH is false. Again let κ be an uncountable cardinal of M, but now let $\mathbb{P} = \mathrm{Fn}(\kappa \times \omega, 2)$, so, if G is $\mathbb{P}$-generic over M, then $\bigcup G : \kappa \times \omega \to 2$. We may think of G as coding a κ-sequence of functions from ω into 2; namely, let $f_\alpha(n) = (\bigcup G)(\alpha, n)$ for $\alpha < \kappa$, $n < \omega$. By absoluteness, the sequence $\langle f_\alpha : \alpha < \kappa \rangle$ (i.e., the function which assigns, to each α, f_α) is in $M[G]$. Furthermore, the f_α are all distinct; to see this, if $\alpha \neq \beta$, let

$$D_{\alpha\beta} = \{p \in \mathbb{P} : \exists n \in \omega (\langle \alpha, n \rangle \in \mathrm{dom}(p) \wedge \langle \beta, n \rangle \in \mathrm{dom}(p) \wedge p(\alpha, n) \neq p(\beta, n))\};$$

$D_{\alpha\beta}$ is dense and in M, so $G \cap D_{\alpha\beta} \neq 0$, which implies $f_\alpha \neq f_\beta$. Thus, $M[G]$ contains a κ-sequence of distinct functions from ω into 2, so the following is obtained.

5.3. LEMMA. *If* $\kappa \in M$ *and* G *is* $\mathrm{Fn}(\kappa \times \omega, 2)$*-generic over* M, *then* $(2^\omega \geq |\kappa|)^{M[G]}$. □

Taking $\kappa = (\omega_2)^M$, this would seem to imply that $2^\omega \geq \omega_2$ in $M[G]$, i.e., CH fails in $M[G]$. *But*, we must first check that κ is also $(\omega_2)^{M[G]}$; this is not immediate since we have just seen, with a slightly different partial order, that an uncountable cardinal of M could become a countable ordinal in $M[G]$. That this does not happen with $\mathrm{Fn}(\kappa \times \omega, 2)$ involves the fact that, as we shall show, this partial order has the countable chain condition (c.c.c.) *in* M (it trivially has c.c.c. in $\mathbf{V}$, since M is countable, but that is irrelevant).

The fact that $(\mathrm{Fn}(\kappa \times \omega, 2)$ has c.c.c.$)^M$ follows from the following slightly more general result, relativized to M.

5.4. LEMMA. *If* I *is arbitrary and* J *is countable, then* $\mathrm{Fn}(I, J)$ *has* c.c.c.

PROOF. Let $p_\alpha \in \mathrm{Fn}(I, J)$ for $\alpha < \omega_1$ and let $a_\alpha = \mathrm{dom}(p_\alpha)$. By the Δ-system lemma (II 1.5), there is an uncountable $X \subset \omega_1$ such that $\{a_\alpha : \alpha \in X\}$ forms a Δ-system, with some root r. Since J is countable, so is J^r, so there are only countably many possibilities for $p_\alpha \upharpoonright r$. It follows that there is an uncountable $Y \subset X$ such that the $p_\alpha \upharpoonright r$ for $\alpha \in Y$ are all the same. But then the p_α for $\alpha \in Y$ are all compatible. Thus, there can never be a family $\{p_\alpha : \alpha < \omega_1\}$ of incompatible conditions. □

There are many more examples of c.c.c. p.o's (see the discussion of MA in II §2). The importance of c.c.c. in forcing is the following lemma, which gives us a way of approximating, within M, any function which appears in $M[G]$.

5.5. LEMMA. *Assume* $\mathbb{P} \in M$, $(\mathbb{P}$ *is* c.c.c.$)^M$, *and* $A, B \in M$; *let* G *be* $\mathbb{P}$*-generic over* M, *and let* $f \in M[G]$, *with* $f : A \to B$. *Then there is a map* $F : A \to \mathscr{P}(B)$ *with* $F \in M$, $\forall a \in A\, (f(a) \in F(a))$ *and* $\forall a \in A\, ((|F(a)| \leq \omega)^M)$.

PROOF. Fix $\tau \in M^{\mathbb{P}}$ with $f = \tau_G$. Since any statement true in $M[G]$ is forced, there is a $p \in G$ such that

$$p \Vdash \tau \text{ is a function from } \check{A} \text{ into } \check{B}.$$

Formally, we are applying here Theorem 3.6(2) to a formula $\phi(x, y, z)$ which asserts that x is a function from y into z; exactly which ϕ we use is irrelevant by Lemma 4.5.

Define

$$F(a) = \{b \in B : \exists q \leq p\, (q \Vdash \tau(\check{a}) = \check{b})\}.$$

$F \in M$ by definability of $\Vdash$ (see Theorem 3.6 and following discussion).

Fix $a \in A$. To see that $f(a) \in F(a)$, let $b = f(a)$. Then there is an $r \in G$ such that $r \Vdash \tau(\check{a}) = \check{b}$, and r and p have a common extension, q. Then $q \Vdash \tau(\check{a}) = \check{b}$, so $b \in F(a)$.

To see that $(|F(a)| \leq \omega)^M$, apply AC in M to find a function $Q \in M$ such that $Q : F(a) \to \mathbb{P}$ and, for $b \in F(a)$, $Q(b) \leq p$ and $Q(b) \Vdash \tau(\check{a}) = \check{b}$. If $b, b' \in F(a)$ and $b \neq b'$, then $Q(b) \perp Q(b')$, since they force inconsistent statements; more precisely, if $Q(b)$ and $Q(b')$ were compatible, there would be a generic H containing both of them, and in $M(H)$, $\tau_H : A \to B$, $\tau_H(a) = b$, and $\tau_H(a) = b'$. Thus, $\{Q(b) : b \in F(a)\}$ is an antichain in $\mathbb{P}$, so, since $Q \in M$ and $(\mathbb{P}$ is c.c.c.$)^M$, $(|F(a)| \leq \omega)^M$. □

We now discuss the relevance of the c.c.c. to absoluteness of cardinals.

5.6. DEFINITION. If $\mathbb{P} \in M$, $\mathbb{P}$ *preserves cardinals* iff whenever G is $\mathbb{P}$-generic over M,

$$\forall \beta \in o(M)\, ((\beta \text{ is a cardinal})^M \leftrightarrow (\beta \text{ is a cardinal})^{M[G]}). \quad \square$$

Note that since ω is absolute, preservation of cardinals is only problematic for $\beta > \omega$. Also, if β is a cardinal of $M[G]$, it is automatically a cardinal of M since any function in M from a smaller ordinal onto β would be in $M[G]$ also. Thus, $\mathbb{P}$ preserves cardinals iff

$$\forall \beta \in o(M)\,[(\beta > \omega \wedge (\beta \text{ is a cardinal})^M) \rightarrow (\beta \text{ is a cardinal})^{M[G]}].$$

It is now easily seen from Lemma 5.5 that if $(\mathbb{P}$ is c.c.c.$)^M$, then $\mathbb{P}$ preserves cardinals (take $B = \beta$ and A an ordinal $<\beta$). In fact, $\mathbb{P}$ preserves cofinalities as well, which is a slightly stronger assertion.

5.7. DEFINITION. If $\mathbb{P} \in M$, $\mathbb{P}$ *preserves cofinalities* iff whenever G is $\mathbb{P}$-generic over M and γ is a limit ordinal in M,

$$\mathrm{cf}(\gamma)^M = \mathrm{cf}(\gamma)^{M[G]}. \quad \square$$

5.8. LEMMA. *If* $\mathbb{P}$ *preserves cofinalities, then* $\mathbb{P}$ *preserves cardinals.*

PROOF. Assume $\mathbb{P}$ preserves cofinalities. If $\alpha \geq \omega$ is a regular cardinal of M, then $\mathrm{cf}(\alpha)^{M[G]} = \mathrm{cf}(\alpha)^M = \alpha$, so α is a regular cardinal of $M[G]$. If $\beta > \omega$ is a limit cardinal of M, then the regular (in fact successor) cardinals of M are unbounded in β; since these remain regular in $M[G]$, β is a limit cardinal in $M[G]$ as well. Since every infinite cardinal is either regular or a limit cardinal (or both), every infinite cardinal of M is a cardinal of $M[G]$. $\square$

There are examples of $\mathbb{P}$ which preserve cardinals without preserving cofinalities (see [Prikry 1970]), but we shall not discuss them in this book.

The following simplifies what needs to be checked for preservation of cofinalities.

5.9. LEMMA. *Assume* $\mathbb{P} \in M$ *and whenever* G *is* $\mathbb{P}$*-generic over* M *and* κ *is a regular uncountable cardinal of* M, $(\kappa$ *is regular*$)^{M[G]}$. *Then* $\mathbb{P}$ *preserves cofinalities.*

PROOF. Let γ be a limit ordinal in M, and let $(\kappa = \mathrm{cf}(\gamma))^M$; then there is an $f \in M$ such that f maps κ into γ cofinally and f is strictly increasing (applying I 10.31 within M). Since $(\kappa$ is regular$)^M$, $(\kappa$ is regular$)^{M[G]}$ (applying absoluteness of ω if $\kappa = \omega$). Since $f \in M[G]$, $(\kappa = \mathrm{cf}(\gamma))^{M[G]}$ (applying I 10.32 within $M[G]$). $\square$

5.10. THEOREM. *If* $\mathbb{P} \in M$ *and* $(\mathbb{P}$ *has* c.c.c.$)^M$, *then* $\mathbb{P}$ *preserves cofinalities (and hence cardinals).*

PROOF. If not, then by Lemma 5.9, there is a $\kappa \in M$ with $\kappa > \omega$, $(\kappa$ regular$)^M$, and $(\kappa$ not regular$)^{M[G]}$. Thus, there is an $\alpha < \kappa$ and an $f \in M[G]$ such that f maps α cofinally into κ. By Lemma 5.5, let F be in M, with $F : \alpha \rightarrow \mathscr{P}(\kappa)$,

$\forall \xi < \alpha(f(\xi) \in F(\xi))$, and $\forall \xi < \alpha(|F(\xi)| \leq \omega)^M$. Let $S = \bigcup_{\xi<\alpha} F(\xi)$. Then, $S \in M$ and S is an unbounded subset of κ. Applying in M the fact that the union of $|\alpha|$ countable sets has cardinality $|\alpha|$, $(|S| = |\alpha| < \kappa)^M$, contradicting that ($\kappa$ is regular)^M. □

This completes everything needed to produce a model of $\neg$CH. Let $\mathbb{P} = \mathrm{Fn}(\omega_2^M \times \omega, 2)$; then $\mathbb{P}$ has c.c.c. in M and thus preserves cardinals, so $\omega_2^M = \omega_2^{M[G]}$. It follows by Lemma 5.3 that $(2^\omega \geq \omega_2)^{M[G]}$.

The next question is whether 2^ω can be exactly ω_2. It is easy to see that if $(2^\omega \geq \omega_3)^M$, the same would be true in any cardinal-preserving extension of M. However, we shall show that if M is a model for GCH, forcing with $\mathrm{Fn}(\omega_2^M \times \omega, 2)$ makes 2^ω exactly ω_2 in $M[G]$. More generally, we shall use the values of cardinal exponents in M to put an upper bound on cardinal exponents in $M[G]$.

We obtain upper bounds by doing the proof of the Power Set Axiom in $M[G]$ slightly more carefully. For Power Set it was sufficient, given a $\sigma \in M^{\mathbb{P}}$, to obtain in M some set S of names which represented all possible subsets of σ. Now, we try to obtain such an S of small cardinality.

5.11. DEFINITION. If $\sigma \in \mathbf{V}^{\mathbb{P}}$, a *nice name* for a subset of σ is $\tau \in \mathbf{V}^{\mathbb{P}}$ of the form $\bigcup\{\{\pi\} \times A_\pi \colon \pi \in \mathrm{dom}(\sigma)\}$, where each A_π is an antichain in $\mathbb{P}$. □

As usual, we plan to use this notion within M, but the property of being a nice name is absolute.

5.12. LEMMA. *If $\mathbb{P} \in M$ and $\sigma, \mu \in M^{\mathbb{P}}$, then there is a nice name $\tau \in M^{\mathbb{P}}$ for a subset of σ such that*

$$\mathbb{1} \Vdash (\mu \subset \sigma \rightarrow \mu = \tau).$$

PROOF. For each $\pi \in \mathrm{dom}(\sigma)$, let $A_\pi \subset \mathbb{P}$ be such that:

(1) $\forall p \in A_\pi (p \Vdash \pi \in \mu)$,

(2) A_π is an antichain in $\mathbb{P}$, and

(3) A_π is maximal with respect to (1) and (2).

We may assume $\langle A_\pi \colon \pi \in \mathrm{dom}(\sigma)\rangle \in M$ by definability of $\Vdash$ and Zorn's Lemma applied within M. Let

$$\tau = \bigcup\{\{\pi\} \times A_\pi \colon \pi \in \mathrm{dom}(\sigma)\}.$$

To show that $\mathbb{1} \Vdash (\mu \subset \sigma \rightarrow \mu = \tau)$, we show that whenever G is $\mathbb{P}$-generic over M, $\mu_G \subset \sigma_G \rightarrow \mu_G = \tau_G$. Assume $\mu_G \subset \sigma_G$.

To show $\mu_G \subset \tau_G$, fix $a \in \mu_G$. Since $\mu_G \subset \sigma_G$, $a = \pi_G$ for some $\pi \in \mathrm{dom}(\sigma)$. If $A_\pi \cap G \neq 0$, fix $p \in A_\pi \cap G$; then $\langle \pi, p\rangle \in \tau$ and $p \in G$, so $a = \pi_G \in \tau_G$. However, if $A_\pi \cap G = 0$, let $q \in G$ be such that $\forall p \in A\, (p \perp q)$ (see Lemma

2.20(a)). Let $q' \in G$ be such that $q' \Vdash \pi \in \mu$, and let r be a common extension of q and q'; then $A_\pi \cup \{r\}$ satisfies (1) and (2) above, contradicting maximality of A_π.

To show $\tau_G \subset \mu_G$, fix $a \in \tau_G$; then $a = \pi_G$, where $\langle \pi, p \rangle \in \tau$ for some $p \in G$. By definition of τ, $p \Vdash \pi \in \mu$, so $a = \pi_G \in \mu_G$. □

If τ is a nice name for a subset of σ, it need not in general be true that $\tau_G \subset \sigma_G$, but that is irrelevant. The important fact is that every subset of σ does get represented by a nice name.

5.13. Lemma. *Assume that* $\mathbb{P} \in M$ *and that in* M, $\mathbb{P}$ *is* c.c.c., $|\mathbb{P}| = \kappa \geq \omega$, λ *is an infinite cardinal, and* $\theta = \kappa^\lambda$ *(i.e., the preceding holds relativized to* M*). Let* G *be* $\mathbb{P}$*-generic over* M*. Then in* $M[G]$, $2^\lambda \leq \theta$.

Proof In M, every antichain in $\mathbb{P}$ is countable, so there are at most κ^ω such antichains. Since $\mathrm{dom}(\check{\lambda}) = \{\check{\xi} \colon \xi < \lambda\}$ has cardinality λ, there are at most $(\kappa^\omega)^\lambda = \kappa^\lambda = \theta$ nice names for subsets of $\check{\lambda}$. Let $\tau_\alpha (\alpha < \theta)$ enumerate, in M, all nice names for subsets of $\check{\lambda}$.

In $M[G]$, there is a function f with domain θ such that $f(\alpha) = \mathrm{val}(\tau_\alpha, G)$ for each $\alpha < \theta$; namely, $f = \pi_G$, where $\pi = \{\langle \mathrm{op}(\check{\alpha}, \tau_\alpha), \mathbb{1} \rangle \colon \alpha < \theta\}$. But by Lemma 5.12, $\mathscr{P}(\lambda)^{M[G]} \subset \mathrm{ran}(f)$, so $(2^\lambda \leq \theta)^{M[G]}$. □

This may be applied to show that it is consistent that 2^ω can be almost anything.

5.14. Lemma. *Let* κ *be an infinite cardinal of* M *such that* $(\kappa^\omega = \kappa)^M$, *and let* $\mathbb{P} = \mathrm{Fn}(\kappa \times \omega, 2)$. *Let* G *be* $\mathbb{P}$*-generic over* M*. Then* $(2^\omega = \kappa)^{M[G]}$.

Proof. Applying Lemma 5.13 with $\lambda = \omega$ yields $2^\omega \leq \kappa$ in M. But by Lemma 5.3, $2^\omega \geq \kappa$ in M; κ is still a cardinal in $M[G]$ since $\mathbb{P}$ has c.c.c. in M. □

In particular, if M satisfies GCH, then in M, $\kappa^\omega = \kappa$ whenever $\mathrm{cf}(\kappa) > \omega$ (see I 10.42). It follows that it is consistent for the continuum to be anything not cofinal with ω (by König's Lemma (I 10.41), $\mathrm{cf}(2^\omega) > \omega$). E.g., we may prove the following.

5.15. Corollary. Con(ZFC) *implies*

(a) $\mathrm{Con}(\mathrm{ZFC} + 2^\omega = \omega_2)$,

(b) $\mathrm{Con}(\mathrm{ZFC} + 2^\omega = \omega_{\omega_1})$, *etc.*

Proof. The fact that our method of generic extensions yields relative consistency proofs was discussed in §1. We may start with M satisfying

ZFC + GCH since in ZFC we can prove the existence of a c.t.m. for any finite number of axioms of ZFC + $\mathbf{V} = \mathbf{L}$ (see VI 4.10), and $\mathbf{V} = \mathbf{L}$ implies GCH.

Thus, to obtain (b), start with M satisfying GCH and apply Lemma 5.14 with $(\kappa = \omega_{\omega_1})^M$. Then in $M[G]$, $2^\omega = \kappa$. Since $\mathbb{P}$ preserves cardinals $\kappa = \omega_{\omega_1}$ in $M[G]$. □

The continuum can also be weakly inaccessible.

5.16. Corollary. *The following four theories are equiconsistent; i.e.,*

$$\mathrm{Con}(T_1) \leftrightarrow \mathrm{Con}(T_2) \leftrightarrow \mathrm{Con}(T_3) \leftrightarrow \mathrm{Con}(T_4),$$

where

T_1 *is* ZFC + GCH + $\exists\kappa$ (κ *is strongly inaccessible*).

T_2 *is* ZFC + $\exists\kappa$ (κ *is weakly inaccessible*).

T_3 *is* ZFC + 2^ω *is weakly inaccessible.*

T_4 *is* ZFC + $\exists\kappa < 2^\omega$(κ *is weakly inaccessible*).

Proof. Both $\mathrm{Con}(T_3)$ and $\mathrm{Con}(T_4)$ obviously imply $\mathrm{Con}(T_2)$. To see that $\mathrm{Con}(T_2) \to \mathrm{Con}(T_1)$, observe that as a theorem of ZFC, if κ is weakly inaccessible, then κ is weakly inaccessible in $\mathbf{L}$ and hence (by GCH in $\mathbf{L}$), strongly inaccessible in $\mathbf{L}$. Thus, within T_2 we can prove that $\mathbf{L}$ is an inner model for T_1.

To prove that $\mathrm{Con}(T_1)$ implies $\mathrm{Con}(T_3)$ and $\mathrm{Con}(T_4)$, let M be a c.t.m. for T_1. If $\mathbb{P}$ is c.c.c. in M and κ is weakly inaccessible in M, then, by preservation of cofinalities, κ will be both regular and a limit cardinal in $M[G]$, and hence κ will remain weakly inaccessible in $M[G]$. Thus, if $\lambda > \kappa$ and (λ is a cardinal)^M, then forcing with $\mathbb{P} = \mathrm{Fn}(\lambda \times \omega, 2)$ makes $M[G]$ a model for T_4. If κ is strongly inaccessible in M, then $(\kappa^\omega = \kappa)^M$, so forcing with $\mathbb{P} = \mathrm{Fn}(\kappa \times \omega, 2)$ makes $(2^\omega = \kappa)^{M[G]}$, whence $M[G]$ satisfies T_3.

Formally, to see that the previous paragraph yields a finitistic relative consistency proof of $\mathrm{Con}(T_1) \to \mathrm{Con}(T_3)$ (or of $\mathrm{Con}(T_1) \to \mathrm{Con}(T_4)$), we apply the discussion in §1 with T_1 as the basic theory instead of ZFC. Thus, within T_1, we may prove the existence of a c.t.m., M, for any desired finite list of axioms of T_1 (see IV 7.11), and then, by forcing, produce a c.t.m. $M[G]$ for any finite list of axioms of T_3. □

By the Gödel Incompleteness Theorem, we cannot expect to produce relative consistency proofs of the form $\mathrm{Con}(\mathrm{ZFC}) \to \mathrm{Con}(T_1)$; see IV §10.

It is also possible to calculate powers of uncountable cardinals in extensions by $\mathrm{Fn}(\kappa \times \omega, 2)$; see Exercise G1. A special case of this, when $\kappa = 1$, yields a quotable relative consistency result.

5.17. COROLLARY. $\mathrm{Con(ZFC)} \rightarrow \mathrm{Con(ZFC + GCH + \mathbf{V} \neq \mathbf{L})}$.

PROOF. Start with M satisfying GCH. Let $\mathbb{P} = \mathrm{Fn}(\omega, 2)$. As pointed out in the proof of Corollary 4.3, $M[G]$ satisfies $\mathbf{V} \neq \mathbf{L}$. If λ is an infinite cardinal of M, let $\theta = (\lambda^+)^M = (\omega^\lambda)^M$. By Lemma 5.13, $(2^\lambda \leq \theta)^{M[G]}$. Thus, $\forall \lambda \geq \omega (2^\lambda \leq \lambda^+)^{M[G]}$ so GCH holds in $M[G]$. □

Finally, we remark that extensions by $\mathrm{Fn}(I, 2)$ cannot be used to get a model of MA + $\neg$CH (Exercise G7); this will require a much more complicated p.o. (see VIII §6).

§6. Forcing with partial functions of larger cardinality

The p.o.'s considered here enable us to violate GCH at larger cardinals without violating CH. Again M is always a fixed c.t.m. for ZFC.

6.1. DEFINITION. For any infinite cardinal λ,

$$\mathrm{Fn}(I, J, \lambda) = \{p \colon |p| < \lambda \wedge p \text{ is a function} \wedge \mathrm{dom}(p) \subset I \wedge \mathrm{ran}(p) \subset J\}.$$

Order $\mathrm{Fn}(I, J, \lambda)$ by: $p \leq q \leftrightarrow q \subset p$. □

Thus, $\mathrm{Fn}(I, J) = \mathrm{Fn}(I, J, \omega)$. As with $\lambda = \omega$, $\mathrm{Fn}(I, J, \lambda)$ is a p.o. with largest element $\mathbb{1} = 0$.

When $\lambda > \omega$, $\mathrm{Fn}(I, J, \lambda)$ is *not* absolute for M. In forcing, we always use $\mathrm{Fn}(I, J, \lambda)^M$ where (λ is a cardinal)^M. Interesting results are only obtained when also (λ is regular)^M, but this restriction does not appear in the elementary discussion.

Analogously to Lemma 5.2 we have the following.

6.2. LEMMA. *If* $I, J, \lambda \in M$, *(*λ *is a cardinal)*^{M}, $J \neq 0$ $(|I| \geq \lambda)^M$, *and* G *is* $\mathrm{Fn}(I, J, \lambda)^M$*-generic over* M, *then* $\bigcup G$ *is a function from* I *onto* J. □

Continuing to replace ω by λ in the discussion of §5, we see that when $I = \kappa \times \lambda$ and $J = 2$, we may think of $\bigcup G$ as coding a κ-sequence of distinct functions from λ into 2, so, analogously to Lemma 5.3, we have the following.

6.3. LEMMA. *If (*λ *is a cardinal)*^{M}; $\kappa \in M$, *and* G *is* $\mathrm{Fn}(\kappa \times \lambda, 2, \lambda)^M$*-generic over* M, *then* $(2^{|\lambda|} \geq |\kappa|)^{M[G]}$. □

As in §5, the difficult part of the discussion involves checking that cardinals are preserved. Some new ideas will be needed here, since if $\lambda > \omega$, then $\mathrm{Fn}(I, J, \lambda)$ has the c.c.c. only in trivial cases (namely $|I| < \omega$ or $|J| \le 1$).

Our argument will split into two parts. First, we modify the c.c.c. argument to check that cardinals $>\lambda$ are preserved. Next, we introduce a new idea to check that cardinals $\le\lambda$ in M remain cardinals in $M[G]$; this fact was trivial when λ was ω. For our arguments to work, we shall eventually need that λ is regular and $2^{<\lambda} = \lambda$ in M.

As in §5, we shall verify preservation of cardinals by verifying preservation of cofinalities. In analogy with 5.6–5.9,

6.4. Definition. Assume that $\mathbb{P} \in M$ and θ is an infinite cardinal of M.

(1) $\mathbb{P}$ *preserves cardinals* $\ge\theta$ (or $\le\theta$) iff whenever G is $\mathbb{P}$-generic over M, $\beta \in o(M)$, and $\beta \ge \theta$ (resp., $\beta \le \theta$),

$$(\beta \text{ is a cardinal})^M \leftrightarrow (\beta \text{ is a cardinal})^{M[G]}.$$

(2) $\mathbb{P}$ *preserves cofinalities* $\ge\theta$ (or $\le\theta$) iff whenever G is $\mathbb{P}$-generic over M, γ is a limit ordinal in M, and $\mathrm{cf}(\gamma)^M \ge \theta$ (resp., $\mathrm{cf}(\gamma)^M \le \theta$), then

$$\mathrm{cf}(\gamma)^M = \mathrm{cf}(\gamma)^{M[G]}. \quad \square$$

6.5. Lemma. *Under the assumptions of Definition* 6.4, *if* $\mathbb{P}$ *preserves cofinalities* $\le\theta$, *then* $\mathbb{P}$ *preserves cardinals* $\le\theta$. *If* $\mathbb{P}$ *preserves cofinalities* $\ge\theta$ *and* (θ *is regular*)^M, *then* $\mathbb{P}$ *preserves cardinals* $\ge\theta$. $\square$

6.6. Lemma. *Under the assumptions of Definition* 6.4, *assume also that whenever* κ *is a regular cardinal of* M, $\kappa \ge \theta$, *and* G *is* $\mathbb{P}$-*generic over* M, *then* (κ *is regular*)$^{M[G]}$. *Then* $\mathbb{P}$ *preserves cofinalities* $\ge\theta$. *Likewise with* $\le\theta$ *replacing* $\ge\theta$. $\square$

If, in the definition of c.c.c., we weaken "countable" to "$<\theta$", then we preserve cofinalities $\ge\theta$.

6.7. Definition. $\mathbb{P}$ has the θ-chain condition (θ-c.c.) iff every antichain in $\mathbb{P}$ has cardinality $<\theta$. $\square$

Thus, the c.c.c. is the ω_1-c.c. Exactly as in 5.5 and 5.10, we have the following.

6.8. Lemma. *Assume* $\mathbb{P} \in M$, $A, B \in M$, *and, in* M, θ *is a cardinal and* $\mathbb{P}$ *is* θ-*c.c. Let* G *be* $\mathbb{P}$-*generic over* M, *and let* $f \in M[G]$, *with* $f : A \to B$. *Then there is a map* $F : A \to \mathscr{P}(B)$ *with* $F \in M$, $\forall a \in A\,(f(a) \in F(a))$, *and* $\forall a \in A\,(|F(a)| < \theta)^M$. $\square$

6.9. LEMMA. *Assume* $\mathbb{P} \in M$, θ *is a cardinal of* M, *and* ($\mathbb{P}$ *has the* θ-c.c.)^M. *Then* $\mathbb{P}$ *preserves cofinalities* $\geq \theta$. *Hence, if also* (θ *is regular*)^M, $\mathbb{P}$ *preserves cardinals* $\geq \theta$. □

We remark on which chain conditions will occur in practice. Let c.c.($\mathbb{P}$) be the least θ such that $\mathbb{P}$ has the θ-c.c. By a theorem of Tarski (see Exercise F4), c.c.($\mathbb{P}$) is finite or regular; it follows that the assumption (θ is regular)^M may be dropped in Lemma 6.9. Also c.c.($\mathbb{P}$) cannot be ω (Exercise F1). If c.c.($\mathbb{P}$) $< \omega$, then $\mathbb{P}$ is uninteresting for forcing (Exercise F2); but if $0 < n < \omega$ then there is a $\mathbb{P}$ with c.c.($\mathbb{P}$) $= n$—namely, $\text{Fn}(1, n-1)$. If θ is weakly inaccessible, then there is an important example of a $\mathbb{P}$ with c.c.($\mathbb{P}$) $= \theta$—namely, the Lévy order (see §8).

Finally, assume $\theta = \lambda^+$. $\text{Fn}(1, \lambda)$ is a trivial example of a $\mathbb{P}$ with c.c.($\mathbb{P}$) $= \theta$. More important, if $\mathbb{P} = \text{Fn}(I, 2, \lambda)$, then under GCH, c.c.($\mathbb{P}$) $= \lambda^+$ if $|I| \geq \lambda$; without GCH, c.c.($\mathbb{P}$) $= (2^{<\lambda})^+$. We leave the fact that c.c.($\mathbb{P}$) $\geq (2^{<\lambda})^+$ as an exercise (F5), but we prove that c.c.($\mathbb{P}$) $\leq (2^{<\lambda})^+$ since that is important for preservation of cardinals.

6.10. LEMMA. $\text{Fn}(I, J, \lambda)$ *has the* $(|J|^{<\lambda})^+$-c.c.

PROOF. Let $\theta = (|J|^{<\lambda})^+$, and suppose that $\{p_\xi : \xi < \theta\}$ formed an antichain. First, assume λ is regular. Then $(|J|^{<\lambda})^{<\lambda} = |J|^{<\lambda}$, so $\forall \alpha < \theta (|\alpha^{<\lambda}| < \theta)$, so by the Δ-system lemma (see II 1.6) there is an $X \subset \theta$ with $|X| = \theta$ such that $\{\text{dom}(p_\xi) : \xi \in X\}$ forms a Δ-system with some root r. Since there are less than θ possibilities for $p_\xi \restriction r$, we have a contradiction as in the proof for $\lambda = \omega$ (see Lemma 5.4).

If λ is singular, then since θ is regular and $> \lambda$, we could find a regular $\lambda' < \lambda$ such that $Y = \{\xi : |p_\xi| < \lambda'\}$ has cardinality θ. Then $\{p_\xi : \xi \in Y\}$ contradicts the $(|J|^{<\lambda'})^+$-c.c. which we have just proved for regular λ'. □

6.11. COROLLARY. *Assume* $I, J \in M$, *and, in* M, λ *is regular*, $|J| \leq 2^{<\lambda}$, *and* $\theta = (2^{<\lambda})^+$. *Then* $\text{Fn}(I, J, \lambda)^M$ *preserves cofinalities and cardinals* $\geq \theta$.

PROOF. Applying Lemma 6.10 within M, $\text{Fn}(I, J, \lambda)^M$ has the θ-c.c. in M, since $(|J|^{<\lambda} = 2^{<\lambda})^M$. Now apply Lemma 6.9. □

By a completely different argument, we shall now show that if λ is regular in M, then $(\text{Fn}(I, \kappa, \lambda))^M$ preserves cofinalities and cardinals $\leq \lambda$. Under GCH, $2^{<\lambda} = \lambda$, so using Corollary 6.11, all cofinalities and cardinals will be preserved. However, if in M there are cardinals κ with $\lambda^+ \leq \kappa \leq 2^{<\lambda}$, then except in trivial cases such κ will have cardinality λ, and hence cease to be cardinals, in $M[G]$ (see Exercise G3).

6.12. DEFINITION. A p.o. $\mathbb{P}$ is *λ-closed* iff whenever $\gamma < \lambda$ and $\{p_\xi : \xi < \gamma\}$ is a decreasing sequence of elements of $\mathbb{P}$ (i.e., $\xi < \eta \rightarrow p_\xi \geq p_\eta$), then

$$\exists q \in \mathbb{P}\, \forall \xi < \gamma (q \leq p_\xi). \quad \square$$

6.13. LEMMA. *If λ is regular, then* $\mathrm{Fn}(I, J, \lambda)$ *is λ-closed.*

PROOF. The q of Definition 6.12 is just $\bigcup \{p_\xi : \xi < \gamma\}$. $|q| < \lambda$ since each $|p_\xi| < \lambda$ and λ is regular. $\square$

If λ is singular, $\mathrm{Fn}(\lambda, 2, \lambda)$ is not λ-closed. Also, if $(\lambda \text{ is singular})^M$, $\mathrm{Fn}(\lambda, 2, \lambda)^M$ collapses λ (Exercise G5).

If λ is regular, then the fact that $\mathrm{Fn}(I, J, \lambda)$ is λ-closed will be used to show that cardinals $\leq \lambda$ are preserved.

The following result should be compared with Lemma 6.8. Lemma 6.8 used a chain condition to approximate, in M, functions from A to B in $M[G]$. Theorem 6.14 shows that functions from A to B are in fact in M if A is small enough.

6.14. THEOREM. *Assume* $\mathbb{P} \in M$, $A, B \in M$, *and, in* M, λ *is a cardinal,* $\mathbb{P}$ *is λ-closed, and* $|A| < \lambda$. *Let* G *be* $\mathbb{P}$*-generic over* M *and let* $f \in M[G]$ *with* $f : A \rightarrow B$. *Then* $f \in M$.

PROOF. Observe first that it is sufficient to prove this with A an ordinal, $A = \alpha < \lambda$. For, then, to prove the general result, we let $j \in M$ be a 1–1 map from $\alpha = |A|^M < \lambda$, onto A, and apply the special case with $f \circ j : \alpha \rightarrow B$ to show that $f \circ j$, and hence f, is in M.

Now, let $K = ({}^\alpha B)^M = {}^\alpha B \cap M$, and $f \in {}^\alpha B \cap M[G]$. We wish to show $f \in K$. If not, fix $\tau \in M^{\mathbb{P}}$ with $f = \tau_G$, and then fix $p \in G$ such that

$$p \Vdash (\tau \text{ is a function from } \check{\alpha} \text{ into } \check{B} \wedge \tau \notin \check{K}). \qquad (*)$$

We now forget about f and G and derive a contradiction directly from $(*)$.

Within M: use transfinite recursion plus AC to choose sequences $\{p_\eta : \eta \leq \alpha\}$ from $\mathbb{P}$ and $\{z_\eta : \eta < \alpha\}$ from B so that

(1) $p_0 = p$,

(2) $p_\eta \leq p_\xi$ for all $\xi \leq \eta$, and

(3) $p_{\eta+1} \Vdash \tau(\check{\eta}) = \check{z}_\eta$.

For the successor steps in this recursion, we are given p_η, and we find $p_{\eta+1}$ and z_η as follows: $p_\eta \leq p$, so

$$p_\eta \Vdash \tau \text{ is a function from } \check{\alpha} \text{ into } \check{B},$$

so (since a consequence of a forced statement is forced—see Lemma 4.5(a)),

$$p_\eta \Vdash \exists x \in \check{B}\, (\tau(\check{\eta}) = x).$$

Thus, by Corollary 3.7(d), there is a $z_\eta \in B$ and a $p_{\eta+1} \leq p_\eta$ such that $p_{\eta+1} \Vdash \tau(\check{\eta}) = \check{z}_\eta$. At the limit steps, p_η, for η a limit, may be chosen to satisfy (2) by the definition of λ-closed.

Still in M, let $g = \langle z_\eta : \eta < \alpha \rangle$; i.e., g is the function with domain α such that $g(\eta) = z_\eta$ for each η. Then $g \in K$.

Let H be $\mathbb{P}$-generic over M, with $p_\alpha \in H$, and hence each $p_\eta \in H$. Then $\tau_H(\eta) = z_\eta$ for each $\eta < \alpha$, so $\tau_H = g \in K$. But $p_0 = p \Vdash \tau \notin \check{K}$, so $\tau_H \notin K$, a contradiction. □

6.15. COROLLARY. *Assume* $\mathbb{P} \in M$, *(*λ *is a cardinal)*^{M}, *and (*$\mathbb{P}$ *is* λ*-closed)*^{M}; *then* $\mathbb{P}$ *preserves cofinalities* $\leq \lambda$, *and hence cardinals* $\leq \lambda$.

PROOF. If not, then, by Lemma 6.6, there is a $\kappa \leq \lambda$ such that (κ is regular)^M but (κ is singular)$^{M[G]}$. Thus, there is an $\alpha < \kappa$ and an $f \in M[G]$ which maps α cofinally into κ. By Theorem 6.14, $f \in M$, contradicting (κ is regular)^M. □

6.16. THEOREM. *Let* $\lambda, I, J \in M$, *and assume that in* M, λ *is regular*, $2^{<\lambda} = \lambda$, *and* $|J| \leq \lambda$; *then* $\mathrm{Fn}(I, J, \lambda)^M$ *preserves cofinalities (and hence cardinals).*

PROOF. By regularity of λ, $\mathrm{Fn}(I, J, \lambda)^M$ is λ-closed in M, and hence preserves cofinalities $\leq \lambda$ (see 6.13–6.15). By $2^{<\lambda} = \lambda$, $\mathrm{Fn}(I, J, \lambda)^M$ has the λ^+-c.c. in M and hence preserves cofinalities $\geq (\lambda^+)^M$ (see 6.8–6.11). □

In particular, we may now force with orders of the form $\mathrm{Fn}(\kappa \times \lambda, 2, \lambda)^M$ to violate GCH as badly as we wish at λ. We may use nice names, as in §5, to obtain a precise computation of 2^λ in $M[G]$. Generalizing Corollary 5.15, we have the following.

6.17. THEOREM. *In* M, *assume that* $\lambda < \kappa$, λ *is regular*, $2^{<\lambda} = \lambda$, *and* $\kappa^\lambda = \kappa$. *Let* $\mathbb{P} = \mathrm{Fn}(\kappa \times \lambda, 2, \lambda)^M$. *Then* $\mathbb{P}$ *preserves cardinals and if* G *is* $\mathbb{P}$*-generic over* M, *then* $(2^\lambda = \kappa)^{M[G]}$.

PROOF. We just proved preservation of cardinals, and $(2^\lambda \geq \kappa)^{M[G]}$ is easy (see Lemma 6.3). We must show $(2^\lambda \leq \kappa)^{M[G]}$.

In M, $\mathbb{P}$ has cardinality $\kappa^{<\lambda} = \kappa$, and $\mathbb{P}$ has the λ^+-c.c., so there are at most $\kappa^\lambda = \kappa$ antichains in $\mathbb{P}$. Hence, there are at most $\kappa^\lambda = \kappa$ nice names for subsets of λ. Let $\langle \tau_\alpha : \alpha < \kappa \rangle$ enumerate these, and let

$$\pi = \{\langle \mathrm{op}(\check{\alpha}, \tau_\alpha), \mathbb{1} \rangle : \alpha < \kappa\}.$$

Then, as in the proof of Lemma 5.13, in $M[G]$, π_G is a function, $\mathrm{dom}(\pi_G) = \kappa$, and $\mathscr{P}(\lambda) \subset \mathrm{ran}(\pi_G)$, so $2^\lambda \leq \kappa$. □

One can generalize Theorem 6.17 to compute the powers of all cardinals in $M[G]$ (not only λ) in terms of cardinal arithmetic in M, but it is probably better always to refer back to the method of Theorem 6.17 instead of trying to memorize the most general result.

We may use the method of Theorem 6.17 to violate GCH as desired at any regular cardinal, or even at any finite number of regular cardinals. As examples, we prove the following.

6.18. THEOREM. *If* ZFC *is consistent, so are*:

(a) ZFC + CH + $2^{\omega_1} = \omega_2 + 2^{\omega_2} = \omega_{\omega_8}$.

(b) ZFC + CH + $2^{\omega_1} = \omega_5 + 2^{\omega_2} = \omega_7$.

(c) ZFC + $2^{\omega} = \omega_3 + 2^{\omega_1} = \omega_4 + 2^{\omega_2} = \omega_6$.

PROOF. In all cases, start with M satisfying ZFC + GCH.

For (a) Let $\mathbb{P} = (\mathrm{Fn}(\omega_{\omega_8} \times \omega_2, 2, \omega_2))^M$. By Theorem 6.17, $\mathbb{P}$ preserves cardinals and if G is $\mathbb{P}$-generic over M, $(2^{\omega_2} = \omega_{\omega_8})^{M[G]}$. The fact that $2^{\omega_1} = \omega_2$ holds in $M[G]$ follows from the fact that $({}^{(\omega_1)}2)^M = ({}^{(\omega_1)}2)^{M[G]}$ by Theorem 6.14. Thus, if $F \in M$ and (F maps ω_2 onto ${}^{(\omega_1)}2)^M$, then (F maps ω_2 onto ${}^{(\omega_1)}2)^{M[G]}$. Likewise, $(2^{\omega} = \omega_1)^{M[G]}$.

For (b) we force twice. Let $\mathbb{P} = (\mathrm{Fn}(\omega_7 \times \omega_2, 2, \omega_2))^M$, let G be $\mathbb{P}$-generic over M, and let $N = M[G]$; then as in (a),

$$(2^{\omega} = \omega_1 \wedge 2^{\omega_1} = \omega_2 \wedge 2^{\omega_2} = \omega_7)^N.$$

Furthermore, $(\kappa^{\omega_1} = \kappa)^N$ whenever $(\kappa \geq \omega_2 \wedge \kappa \text{ is regular})^N$, since this is true in M by $(\mathrm{GCH})^M$, and $({}^{(\omega_1)}\kappa)^M = ({}^{(\omega_1)}\kappa)^N$. We now apply our results on forcing with N as the ground model instead of M. Let

$$\mathbb{Q} = (\mathrm{Fn}(\omega_5 \times \omega_1, 2, \omega_1))^N.$$

By $(2^{<\omega_1} = \omega_1)^N$, $\mathbb{Q}$ preserves cardinals. Let H be $\mathbb{Q}$-generic over N. $(\mathrm{CH})^{N[H]}$ is proved as in (a). $(2^{\omega_2} \geq \omega_7)^{N[H]}$ follows from $(2^{\omega_2} \geq \omega_7)^N$. To see that in fact $(2^{\omega_2} = \omega_7)^{N[H]}$, use the method of Theorem 6.17; namely, in N, $\mathbb{Q}$ has the ω_2-c.c. and $|\mathbb{Q}| = \omega_5^{\omega_1} = \omega_5$, so there are only $((\omega_5)^{\omega_1})^{\omega_2} = \omega_7$ nice names for subsets of ω_2. To see that $(2^{\omega_1} = \omega_5)^{N[H]}$, apply Theorem 6.17 directly, plus the fact that $(\omega_5^{\omega_1} = \omega_5)^N$.

For (c), force three times, and construct $M \subset N_1 \subset N_2 \subset N_3$. N_1 satisfies $2^{\omega} = \omega_1 \wedge 2^{\omega_1} = \omega_2 \wedge 2^{\omega_2} = \omega_6$, N_2 satisfies $2^{\omega} = \omega_1 \wedge 2^{\omega_1} = \omega_4 \wedge 2^{\omega_2} = \omega_6$, and N_3 satisfies (c). □

In proving (b) and (c), it is very important that we proceed *backwards*, dealing with the largest cardinal first. For example, if we tried to prove (b) by letting $\mathbb{P} = (\mathrm{Fn}(\omega_5 \times \omega_1, 2, \omega_1))^M$ and $N = M[G]$, where G is $\mathbb{P}$-generic over M, then N would satisfy $2^{\omega_1} = \omega_5$. Thus, $(2^{<\omega_2} \neq \omega_2)^N$, so if we set $\mathbb{Q} = (\mathrm{Fn}(\omega_7 \times \omega_2, 2, \omega_2))^N$, $\mathbb{Q}$ would not preserve cardinals. In fact, if H is $\mathbb{Q}$-generic over N, then $(\omega_5)^N$ would have cardinality ω_2 in $N[H]$ (see Exercise G3), and $N[H]$ would satisfy $2^{\omega_1} = \omega_2$.

We may easily generalize Theorem 6.18 to deal with the powers of any finite number of regular cardinals (see Exercise G6). The following two questions now suggest themselves.

Question 1. How free are we to monkey with the powers of regular cardinals?

The answer, due to Easton, is: as free as we wish, subject, of course, to monotonicity $(\lambda < \lambda' \to 2^\lambda \leq 2^{\lambda'})$ and König's Lemma $(\mathrm{cf}(2^\lambda) > \lambda$; see I 10.41). We discuss this further in VIII §4. We merely remark here that some new idea is needed. Say, e.g., we start with M satisfying GCH and we want $M[G]$ to satisfy $\forall n \in \omega\, (2^{\omega_n} = \omega_{n+3})$. The method of proof of Theorem 6.18 indicates that we should iterate forcing ω times, starting with the *largest* ω_n, which is clearly nonsense. Even if we could turn this around, one cannot naively repeat the forcing process ω times and expect to produce a model of ZFC (see Exercise B6).

Question 2. What about singular cardinals?

If λ is singular in M, p.o.'s of the form $\mathrm{Fn}(I, \kappa, \lambda)^M$ will always collapse λ (Exercise G5), so questions about powers of singular cardinals are not settled by the methods of this section. In fact, there are restrictions beyond monotonicity and König's Lemma on the powers of singular cardinals; see the end of VIII §4 for more details.

§7. Embeddings, isomorphisms, and Boolean-valued models

As before, M is always a fixed c.t.m. for ZFC.

Suppose $\mathbb{P}$ and $\mathbb{Q}$ are p.o.'s in M, $i \in M$, and $i : \mathbb{P} \to \mathbb{Q}$ embeds $\mathbb{P}$ as a sub-order of $\mathbb{Q}$. We shall show how, under suitable restrictions on i, we may use i to embed the whole $\mathbb{P}$ forcing apparatus into the $\mathbb{Q}$ forcing apparatus. This one idea has many diverse and seemingly unrelated applications. For one, the special case where $\mathbb{P} = \mathbb{Q}$ is the key idea behind constructing models of ZF + $\neg$AC (see Exercises E1–5). For another, if $\mathbb{P} \neq \mathbb{Q}$ but i is an isomorphism, we may show that isomorphic p.o.'s lead to identical generic extensions (provided that the isomorphism is in M). A third application, when $\mathbb{P} \subset \mathbb{Q}$ and i is inclusion, is that we may regard larger p.o.'s as yielding larger generic extensions; this will be of vital importance for iterated forcing in VIII. Finally, we shall use these ideas to relate forcing to Boolean-valued models.

For Boolean-valued models, we consider the special case where $\mathbb{Q} = \{b \in \mathscr{B}: b > 0\}$, where ($\mathscr{B}$ is a complete Boolean algebra)^M. For this discussion we shall assume that the reader is familiar with the relationship between p.o.'s and Boolean algebras discussed in II §3. However, the reader who is not interested in Boolean-valued models may simply skip all references to Boolean algebras in this section without loss of continuity.

We begin by examining the third application above. Suppose that $\mathbb{P}$ is a sub-order of $\mathbb{Q}$; i.e., $\mathbb{P} \subset \mathbb{Q}$ and $\leq_{\mathbb{P}} = \leq_{\mathbb{Q}} \cap \mathbb{P} \times \mathbb{P}$, then $\mathbb{Q}$ "should" yield a bigger extension than does $\mathbb{P}$; if H is $\mathbb{Q}$-generic over M, then $H \cap \mathbb{P}$ "should" be $\mathbb{P}$-generic over M, with $M[H \cap \mathbb{P}] \subset M[H]$. However, this is false without some further restrictions on $\mathbb{P}$ and $\mathbb{Q}$.

To appreciate one necessary restriction, suppose that $p_1, p_2 \in \mathbb{P}$ and that p_1 and p_2 are incompatible in $\mathbb{P}$ but are compatible in $\mathbb{Q}$; say $q \in \mathbb{Q}$, $q \leq p_1$, and $q \leq p_2$. If H is $\mathbb{Q}$-generic and $q \in H$, then $p_1, p_2 \in H$, so $H \cap \mathbb{P}$ is not even a filter in $\mathbb{P}$. We must thus require that if p_1 and p_2 are incompatible in $\mathbb{P}$, then they are incompatible in $\mathbb{Q}$ also.

To obtain a second restriction, fix $q \in \mathbb{Q}$, and let $D = \{p \in \mathbb{P}: p \perp q\}$. If H is $\mathbb{Q}$-generic over M and $q \in H$, then $H \cap D = 0$, so if we wish $H \cap \mathbb{P}$ to be $\mathbb{P}$-generic, we had better require that D not be dense in $\mathbb{P}$. Thus, there must be a $p \in \mathbb{P}$ such that

$$\forall p' \in \mathbb{P}\,(p' \leq p \rightarrow p' \text{ and } q \text{ are compatible in } \mathbb{Q}).$$

If these two restrictions hold, we shall say that $\mathbb{P}$ is *completely contained* in $\mathbb{Q}$, or $\mathbb{P} \subset_c \mathbb{Q}$. These restrictions are sufficient, as we shall see in Theorem 7.5. We now present the formal development in the somewhat more general framework of an embedding from one p.o. into another.

7.1. Definition. Let $\langle \mathbb{P}, \leq_{\mathbb{P}}, \mathbb{1}_{\mathbb{P}} \rangle$ and $\langle \mathbb{Q}, \leq_{\mathbb{Q}}, \mathbb{1}_{\mathbb{Q}} \rangle$ be p.o.'s and $i : \mathbb{P} \rightarrow \mathbb{Q}$. i is a *complete embedding* iff

(1) $\forall p, p' \in \mathbb{P}\,(p' \leq p \rightarrow i(p') \leq i(p))$.
(2) $\forall p_1, p_2 \in \mathbb{P}\,(p_1 \perp p_2 \leftrightarrow i(p_1) \perp i(p_2))$.
(3) $\forall q \in \mathbb{Q}\, \exists p \in \mathbb{P}\, \forall p' \in \mathbb{P}\,(p' \leq p \rightarrow (i(p') \text{ and } q \text{ are compatible in } \mathbb{Q}))$.

In (3), we call p *a reduction* of q to $\mathbb{P}$. □

In Definition 7.1, we very quickly dropped the subscripts for $\leq_{\mathbb{P}}$ and $\leq_{\mathbb{Q}}$, since it is clear from context which order is being referenced. Likewise, there should, formally, be subscripts on the $\perp$. Observe in (3) that the reduction, p, of q to $\mathbb{P}$ is not unique; if $p_1 \leq p$, then p_1 is another such reduction.

7.2. Definition. $\langle \mathbb{P}, \leq_{\mathbb{P}}, \mathbb{1}_{\mathbb{P}} \rangle \subset_c \langle \mathbb{Q}, \leq_{\mathbb{Q}}, \mathbb{1}_{\mathbb{Q}} \rangle$, or $\mathbb{P} \subset_c \mathbb{Q}$, iff $\leq_{\mathbb{P}} = \leq_{\mathbb{Q}} \cap \mathbb{P} \times \mathbb{P}$ and the inclusion (identity) map from $\mathbb{P}$ to $\mathbb{Q}$ is a complete embedding. □

If i is an inclusion, as in Definition 7.2, then (1) of Definition 7.1 holds trivially and (2) and (3) of 7.1 are precisely the two restrictions we discussed before stating 7.1. In the more general framework of Definition 7.1, the reader might have expected us to state (1) as

$$\forall p, p' \in \mathbb{P}\,(p' \leq p \leftrightarrow i(p') \leq i(p)),$$

and to add to (1) the requirements that i be 1–1 and that $i(\mathbb{1}_{\mathbb{P}}) = \mathbb{1}_{\mathbb{Q}}$. It is true that in many (but not all) cases of interest, all these additional things hold (see Exercise C2), but the general theory is just as easy to carry out under Definition 7.1 as it stands, and, as we shall see later, this generality is needed in the theory of Boolean-valued models (see also Exercise C9).

We remark that in Definition 7.1(2), the implication from right to left follows from (1), but the implication from left to right says something new.

As a trivial example of Definition 7.1, if all elements of $\mathbb{P}$ are compatible, and $i(p) = \mathbb{1}_{\mathbb{Q}}$ for all $p \in \mathbb{P}$, then i is a complete embedding. More useful examples are given by the following.

7.3. Lemma. (a) *If i is an isomorphism from $\mathbb{P}$ onto $\mathbb{Q}$, then i is a complete embedding.*

(b) *If $I \subset I'$, then* $\mathrm{Fn}(I, J, \kappa) \subset_{\mathrm{c}} \mathrm{Fn}(I', J, \kappa)$.

Proof. For (b), clauses (1) and (2) of Definition 7.1 are clear. For (3), if $q \in \mathrm{Fn}(I', J, \kappa)$, then $q \restriction I$ is a reduction of q to $\mathrm{Fn}(I, J, \kappa)$. □

It should not be presumed that all "naturally occurring" inclusions are complete inclusions. For cxample, $\mathrm{Fn}(\kappa, 2) \subset \mathrm{Fn}(\kappa, 2, \omega_1)$, but this inclusion is not complete if $\kappa \geq \omega$; clauses (1) and (2) of Definition 7.1 hold, but no $q \in \mathrm{Fn}(\kappa, 2, \omega_1)$ with infinite domain has a reduction to $\mathrm{Fn}(\kappa, 2)$. Furthermore, relativizing to a c.t.m. M, $\mathrm{Fn}(\kappa, 2, \omega_1)^M$ cannot be thought of as corresponding to a larger extension than $\mathrm{Fn}(\kappa, 2)$, since $\mathrm{Fn}(\kappa, 2)$ adds new subsets of ω, whereas $\mathrm{Fn}(\kappa, 2, \omega_1)^M$ does not.

We now proceed to show that if $\mathbb{P} \subset_{\mathrm{c}} \mathbb{Q}$, then $\mathbb{Q}$ does yield a bigger extension than does $\mathbb{P}$. As usual, we would expect to relativize all relevant order-theoretic notions to M, but the definitions of "complete embedding" and "$\subset_{\mathrm{c}}$" are easily seen to be absolute for M, so we do not have to relativize these notions.

As a preliminary, we prove the following.

7.4. Lemma. *Suppose $\mathbb{P} \in M$ and $G \subset \mathbb{P}$; then G is $\mathbb{P}$-generic over M iff*

(1) $\forall p, q \in G\, \exists r \in \mathbb{P}\,(r \leq p \wedge r \leq q)$,

(2) $\forall p \in G\, \forall q \in \mathbb{P}\,(q \geq p \rightarrow q \in G)$, *and*

(3) $\forall D \subset \mathbb{P}\,((D \in M \wedge D \text{ dense in } \mathbb{P}) \rightarrow G \cap D \neq 0)$.

Proof. The only difference between (1)–(3) and the definition of generic is that we required G to be a filter, which meant that (1) was strengthened to require r to be in G (see II Definition 2.4). Thus, to prove Lemma 7.4, we assume G satisfies (1)–(3), fix $p, q \in G$, and show that $\exists r \in G\,(r \leq p \wedge r \leq q)$. Let

$$D = \{r \in \mathbb{P}\colon r \perp p \vee r \perp q \vee (r \leq p \wedge r \leq q)\}.$$

D is dense and in M, so by (3) fix $r \in G \cap D$. Since, by (1), elements of G are pairwise compatible, $r \leq p \wedge r \leq q$. □

7.5. Theorem. *Suppose i, $\mathbb{P}$. and $\mathbb{Q}$ are in M, $i : \mathbb{P} \to \mathbb{Q}$, and i is a complete embedding. Let H be $\mathbb{Q}$-generic over M. Then $i^{-1}(H)$ is $\mathbb{P}$-generic over M and $M[i^{-1}(H)] \subset M[H]$.*

Proof. We check first that $i^{-1}(H)$ is generic. Clauses (1) and (2) of Lemma 7.4 are easily verified using clauses (2) and (1), respectively, of Definition 7.1. For (3), fix $D \in M$ with D dense in $\mathbb{P}$. If $i^{-1}(H) \cap D = 0$, then $H \cap i''D = 0$, so there is a $q \in H$ such that $\forall q' \in i''D\,(q' \perp q)$ (see Lemma 2.20(a)), so $\forall p' \in D\,(i(p') \perp q)$. If p is a reduction of q to $\mathbb{P}$, then for all $p' \leq p$, $\neg(i(p') \perp q)$, so $p' \notin D$, which is impossible if D is dense.

Since $i \in M \subset M[H]$ and $H \in M[H]$, we have $i^{-1}(H) \in M[H]$, whence $M[i^{-1}(H)] \subset M[H]$ by minimality of $M[i^{-1}(H)]$ (see Lemma 2.9). □

Also, if we are given a $G \subset \mathbb{P}$ which is $\mathbb{P}$-generic over M, then we can always find a $\mathbb{Q}$-generic H such that $G = i^{-1}(H)$. Furthermore, there is a p.o. $\mathbb{R} \in M[G]$ such that $M[H] = M[G][K]$ for some K which is $\mathbb{R}$-generic over $M[G]$ (see Exercises D3–D6). The relationship between complete embeddings and iterated forcing extensions will be taken up again in VIII.

If $i : \mathbb{P} \to \mathbb{Q}$ is in fact an isomorphism then we may apply Theorem 7.5 both to i and its inverse to show that p.o.'s isomorphic by an isomorphism in M yield the same extensions.

7.6. Corollary. *In Theorem 7.5, suppose also that i is an isomorphism. Let $G \subset \mathbb{P}$. Then G is $\mathbb{P}$-generic over M iff $i''G$ is $\mathbb{Q}$-generic over M, and in that case, $M[G] = M[i''G]$.* □

Thus, for example, if $\kappa \geq \omega$ and $\kappa \in M$, then $\mathrm{Fn}(\kappa, 2)$ and $\mathrm{Fn}(\kappa \times \omega, 2)$ yield the same extensions, since $(|\kappa| = |\kappa \times \omega|)^M$, so that the two p.o.'s are isomorphic in M. We used $\mathrm{Fn}(\kappa \times \omega, 2)$, which made it simpler to describe the κ-sequence of elements of ${}^{\omega}2$ added, but $\mathrm{Fn}(\kappa, 2)$ is quicker to write down and is the order usually referred to in the literature. If $(\kappa \geq \omega_1)^M$, then $\mathrm{Fn}(\kappa, 2)$ and $\mathrm{Fn}(\omega, 2)$ are still isomorphic in $\mathbf{V}$, but *not* in M, and they need *not* yield the same extensions, since CH may be true in one extension but false in another.

The orders $\mathrm{Fn}(\omega, 2)$, $\mathrm{Fn}(\omega, 3)$ and $\mathrm{Fn}(\omega, \omega)$ are *not* isomorphic in M or in $\mathbf{V}$, but they do yield the same extensions. This can be proved (Exercise C3) using the concept of *dense embedding*, which we take up next.

7.7. Definition. Let $\mathbb{P}$, and $\mathbb{Q}$, be p.o.'s and $i : \mathbb{P} \to \mathbb{Q}$. i is a *dense embedding* iff

(1) $\forall p, p' \in \mathbb{P}\,(p' \leq p \to i(p') \leq i(p))$.

(2) $\forall p_1, p_2 \in \mathbb{P}\,(p_1 \perp p_2 \to i(p_1) \perp i(p_2))$.

(3) $i``\mathbb{P}$ is dense in $\mathbb{Q}$. □

7.8. Lemma. *Every dense embedding is a complete embedding.*

Proof. If $q \in \mathbb{Q}$, any $p \in \mathbb{P}$ with $i(p) \leq q$ is a reduction of q to $\mathbb{P}$. □

An important special case is the following.

7.9. Corollary. *If $\mathbb{P}$ is a sub-order of $\mathbb{Q}$ and $\mathbb{P}$ is dense in $\mathbb{Q}$, then the identity on $\mathbb{P}$ is a dense embedding into $\mathbb{Q}$.* □

If $i : \mathbb{P} \to \mathbb{Q}$ is a dense embedding, then $\mathbb{P}$ and $\mathbb{Q}$ yield the same generic extensions, as we show now after a preliminary lemma.

7.10. Lemma. *Suppose $\mathbb{P} \in M$, G_1 and G_2 are $\mathbb{P}$-generic over M, and $G_1 \subset G_2$. Then $G_1 = G_2$.*

Proof. Suppose $p \in G_2$ but $p \notin G_1$. Since $G_1 \cap \{p\} = 0$, there is a $q \in G_1$, with $q \perp p$ (applying Lemma 2.20(a) to G_1 and $\{p\}$), which is impossible since G_2 is a filter. □

7.11. Theorem. *Suppose i, $\mathbb{P}$, and $\mathbb{Q}$ are in M, $i : \mathbb{P} \to \mathbb{Q}$, and i is a dense embedding. If $G \subset \mathbb{P}$, let $\tilde{\imath}(G) = \{q \in \mathbb{Q} : \exists p \in G\,(i(p) \leq q)\}$.*

(a) *If $H \subset \mathbb{Q}$ is $\mathbb{Q}$-generic over M, then $i^{-1}(H)$ is $\mathbb{P}$-generic over M and $H = \tilde{\imath}(i^{-1}(H))$.*

(b) *If $G \subset \mathbb{P}$ is $\mathbb{P}$-generic over M, then $\tilde{\imath}(G)$ is $\mathbb{Q}$-generic over M and $G = i^{-1}(\tilde{\imath}(G))$.*

(c) *In* (a) *or* (b), *if $G = i^{-1}(H)$ (or, equivalently, if $H = \tilde{\imath}(G)$), then $M[G] = M[H]$.*

Proof. We first verify genericity of $i^{-1}(H)$ in (a) and $\tilde{\imath}(G)$ in (b). Since every dense embedding is a complete embedding genericity of $i^{-1}(H)$ in (a) follows from Theorem 7.5. In (b), $\tilde{\imath}(G)$ is easily seen to be a filter in $\mathbb{Q}$. To see that it is generic, let $D \in M$ be dense in $\mathbb{Q}$. Let

$$D^* = \{p \in \mathbb{P}\colon \exists q \in D\,(i(p) \leq q)\}.$$

If $D^* \cap G \neq 0$, then $D \cap \tilde{\imath}(G) \neq 0$, and $D^* \cap G \neq 0$ will follow if we can show that D^* is dense in $\mathbb{P}$. To see this fix $p \in \mathbb{P}$; now let $q \in D$ be such that $q \leq i(p)$, and let $p' \in \mathbb{P}$ be such that $i(p') \leq q$; then $i(p') \leq i(p)$, so $i(p')$ and $i(p)$ are compatible, and hence p' and p are compatible. Let $p'' \in \mathbb{P}$ be such that $p'' \leq p$ and $p'' \leq p'$; then $p'' \in D^*$ (since $i(p'') \leq q \in D$), and $p'' \leq p$.

To see that $G = i^{-1}(\tilde{\imath}(G))$ in (b), we have just seen that $\tilde{\imath}(G)$ is $\mathbb{Q}$-generic over M, and hence that $i^{-1}(\tilde{\imath}(G))$ is $\mathbb{P}$-generic over M. Since $G \subset i^{-1}(\tilde{\imath}(G))$ is immediate from the definitions, equality follows by Lemma 7.10. Likewise in (a), $\tilde{\imath}(i^{-1}(H)) \subset H$ follows directly from the definitions, so equality holds by Lemma 7.10.

Finally, for (c), we have seen (Theorem 7.5) that $M[G] \subset M[H]$. The same proof now shows that $M[H] \subset M[G]$; namely $H \in M[G]$ and $M \subset M[G]$, so $M[H] \subset M[G]$ by minimality of $M[H]$ (Lemma 2.9). □

We may also use our $i : \mathbb{P} \to \mathbb{Q}$ to associate to every $\mathbb{P}$-name, a $\mathbb{Q}$-name for the same object.

7.12. Definition. If $i : \mathbb{P} \to \mathbb{Q}$, define, by recursion on $\tau \in V^{\mathbb{P}}$,

$$i_*(\tau) = \{\langle i_*(\sigma), i(p)\rangle\colon \langle \sigma, p\rangle \in \tau\}. \quad \square$$

It is easily seen that $i_*(\tau) \in V^{\mathbb{Q}}$ and that the definition of i_* is absolute for M, so that, if $\mathbb{P}$, $\mathbb{Q}$ and i are in M, then $i_* : M^{\mathbb{P}} \to M^{\mathbb{Q}}$.

7.13. Lemma. *Suppose i, $\mathbb{P}$, and $\mathbb{Q}$ are in M, $i : \mathbb{P} \to \mathbb{Q}$, and i is a complete embedding, then:*

(a) *If H is $\mathbb{Q}$-generic over M, then* $\mathrm{val}(\tau, i^{-1}(H)) = \mathrm{val}(i_*(\tau), H)$ *for each* $\tau \in M^{\mathbb{P}}$.

(b) *If $\phi(x_1, \ldots, x_n)$ is a formula which is absolute for transitive models of* ZFC, *then*

$$p \Vdash_{\mathbb{P}} \phi(\tau_1, \ldots, \tau_n) \quad \textit{iff} \quad i(p) \Vdash_{\mathbb{Q}} \phi(i_*(\tau_1), \ldots, i_*(\tau_n)).$$

(c) *If i is a dense embedding and $\phi(x_1, \ldots, x_n)$ is any formula, then*

$$p \Vdash_{\mathbb{P}} \phi(\tau_1, \ldots, \tau_n) \quad \textit{iff} \quad i(p) \Vdash_{\mathbb{Q}} \phi(i_*(\tau_1), \ldots, i_*(\tau_n)).$$

Proof. (a) is a straightforward induction on τ, and does not actually require that H be generic or that i be complete.

To prove the implication from left to right in (b) and (c), suppose

$p \Vdash_{\mathbb{P}} \phi(\tau_1, \ldots, \tau_n)$. Fix $H \subset \mathbb{Q}$ with $i(p) \in H$ and H $\mathbb{Q}$-generic over M, then $p \in i^{-1}(H)$, so by definition of $\Vdash_{\mathbb{P}}$,

$$\phi(\mathrm{val}(\tau_1, i^{-1}(H)), \ldots, \mathrm{val}(\tau_n, i^{-1}(H)))^{M[i^{-1}(H)]}.$$

Since $\mathrm{val}(\tau_i, i^{-1}(H)) = \mathrm{val}(i_*(\tau_i), H)$, and $M[i^{-1}(H)] \subset M[H]$, we have

$$\phi(\mathrm{val}(i_*(\tau_1), H), \ldots, \mathrm{val}(i_*(\tau_n), H))^{M[H]}$$

(applying absoluteness in (b) and $M[i^{-1}(H)] = M[H]$ in (c)). Thus, by definition of $\Vdash_{\mathbb{Q}}$,

$$i(p) \Vdash_{\mathbb{Q}} \phi(i_*(\tau_1), \ldots, i_*(\tau_n)).$$

To prove (b) and (c) from right to left, assume $\neg(p \Vdash_{\mathbb{P}} \phi(\tau_1, \ldots, \tau_n))$; then there is a $p' \leq p$ such that $p' \Vdash_{\mathbb{P}} \neg\phi(\tau_1, \ldots, \tau_n)$, whence, as we have just seen, $i(p') \Vdash_{\mathbb{Q}} \neg\phi(i_*(\tau_1), \ldots, i_*(\tau_n))$. Since $i(p') \leq i(p)$,

$$\neg(i(p) \Vdash_{\mathbb{Q}} \phi(i_*(\tau_1), \ldots, i_*(\tau_n))). \quad \square$$

We turn now to Boolean-valued models. Our discussion parallels the treatment in II §3 relating MA to Boolean algebras.

If $\mathscr{B}$ is a Boolean algebra, we shall, as in II §3, abuse notation somewhat and apply our forcing terminology to $\mathscr{B}$ when we really mean $\mathscr{B} \setminus \{\mathbb{0}\}$. Thus, $M^{\mathscr{B}}$ is really $M^{\mathbb{P}}$ where $\mathbb{P} = \mathscr{B} \setminus \{\mathbb{0}\}$, and $p \Vdash \phi$ has been defined only when $p \neq \mathbb{0}$. However, it is consistent with our terminology to take $\mathbb{0} \Vdash \phi$ to be always true (vacuously), since no filter contains $\mathbb{0}$.

Now recall that by II 3.3, for every p.o. $\mathbb{P}$ there is a dense embedding i, of $\mathbb{P}$ into some complete Boolean algebra, $\mathscr{B}$. $\mathscr{B}$ is called *the completion* of $\mathbb{P}$ and is unique up to isomorphism (see II Exercise 18). i is not in general 1–1 (see II §3 and Exercises C8 and D3 of this chapter). Applying II 3.3 relativized to M, together with Theorem 7.11, we see that any generic extension of M can be obtained by forcing over M with a $\mathscr{B} \in M$ such that ($\mathscr{B}$ is a complete Boolean algebra)^M. This fact suggests an alternate approach to the exposition of forcing. One may go through the abstract development in §§2–4 only for the very special case of forcing with complete Boolean algebras of M; in this special case, many of the basic definitions are simpler, and it is easier to grasp intuitively what is going on. Then, when, as in §§5–6, we wish to apply a specific p.o. $\mathbb{P}$, which is probably not a Boolean algebra, we simply force not with $\mathbb{P}$, but with the $\mathscr{B} \in M$ such that ($\mathscr{B}$ is the completion of $\mathbb{P}$)^M.

We did not take this approach in this book because we did not wish to make familiarity with Boolean algebras a prerequisite for understanding forcing. Also, the general theory of forcing can easily be adapted to produce generic extensions of models of ZF − P (Exercise B10) or even weaker theories. This has applications in model theory and recursion theory (see

[Keisler 1973] and [Sacks 1971]). Since constructing the completion requires the Power Set Axiom, one cannot, when forcing over these models, reduce the general theory to forcing with complete Boolean algebras of M.

We now outline more specifically what simplifications take place when forcing with complete Boolean algebras of M.

7.14. Definition. If $\mathscr{B} \in M$, ($\mathscr{B}$ is a complete Boolean algebra)^M, and $\tau_1, \ldots, \tau_n \in M$, then

$$[\![\phi(\tau_1, \ldots, \tau_n)]\!] = \bigvee \{p \in \mathscr{B}: p \Vdash \phi(\tau_1, \ldots, \tau_n)\}.$$

$[\![\phi(\tau_1, \ldots, \tau_n)]\!]$ is called the *truth value* of $\phi(\tau_1, \ldots, \tau_n)$. $\square$

Of course, "complete" is not absolute, and $\mathscr{B}$ may well fail to be complete in $\mathbf{V}$; in fact, every countable complete Boolean algebra is finite (Exercise F6). Nevertheless, the definition of $[\![\phi]\!]$ makes sense by the definability of $\Vdash$; $\{p \in \mathscr{B}: p \Vdash \phi\}$ is in M, so its supremum exists.

Intuitively, we may think of the people living in M as defining a *Boolean-valued model of set theory*, where the truth values, $[\![\phi]\!]$ (or $[\![\phi(\tau_1, \ldots, \tau_n)]\!]$) may be $\mathbb{1}$ (true) or may be $\mathbb{0}$ (false), but may also have some value in $\mathscr{B}$ intermediate between $\mathbb{0}$ and $\mathbb{1}$. If ϕ is true in all $\mathscr{B}$-generic extensions, then $\mathbb{1} \Vdash \phi$, so $[\![\phi]\!] = \mathbb{1}$, and if ϕ is false in all $\mathscr{B}$-generic extensions, then only $\mathbb{0} \Vdash \phi$, so $[\![\phi]\!] = \mathbb{0}$, but if ϕ is true in some extensions and false in others, then $\mathbb{0} < [\![\phi]\!] < \mathbb{1}$. By definability of $\Vdash$, the M-people are able to define the Boolean truth value of ϕ without ever being able to construct a real (2-valued) generic filter.

(a) of the next lemma says that $[\![\phi]\!]$ is the largest condition which forces ϕ.

7.15. Lemma. *Under the assumptions of Definition* 7.14,

(a) $\forall p \in \mathscr{B}\,(p \Vdash \phi(\tau_1, \ldots, \tau_n) \leftrightarrow p \leq [\![\phi(\tau_1, \ldots, \tau_n)]\!])$.

(b) $[\![\phi(\tau_1, \ldots, \tau_n) \wedge \psi(\tau_1, \ldots, \tau_n)]\!] = [\![\phi(\tau_1, \ldots, \tau_n)]\!] \wedge [\![\psi(\tau_1, \ldots, \tau_n)]\!]$.

(c) $[\![\neg\phi(\tau_1, \ldots, \tau_n)]\!] = [\![\phi(\tau_1, \ldots, \tau_n)]\!]'$.

(d) $[\![\exists x\, \phi(x, \tau_1, \ldots, \tau_n)]\!] = \bigvee \{[\![\phi(\sigma, \tau_1, \ldots, \tau_n)]\!]: \sigma \in M^{\mathscr{B}}\}$.

Proof. In all cases, we drop explicit mention of $\tau_1, \ldots, \tau_n$.

For (a), $p \Vdash \phi$ implies $p \leq [\![\phi]\!]$ by the definition of $[\![\phi]\!]$. Now, assume $p \leq [\![\phi]\!]$. Either $p \Vdash \phi$ or there is a (non-$\mathbb{0}$) $q \leq p$ such that $q \Vdash \neg\phi$. But for such a q, $\forall r\,(r \Vdash \phi \rightarrow q \wedge r = \mathbb{0})$, so $q \wedge [\![\phi]\!] = \mathbb{0}$, contradicting $\mathbb{0} < q \leq p \leq [\![\phi]\!]$.

(b)–(d) are easy exercises using (a) and basic properties of $\Vdash$. We do (b) as an example. $[\![\phi]\!] \Vdash \phi$ by (a), and $[\![\phi]\!] \wedge [\![\psi]\!] \leq [\![\phi]\!]$, so

$$[\![\phi]\!] \wedge [\![\psi]\!] \Vdash \phi.$$

Likewise, $[\![\phi]\!] \wedge [\![\psi]\!] \Vdash \psi$, so $[\![\phi]\!] \wedge [\![\psi]\!] \Vdash \phi \wedge \psi$, whence

$$[\![\phi]\!] \wedge [\![\psi]\!] \leq [\![\phi \wedge \psi]\!].$$

Also, $[\![\phi \wedge \psi]\!] \Vdash \phi \wedge \psi$, and ZFC $\vdash \phi \wedge \psi \rightarrow \phi$, so $[\![\phi \wedge \psi]\!] \Vdash \phi$ (by Lemma 4.5). Thus, $[\![\phi \wedge \psi]\!] \leq [\![\phi]\!]$. Likewise, $[\![\phi \wedge \psi]\!] \leq [\![\psi]\!]$, so

$$[\![\phi \wedge \psi]\!] \leq [\![\phi]\!] \wedge [\![\psi]\!]. \quad \square$$

Lemma 715 (b)–(d) say that in computing Boolean truth values, the logical operations are mirrored by the corresponding Boolean operations. By Lemma 7.15 (a), $\Vdash$ can be defined in terms of $[\![\ldots]\!]$.

Lemma 7.15 suggests a substantial simplification in the treatment of $\Vdash^*$ in §3. Given a complete Boolean algebra, $\mathscr{B}$, we consider $\Vdash^*$ to be an auxiliary notion, introduced by the definition:

$$p \Vdash^* \phi(\tau_1, \ldots, \tau_n) \leftrightarrow p \leq [\![\phi(\tau_1, \ldots, \tau_n)]\!]^*.$$

Thus, $[\![\ldots]\!]^*$ is considered to be the basic notion, and its definition is similar to that of $\Vdash^*$ in Definition 3.3—but much simpler. In the induction steps for $\wedge$, $\neg$, and $\exists$, we use the clauses of Lemma 7.15(b)–(d) as definitions; thus, $[\![\neg\phi]\!]^*$ is defined to be $([\![\phi]\!]^*)'$. The two inductive clauses for atomic formulas are also simplified (see Exercise D7). In all cases in the definition of $[\![\ldots]\!]^*$, logical operations are reflected by Boolean operations in a natural way, and we avoid the use of auxiliary concepts such as "dense below". As with $\Vdash^*$, we may think of the definition of $[\![\ldots]\!]^*$ as taking place within **V**, but we relativize it to M when discussing generic extensions, using a $\mathscr{B} \in M$ such that ($\mathscr{B}$ is a complete Boolean algebra)^M.

Boolean-valued models also simplify the treatment of nice names (Definition 5.11). It is convenient here to allow $\mathbb{0}$ as a second co-ordinate in names; i.e., re-define $\tau \in V^{\mathscr{B}} \leftrightarrow \tau \subset V^{\mathscr{B}} \times \mathscr{B}$ (not $V^{\mathscr{B}} \times (\mathscr{B} \setminus \{\mathbb{0}\})$, as we would have if we are forcing with $\mathscr{B} \setminus \{\mathbb{0}\}$). It is easily seen that this entails no essential change in the development of §§2–4, and it is more natural to treat all elements of $\mathscr{B}$ equally. Now, if

$$\tau = \bigcup \{\{\pi\} \times A_\pi : \pi \in \mathrm{dom}(\sigma)\}$$

is a nice name for a subset of σ, let $b_\pi = \bigvee A_\pi$ (so $b_\pi = \mathbb{0}$ if $A_\pi = 0$), and let $\tilde{\tau} = \{\langle \pi, b_\pi \rangle : \pi \in \mathrm{dom}(\tau)\}$. Then $\tilde{\tau}$ is a function from $\mathrm{dom}(\sigma)$ into $\mathscr{B}$. Call such a $\tilde{\tau}$ a *very nice name*. It is easily seen that $[\![\tau = \tilde{\tau}]\!] = \mathbb{1}$, so that corresponding to Lemma 5.12, every subset of σ in $M[G]$ is represented by a very nice name. Beside eliminating the somewhat ad hoc introduction of antichains, the use of such names makes it intuitively clearer what the elements of $M[G]$ "are". For example, a very nice name for a subset of $\check{\omega}$ is a function from $\{\check{n} : n \in \omega\}$ into $\mathscr{B}$, which we may indentify with a function $f : \omega \rightarrow \mathscr{B}$. We may think of f as a characteristic function of a Boolean-valued subset of ω.

Another simplification occurs in the notion of complete embedding (Definition 7.1). Instead of another seemingly ad hoc definition, this notion, when restricted to complete Boolean algebras reduces to the natural notion of complete injective homomorphism (Exercise C7).

Thus, we see that restricting our p.o.'s to be complete Boolean algebras of M makes the abstract theory somewhat simpler and more natural, although this approach then requires us to go through the additional step of embedding arbitrary p.o.'s into complete Boolean algebras.

For more on Boolean-valued models, see §9.

§8. Further results

In this section we collect some technical results which did not seem to fit in earlier, together with three more examples of forcing. As always, M is a fixed c.t.m. for ZFC.

8.1. LEMMA. *Suppose that in M: A is an antichain in $\mathbb{P}$ and, for each $q \in A$, σ_q is a $\mathbb{P}$-name. Then there is a $\pi \in M^{\mathbb{P}}$ such that $q \Vdash \pi = \sigma_q$ for each $q \in A$.*

PROOF. In M, let

$$\pi = \bigcup_{q \in A} \{\langle \tau, r\rangle : (r \leq q) \wedge (r \Vdash \tau \in \sigma_q) \wedge (\tau \in \operatorname{dom}(\sigma_q))\}.$$

Fix $q \in A$ and fix a generic G with $q \in G$. We show $\pi_G = \operatorname{val}(\sigma_q, G)$.

Any element of π_G is of the form τ_G where $\langle \tau, r\rangle \in \pi$ for some $r \in G$. Then for some $q' \in A$, $r \leq q'$ and $r \Vdash \tau \in \sigma_{q'}$. But A is an antichain and $q \in G$, so $q' = q$. Thus, $r \Vdash \tau \in \sigma_q$, so $\tau_G \in \operatorname{val}(\sigma_q, G)$.

Any element of $\operatorname{val}(\sigma_q, G)$ is of the form τ_G for some $\tau \in \operatorname{dom}(\sigma_q)$. Fix $p \in G$ with $p \Vdash \tau \in \sigma_q$, and let $r \in G$ be a common extension of p and q. Then $\langle \tau, r\rangle \in \pi$, so $\tau_G \in \pi_G$. □

As motivation for the next result, observe that in proving that $M[G]$ satisfies ZFC, every time that we verified an existential statement we produced a name which witnessed the existence independently of G. For example, to verify the Pairing Axiom, it would have been sufficient to show that for each $\sigma, \tau \in M^{\mathbb{P}}$, $\mathbb{1} \Vdash \exists x (\sigma \in x \wedge \tau \in x)$; equivalently for every generic G there is a π such that $\sigma_G \in \pi_G \wedge \tau_G \in \pi_G$. But in fact, we found a π, namely $\operatorname{up}(\sigma, \tau)$, which was independent of G (see Definition 2.16), such that $\mathbb{1} \Vdash (\sigma \in \pi \wedge \tau \in \pi)$. The fact that this can be done is part of a more general fact, known as the *maximal principle*.

8.2. THEOREM. *If $\mathbb{P} \in M$, $\tau_1, \ldots, \tau_n \in M^{\mathbb{P}}$, and $p \Vdash \exists x\, \phi(x, \tau_1, \ldots, \tau_n)$, then there is a $\pi \in M^{\mathbb{P}}$ such that $p \Vdash \phi(\pi, \tau_1, \ldots, \tau_n)$.*

PROOF. We suppress throughout mention of $\tau_1, \ldots, \tau_n$. Using Zorn's Lemma in M, let $A \in M$ be such that

(1) A is an antichain in $\mathbb{P}$.

(2) $\forall q \in A \left(q \leq p \wedge \exists \sigma \in M^{\mathbb{P}} \left(q \Vdash \phi(\sigma)\right)\right)$.

(3) A is a maximal with respect to (1) and (2).

By AC in M, pick $\sigma_q \in M^{\mathbb{P}}$ for $q \in A$ so that $q \Vdash \phi(\sigma_q)$, and by Lemma 8.1, let $\pi \in M^{\mathbb{P}}$ be such that $q \Vdash \pi = \sigma_q$ for each $q \in A$. So, for $q \in A$,

$$q \Vdash (\phi(\sigma_q) \wedge \pi = \sigma_q),$$

so $q \Vdash \phi(\pi)$.

We now show that $p \Vdash \phi(\pi)$. If not, let $r \leq p$ be such that $r \Vdash \neg\phi(\pi)$. Since $p \Vdash \exists x\, \phi(x)$, $\{q\colon \exists \sigma \in M^{\mathbb{P}} \left(q \Vdash \phi(\sigma)\right)\}$ is dense below p (see Corollary 3.7), so fix $q_0 \leq r$ with $\exists \sigma \in M^{\mathbb{P}} \left(q_0 \Vdash \phi(\sigma)\right)$. For each $q \in A$, we have $q \Vdash \phi(\pi)$ and $q_0 \Vdash \neg\phi(\pi)$, so $q \perp q_0$. Thus, $A \cup \{q_0\}$ satisfies (1) and (2) above, contradicting maximality of A. □

Theorem 8.2 is best appreciated in the framework of Boolean-valued models (see §7). By Lemma 7.15,

$$[\![\exists x\, \phi(x)]\!] = \bigvee \{ [\![\phi(\sigma)]\!] \colon \sigma \in M^{\mathbb{P}} \},$$

but Theorem 8.2 says that we can in fact find a $\sigma \in M^{\mathbb{P}}$ such that $[\![\phi(\sigma)]\!]$ is the maximum possible value, $[\![\exists x\, \phi(x)]\!]$.

We now turn to some more applications of forcing.

Many of the combinatorial principles known to be true in **L** can also be proved consistent by forcing. The easiest of these is $\diamondsuit$ (see II 7.1). This argument uses countable partial functions, but has a different flavor from the methods of §6.

8.3. THEOREM. *Let* $\mathbb{P} = \mathrm{Fn}(\omega_1, 2, \omega_1)^M$. *If* G *is* $\mathbb{P}$*-generic over* M, *then* $\mathscr{P}(\omega) \cap M = \mathscr{P}(\omega) \cap M[G]$, $\omega_1^M = \omega_1{}^{M[G]}$, *and* $\diamondsuit$ *holds in* $M[G]$.

PROOF. The first two statements are immediate from $\mathbb{P}$ being ω_1-closed in M (see 6.14 and 6.15). The argument for $\diamondsuit$ will be more transparent if we use a different p.o. isomorhic to $\mathbb{P}$ in M. Let $I = \{\langle \alpha, \xi \rangle \colon \xi < \alpha < \omega_1^M\}$, and let $\mathbb{Q} = \mathrm{Fn}(I, 2, \omega_1)^M$. Since $|I| = \omega_1$ in M, $\mathbb{P}$ and $\mathbb{Q}$ are isomorphic in M, so (by Corollary 7.6) it is sufficient to check that whenever G is $\mathbb{Q}$-generic over M, $\diamondsuit$ holds in $M[G]$.

If A is any function from I into 2, we let $A_\alpha\colon \alpha \to 2$ be defined by $A_\alpha(\xi) = A(\alpha, \xi)$ (for $\alpha < \omega_1^M$). If G is $\mathbb{Q}$-generic over M, then $\bigcup G$ is a function from I into 2. We shall prove $(\diamondsuit)^{M[G]}$ by showing that in $M[G]$, $\langle (\bigcup G)_\alpha \colon \alpha < \omega_1^M \rangle$ is a $\diamondsuit$-sequence if we identify sets with their characteristic functions. Equivalently, we shall show that if $B \in M[G]$ and $B : \omega_1^M \to 2$, then $\{\alpha \colon B \restriction \alpha = (\bigcup G)_\alpha\}$ is stationary in $M[G]$. Assume that this is false. Then there is a name $\tau \in M^{\mathbb{Q}}$ (for B), a name $\sigma \in M^{\mathbb{Q}}$ (for a c.u.b.), and a $p \in G$ such that

$$p \Vdash [(\tau \subset \omega_1) \wedge (\sigma \subset \omega_1) \wedge (\sigma \text{ is c.u.b.}) \wedge \forall \alpha \in \sigma (\tau \upharpoonright \alpha \neq (\bigcup \Gamma)_\alpha)], \quad (*)$$

where $\Gamma = \{\langle \check{q}, q \rangle : q \in \mathbb{Q}\}$ is the $\mathbb{Q}$-name for the $\mathbb{Q}$-generic filter (see Definition 2.12). We now forget about G and B, and derive a contradiction directly from $(*)$.

The following paragraph takes place in M. For any $q \in \mathbb{Q}$, let $\text{supt}(q)$ be the least $\beta < \omega_1$ such that $\text{dom}(q) \subset \{\langle \alpha, \xi \rangle : \xi < \alpha < \beta\}$. Now, define inductively p_n $(n \in \omega)$ along with β_n, δ_n, and b_n so that

(1) $p_0 = p$.
(2) $\beta_n = \text{supt}(p_n)$.
(3) $\delta_n > \beta_n$.
(4) $p_{n+1} \leq p_n$.
(5) $p_{n+1} \Vdash \check{\delta}_n \in \sigma$.
(6) $\text{supt}(p_{n+1}) > \delta_n$.
(7) $b_n : \beta_n \to 2$ and $p_{n+1} \Vdash (\tau \upharpoonright \check{\beta}_n = \check{b}_n)$.

Let us check that this induction can be accomplished. Given p_n, β_n is defined. Since $p_n \leq p$, $p_n \Vdash (\sigma$ is c.u.b.), so $p_n \Vdash \exists x \in \check{\omega}_1 (x > \check{\beta}_n \wedge x \in \sigma)$. Thus, applying Corollary 3.7(d), there is a $q \leq p_n$ and a $\delta_n \in \omega_1$ such that $q \Vdash (\check{\delta}_n > \check{\beta}_n \wedge \check{\delta}_n \in \sigma)$, so $\delta_n > \beta_n$ and $q \Vdash \check{\delta}_n \in \sigma$. Let r be any extension of q with $\text{supt}(r) > \delta_n$. Finally, to handle (7), let $F = {}^{(\beta_n)}2$. Then $r \Vdash \tau \upharpoonright \check{\beta}_n \in \check{F}$, since an ω_1-closed p.o. adds no new functions from β_n into 2 (see Theorem 6.14). Thus, there is a $b_n \in F$ and a $p_{n+1} \leq r$ such that $p_{n+1} \Vdash \tau \upharpoonright \check{\beta}_n = \check{b}_n$ (see Corollary 3.7(d)).

Still within M, we have $\beta_0 < \delta_0 < \beta_1 < \delta_1 < \cdots$. Let

$$\gamma = \sup\{\beta_n : n \in \omega\} = \sup\{\delta_n : n \in \omega\}.$$

Let $p_\omega = \bigcup_n p_n$; then $\text{supt}(p_\omega) = \gamma$. For each $n < \omega$, $p_\omega \leq p_{n+1}$ so $p_\omega \Vdash (\tau \upharpoonright \check{\beta}_n = \check{b}_n)$. Thus, the b_n for $n \in \omega$ agree on their common domains, $b_\omega = \bigcup_n b_n$ is a function from γ to 2, and $p_\omega \Vdash (\tau \upharpoonright \check{\gamma} = \check{b}_\omega)$. Now there are no pairs $\langle \gamma, \xi \rangle \in \text{dom}(p_\omega)$, so we may extend p_ω to an s such that $s(\gamma, \xi) = b_\omega(\xi)$ for each $\xi < \gamma$. Then $s \Vdash [(\bigcup \Gamma)_{\check{\gamma}} = \check{b}_\omega]$, so $s \Vdash [\tau \upharpoonright \check{\gamma} = (\bigcup \Gamma)_{\check{\gamma}}]$. But also, $s \Vdash (\check{\gamma} \in \sigma)$ since $s \Vdash (\sigma$ is closed) and $s \Vdash (\check{\delta}_n \in \sigma)$ for each n. Thus,

$$s \Vdash [\exists \alpha \in \sigma (\tau \upharpoonright \alpha = (\bigcup \Gamma)_\alpha)],$$

which, since $s \leq p$, is a contradiction. □

Since $\diamondsuit \to$ CH, CH holds in $M[G]$ regardless of whether it holds in M; what happens is that if CH fails in M, the cardinal $(2^\omega)^{(M)}$ gets collapsed in $M[G]$.

We cannot use the method of Theorem 8.3 to prove the consistency of $\diamondsuit^+$, since $\diamondsuit^+$ may fail in $M[G]$ (see VIII Exercises J5–J7). If M is chosen carefully, $\diamondsuit^+$ will hold in $M[G]$. For example, if M satisfied $\mathbf{V} = \mathbf{L}$, then $M[G]$ would satisfy $\exists X \subset \omega_1 (\mathbf{V} = \mathbf{L}(X))$, which implies $\diamondsuit^+$ (see VI Exercise 8). Of course, M also satisfies $\diamondsuit^+$, so this forcing argument establishes nothing new. One can, in the spirit of Theorem 8.3, prove the consistency of $\diamondsuit^+$ by forcing, but the partial order is more complicated (see Exercises H18–H20).

Theorem 8.3 does tell us something that $\mathbf{L}$ did not; namely, that $\diamondsuit$ does not imply $2^{\omega_1} = \omega_2$. To see this, start with M satisfying CH + $2^{\omega_1} > \omega_2$; there is such an M by Theorem 6.18. Then $\mathbb{P}$ will not collapse cardinals, so $M[G]$ will satisfy $\diamondsuit + 2^{\omega_1} > \omega_2$. Likewise, by Exercise H20, $\diamondsuit^+$ does not imply $2^{\omega_1} = \omega_2$.

Since $\diamondsuit$ implies that there is an ω_1-Suslin tree, there is one in the $M[G]$ of Theorem 8.3. Historically, Jech first showed that one can add a Suslin tree in a generic extension, and with the advent of $\diamondsuit$, his argument was easily modified to yield Theorem 8.3. His original argument is still of some interest, since it can be generalized to add a κ-Suslin tree for any regular κ of M (see Exercises H11–H14), even though the existence of such a tree does not follow from the analogue of $\diamondsuit$ on κ.

It is also easy to destroy a Suslin tree; in fact if we stand the tree upside down it will destroy itself.

8.4. THEOREM. *Let $\kappa > \omega$ be a regular cardinal of M and T a κ-Suslin tree of M. Then there is a* p.o. $\mathbb{P} \in M$, *such that* ($\mathbb{P}$ *has* κ-c.c.)^M, $\mathbb{P}$ *preserves cofinalities, and, whenever G is $\mathbb{P}$-generic over M,*

(1) *T is not a κ-Suslin tree in $M[G]$, and*

(2) *If $\alpha < \kappa$ and $B \in M$, then ${}^{\alpha}B \cap M[G] = {}^{\alpha}B \cap M$.*

PROOF. We begin by imitating the proof that MA(ω_1) implies that there are no ω_1-Suslin trees (see II 5.14). Within M, let T' be a well-pruned κ-sub-tree of T (see II 5.11). Let $\lhd$ be the tree order on T', and let $\mathbb{P}$ be T' with the reverse order; $p <_{\mathbb{P}} q$ iff $q \lhd p$; then $\mathbb{P}$ is a p.o., with $\mathbb{1}_{\mathbb{P}}$ the least element of the tree T'. Since T' is well-pruned, $D_\alpha = \{p\colon \operatorname{ht}(p, T') > \alpha\}$ is dense in $\mathbb{P}$ for each $\alpha < \kappa$. If G is a filter in $\mathbb{P}$, then G is a chain in T'. If G is $\mathbb{P}$-generic over M, then $G \cap D_\alpha \neq 0$ for all $\alpha < \kappa$, so G contains elements of T' of arbitrarily large height below κ. Thus, T' is not even Aronszajn in $M[G]$.

Since T' is a κ-Suslin tree in M, ($\mathbb{P}$ has the κ-c.c.)^M, so $\mathbb{P}$ preserves cofinalities $\geq\kappa$ (see Lemma 6.9). We shall thus be done if we can check (2), since (2) implies that $\mathbb{P}$ preserves cofinalities $\leq\kappa$ (as in the proof of Corollary 6.15 from Theorem 6.14).

To prove (2), suppose $f : \alpha \to B, \alpha < \kappa, B \in M$, and $f \in M[G]$. Say $f = \tau_G$. For each $\xi < \alpha$, there is a $p_\xi \in G$ such that $p_\xi \Vdash \tau(\xi) = (f(\xi))\check{}$. Furthermore, the sequence $\langle p_\xi \colon \xi < \alpha \rangle$ may be chosen in $M[G]$ by using AC in $M[G]$ and the fact that

$$\{\langle p, \xi, y \rangle \colon p \in \mathbb{P} \wedge y \in B \wedge \xi \in \alpha \wedge p \Vdash \tau(\check{\xi}) = \check{y}\}$$

is in M and hence in $M[G]$. Now G is a chain, $\{p_\xi \colon \xi < \alpha\} \subset G$, and κ is regular in $M[G]$, so there is a $q \in G$ above all the p_ξ in T'; i.e., $\forall \xi < \alpha\, (q \leq p_\xi)$. Then $q \Vdash \tau(\check{\xi}) = (f(\xi))\check{}$ for each $\xi < \alpha$, so

$$f = \{\langle \xi, y \rangle \in \alpha \times B \colon q \Vdash \tau(\check{\xi}) = \check{y}\} \in M. \quad \square$$

We see in Theorem 8.4 our first application in forcing of a p.o. which did not arise from partial functions (although other such p.o.'s have already been used when applying MA in II). Theorem 8.4 also illustrates that 8.4(2) can hold for $\mathbb{P}$ without $\mathbb{P}$ being κ-closed in M. A well-pruned κ-Suslin tree can never be κ-closed, since if it were, one could inductively pick a path through it. For any $\mathbb{P}$. the conclusion 8.4(2) is equivalent to $\mathbb{P}$ having the property of being κ-Baire in M (see Exercise B4). It is possible to show directly that a κ-Suslin tree is κ-Baire (Exercise H16), without mentioning forcing, but the proof given of Theorem 8.4(2) is somewhat easier.

The ease with which Theorem 8.4 was proved should not mislead one to thinking that it is easy to prove the consistency of SH, since for this one needs to iterate forcing a transfinite number of times to destroy all the ω_1-Suslin trees in M. We shall do such an iteration in VIII to prove the consistency of MA + $\neg$CH, and hence of SH.

Since Theorem 8.4 says that one may destroy an ω_1-Suslin tree without enlarging ${}^{\omega}2$, it might be expected that SH + CH is consistent. It is, but the iterated forcing argument (due to Jensen; see [Devlin–Johnsbråten 1974]) is more difficult than the ones considered in this book.

When $\kappa > \omega_1$, it becomes more difficult to destroy all κ-Suslin trees, and there are still a number of open questions regarding the consistency of the κ-Suslin Hypothesis. Most notably, it is unknown whether GCH + ω_2-SH is consistent; it "should" be in view of Theorem 8.4, but [Gregory 1976] shows that GCH + ω_2-SH implies that ω_2 is Mahlo in L. For more on the κ-SH, see II §5.

To introduce our next example of forcing, we ask: what is $\omega_1^{M[G]}$? $\omega_1^{M[G]}$ must be regular in M, and we shall see that by suitably defining $\mathbb{P}$, we can arrange for $\omega_1^{M[G]}$ to be any regular cardinal of M except ω. We first remark that we have already taken care of successor cardinals.

8.5. Lemma. *Suppose that in M, λ is an infinite cardinal and $\kappa = \lambda^+$. Let $\mathbb{P} = \mathrm{Fn}(\omega, \lambda)$. Let G be $\mathbb{P}$-generic over M. Then $\omega_1^{M[G]} = \kappa$.*

Proof. If $\alpha < \kappa$, then in M there is a map from λ onto α. Since $\bigcup G$ maps ω onto λ, (α is countable)$^{M[G]}$. Thus, $\omega_1^{M[G]} \geq \kappa$. However, $(|\mathbb{P}| = \lambda)^M$, so

($\mathbb{P}$ has the κ-c.c.)^M, so $\mathbb{P}$ preserves cardinals $\geq \kappa$. Thus, κ remains a cardinal in $M[G]$, so $\omega_1^{M[G]} = \kappa$. □

If κ is weakly inaccessible in M, we force with finite functions in a slightly different way, collapsing every ordinal below κ, but preserving κ. We wish to add, for every $\alpha < \kappa$, a map f_α: $\omega \to \alpha$. For notational convenience, we code $\langle f_\alpha: \alpha < \kappa \rangle$ as on function with domain $\kappa \times \omega$.

8.6. Definition. For any κ, the *Lévy collapsing order* for κ, $\mathrm{Lv}(\kappa)$, is

$$\{p: |p| < \omega \wedge p \text{ is a function } \wedge \mathrm{dom}(p) \subset \kappa \times \omega \wedge$$
$$\wedge \forall \langle \alpha, n \rangle \in \mathrm{dom}(p) (p(\alpha, n) \in \alpha)\}.$$

$\mathrm{Lv}(\kappa)$ is ordered by reverse inclusion; $p \leq q$ iff $q \subset p$. □

We shall show now that if κ is a regular uncountable cardinal of M, then forcing with $\mathrm{Lv}(\kappa)$ makes $\kappa = \omega_1^{M[G]}$. This is usually only of interest when κ is weakly inaccessible in M, since successor cardinals were handled by the easier argument of Lemma 8.5.

8.7. Lemma. *If κ is regular and uncountable, then* $\mathrm{Lv}(\kappa)$ *has the* κ-c.c.

Proof. Fix $p_\mu \in \mathrm{Lv}(\kappa)$ for $\mu < \kappa$. By the Δ-system lemma (II 1.6), there is a set $B \subset \kappa$ such that $|B| = \kappa$, and $\{\mathrm{dom}(p_\mu): \mu \in B\}$ forms a Δ-system with some root r. Since κ is regular and there are less than κ possibilities for the $p_\mu \restriction r$, there is a $C \subset B$ with $|C| = \kappa$ and the $p_\mu \restriction r$ for $\mu \in C$ all the same; then the p_μ for $\mu \in C$ are pairwise compatible. In particular, the p_μ for $\mu < \kappa$ could not have been pairwise incompatible. □

8.8. Theorem. *Suppose that in M, κ is regular and uncountable. Let G be* $\mathrm{Lv}(\kappa)$*-generic over M. Then $\kappa = \omega_1^{M[G]}$.*

Proof. Applying Lemma 8.7 in M, ($\mathrm{Lv}(\kappa)$ has the κ-c.c.)^M, so κ remains a cardinal in $M[G]$. However, if $0 < \alpha < \kappa$, then standard density arguments show that

$$(\bigcup G)_\alpha = \{\langle n, \xi \rangle: \langle \alpha, n, \xi \rangle \in \bigcup G)$$

is a function from ω onto α, so α is countable in $M[G]$. □

[Solovay 1970] uses the Lévy collapsing order to get a model of ZF in which all subsets of $\mathbb{R}$ are Lebesgue measurable. More precisely, he starts with κ strongly inaccessible in M. In $M[G]$ it is true that if $X \subset \mathbb{R}$ and $X \in \mathbf{HOD}(\mathscr{P}(\omega))$ (defined in V Exercise 9), then X is Lebesgue measurable.

Of course, $M[G]$ satisfies AC and thus has a non-measurable set, but the inner model $\mathbf{HOD}(\mathscr{P}(\omega))$ (defined within $M[G]$) satisfies ZF + "all sets of reals are Lebesgue measurable." It is unknown whether one can prove from Con(ZFC) alone the consistency of ZF + "all sets of reals are Lebesgue measurable."[1]

The Lévy partial order is also relevant to Kurepa's Hypothesis, KH (see II 5.15).

8.9. COROLLARY. *Suppose that in M, κ is strongly inaccessible. Let G be $\mathrm{Lv}(\kappa)$-generic over M. Then* $(\mathrm{KH})^{M[G]}$.

PROOF. In M, let T be the complete binary tree of height κ. Since cardinals $\geq\kappa$ do not get collapsed, and since in M T has $\geq\kappa^+$ paths, (T is an ω_1-Kurepa tree)$^{M[G]}$. □

However, one does not need inaccessibles to get models of KH. To prove Con(ZFC) → Con(ZFC + KH), one can either use the fact that KH follows from $\diamondsuit^+$ and thus holds in **L** (see II 7.11 and VI 5.2), or one can use a different forcing argument which does not require inaccessibles (see Exercise H19).

Ironically, a minor variant of $\mathrm{Lv}(\kappa)$ can be used to construct a model of $\neg$KH. Again, κ is strongly inaccessible in M, but we force with countable (in M) partial functions to add, for $\alpha < \kappa$, a map from ω_1^M onto α. The argument in Theorem 8.8 is easily modified to show that $\omega_1^M = \omega_1^{M[G]}$ and $\kappa = \omega_2^{M[G]}$. However, to prove $(\neg\mathrm{KH})^{M[G]}$, we need the technique of iterated forcing, so we defer the further discussion of this order until VIII §3.

Unlike KH, $\neg$KH needs an inaccessible, since $\neg$KH implies that ω_2 is inaccessible in **L** (Exercise B9).

§9. Appendix: Other approaches and historical remarks

There are several different ways of presenting forcing. They all yield precisely the same consistency proofs, but they differ in their metamathematical conception. We survey here the various approaches.

Approach 1: via countable transitive models. This is usually the approach favored by non-logicians, since we handle models and their extensions in a rather straightforward mathematical way. An analogy is often drawn between generic extensions and field extensions in algebra, where one also uses names (polynomials) for objects in the extension field. One can always skim over, or ignore, the rather tedious details in §§3 and 4 of showing that the procedure really works. One can likewise ignore the logical unpleas-

[1](Added in proof.) One cannot, by a recent result of Shelah.

antries associated with the fact that in ZFC we cannot actually produce a c.t.m. for ZFC. These unpleasantries may be handled in one of the following three ways.

(1a) The approach in this book. We show that, given any finite list, $\phi_1, \ldots, \phi_n$, of axioms of, say, ZFC + $\neg$CH, we can prove in ZFC that there is a c.t.m. for $\phi_1, \ldots, \phi_n$. The procedure involves finding (in the metatheory) another finite list $\psi_1, \ldots, \psi_m$ of axioms of ZFC, and proving in ZFC that given a c.t.m. M for $\psi_1, \ldots, \psi_m$, there is a generic extension, $M[G]$, satisfying $\phi_1, \ldots, \phi_n$. The inelegant part of this argument is that the procedure for finding $\psi_1, \ldots, \psi_m$, although straightforward, completely effective, and finitistically valid, is also very tedious. We must list in $\psi_1, \ldots, \psi_m$ not only the axioms of ZFC "obviously" used in checking that $\phi_1, \ldots, \phi_n$ hold in $M[G]$ (e.g., if ϕ_1 is the Power Set Axiom, then ϕ_1 should be listed among $\psi_1, \ldots, \psi_m$), but also all the axioms needed to verify that various concepts are absolute for M ("finite", "p.o.", etc.), as well as the axioms needed to show that certain mathematical results, such as the Δ-system lemma, hold in M. Of course, for the relative consistency proof,

$$\mathrm{Con}(\mathrm{ZFC}) \to \mathrm{Con}(\mathrm{ZFC} + \neg\mathrm{CH}),$$

it is not necessary to display explicitly $\psi_1, \ldots, \psi_m$; it is sufficient to convince the reader that $\psi_1, \ldots, \psi_m$ may be found.

(1b) This is a way of avoiding dealing with an unspecified list, $\psi_1, \ldots, \psi_m$. Let $\mathscr{L}$ be the language with basic non-logical symbols $\in$ and c, where c is a constant symbol. Let T be the theory in $\mathscr{L}$ consisting of ZFC (written using just $\in$), plus the sentence "c is countable and transitive," plus the sentence ϕ^c for each axiom ϕ of ZFC. T may be seen (finitistically) to be a conservative extension of ZFC, since, by the Reflection Theorem (IV 7.11) any finite sub-theory of T may be interpreted within ZFC. Thus, Con(ZFC) $\to$ Con(T). Within T, one can produce an extension $c[G]$ and prove $\phi^{c[G]}$ for each axiom ϕ of ZFC + $\neg$CH. Thus,

$$\mathrm{Con}(T) \to \mathrm{Con}(\mathrm{ZFC} + \neg\mathrm{CH}).$$

(1c) One can formalize logic within ZFC and then write a predicate, $M \models \ulcorner\mathrm{ZFC}\urcorner$, in the free variable M (see IV §10). It is then a formal theorem of ZFC that

$$\forall M\,(M \text{ is a c.t.m. for } \ulcorner\mathrm{ZFC}\urcorner \to \exists N \supset M\,(N \text{ is a c.t.m. for } \ulcorner\mathrm{ZFC} + \neg\mathrm{CH}\urcorner)).$$

The mathematics of the proof of this theorem is precisely the material in §§2–5, although the logical interpretation is different. This approach does not yield a finitistic proof of Con(ZFC) $\to$ Con(ZFC + $\neg$CH), since in ZFC one cannot prove that

$$\exists M\,(M \text{ is a c.t.m. for } \ulcorner\mathrm{ZFC}\urcorner),$$

but it should convince the confirmed Platonist that Con(ZFC + $\neg$CH) is "true", since the existence of such an M is "true" (see IV §7).

Approach 2: via syntactical models, or forcing over **V**. Here one never discusses set models at all.

(2a) In §3, we defined an auxiliary notion, $\Vdash^*$. The definition of $\Vdash^*$ did not refer to models, although our intent was to relativize it to M to prove that $(p \Vdash \phi) \leftrightarrow (p \Vdash^* \phi)^M$. After §3, we essentially forgot about $\Vdash^*$, although we frequently used the result that $p \Vdash \phi$ was equivalent to some (it did not matter which) formula relativized to M. However, in the syntactical model approach, we forget about $\Vdash$, do $\Vdash^*$ in **V**, and never relativize it to anything. One must check that all the facts that we developed about $\Vdash$ may in fact be proved in **V** about $\Vdash^*$; see Exercises B12–B15. Eventually one checks that $\mathbb{1} \Vdash^* \phi$ whenever ϕ is an axiom for ZFC, and that, with the right $\mathbb{P}$, $\mathbb{1} \Vdash^* \neg$CH. It is also necessary to verify the following.

9.1. LEMMA. *If* $\phi_1, \ldots, \phi_n \vdash \psi$ *and* $p \Vdash^* \phi_1, \ldots, \phi_n$, *then* $p \Vdash^* \psi$. □

Using Lemma 9.1, if we succeed in finding a contradiction in ZFC + $\neg$CH, say

$$\phi_1, \ldots, \phi_n \vdash \psi \wedge \neg\psi,$$

where $\phi_1, \ldots, \phi_n$ are axioms of ZFC + $\neg$CH, then in ZFC we could prove $\mathbb{1} \Vdash^* \psi \wedge \neg\psi$, which is easily seen to contradict the definition of $\Vdash^*$. Thus, we have defined a procedure for finding a contradiction in ZFC, given one in ZFC + $\neg$CH, so Con(ZFC) $\rightarrow$ Con(ZFC + $\neg$CH). This approach is unpalatable to some, since a model in the traditional sense for ZFC + $\neg$CH is never constructed. We may think of this approach as putting ourselves (in **V**) in the place of the M-people of the c.t.m. approach; so we are making up names for, and talking about, objects in some generic extension of **V** which does not exist at all (to us).

(2b) The Boolean-valued model approach (over **V**) is the special case of (2a) in which we consider only p.o.'s which are complete Boolean algebras. As in §7, this special case will produce all the independence proofs that can be done using arbitrary p.o.'s. This special case has, perhaps, a clearer intuitive motivation, since we may think that we really are creating a model, $\mathbf{V}^{\mathscr{B}}$, for ZFC + $\neg$CH, except that the model is in many-valued logic, with truth values lying in some complete Boolean algebra. Lemma 9.1 becomes now the following lemma, which asserts that Boolean valued logic, like 2-valued logic, is valid for classical proof theory.

9.2. LEMMA. *If* $\phi_1, \ldots, \phi_n \vdash \psi$, *then* $[\![\phi_1]\!]^* \wedge \cdots \wedge [\![\phi_n]\!]^* \leq [\![\psi]\!]^*$. □

Intuitively, we may think of $\mathbf{V}$ as a sub-class of $\mathbf{V}^{\mathscr{B}}$, although formally we are defining an embedding, $\check{}$, from $\mathbf{V}$ into $\mathbf{V}^{\mathscr{B}}$. For a detailed development of this approach, see [Rosser 1969] or [Bell 1977].

(2c) Two-valued class models. This is just a curiosity, but in (2b) if G is *any* ultrafilter on $\mathscr{B}$, we may form a two-valued relational system, $\mathbf{V}^{\mathscr{B}}/G$ as follows: $\mathbf{V}^{\mathscr{B}}/G$ has $\mathbf{V}^{\mathscr{B}}$ as its base class, but interprets $\in$ as (2-valued) binary relation E, and interprets $=$ as another binary relation $\equiv$ (instead of as real equality, as is more common in model theory). We define $\tau\, E\, \sigma$ iff $[\![\tau \in \sigma]\!]^* \in G$ and $\tau \equiv \sigma$ iff $[\![\tau = \sigma]\!]^* \in G$. We then show, by induction on ϕ (in the metatheory), that

$$[\![\phi(\tau_1, \ldots, \tau_n)]\!]^* \in G \leftrightarrow V^{\mathscr{B}}/G \models \phi[\tau_1, \ldots, \tau_n].$$

The induction is straightforward except in the step for $\exists$, where essential use is made of the maximal principle (see Theorem 8.2). Thus, with a suitable $\mathscr{B}$, $V^{\mathscr{B}}/G$ will be a 2-valued model for ZFC + $\neg$CH.

$V^{\mathscr{B}}/G$ is almost never well founded, and thus cannot be identified with a transitive model for set theory. See Exercise B18 for more details.

We now briefly survey the history of forcing. Forcing was invented by Cohen, who used it to establish the consistency of ZFC + $\neg$CH, ZF + $\neg$AC, and ZFC + GCH + $\mathbf{V} \neq \mathbf{L}$. Cohen conceived of forcing via the syntactical model approach (2a), but developed the presentations (1a) and (1c) in his published works (see [Cohen 1963, 1964, 1966]), so as to deal with real models. The modification (1b) is due to Shoenfield.

Cohen's original treatment made forcing seem very much related to the constructible hierarchy. His M was always a model for $\mathbf{V} = \mathbf{L}$, so $M = L(\gamma)$, where $\gamma = o(M)$, and his $M[G]$ was defined as $L(\gamma, G)$ (as in VI Exercise 6); this is in fact an equivalent definition in this case (see Exercise B10). He also did not have the idea of working with an abstract p.o., but thought of his conditions as associated with sets of statements in a formal language. Scott and Solovay developed the approach (2b), using an arbitrary complete Boolean algebra, and they realized that the $L(\alpha)$ construction really had nothing to do with forcing, but that it was the $R(\alpha)$ construction that was relevant; in fact, $\mathbf{V}^{\mathscr{B}}$ may be thought of as constructed by iterating the $\mathscr{B}$-valued power set operation (see Exercise B17). They also saw that one could embed any p.o. densely into a complete Boolean algebra, so that the Boolean algebra approach is completely general.

Modern expositions of forcing owe much to Shoenfield, who realized that one could do the Scott–Solovay construction directly from a p.o., without embedding it first into a Boolean algebra. He also invented our definition of $\Vdash$. Previously, there was only $\Vdash^*$. Expositions defined $\Vdash^*$, and proved the basic fact that

$$\phi^{M[G]} \leftrightarrow \exists p \in G\,((p \Vdash^* \phi)^M)$$

(our 3.5). Once this is done, it follows easily that

$$(p \Vdash^* \phi)^M \leftrightarrow \forall \text{generic } G\,(p \in G \to \phi^{M[G]}),$$

but there is really no need to introduce the notion of $\Vdash$, since we may always refer to $\Vdash^*$. The advantage of introducing $\Vdash$ first by the definition

$$p \Vdash \phi \leftrightarrow \forall \text{generic } G\,(p \in G \to \phi^{M[G]}),$$

as we have done (following [Shoenfield 1971a]), is that the reader may gain some insight into what is going on before plunging into the details of $\Vdash^*$.

In the literature, the notations $\Vdash^*$ and $[\![\phi]\!]^*$ are not used. Thus, $p \Vdash \phi$ may mean, in our notation, $p \Vdash \phi$, $p \Vdash^* \phi$, or $(p \Vdash^* \phi)^M$, depending on context. Once the basics are understood, this ambiguity never causes confusion.

Actually, our forcing, $\Vdash$, was what was once called "weak forcing". Cohen defined strong forcing, $\Vdash_s$, as the basic concept. Unlike $\Vdash$, $\Vdash_s$ did not respect logical equivalence; for example, $p \Vdash_s \neg\neg\phi$ did not imply $p \Vdash \phi$. Weak forcing was defined in terms of $\Vdash_s$ by: $p \Vdash \phi$ iff $p \Vdash_s \neg\neg\phi$. $\Vdash_s$ may now seem like a historical anachronism, but it is still relevant to intuitionistic logic, where $\neg\neg\phi$ is not equivalent to ϕ.

There are two important precursors to the modern theory of forcing: one in recursion theory and one in model theory.

In recursion theory, many classical results may be viewed, in hindsight, as forcing arguments. Consider, for example, the Kleene–Post theorem that there are incomparable Turing degrees. Let $\mathbb{P} = \mathrm{Fn}(2 \times \omega, 2)$, let G be $\mathbb{P}$-generic over M, and think of G as coding $f_0, f_1 \in 2^\omega$, where $f_i(n) = \bigcup G(i, n)$. Then, f_0 and f_1 are recursively incomparable (see Exercise G8). Furthermore, to conclude recursively incomparability of f_0 and f_1, it is not necessary that G be generic over all of M; it is sufficient that G intersect only a few of the arithmetically defined dense sets of M; so few that in fact G, and hence also f_0 and f_1, may be taken to be recursive in $0'$. This forcing argument for producing incomparable degrees below $0'$ is in fact precisely the original Kleene–Post argument, with a slight change in notation. See [Sacks 1971] for some deeper applications of forcing to recursion theory and a comparison of these methods with earlier (pre-forcing) techniques.

In model theory, it was well-known that models with truth values taken in an arbitrary Boolean algebra (instead of $\{\mathbb{0}, \mathbb{1}\}$) were correct for classical logic (in the sense of Lemma 9.2 above). Indeed, one proof of the Gödel Completeness Theorem, due to [Rasiowa–Sikorski 1963], involved first constructing a Boolean valued model for a theory, and then applying the Rasiowa–Sikorski Theorem (Exercise D2) to get a suitably "generic" homomorphism into $\{\mathbb{0}, \mathbb{1}\}$ which produced a 2-valued model. This approach also yields a completeness theorem for logic with *infinite* formulas of countable length. However, using Boolean valued models, the analogous completeness theorem for uncountable languages was known to fail. For example, if ϕ is the infinite distributive law,

$$\bigwedge_{n<\omega} \bigvee_{i<2} P_{n,i} \rightarrow \bigvee_{f\in{}^{\omega}2} \bigwedge_{n<\omega} P_{n,f(n)}$$

(ϕ is a sentence of length 2^{ω}, in proposition letters P_{ni}), then ϕ is valid, but is not derivable using the ordinary infinitary proof rules, since ϕ has truth value $\mathbb{0}$ in a suitable Boolean interpretation—namely, in the regular open algebra of 2^{ω} (which is the completion of $\mathrm{Fn}(\omega, 2)$; see Exercise G9), assign P_{ni} truth value $\{f \in 2^{\omega}\colon f(n) = i\}$. In modern times, ϕ fails to be valid in the extension $\mathbf{V}^{\mathscr{B}}$ because in the disjunction $\bigvee_{f\in{}^{\omega}2}$, f ranges only over functions in $\mathbf{V}$. For more on Boolean valued methods in infinitary logic, see [Karp 1964].

Ironically, Cohen was not at first aware of these precursors to his work, and the relationships discussed above only became clear as the theory of forcing was developed further.

EXERCISES

In the following, unless we state otherwise: M represents a c.t.m. for ZFC and $\mathbb{P}$ is a p.o.; furthermore, if $\mathbb{P} \in M$, then G is a filter which is $\mathbb{P}$-generic over M.

A. Warming-up exercises

(A1) $p \in \mathbb{P}$ is called an *atom* iff

$$\neg\exists q, r \in \mathbb{P}\,(q \leq p \wedge r \leq p \wedge q \perp r).$$

$\mathbb{P}$ is *non-atomic* iff $\mathbb{P}$ has no atoms. Show that if $\mathbb{P} \in M$ and p is an atom of $\mathbb{P}$, then there is a filter $G \in M$ such that $p \in G$ and G intersects *all* dense subsets of $\mathbb{P}$. *Remark.* This is a converse to Lemma 2.4.

(A2) Assume that $\mathbb{P} \in M$ and $\mathbb{P}$ infinite. Show that there is an $H \subset \mathbb{P}$ such that $M[H]$ is not a model of ZF $-$ P. *Hint.* Fix $f \in M$ such that f maps $\omega \times \omega$ 1–1 into $\mathbb{P}$. Choose H so that $f^{-1}(H)$ is a well-order of ω in type $> \mathrm{o}(M)$.

(A3) In Exercise A2, assume also that $\mathbb{P}$ is non-atomic. Show that H may be chosen to be a filter.

(A4) Suppose that $\mathbb{P} \in M$ and $\mathbb{P}$ is non-atomic. Show that $\{G\colon G \text{ is } \mathbb{P}\text{-generic over } M\}$ has cardinality 2^{ω}.

(A5) If $\sigma, \tau \in M^{\mathbb{P}}$, show that $\sigma_G \cup \tau_G = (\sigma \cup \tau)_G$. *Remark.* This does not require G to be generic.

(A6) If $\tau \in M\ \mathbb{P}$, let

$$\pi = \{\langle \rho, p\rangle\colon \exists\langle \sigma, q\rangle \in \tau\, \exists r\, (\langle \rho, r\rangle \in \sigma \wedge p \leq r \wedge p \leq q\}.$$

Show that $\pi_G = \bigcup(\tau_G)$. *Remark.* This requires G to be a filter, but G need not be generic.

(A7) If $\tau, \sigma \in M^{\mathbb{P}}$ and $\operatorname{dom}(\tau), \operatorname{dom}(\sigma) \subset \{\check{n}\colon n \in \omega\}$, let

$$\pi = \{\langle \check{n}, p\rangle\colon \exists q, r\, (p \leq q \wedge p \leq r \wedge \langle \check{n}, q\rangle \in \tau \wedge \langle \check{n}, r\rangle \in \sigma)\}.$$

Show that $\pi_G = \tau_G \cap \sigma_G$. *Remark.* This requires G to be a filter, but G need not be generic.

(A8) Suppose $\tau \in M^{\mathbb{P}}$ and $\operatorname{dom}(\tau) \subset \{\check{n}\colon n \in \omega\}$. Let

$$\sigma = \{\langle \check{n}, p\rangle\colon \forall q \in \mathbb{P}\, (\langle \check{n}, q\rangle \in \tau \rightarrow p \perp q)\}.$$

Show that $\sigma_G = \omega \setminus \tau_G$. *Hint.*

$$\{r\colon \exists p \geq r\, (\langle \check{n}, p\rangle \in \sigma \vee \langle \check{n}, p\rangle \in \tau)\}$$

is dense. *Remark.* To show $\forall \tau \in M^{\mathbb{P}}\ ((\omega \setminus \tau_G) \in M[G])$ requires forcing.

(A9) Assume that $\mathbb{P} \in M$, $p \in \mathbb{P}$, and $\exists q \in \mathbb{P}\, (p \perp q)$. Show that

$$\{\tau \in M^{\mathbb{P}}\colon p \Vdash \tau = \check{0}\}$$

is a proper class in M. *Remark.* Thus, if $p \in G$, $\{\tau \in M^{\mathbb{P}}\colon \tau_G = 0\}$ is not a subset of any set of M.

(A10) Assume $\mathbb{P} \in M$ and that $\mathbb{P}$ is separative (see II Exercise 15). Show that

$$p \Vdash \{\langle\{\langle 0, q\rangle\}, r\rangle\} = \check{1},$$

iff $p \leq r$ and $p \perp q$.

(A11) Assume $\mathbb{P} \in M$ and $p \perp q$ for some $p, q \in \mathbb{P}$. Show that

$$\{\tau \in M^{\mathbb{P}}\colon \mathbb{1} \Vdash \tau = \check{1}\}$$

is a proper class of M. *Hint.* Consider, for any $\sigma \in M^{\mathbb{P}}$,

$$\{\langle\{\langle \sigma, p\rangle\}, q\rangle, \langle \check{0}, \mathbb{1}\rangle\}.$$

(A12) Assume $\mathbb{P} \in M$ and G is a filter in $\mathbb{P}$. Show that the following are equivalent.

(a) $G \cap D \neq 0$, whenever $D \in M$ and D is dense in $\mathbb{P}$.

(b) $G \cap A \neq 0$, whenever $A \in M$ and A is a maximal antichain in $\mathbb{P}$.

(c) $G \cap E \neq 0$, whenever $E \in M$ and $\forall p \in \mathbb{P}\, \exists q \in E(p$ and q are compatible).

Furthermore, show that in the definition of filter, we may weaken the requirement

(1) $\forall p, q \in G\, \exists r \in G\, (r \leq p \wedge r \leq q)$ to

(1′) $\forall p, q \in G\, \exists r \in \mathbb{P}\, (r \leq p \wedge r \leq q)$.

Thus, this Exercise provides $3 \cdot 2 = 6$ equivalent definitions of "generic".

B. Miscellaneous results

(B1) Show that if we redefine $\check{x}$ by

$$\check{x} = \{\langle \check{y}, p \rangle : y \in x \wedge p \in \mathbb{P}\},$$

then we may drop the assumption that $\mathbb{P}$ has a largest element for all the results in this chapter. *Remark.* The usefulness of $\mathbb{1}_{\mathbb{P}}$ will be more apparent in VIII.

(B2) Suppose $\langle \mathbb{P}, \leq \rangle$ is a partial order in M which may or may not have a largest element. In M, fix $\mathbb{1} \notin \mathbb{P}$, and define the p.o. $\langle \mathbb{Q}, \leq, \mathbb{1} \rangle$ by: $\mathbb{Q} = \mathbb{P} \cup \{\mathbb{1}\}$, where $\mathbb{P}$ retains the same order and $\forall p \in \mathbb{P}\, (p < \mathbb{1})$. Show that if $G \subset \mathbb{P}$, G is $\mathbb{P}$-generic over M iff $G \cup \{\mathbb{1}\}$ is $\mathbb{Q}$-generic over M, and $M[G]$ (defined as a $\mathbb{P}$-extension) is the same as $M[G \cup \{\mathbb{1}\}]$ (defined as a $\mathbb{Q}$-extension).

(B3) Define $\Vdash'$ by the following clauses:

(a) $p \Vdash' \tau_1 = \tau_2$ iff

$$\forall \pi \in \operatorname{dom}(\tau_1) \cup \operatorname{dom}(\tau_2)\, \forall q \leq p\, ((q \Vdash' \pi \in \tau_1) \leftrightarrow (q \Vdash' \pi \in \tau_2)).$$

(b)–(e) As in the definition of $\Vdash^*$ (3.3). Show that this definition makes sense. *Remark* and *hint*. Clauses (a) and (b) define $\Vdash'$ for both kinds of atomic formulas by a simultaneous recursion. Say, in (a),

$$\operatorname{rank}(\tau_1) < \operatorname{rank}(\pi) < \operatorname{rank}(\tau_2).$$

Then $\langle \pi, \tau_2 \rangle$ is "bigger" than $\langle \tau_1, \tau_2 \rangle$. However, clause (b) now reduces the forcing of $\pi \in \tau_2$ to the forcing of $\pi = \pi'$ for $\pi' \in \operatorname{dom}(\tau_2)$, and

$$\max(\operatorname{rank}(\pi), \operatorname{rank}(\pi')) < \max(\operatorname{rank}(\tau_1), \operatorname{rank}(\tau_2)).$$

(B4) Prove Enunciations 3.4–3.6 using $\Vdash'$ instead of $\Vdash^*$.

(B5) Assume $f: A \to M$ and $f \in M[G]$. Show that there is a $B \in M$ such that $f: A \to B$. *Hint.* Let

$$B = \{b\colon \exists p \in \mathbb{P}\,(p \Vdash \check{b} \in \operatorname{ran}(\tau))\},$$

where $f = \tau_G$.

(B6) Assume $\mathbb{P} \in M$ and α is a cardinal of M. Show that the following are equivalent.

(1) Whenever $B \in M$, ${}^{\alpha}B \cap M = {}^{\alpha}B \cap M[G]$.

(2) ${}^{\alpha}M \cap M = {}^{\alpha}M \cap M[G]$.

(3) In M: The intersection of α dense open subsets of $\mathbb{P}$ is dense.

See the proof of II 3.3 for the topology on $\mathbb{P}$. *Remark.* A p.o. satisfying (3) is called α^+-*dense* or α^+-*Baire*. κ-Baire means that the intersection of *less than* κ dense open sets is dense.

(B7) Show that if $\mathbb{P}$ is λ-closed and λ is singular then $\mathbb{P}$ is λ^+-closed.

(B8) Let $\mathbb{P} \in M$ be non-atomic (see Exercise A1). Let

$$M = M_0 \subset M_1 \subset \cdots \subset M_n \subset \cdots \qquad (n \in \omega).$$

such that $M_{n+1} = M_n[G_n]$ for some G_n which is $\mathbb{P}$-generic over M_n. Show that $\bigcup_n M_n$ cannot satisfy the Power Set Axiom. Furthermore, show that the G_n may be chosen so that there is no c.t.m. N for ZFC with $\langle G_n\colon n \in \omega\rangle \in N$ and $o(N) = o(M)$. *Hint.* $\{n\colon p \in G_n\}$ can code $o(M)$.

(B9) Show that $\neg$KH implies that ω_2 is inaccessible in $\mathbf{L}$. *Hint.* Suppose ω_2 is $(\lambda^+)^{\mathbf{L}}$, where λ is a cardinal of $\mathbf{L}$. Let $X \subset \omega_1$ be such that $\omega_1^{\mathbf{L}(X)} = \omega_1$ and $\omega_2^{\mathbf{L}(X)} = \omega_2$. $\diamondsuit^+$, and hence KH, hold in $\mathbf{L}(X)$ (see VI Exercise 8), and a Kurepa tree of $\mathbf{L}(X)$ remains such in $\mathbf{V}$.

(B10) Suppose M satisfies $\mathbf{V} = \mathbf{L}$, so $M = L(\gamma)$, where $\gamma = o(M)$. Show that $M[G] = L(\gamma, G)$ in the sense of VI Exercise 6.

(B11) Verify that one may do forcing over a c.t.m. M for ZF* − P to produce $M[G]$ satisfying ZF* − P. Show that $\mathrm{AC}^M \to \mathrm{AC}^{M[G]}$ and that $M[G]$ will satisfy the Power Set Axiom if M does. *Remark.* See III Exercise 19 for ZF*. AR^* is needed in M to prove AR (and AR^*) in $M[G]$.

(B12) Express Exercises A9–A11 as exercises about $\Vdash^*$ in $\mathbf{V}$ (without any mention of models), and work them using the definition of $\Vdash^*$. For example, the conclusion of Exercise A9 says that

$$\{\tau \in \mathbf{V}^{\mathbb{P}} : p \Vdash^* \tau = \check{0}\}$$

is a proper class.

(B13) Let $\mathbb{P}$ be a p.o. Show that for any x, y,

$$x = y \to \mathbb{1} \Vdash^* \check{x} = \check{y},$$

$$x \neq y \to \mathbb{1} \Vdash^* \neg(\check{x} = \check{y})$$

$$x \in y \to \mathbb{1} \Vdash^* \check{x} \in \check{y},$$

$$x \notin y \to \mathbb{1} \Vdash^* \neg(\check{x} \in \check{y}).$$

Remark. This exercise does not mention models.

(B14) Show that all the relative consistency results of this chapter can be done in the syntactical model approach ((2a) of §9). As a start, one must show, without reference to models, that $\mathbb{1} \Vdash^* \phi$ for each axiom ϕ of ZFC. Interpret Exercise B6 as a characterization of κ-Baire in $\mathbf{V}$.

(B15) Do Exercise B14 using reflection. Thus, if $\neg(\mathbb{1} \Vdash^* \phi)$, for ϕ an axiom of ZFC, there would be c.t.m.'s, M, for arbitrary finite fragments of ZFC, such that $\neg(\mathbb{1} \Vdash^* \phi)^M$, and hence $\neg(\mathbb{1} \Vdash \phi)$.

(B16) Let $\mathscr{B}$ be a complete Boolean algebra. A *$\mathscr{B}$-valued structure* for the language of set theory is a triple $\langle M, f_=, f_\in \rangle$, where $f_=, f_\in : M \times M \to \mathscr{B}$ and the axioms of predicate calculus with equality are valid; for example, we require

$$[\![\forall x, y, z\,((x = y \wedge y \in z) \to x \in z)]\!] = \mathbb{1},$$

where we evaluate $[\![\phi]\!]$ by setting $[\![x \in y]\!] = f_\in(x, y)$, $[\![x = y]\!] = f_=(x, y)$, and interpreting the logical connectives by their corresponding Boolean operations. Show that if $\vdash \phi$, then $[\![\phi]\!] = \mathbb{1}$ in any such structure.

(B17) Verify that one can construct $\mathbf{V}^{\mathscr{B}}$ by iterating the Boolean power set operation. Thus, for each α, define $\mathscr{B}$-valued structures, $\langle R^{\mathscr{B}}(\alpha), f_=(\alpha), f_\in(\alpha)\rangle$ (see Exercise B16). $R^{\mathscr{B}}(\alpha + 1)$ is $R^{\mathscr{B}}(\alpha)$ plus the set of *extensional* functions, h, from $R^{\mathscr{B}}(\alpha)$ into $\mathscr{B}$; where extensional means

$$\forall x, y \in R^{\mathscr{B}}(\alpha)\,(([\![x = y]\!] \wedge h(y)) \leq h(x)).$$

Let $\mathbf{WF}^{\mathscr{B}} = \bigcup_{\alpha \in \mathbf{ON}} R^{\mathscr{B}}(\alpha)$. There is a Boolean isomorphism, $\mathbf{I}$, between $\mathbf{V}^{\mathscr{B}}$ (in the sense of approach (2b) of §9) and $\mathbf{WF}^{\mathscr{B}}$; thus, $\mathbf{I}$: $\mathbf{V}^{\mathscr{B}} \times \mathbf{WF}^{\mathscr{B}} \to \mathscr{B}$, and satisfies

$$(\llbracket \sigma \in \tau \rrbracket^* \wedge \mathbf{I}(\sigma, x) \wedge \mathbf{I}(\tau, y)) \leq \llbracket x \in y \rrbracket,$$
$$\bigvee \{\mathbf{I}(\sigma, x) \colon x \in \mathbf{WF}^{\mathscr{B}}\} = \mathbb{1}, \text{etc.}$$

(B18) In Approach (2c) of §9, show that $\mathbf{V}^{\mathscr{B}}/G$ is well founded iff G is countably complete. Furthermore, if $\mathscr{B}$ is non-atomic, G cannot be countably complete unless $|\mathscr{B}|$ is at least as large as the first 2-valued measurable cardinal.

C. Complete embedding and complete Boolean algebras

(C1) Show that a composition of complete embeddings is complete. That is, if $i : \mathbb{P} \to \mathbb{Q}$ and $j : \mathbb{Q} \to \mathbb{R}$ are complete, so is $j \circ i : \mathbb{P} \to \mathbb{R}$.
(C2) Show that if $\mathbb{P}$ and $\mathbb{Q}$ are separative (see II Exercise 15), and $i : \mathbb{P} \to \mathbb{Q}$ is a complete embedding, then i is 1–1, $i(\mathbb{1}_{\mathbb{P}}) = \mathbb{1}_{\mathbb{Q}}$, and for all $p, p' \in \mathbb{P}$, $p \leq p'$ iff $i(p) \leq i(p')$.

(C3) Let $i : \mathbb{P} \to \mathbb{Q}$. Show that if i is a complete embedding, then

$$\mathbb{P} \text{ is non-atomic} \to \mathbb{Q} \text{ is non-atomic},$$

and if i is a dense embedding, then

$$\mathbb{P} \text{ is non-atomic} \leftrightarrow \mathbb{Q} \text{ is non-atomic}.$$

(C4) Let $\mathbb{P}$ be a countable non-atomic p.o. Show that there is a dense embedding from $\{p \in \mathrm{Fn}(\omega, \omega) \colon \mathrm{dom}(p) \in \omega\}$ into $\mathbb{P}$. *Hint.* If $\mathbb{P} = \mathrm{Fn}(\omega, \omega)$, inclusion works. In general, map $\{p \colon \mathrm{dom}(p) = 1\}$ onto an infinite antichain in $\mathbb{P}$, now handle $\{p \colon \mathrm{dom}(p) = 2\}$, etc. *Remark.* Hence, $\mathbb{P}$, $\mathrm{Fn}(\omega, 2)$, $\mathrm{Fn}(\omega, \omega)$, and $\mathrm{Fn}(\omega \times \omega, 18)$ all yield the same generic extentions.

(C5) If $\mathbb{P}$ is the non-$\mathbb{0}$ elements of a Boolean algebra $\mathscr{B}$, show that $\mathbb{P}$ is non-atomic by the definition in Exercise A1 iff $\mathscr{B}$ is non-atomic in the usual sense for Boolean algebras:

$$\forall b > \mathbb{0}\, \exists c\, (\mathbb{0} < c < b).$$

(C6) Let $\mathscr{A}$ and $\mathscr{B}$ be non-atomic countable Boolean algebras. Show that $\mathscr{A}$ and $\mathscr{B}$ are isomorphic. *Remark.* Via Stone spaces, this implies that all compact 0-dimensional second-countable Hausdorff spaces with no isolated points are homomorphic to the Cantor set.

(C7) If $\mathscr{A}$ and $\mathscr{B}$ are Boolean algebras, a homomorphism $i : \mathscr{A} \to \mathscr{B}$ is *complete* iff whenever $S \subset \mathscr{A}$ and $\bigvee S$ exists, then $\bigvee (i''S)$ exists in $\mathscr{B}$ and

$\bigvee(i''S) = i(\bigvee S)$. Show that $i : \mathscr{A} \to \mathscr{B}$ is a complete embedding (in the sense of Definition 7.1) iff i is a complete injective homomorphism. Furthermore, show that in this case, if, for $b \in \mathscr{B}$, $h(b) = \bigwedge\{a \in \mathscr{A}: i(a) \geq b\}$, then $h(b)$ is the largest reduction of b to $\mathscr{A}$. $h(b)$ is called the *$\mathscr{A}$-hull* of b.

(C8) Let $i : \mathbb{P} \to \mathbb{Q}$ be a complete embedding, $\mathscr{A}$ = the completion of $\mathbb{P}$, and $\mathscr{B}$ = the completion of $\mathbb{Q}$. Show how i defines a complete embedding, j, from $\mathscr{A}$ into $\mathscr{B}$. If i is a dense embedding, show that j is an isomorphism.

(C9) Let $\mathbb{P}$ be the p.o. used in proving the $<c$ additivity of Lebesgue measure from MA (see II 2.21), and let $i : \mathbb{P} \to \mathscr{B}$ be the completion of $\mathbb{P}$. Show that there are $p, q \in \mathbb{P}$ such that $p \not\leq q$, $q \not\leq p$, and $i(p) = i(q)$.

(C10) Let $\mathscr{B}$ be a complete Boolean algebra. Show that $\mathscr{B}$ is ω_1-Baire (see Exercise B6) iff $\mathscr{B}$ is (ω, ∞) distributive, i.e., for each κ, the equation

$$\bigwedge_{n\in\omega} \bigvee_{\alpha\in\kappa} b_{n,\alpha} = \bigvee_{f\in\kappa^\omega} \bigwedge_{n\in\omega} b_{n,f(n)}$$

holds.

(C11) Show that $\neg$SH is equivalent to the existence of a non-atomic, c.c.c., complete Boolean algebra, $\mathscr{B}$, which is (ω, ∞) distributive. Such a $\mathscr{B}$ is called a *Suslin algebra*.

D. Relations of C to forcing

(D1) Let $\mathscr{B}$ be a complete Boolean algebra of M and $F \subset \mathscr{B}$. Show that F is $\mathscr{B}$-generic over M iff F is an ultrafilter and the associated homomorphism h of $\mathscr{B}$ into the 2-element algebra preserves all sups in M—i.e., for all $S \in M$ with $S \subset \mathscr{B}$, $h(\bigvee S) = \bigvee\{h(b): b \in S\}$.

(D2) ([Rasiowa–Sikorsky 1963]). Let $\mathscr{B}$ be any Boolean algebra, and, for $n \in \omega$, $S_n \subset \mathscr{B}$ such that the supremum of S_n exists and $\bigvee S_n = b_n$. Show that there is a homomorphism h of $\mathscr{B}$ into the 2-element algebra preserving each $\bigvee S_n$ (i.e., $h(b_n) = \bigvee\{h(b): b \in S_n\}$). *Remark.* By Exercise D1, this Rasiowa–Sikorski theorem is the generic filter existence theorem for Boolean-valued models.

(D3) Assume that i, $\mathbb{P}$, and $\mathbb{Q}$ are in M, and $i : \mathbb{P} \to \mathbb{Q}$ is a complete embedding. For any $G \subset \mathbb{P}$, define

$$\mathbb{Q}/G = \{q \in \mathbb{Q}: \forall p \in G\,(q \text{ is compatible with } i(p))\}.$$

Show that p is a reduction of q to $\mathbb{P}$ iff $p \Vdash \check{q} \in \check{\mathbb{Q}}/\Gamma$.

(D4) Suppose that in M, $i: \mathbb{P} \to \mathbb{Q}$ is a complete embedding. Let G be $\mathbb{P}$-generic over M, and let K be $\mathbb{Q}/G$ generic over $M[G]$. Show that K is $\mathbb{Q}$-generic over M and that $M[K]_{\mathbb{Q}} = M[G][K]_{\mathbb{Q}/G}$. Here $M[K]_{\mathbb{Q}} = \{\tau_K : \tau \in M^{\mathbb{Q}}\}$, while $M[G][K]_{\mathbb{Q}/G} = \{\tau : \tau \in M[G]^{\mathbb{Q}/G}\}$; this notation is needed since K is a filter in two different p.o.'s. *Hint*. If D is dense in $\mathbb{Q}$, then $D \cap \mathbb{Q}/G$ is dense in $\mathbb{Q}/G$. To see this, fix $q_0 \in \mathbb{Q}/G$ and fix $p_0 \in G$ with $p_0 \Vdash \check{q}_0 \in \check{\mathbb{Q}}/\Gamma$. Show that the following is dense below p_0:

$$\{p \in \mathbb{P}: \exists q \in \mathbb{Q}\,(q \leq q_0 \wedge q \in D \wedge p \Vdash \check{q} \in \check{\mathbb{Q}}/\Gamma)\}.$$

(D5) Suppose that in M, $i: \mathbb{P} \to \mathbb{Q}$ is a complete embedding. Let H be $\mathbb{Q}$-generic over M and let $G = i^{-1}(H)$. Show that $H \subset \mathbb{Q}/G$, H is $\mathbb{Q}/G$-generic over $M[G]$, and $M[H]_{\mathbb{Q}} = M[G][H]_{\mathbb{Q}/G}$. *Hint* If $D \subset \mathbb{Q}/G$, D is dense, and $D \in M[G]$, let $D = \tau_G$ and let $p_0 \Vdash (\tau$ is dense in $\check{\mathbb{Q}}/\Gamma)$; then

$$\{q: \exists p\, \exists q_1 [(p \Vdash \check{q}_1 \in \tau) \wedge q \leq i(p) \wedge q \leq q_1]\}$$

is dense below $i(p_0)$. *Remark*. Exercises D4 and D5 show that a one-step extension by $\mathbb{Q}$ is equivalent to extending via $\mathbb{P}$ and then $\mathbb{P}/G$.

(D6) Assume that in M, $i: \mathscr{A} \to \mathscr{B}$ is a complete Boolean embedding of complete Boolean algebras. Let G be $\mathscr{A}$-generic over M, and, in $M[G]$, let

$$\mathscr{I} = \{b \in \mathscr{B}: \exists a \in G\,(b \perp i(a))\}.$$

Show that $\mathscr{I}$ is an ideal in $\mathscr{B}$, the quotient algebra $\mathscr{B}/\mathscr{I}$ is complete in $M[G]$, and $\mathscr{I} = \{b: (h(b))' \in G\}$, where h is as in Exercise C7. Furthermore, if $\mathscr{B}/G$ is as in Exercise D3, and $i(b) = [b]_{\mathscr{I}}$, then $i: \mathscr{B}/G \to \mathscr{B}/\mathscr{I}$, is a dense embedding, so that $\mathscr{B}/\mathscr{I}$ is the completion of $\mathscr{B}/G$ in $M[G]$.

(D7) If $\mathscr{B}$ is a complete Boolean algebra of M, show that
(a) $[\![\tau_1 = \tau_2]\!] = \bigwedge\{[\![\pi \in \tau_1]\!] \leftrightarrow [\![\pi \in \tau_2]\!]: \pi \in \mathrm{dom}(\tau_1) \cup \mathrm{dom}(\tau_2)\}$, and
(b) $[\![\tau_1 \in \tau_2]\!] = \bigvee\{[\![\pi = \tau_1]\!] \wedge b: \langle \pi, b\rangle \in \tau_2\}$,
where, for $a, b \in \mathscr{B}$, $(a \leftrightarrow b) = (a' \vee b) \wedge (b' \vee a)$.
Remark. In the spirit of B3, (a) and (b) can be taken as clauses in an inductive definition of $[\![\ldots]\!]^*$.

(D8) In M, let $i: \mathbb{P} \to \mathscr{B}$ be the completion of $\mathbb{P}$. Show that $i(p) = i(q)$ iff for all G which are $\mathbb{P}$-generic over $M, p \in G \leftrightarrow q \in G$.

E. Automorphisms and AC

(E1) If $\mathbb{P}$ (i.e., $\langle \mathbb{P}, \leq, \mathbb{1}\rangle$) is a p.o., an *automorphism* of $\mathbb{P}$ is a 1–1 i from $\mathbb{P}$ onto $\mathbb{P}$ which preserves $\leq$ and satisfies $i(\mathbb{1}) = \mathbb{1}$; thus also $i_*(\check{x}) = \check{x}$ for each x. $\mathbb{P}$ is called *almost homogeneous* iff for all $p, q \in \mathbb{P}$, there is an automorphism i of $\mathbb{P}$ such that $i(p)$ and q are compatible. Suppose that $\mathbb{P} \in M$ and $\mathbb{P}$ is almost homogeneous in M. Show that if $p \Vdash \phi(\check{x}_1, \dots, \check{x}_n)$, then $\mathbb{1} \Vdash \phi(\check{x}_1, \dots, \check{x}_n)$; thus, either $\mathbb{1} \Vdash \phi(\check{x}_1, \dots, \check{x}_n)$ or $\mathbb{1} \Vdash \neg\phi(\check{x}_1, \dots, \check{x}_n)$.

(E2) Show that any $\mathrm{Fn}(I, J, \kappa)$ is almost homogeneous.

(E3) (A. Miller). In M, let I and J be uncountable, $\mathbb{P} = \mathrm{Fn}(I, 2)$, and $\mathbb{Q} = \mathrm{Fn}(J, 2)$. Let $\phi(x)$ be a formula. Show that

$$\mathbb{1} \Vdash_{\mathbb{P}} (\phi(\check{\alpha})^{\mathbf{L}(\mathscr{P}(\omega))}) \quad \text{iff} \quad \mathbb{1} \Vdash_{\mathbb{Q}} (\phi(\check{\alpha})^{\mathbf{L}(\mathscr{P}(\omega))}).$$

Hint. If $(|I| = |J|)^M$ apply Lemma 7.13. More generally, say $(|I| \le |J|)^M$ and let H be $\mathrm{Fn}(I, J, \omega_1)^M$-generic over M; then $(|I| = |J|)^{M[H]}$. If G is $\mathbb{P}$-generic over $M[H]$, then G is $\mathbb{P}$-generic over M and $\mathscr{P}(\omega) \cap M[H][G] = \mathscr{P}(\omega) \cap M[G]$. Thus, applying E1,

$$\mathbb{1} \Vdash_{\mathbb{P}, M} (\phi(\check{\alpha})^{\mathbf{L}(\mathscr{P}(\omega))}) \quad \text{iff} \quad \mathbb{1} \Vdash_{\mathbb{P}, M[H]} (\phi(\check{\alpha})^{\mathbf{L}(\mathscr{P}(\omega))}).$$

Likewise with $\mathbb{Q}$.

(E4) Show $\mathrm{Con}(\mathrm{ZF}) \to \mathrm{Con}(\mathrm{ZF} + \neg\mathrm{AC})$. In fact, show the consistency of $\neg\mathrm{AC}$ with $\mathbf{V} = \mathbf{L}(\mathscr{P}(\omega)) = \mathbf{HOD}(\mathscr{P}(\omega))$. *Hint*. (A. Miller). Let $(|I| \ge \omega_1)^M$, Let G be $\mathrm{Fn}(I, 2)$-generic over M, and let $N = \mathbf{L}(\mathscr{P}(\omega))^{M[G]}$. If AC holds in N, let $(\kappa = |\mathscr{P}(\omega)|)^N$, then $\mathbb{1} \Vdash ((\check{\kappa} = |\mathscr{P}(\omega)|)^{\mathbf{L}(\mathscr{P}(\omega))})$. Now, taking J such that $(|J| > \kappa)^M$ will contradict E3.

(E5) Show the consistency of ZFC + GCH with $\mathbf{V} \ne \mathbf{OD}$. *Hint*. Use $\mathrm{Fn}(I, 2)$, and apply E1.

F. Chain conditions

(F1) Let $\mathrm{c.c.}(\mathbb{P})$ be the least θ such that $\mathbb{P}$ has the θ-c.c. Show that $\mathrm{c.c.}(\mathbb{P}) \ne \omega$. *Hint*. Consider a maximal incompatible family of atoms.

(F2) If $\mathbb{P} \in M$ and $\mathrm{c.c.}(\mathbb{P}) < \omega$, show that every G which is $\mathbb{P}$-generic over M is in M. Do likewise when $(\mathrm{c.c.}(\mathbb{P}) = \kappa \wedge \mathbb{P}$ is κ-closed$)^M$.

(F3) If $\mathbb{P} \in M$, show that $\mathrm{c.c.}(\mathbb{P}) < \omega$ iff $(\mathrm{c.c.}(\mathbb{P}) < \omega)^M$. *Hint*. Resolve $\mathbb{P}$ into atoms in M.

*(F4) (Tarski) Show that if $\mathrm{c.c.}(\mathbb{P}) \ge \omega$ then $\mathrm{c.c.}(\mathbb{P})$ is regular.

(F5) Show that if $|I| \geq \lambda$ and $|J| \geq 2$, then $\mathrm{Fn}(I, J, \lambda)$ has an antichain of size $2^{<\lambda}$. *Remark.* This is easier when λ is a successor cardinal.

(F6) Show that every countable complete Boolean algebra is finite. *Hint.* See Exercise F1.

G. Fn(I, J, λ)

(G1) Let κ, λ, θ be infinite cardinals of M, and let $\mathbb{P} = \mathrm{Fn}(\kappa \times \omega, 2)$. Show that

$$(\lambda^{\theta})^{M[G]} = ((\max(\kappa, \lambda))^{\theta})^{M}.$$

(G2) Suppose $I \in M$ is infinite, and let $2 \leq \alpha \leq \beta \leq \omega$. Show that $\mathrm{Fn}(I, \alpha)$ and $\mathrm{Fn}(I, \beta)$ yield the same generic extensions. *Hint.* Let $I = \kappa \times \omega$, and apply Exercise C4 co-ordinatewise.

(G3) Assume λ is regular, and let $\mathbb{P}(J) = \{p \in \mathrm{Fn}(\lambda, J, \lambda)\colon \mathrm{dom}(p) \in \lambda\}$. Show that $\mathbb{P}(J)$ is densely included in $\mathrm{Fn}(\lambda, J, \lambda)$. Furthermore, show that if $2 \leq |J| \leq 2^{<\lambda}$, then $\mathbb{P}(2^{<\lambda})$ densely embeds into $\mathrm{Fn}(\lambda, J, \lambda)$. *Remark.* Applying this within M show that $\mathrm{Fn}(\lambda, 2, \lambda)$ adds a function from λ onto $2^{<\lambda}$, since $\mathrm{Fn}(\lambda, 2^{<\lambda}, \lambda)$ obviously does.

(G4) Let $(\mathbb{P} = \mathrm{Fn}(I, 2, \omega_1))^M$, where $(|I| \geq \omega_1)^M$. Show that $M[G]$ satisfies CH, regardless of whether M does.

(G5) Suppose, in M, $\theta = \mathrm{cf}(\lambda) < \lambda$. Show that $\mathrm{Fn}(\lambda, 2, \lambda)$ adds a map from θ onto λ^+. *Hint.* Say $\lambda = (\omega_\omega)^M$. Let $f(n)$ be the ω_n-th element of $\{\alpha\colon \bigcup G(\alpha) = 1\}$. Let $g(n)$ be the $\beta < \omega_n$ such that $f(n)$ is of the form $\omega_n \cdot \delta + \beta$. Then g maps ω onto λ. Now show that $\bigcup G$ codes every function in M from ω into λ.

(G6) Suppose M satisfies GCH. Let $\kappa_1 < \cdots < \kappa_n$ be regular cardinals of M, and let $\lambda_1 \leq \cdots \leq \lambda_n$ be cardinals of M such that $(\mathrm{cf}(\lambda_i) > \kappa_i)^M$. Force n times to construct a c.t.m. $N \supset M$ with the same cardinals such that for each i, $(2^{\kappa_i} = \lambda_i)^N$.

(G7) Suppose M satisfies CH and $\mathbb{P} = \mathrm{Fn}(I, 2)$, where $(|I| \geq \omega_2)^M$. Show that MA fails in $M[G]$. *Hint.* Show that if $f \in \omega^\omega \cap M[G]$, there is a $g \in \omega^\omega \cap M$ such that $\{n\colon g(n) \geq f(n)\}$ is infinite. Now apply II Exercise 8.

(G8) If $\mathbb{P} = \mathrm{Fn}(\kappa \times \omega, 2)$, and $f_\alpha(n) = \bigcup G(\alpha, n)$, show that the f_α for $\alpha < \kappa$ are recursively incomparable.

(G9) Show that the completion of the p.o. $\mathrm{Fn}(I, 2)$ is isomorphic to the regular open algebra of the space 2^I (where $2 = \{0, 1\}$ has the discrete topology and 2^I has the product topology).

H. Specific forcing constructions

(H1) Assume in M that $\kappa > \omega$, κ is regular, and $\mathbb{P}$ has the κ-c.c. In $M[G]$, let $C \subset \kappa$ and C c.u.b. Show that there is a $C' \in M$ such that $C' \subset C$ and C' is c.u.b. in κ. *Hint*. In $M[G]$, let $f: \kappa \to \kappa$ be such that

$$\forall \alpha < \kappa \, (\alpha < f(\alpha) \in C),$$

and apply Lemma 6.8.

(H2) Suppose that in M: $\kappa > \omega$, κ is regular, $S \subset \kappa$ is stationary and $\mathbb{P}$ is either κ-c.c. or κ-closed. Show that S remains stationary in $M[G]$.

(H3) A cardinal κ is called *weakly Mahlo* iff κ is regular and $\{\alpha < \kappa\colon \alpha$ is regular$\}$ is stationary in κ. Show that if κ is weakly Mahlo, then

$$\{\alpha < \kappa\colon \alpha \text{ is weakly inaccessible}\}$$

is stationary in κ.

(H4) Assume in M that $\kappa > \omega$ and $\mathbb{P}$ is either κ-closed or has the λ-c.c. for some $\lambda < \kappa$. Show that if κ is weakly Mahlo in M, then κ is weakly Mahlo in $M[G]$. Do the same for weakly hyper-Mahlo, where κ is *weakly hyper-Mahlo* iff κ is regular and

$$\{\alpha < \kappa\colon \alpha \text{ is weakly Mahlo}\}$$

is stationary in κ.

(H5) Show that 5.16 remains true if "inaccessible" is everywhere replaced by "Mahlo". Do likewise for "hyper-Mahlo". See II Exercises 48 and 50 for the definitions of strongly Mahlo and strongly hyper-Mahlo. Furthermore, show that it is relatively consistent to have a Mahlo or hyper-Mahlo κ such that $2^\kappa > \kappa^+$ and GCH holds below κ.

(H6) Assume that in M, $\mathscr{A}, \mathscr{C} \subset \mathscr{P}(\omega)$ and for all $y \in \mathscr{C}$ and all finite $F \subset \mathscr{A}$ $|y \setminus \bigcup F| = \omega$. Show that there is a $\mathbb{P} \in M$ such that $(\mathbb{P}$ is c.c.c.$)^M$ and whenever G is $\mathbb{P}$-generic over M, $M[G]$ contains a $d \subset \omega$ such that

$$\forall x \in \mathscr{A} \, (|d \cap x| < \omega) \quad \text{and} \quad \forall y \in \mathscr{C} \, (|d \cap y| = \omega).$$

Hint. See II 2.15.

(H7) Let $\langle X, <\rangle$ be a total order in M. Show that there is a $\mathbb{P} \in M$ such that ($\mathbb{P}$ has the c.c.c.)^M and whenever G is $\mathbb{P}$-generic over M, there are $a_x \subset \omega\,(x \in X)$ such that $x < y \rightarrow a_x \subset^* a_y$. *Hint*. See II Exercise 22.

(H8) Assume ($\mathbb{P}$ is c.c.c. $\wedge$ $|\mathbb{P}| \leq \omega_1)^M$ and $\diamondsuit$ holds in M. Show that $\diamondsuit$ holds in $M[G]$. *Hint*. Use $\diamondsuit$ in M to "capture" all nice names for subsets of ω_1.

(H9) Assume ($\mathbb{P}$ is c.c.c.)^M and $\diamondsuit$ holds in $M[G]$. Show that $\diamondsuit$ holds in M. *Hint*. It is sufficient to verify $\diamondsuit^-$ in M; see II 7.13, 7.14.

(H10) Suppose that in M, $\mathbb{P}$ has ω_1 as a precaliber (see II Exercise 26). Show that if T is an ω_1-Suslin tree in M, then T remains a Suslin tree in $M[G]$. Do likewise if $\mathbb{P}$ is ω_1-closed in M.

(H11) Let $\kappa > \omega$ be regular. Let $\mathbb{P}$ be the Jech p.o. for adding a κ-Suslin tree. Elements of $\mathbb{P}$ are either the empty tree, $\mathbb{1}$, or subtrees, p, of $2^{<\kappa}$ such that $\mathrm{ht}(p) = \alpha + 1$ for some limit $\alpha < \kappa$,

$$\forall s \in p\,(\mathrm{dom}(s) < \alpha \rightarrow (s^\frown\langle 0\rangle \in p \wedge s^\frown\langle 1\rangle \in p)),$$

$$\forall \xi \leq \alpha\,(|\mathrm{Lev}_\xi(p)| < \kappa), \quad \text{and} \quad \forall s \in p\,\exists t \in p\,(\mathrm{dom}(t) = \alpha \wedge s \subset t).$$

$p \leq q$ iff $q = \{s \in p\colon \mathrm{dom}(s) < \mathrm{ht}(q)\}$; i.e., q is obtained by sawing off p parallel to the ground. Show that $\mathbb{P}$ is ω_1-closed and not ω_2-closed. *Hint*. If $p_0 > p_1 > \cdots$, $\bigcup_n p_n$ is a tree of some limit height, γ, but since $\mathrm{cf}(\gamma) = \omega$, it is easy to extend $\bigcup_n p_n$ to a tree of height $\gamma + 1$ in $\mathbb{P}$. However, $\bigcup_{\xi<\omega_1} p_\xi$ could be an ω_1-Aronszajn tree.

(H12) Show that if $\mathbb{P}$ is as in Exercise H11 and $\alpha < \kappa$, $D_\alpha = \{p \in \mathbb{P}\colon \mathrm{ht}(p) > \alpha\}$ is dense in $\mathbb{P}$. *Remark*. This is where it is important that conditions have successor height. *Hint*. To see that $D_\alpha \neq 0$, let

$$p = \{s \in 2^{<\alpha+1}\colon |\{\xi\colon s(\xi) = 1\}| < \omega\}.$$

(H13) Show that the $\mathbb{P}$ of Exercise H11 is κ-Baire (see Exercise B6), even though it is not κ-closed if $\kappa > \omega_1$. *Hint*. Suppose $\lambda < \kappa$ and $D_\xi(\xi < \lambda)$ are dense and open. Find a decreasing sequence, $p_\xi\,(\xi \leq \lambda)$, with $p_\xi \in D_\xi$ for each $\xi < \lambda$ (so then $p_\lambda \in \bigcap_{\xi<\lambda} D_\xi$). $\mathrm{ht}(p_\xi)$ will be $\alpha_\xi + 1$. For limit γ, $\alpha_\gamma = \sup\{\alpha_\xi\colon \xi < \gamma\}$. To ensure $\bigcup_{\xi<\gamma} p_\xi$ can be extended when $\mathrm{cf}(\gamma) > \omega$, choose, along with each p_ξ, an $f(x, \xi) \in \mathrm{Lev}_{\alpha_\xi}(p_\xi)$ for each $s \in p_\xi$ such that $s \subset f(s, \xi)$ and $\xi < \eta \rightarrow f(s, \xi) \subset f(s, \eta)$. *Remark*. For $\kappa > \omega_1$, this argument is due to Prikry and Silver.

(H14) Let $\kappa > \omega$ be a regular cardinal of M, and let $\mathbb{P}$ be defined within

M as in Exercise H11. Show that in $M[G]$, $\bigcup G$ is a κ-Suslin tree, and that $\mathbb{P}$ preserves cofinalities if $2^{<\kappa} = \kappa$ in M. *Hint.* If

$$p \Vdash (\tau \text{ is a maximal antichain in } \Gamma),$$

find $p_0 = p > p_1 > p_2 > \cdots$ such that

$$\forall n\, \forall s \in p_n((p_{n+1} \Vdash \check{s} \in \tau) \vee (p_{n+1} \Vdash \check{s} \notin \tau)).$$

Let $q = \bigcup_n p_n$ (note $q \notin \mathbb{P}$), and let $A = \{s \in q\colon \exists n(p_n \Vdash \check{s} \in \tau)\}$. Make sure that the p_n are defined so that A is a maximal antichain in q. Now, extend q to a q' such that $q' \Vdash \tau = A$. This is like the proof of $\diamondsuit \to \neg$SH (see II 7.8).

(H15) Show that $\mathrm{Fn}(\omega_1, 2^\omega, \omega_1)$ and the Jech p.o. for adding an ω_1-Suslin tree have isomorphic completions. *Hint.*

$$\{p \in \mathrm{Fn}(\omega_1, 2^\omega, \omega_1)\colon \mathrm{dom}(p) \text{ is a successor ordinal}\}$$

can be densely embedded in both of them.

(H16) Let T be a κ-Suslin tree; concepts like "dense" refer to the reverse order, $<$, on T, as in Theorem 8.4. Show that if $X \subset T$, then there is an $\alpha < \kappa$ such that for all $p \in \mathrm{Lev}_\alpha(T)$, either X is dense below p or empty below p (i.e. $\neg\exists q < p\,(q \in X)$). Use this to show that T is κ-Baire.

(H17) Suppose that in M, $\mathbb{P}$ is c.c.c. and T is a κ-Suslin tree for some $\kappa > \omega_1$. Show that T remains Suslin in $M[G]$. *Hint.* In M, if $X \subset T$ and $|X| = \kappa$, then X contains an uncountable chain and an uncountable antichain.

(H18) Jensen's $\diamondsuit^+$ order is the set $\mathbb{P}$ of pairs $\langle p, \mathscr{S}\rangle$, where p is a condition in the Jech p.o. (H11, with $\kappa = \omega_1$), $\mathscr{S} \subset 2^{\omega_1}$, $|\mathscr{S}| \le \omega$, and

$$(f \restriction (\mathrm{ht}(p) - 1)) \in p$$

for each $f \in \mathscr{S}$. $\langle p, \mathscr{S}\rangle \le \langle p', \mathscr{S}'\rangle$ iff $p \le p'$ and $\mathscr{S}' \subset \mathscr{S}$. Assuming CH, show that $\mathbb{P}$ is ω_1-closed and ω_2-c.c.

(H19) Let $\mathbb{P}$ be the $\diamondsuit^+$-order in M (see H18), and assume CH in M. In $M[G]$, let $T = \bigcup\{p\colon \exists \mathscr{S}(\langle p, \mathscr{S}\rangle \in G)\}$, and show that T is a Kurepa tree.

(H20) Let $\mathbb{P}$ be the $\diamondsuit^+$-order in M and assume CH in M. Show that $\diamondsuit^+$ holds in $M[G]$. *Hint.* See VI Exercise 9. In $M[G]$, fix $h \in 2^{\omega_1}$ such that $T \in \mathbf{L}(h)$ and $\mathscr{P}(\omega) \in \mathbf{L}(h)$; then show, in $M[G]$, that
 (a) $\forall f \in 2^{\omega_1} \cap M\, \exists g \in 2^{\omega_1}\, (f \in \mathbf{L}(g, h) \wedge g \text{ is a path through } T)$, and
 (b) $\forall f \in 2^{\omega_1}\, \exists g \in 2^{\omega_1} \cap M(f \in \mathbf{L}(g, h))$.
For (b), observe that $G \cap \mathbf{L}(g, h) \in \mathbf{L}(g, h)$.

(H21) In M, let $\mathscr{B}$ be the measure algebra of 2^I; that is, $2 = \{0, 1\}$ is given a measure by defining $\{0\}$ and $\{1\}$ to have measure $\frac{1}{2}$. 2^I has the product measure, and $\mathscr{B}$ is the algebra of measurable sets modulo measure 0 sets. Show that $\mathscr{B}$ has the c.c.c. in M, and, if G is $\mathscr{B}$-generic over M, then $2^\omega \geq |I|$ in $M[G]$. *Hint.* G defines a function $f : I \to 2$. If $I = \kappa \times \omega$, and $f_\alpha(n) = f(\alpha, n)$, then the f_α are distinct.

(H22) In Exercise H21, assume also that in M, $|I| = \kappa$ and $\kappa^\omega = \kappa$. Show that $2^\omega = \kappa$ in $M[G]$. *Remark.* In M, $\mathscr{B}$ is isomorphic to the Baire sets in 2^I modulo measure 0 sets, whereas the completion of $\mathrm{Fn}(I, 2)$ is the regular open algebra of 2^I, which is the same (for 2^I) as the Baire sets modulo first category. For more on the analogy between measure and category in generic extensions, see [Kunen 1900].

(H23) In Exercise H21 (or with $\mathscr{B}$ any measure algebra in M), let $f \in \omega^\omega \cap M[G]$. Show that there is a $g \in \omega^\omega \cap M$ such that $\forall n\,(f(n) \leq g(n))$. Show that this fails in extension by $\mathrm{Fn}(I, 2)$. *Remark.* Thus, in contrast with Exercise H22, not all properties of these extensions are the same. *Hint.* If $\mathbb{1} \Vdash \tau : \check{\omega} \to \check{\omega}$, and $\varepsilon > 0$, let $g(n)$ be an m such that $\mu([\![\tau(\check{n}) \leq \check{m}]\!]) \geq 1 - \varepsilon/2^n$. For $\mathrm{Fn}(I, 2)$, consider instead $\mathrm{Fn}(\omega, \omega)$.

(H24) (Prikry). In M, assume that κ is 2-valued measurable and that $\mathbb{P}$ is c.c.c. Show that $\exists \mathscr{I}\, S(\kappa, \omega_1, \mathscr{I})$ holds in $M[G]$ (see II Exercise 60). *Hint.* If $S(\kappa, 2, \mathscr{J})$ holds in M, define

$$\mathscr{I} = \{X \in \mathscr{P}(\kappa) \cap M[G] : \exists Y \in \mathscr{J}\,(X \subset Y)\}.$$

Remark. If $\mathbb{P}$ is a measure algebra then κ is real-valued measurable in $M[G]$; see [Solovay 1971].

*(H25) (Jensen). In M, let $T \subset \omega_1$ be stationary, and let $\mathbb{P}$ be the set of all $p \subset T$ such that p is closed in ω_1 and countable. $p \leq q$ iff $q \subset p$ and $q = p \cap (\max(q) + 1)$. Show that $\mathbb{P}$ is ω_1-Baire and that in $M[G]$, T contains a c.u.b. *Remark.* Thus, $S = \omega_1 \setminus T$ is not stationary in $M[G]$, although it may be stationary in M; compare this with H2.

CHAPTER VIII

ITERATED FORCING

Throughout this chapter, M is a fixed c.t.m. for ZFC, although we shall apply the results of VII using not only M, but also various extensions of M, as the ground model.

The idea behind iterated forcing is to repeat the generic extension process α times, for some ordinal α, to obtain a chain of models,

$$M = M_0 \subset M_1 \subset \cdots \subset M_\xi \subset \cdots \subset M_\alpha,$$

where $M_{\xi+1} = M_\xi[G_\xi]$, with G_ξ $\mathbb{P}_\xi$-generic over M_ξ for some p.o. $\mathbb{P}_\xi \in M_\xi$. M_α will be the model constructed from M by adjoining the sequence $\langle G_\xi \colon \xi < \alpha \rangle$. The discussion of iterated forcing breaks into two parts: two stage iteration and limit iteration.

Two stage iteration (i.e., $\alpha = 2$) at first sight seems trivial. $M_1 = M_0[G_0]$ is another c.t.m. for ZFC, and one can simply apply the forcing process to any p.o. $\mathbb{P}_1 \in M_1$ to obtain $M_2 = M_1[G_1]$. In fact, we have already done precisely this in obtaining models violating GCH at various cardinals (see VII 6.18).

Once two stage iteration is understood, it is clear, by induction, how to do n stage iteration for any $n \in \omega$. Presumably, in transfinite iterations, the two stage process will get us from M_ξ to $M_{\xi+1}$; i.e., we obtain $M_{\xi+1}$ by forcing over M_ξ.

Limit iteration, on the other hand, seems to present a problem. Once we have M_n for $n < \omega$, what is M_ω? We cannot just set $M_\omega = \bigcup_n M_n$, since this will usually not contain $\{G_n \colon n \in \omega\}$, and it is easy to find examples where there is no c.t.m. N at all with $\langle G_n \colon n \in \omega \rangle \in N$ and $o(N) = o(M)$ (see VII Exercise B8). Intuitively, we might avoid this pathology by choosing the whole sequence $\langle G_n \colon n \in \omega \rangle$ generically, but what does that actually mean?

To solve these problems, we go back to two stage iteration and make it harder. We shall show how to do a two stage iteration in one step. Thus, we shall find a p.o. $\mathbb{Q}_1 \in M$ such that M_2, which was $M[G_0][G_1]$, is also $M[H_1]$ for an H_1 which is $\mathbb{Q}_1$-generic over M. Furthermore, we shall have $\mathbb{P}_0 \subset_c \mathbb{Q}_1$ (see VII Definition 7.2). Likewise, an n stage iteration will correspond to a chain,

$$\mathbb{P}_0 = \mathbb{Q}_0 \subset_c \mathbb{Q}_1 \subset_c \cdots \subset_c \mathbb{Q}_n$$

in M, and M_n will be $M[H_n]$ for some H_n which is $\mathbb{Q}_n$-generic over M.

In general, an α-stage iteration will correspond to building, in M, a chain of p.o.'s,

$$\mathbb{P}_0 = \mathbb{Q}_0 \subset_c \mathbb{Q}_1 \subset_c \cdots \subset_c \mathbb{Q}_\xi \subset_c \cdots \subset_c \mathbb{Q}_\alpha.$$

Thus, in this approach, we do not begin by constructing a chain of models. Rather, our efforts are concentrated on constructing the $\mathbb{Q}_\xi$ in M. Once we are done, we may define $M_\xi = M[H_\alpha \cap \mathbb{Q}_\xi]$, and obtain a chain of models after all, but this is of secondary importance.

In our building the chain of p.o.'s, there is still a problem of what to do at limits. The question, "what is M_ω?" now becomes, "what is $\mathbb{Q}_\omega$?" This is indeed a problem in some iterated forcing constructions, but not in all. For example, in proving Con(MA + $\neg$CH), the most obvious answer, $\mathbb{Q}_\omega = \bigcup_n \mathbb{Q}_n$ works. Thus, most of this chapter will be taken up with what to do at the successor stages, and this easily reduces to two-stage iterations, or how to define $\mathbb{Q}_1$, given $\mathbb{P}_0 \in M$ and $\mathbb{P}_1 \in M_1 = M[G_0]$.

A special case of this, when $\mathbb{P}_1$ is in fact in M, is much simpler than the general case and has many applications (although it will not suffice to handle Con(MA + $\neg$CH)). If $\mathbb{P}_1 \in M$, then we may simply take $\mathbb{Q}_1 = \mathbb{P}_0 \times \mathbb{P}_1$, as we shall see in §1. In §§2–4, we shall apply this special case to prove further results about the Cohen extensions ($\text{Fn}(\kappa, 2)$), prove Con($\neg$KH), and violate GCH at infinitely many regular cardinals.

In §5 we shall describe the general theory, when $\mathbb{P}_1$ need not be in M. In §6 we shall apply this to show Con(MA + $\neg$CH).

§1. Products

1.1. Definition. If $\langle \mathbb{P}_0, \leq_0, \mathbb{1}_0 \rangle$ and $\langle \mathbb{P}_1, \leq_1, \mathbb{1}_1 \rangle$ are p.o.'s, then the product p.o.,

$$\langle \mathbb{P}_0, \leq_0, \mathbb{1}_0 \rangle \times \langle \mathbb{P}_1, \leq_1, \mathbb{1}_1 \rangle = \langle \mathbb{P}_0 \times \mathbb{P}_1, \leq, \mathbb{1} \rangle$$

is defined by

$$\langle p_0, p_1 \rangle \leq \langle q_0, q_1 \rangle \quad \text{iff} \quad p_0 \leq q_0 \wedge p_1 \leq q_1,$$

and $\mathbb{1} = \langle \mathbb{1}_0, \mathbb{1}_1 \rangle$. Furthermore, define $i_0 : \mathbb{P}_0 \to \mathbb{P}_0 \times \mathbb{P}_1$, $i_1 : \mathbb{P}_1 \to \mathbb{P}_0 \times \mathbb{P}_1$ by $i_0(p_0) = \langle p_0, \mathbb{1}_1 \rangle$, and $i_1(p_1) = \langle \mathbb{1}_0, p_1 \rangle$. □

When discussing $\mathbb{P}_0$ and $\mathbb{P}_1$, we shall usually drop the subscripts on their respective $\mathbb{1}$'s and $\leq$'s.

It is here that it becomes really useful that our p.o.'s have a largest element. That restriction could have been avoided in VII by a minor modification in the definition of $\check{x}$ (see VII Exercise B1). The restriction can still be avoided here, but with more work, since we need $\mathbb{1}_0$ and $\mathbb{1}_1$ to define i_0 and i_1.

1.2. LE&MA. *If i_0, i_1 are as in Definition* 1.1, *then i_0 and i_1 are complete embeddings.*

PROOF. Clauses (1) and (2) of VII Definition 7.1 are immediate. To verify clause (3) for i_0, observe that if $\langle p_0, p_1 \rangle \in \mathbb{P}_0 \times \mathbb{P}_1$, then p_0 is a reduction of $\langle p_0, p_1 \rangle$ to $\mathbb{P}_0$. □

1.3. LEMMA. *Suppose that $\mathbb{P}_0, \mathbb{P}_1 \in M$ and G is $\mathbb{P}_0 \times \mathbb{P}_1$-generic over M. Then $i_0^{-1}(G)$ is $\mathbb{P}_0$-generic over M, $i_1^{-1}(G)$ is $\mathbb{P}_1$-generic over M, and $G = i_0^{-1}(G) \times i_1^{-1}(G)$.*

PROOF. Genericity of $i_0^{-1}(G)$ and $i_1^{-1}(G)$ follows from the completeness of i_0 and i_1 (see VII 7.5). If $\langle p_0, p_1 \rangle \in G$, then $\langle p_0, \mathbb{1} \rangle \in G$ and $\langle \mathbb{1}, p_1 \rangle \in G$, so $\langle p_0, p_1 \rangle \in i_0^{-1}(G) \times i_1^{-1}(G)$. If $\langle p_0, p_1 \rangle \in i_0^{-1}(G) \times i_1^{-1}(G)$, then $\langle p_0, \mathbb{1} \rangle$ and $\langle \mathbb{1}, p_1 \rangle$ are both in G, and thus have a common extension, $\langle q_0, q_1 \rangle \in G$; then $\langle q_0, q_1 \rangle \leq \langle p_0, p_1 \rangle$, so $\langle p_0, p_1 \rangle \in G$. □

Thus, all $\mathbb{P}_0 \times \mathbb{P}_1$-generic filters are of the form $G_0 \times G_1$, where G_0 and G_1 are generic over M. Not every such product is generic, however. For example, in $\mathbb{P} \times \mathbb{P}$, a product of the form $G \times G$ is usually not generic (see Exercise J1). Genericity of $G_0 \times G_1$ requires that G_1 be generic over $M[G_0]$, as we shall see in the following theorem which relates $\mathbb{P}_0 \times \mathbb{P}_1$ forcing with a two stage iterated forcing.

1.4. THEOREM. *Suppose $\mathbb{P}_0 \in M$, $\mathbb{P}_1 \in M$, $G_0 \subset \mathbb{P}_0$, and $G_1 \subset \mathbb{P}_1$; then the following are equivalent:*

(1) *$G_0 \times G_1$ is $\mathbb{P}_0 \times \mathbb{P}_1$-generic over M.*

(2) *G_0 is $\mathbb{P}_0$-generic over M and G_1 is $\mathbb{P}_1$-generic over $M[G_0]$.*

(3) *G_1 is $\mathbb{P}_1$-generic over M and G_0 is $\mathbb{P}_0$-generic over $M[G_1]$.*

Furthermore, if (1)–(3) hold, then $M[G_0 \times G_1] = M[G_0][G_1] = M[G_1][G_0]$.

PROOF. We show (1)$\leftrightarrow$(2). The proof of (1)$\leftrightarrow$(3) is the same.

To show (1)$\rightarrow$(2), assume (1). By Lemma 1.3, $G_0 = i_0^{-1}(G_0 \times G_1)$ is $\mathbb{P}_0$-generic over M. To show G_1 is $\mathbb{P}_1$-generic over $M[G_0]$, fix $D \in M[G_0]$ such that D is dense in $\mathbb{P}_1$. Now fix $\tau \in M^{\mathbb{P}_0}$ such that $D = \tau_{G_0}$, and fix

$p_0 \in G_0$ such that

$$p_0 \Vdash (\tau \text{ is dense in } \check{\mathbb{P}}_1). \tag{a}$$

Let

$$D' = \{\langle q_0, q_1 \rangle : (q_0 \leq p_0) \wedge (q_0 \Vdash \check{q}_1 \in \tau)\} \subset \mathbb{P}_0 \times \mathbb{P}_1.$$

We first show that D' is dense below $\langle p_0, \mathbb{1} \rangle$. Fix $\langle r_0, r_1 \rangle \leq \langle p_0, \mathbb{1} \rangle$. $r_0 \leq p_0$, so

$$r_0 \Vdash \exists x \in \check{\mathbb{P}}_1 (x \in \tau \wedge x \leq \check{r}_1), \tag{b}$$

so there is a $q_1 \in \mathbb{P}_1$ and a $q_0 \leq r_0$, such that

$$q_0 \Vdash (\check{q}_1 \in \tau \wedge \check{q}_1 \leq \check{r}_1);$$

thus, $\langle q_0, q_1 \rangle \leq \langle r_0, r_1 \rangle$ and $\langle q_0, q_1 \rangle \in D'$.

Now, since $\langle p_0, \mathbb{1} \rangle \in G_0 \times G_1$, there is some $\langle q_0, q_1 \rangle \in (G_0 \times G_1) \cap D'$. Then $q_0 \Vdash \check{q}_1 \in \tau$, so $q_1 \in \tau_{G_0} = D$; and $q_1 \in G_1$, so $G_1 \cap D \neq 0$.

To show (2) → (1), assume (2). $G_0 \times G_1$ is easily seen to be a filter. To show that it is generic, fix $D \subset \mathbb{P}_0 \times \mathbb{P}_1$ such that D is dense and in M. Let

$$D^* = \{p_1 : \exists p_0 \in G_0 (\langle p_0, p_1 \rangle \in D)\} \subset \mathbb{P}_1;$$

then $D^* \in M[G_0]$, and

$$D^* \cap G_1 \neq 0 \rightarrow D \cap (G_0 \times G_1) \neq 0,$$

so we shall be done if we can check that D^* is dense in $\mathbb{P}_1$. Fix $r_1 \in \mathbb{P}_1$. Then

$$D_0 = \{p_0 : \exists p_1 \leq r_1 (\langle p_0, p_1 \rangle \in D)\}$$

is dense in $\mathbb{P}_0$ and $D_0 \in M$, so fix $p_0 \in D_0 \cap G_0$, and then fix $p_1 \leq r_1$ with $\langle p_0, p_1 \rangle \in D$. Then, p_1 is an extension of r_1 in D^*.

Finally, $M[G_0 \times G_1] = M[G_0][G_1]$ follows by minimality of the various generic extensions. Thus, $M \subset M[G_0][G_1]$ and

$$G_0 \times G_1 \in M[G_0][G_1],$$

so $M[G_0 \times G_1] \subset M[G_0][G_1]$. Conversely, $M \subset M[G_0 \times G_1]$ and $G_0 \in M[G_0 \times G_1]$, so $M[G_0] \subset M[G_0 \times G_1]$; but also $G_1 \in M[G_0 \times G_1]$, so $M[G_0][G_1] \subset M[G_0 \times G_1]$. □

A careful examination of the preceding argument will reveal several abuses of notation. The $\Vdash$ in (a) should really be $\Vdash_{\mathbb{P}_0}$. Also, it must be de-determined from context which model is being considered to be the ground model; for example, in (a), we are forcing over M, not $M[G_1]$, whereas if we had been discussing the extension of $M[G_1]$ to $M[G_1][G_0]$, then $p_0 \Vdash \phi$ would refer to forcing over $M[G_1]$. Furthermore, the $\leq$ in (b) really refers to the object $\leq_1 \in M$, and should have been written $(\leq_1)\check{}$.

This concludes the theory of products. In §§2–4, we give some applications. Unlike in the general theory of iterations (see §5), in applications of products we do not usually start with $\mathbb{P}_0$ and $\mathbb{P}_1$ and form $\mathbb{P}_0 \times \mathbb{P}_1$. Rather, we start with one $\mathbb{P}$, given ab initio by some explicit definition (such as $\mathbb{P} = \mathrm{Fn}(I, 2)$ in §2), and we prove further results about the usual $\mathbb{P}$ extension $M[G]$ by factoring $\mathbb{P}$ into a product $\mathbb{P}_0 \times \mathbb{P}_1$ in varous ways.

The reader may now, if desired, skip directly to §5 without loss of continuity.

§2. More on the Cohen model

By the Cohen model, we mean generic extensions using p.o.'s of the form $\mathrm{Fn}(I, 2)$, as in Cohen's original proof of consistency of $\neg$CH. In this section we deduce some additional properties of such extensions.

$\mathrm{Fn}(I, 2)$ can be viewed as a product in many ways; namely, for each $I_0 \subset I$, $\mathrm{Fn}(I, 2)$ is isomorphic to $\mathrm{Fn}(I_0, 2) \times \mathrm{Fn}(I \setminus I_0, 2)$. Using this isomorphism, we may translate all of the basic results of §1 into statements about $\mathrm{Fn}(I, 2)$. The following theorem translates as much of §1 as we shall need.

2.1. THEOREM. *Suppose $I = I_0 \cup I_1$, $I_0 \cap I_1 = 0$, and $I_0, I_1 \in M$. Let G be $\mathrm{Fn}(I, 2)$-generic over M. Let $G_0 = G \cap \mathrm{Fn}(I_0, 2)$ and $G_1 = G \cap \mathrm{Fn}(I_1, 2)$. Then G_0 is $\mathrm{Fn}(I_0, 2)$-generic over M, G_1 is $\mathrm{Fn}(I_1, 2)$-generic over $M[G_0]$, and $M[G] = M[G_0][G_1]$.*

PROOF. Define

$$j: \mathrm{Fn}(I_0, 2) \times \mathrm{Fn}(I_1, 2) \to \mathrm{Fn}(I, 2)$$

by: $j(p_0, p_1) = p_0 \cup p_1$. It is easily seen that j is an isomorphism. Thus, $H = j^{-1}(G)$ is $\mathrm{Fn}(I_0, 2) \times \mathrm{Fn}(I_1, 2)$-generic over M and $M[G] = M[H]$ (see VII 7.6). By Lemma 1.3, $H = H_0 \times H_1$, where $H_0 = i_0^{-1}(H)$ and $H_1 = i_1^{-1}(H)$. By Theorem 1.4, H_0 is $\mathrm{Fn}(I_0, 2)$-generic over M, H_1 is $\mathrm{Fn}(I_1, 2)$-generic over $M[H_0]$, and $M[G] = M[H] = M[H_0][H_1]$. But

$$H_0 = \{p_0 \in \mathrm{Fn}(I_0, 2): \langle p_0, \mathbb{1} \rangle \in H\} = \{p_0 \in \mathrm{Fn}(I_0, 2): p_0 \cup 0 \in G\} = G_0,$$

and, likewise, $H_1 = G_1$. $\square$

Theorem 2.1 is useful because any small set in $M[G]$ is in $M[G_0]$ for a small I_0. More precisely, the following holds.

2.2. Lemma. *Suppose* $I, S \in M$. *Let* G *be* $\mathrm{Fn}(I, 2)$-*generic over* M, *and let* $X \subset S$ *with* $X \in M[G]$. *Then* $X \in M[G \cap \mathrm{Fn}(I_0, 2)]$ *for some* $I_0 \subset I$ *such that* $I_0 \in M$ *and* $(|I_0| \le |S|)^M$.

Proof. Assume S is infinite, since $|S| < \omega$ would imply that $X \in M$. Let $X = \tau_G$, where τ is a nice name for a subset of $\check{S}$ (see VII 5.11); then

$$\tau = \bigcup_{s \in S} (\{\check{s}\} \times A_s),$$

where A_s is an antichain in $\mathrm{Fn}(I, 2)$. Let

$$I_0 = \bigcup \{\mathrm{dom}(p) : \exists s \in S\, (p \in A_s)\},$$

and let $G_0 = G \cap \mathrm{Fn}(I_0, 2)$. By the c.c.c. (in M), each $|A_a| \le \omega$, so $|I_0| \le |S|$ in M.

Since each $A_s \subset \mathrm{Fn}(I_0, 2)$, τ is an $\mathrm{Fn}(I_0, 2)$-name, and

$$\tau_G = \{s \in S : \exists p \in G_0\, (p \in A_s)\} = \tau_{G_0} \in M[G_0]. \quad \square$$

It follows that every subset of ω in $M[G]$ is in an extension of M by a p.o. isomorphic to $\mathrm{Fn}(\omega, 2)$ in M. Since extensions by $\mathrm{Fn}(\omega, 2)$ preserve CH, it is possible to use Lemma 2.2 to show that certain consequences of CH hold also in extensions by arbitrary $\mathrm{Fn}(I, 2)$. For example, we show now that if M satisfies CH, then there is a maximal almost disjoint family (see II 1.1) of subsets of ω of size ω_1, a fact which contradicts MA if CH fails (see II 2.16). For an easier way of seeing that MA fails in these extensions, see VII Exercise G7.

2.3. Theorem. *Suppose that* $I \in M$ *and* CH *is true in* M. *Let* G *be* $\mathrm{Fn}(I, 2)$-*generic over* M. *Then in* $M[G]$ *there is a* m.a.d.f. *of size* ω_1.

Proof. Within M, we shall define an m.a.d.f. $\mathscr{A}$ of size ω_1 such that for *all* $I \in M$, $\mathscr{A}$ remains maximal in $M[G]$ whenever G is $\mathrm{Fn}(I, 2)$-generic over M. By Lemma 2.2, it is sufficient to verify maximality of $\mathscr{A}$ in extensions via I_0 which are countable in M, since any $X \subset \omega$ in $M[G]$ is in $M[G \cap \mathrm{Fn}(I_0, 2)]$ for some such I_0. When I_0 is finite, $M[G_0] = M$, and when $(|I_0| = \omega)^M$, $\mathrm{Fn}(I_0, 2)$ is isomorphic to $\mathrm{Fn}(\omega, 2)$ in M. Since p.o.'s which are isomorphic in M lead to the same extensions (see VII 7.6), it is sufficient to define $\mathscr{A}$ such that whenever G is $\mathrm{Fn}(\omega, 2)$-generic over M, no infinite subset of ω in $M[G]$ is a.d. from all elements in $\mathscr{A}$.

From now on, all forcing terminology refers to the p.o. $\mathrm{Fn}(\omega, 2)$. Within M, do the following: By CH, let $\langle p_\xi, \tau_\xi \rangle$ for $\omega \le \xi < \omega_1$ enumerate all pairs $\langle p, \tau \rangle$ such that $p \in \mathrm{Fn}(\omega, 2)$ and τ is a nice name for a subset of $\check{\omega}$. By recursion, pick infinite $A_\xi \subset \omega$ as follows. Let A_n, for $n < \omega$, be any disjoint infinite sets. If $\omega \le \xi < \omega_1$, and we have A_η for $\eta < \xi$, choose A_ξ

so that

(1) $\forall \eta < \xi (|A_\eta \cap A_\xi| < \omega)$, and

(2) If

$$p_\xi \Vdash (|\tau_\xi| = \omega) \quad \text{and} \quad \forall \eta < \xi [p_\xi \Vdash (|\tau_\xi \cap \check{A}_\eta| < \check{\omega})], \tag{$*$}$$

then

$$\forall n \in \omega \, \forall q \leq p_\xi \, \exists r \leq q \, \exists m \geq n (m \in A_\xi \wedge r \Vdash \check{m} \in \tau_\xi).$$

To see that A_ξ may be so chosen, assume ($*$) holds, since if ($*$) fails then only (1) need be considered, and we simply apply the fact that there is no m.a.d.f. of size ω (II 1.2). Let B_i ($i \in \omega$) enumerate $\{A_\eta \colon \eta < \xi\}$ and let $\langle n_i, q_i \rangle$ ($i \in \omega$) enumerate $\omega \times \{q \colon q \leq p_\xi\}$. By ($*$), for each i

$$q_i \Vdash (|\tau_\xi \setminus (\check{B}_0 \cup \cdots \cup \check{B}_i)| = \check{\omega}),$$

so we may choose an $r_i \leq q_i$ and an $m_i \geq n_i$ such that $m_i \notin B_0 \cup \cdots \cup B_i$ and $r_i \Vdash \check{m}_i \in \tau_\xi$. Let $A_\xi = \{m_i \colon i \in \omega\}$.

Now, let $\mathscr{A} = \{A_\xi \colon \xi < \omega_1^M\}$. Let G be $\mathrm{Fn}(\omega, 2)$-generic over M. If $\mathscr{A}$ failed to be maximal in $M[G]$, there would be a $\langle p_\xi, \tau_\xi \rangle$ such that $p_\xi \in G$, $p_\xi \Vdash |\tau_\xi| = \check{\omega}$, and $p_\xi \Vdash \forall X \in \mathscr{A} (|\tau_\xi \cap X| < \check{\omega})$. Thus, ($*$) holds at ξ, but also $p_\xi \Vdash |\tau_\xi \cap \check{A}_\xi| < \omega$, so there is a $q \leq p$ and an $n \in \omega$ such that $q \Vdash (\tau_\xi \cap \check{A}_\xi \subset \check{n})$, contradicting that

$$\exists r \leq q \, \exists m \geq n (m \in A_\xi \wedge r \Vdash \check{m} \in \tau_\xi). \quad \square$$

Some consequences of MA do hold in the Cohen extension. For example, every maximal independent family must have size 2^ω.

2.4. Definition. If $\mathscr{A} \subset \mathscr{P}(\omega)$, $\mathscr{A}$ is an *independent family* (i.f.) iff whenever $m, n \in \omega$ and $a_1, \ldots, a_m, b_1, \ldots, b_n$ are distinct members of $\mathscr{A}$,

$$|a_1 \cap \cdots \cap a_m \cap (\omega \setminus b_1) \cap \cdots \cap (\omega \setminus b_n)| = \omega.$$

$\mathscr{A}$ is a *maximal independent family* (m.i.f.) iff $\mathscr{A}$ is an i.f. and is maximal with respect to that property. $\square$

As with almost disjoint sets, there is always an i.f. of size 2^ω (see Exercise A6), and hence an m.i.f. of size 2^ω. Thus, there remains only the question of whether *every* m.i.f. must have size 2^ω. By an easy diagonal argument (Exercise A15), every m.i.f. has size $\geq \omega_1$, so we are led to consider models of $\neg$CH. As we have just seen with m.a.d.f.'s, it is consistent with $\neg$CH that there is an m.i.f. of size ω_1, but the argument involves an iterated forcing extension (Exercise A13). As we shall show next (Theorem 2.6), in the Cohen extension every m.i.f. must have size 2^ω.

It also follows from MA that every m.i.f. has size 2^ω; this is an easy exercise (A15) applying MA to the p.o. $\mathrm{Fn}(\omega, 2)$, and is exactly analogous to the proof of Lemma 2.5. In general, the Cohen extension satisfies those consequences of MA which may be proved using just p.o.'s of the form $\mathrm{Fn}(I, 2)$. To obtain a model of MA + $\neg$CH (see §6), we must iterate forcing with all possible c.c.c. p.o.'s.

2.5. Lemma. *Assume that* $\mathscr{A} \subset \mathscr{P}(\omega)$, $\mathscr{A} \in M$, $I \in M$, *and* I *is infinite; let* G *be* $\mathrm{Fn}(I, 2)$*-generic over* M; *then* $\mathscr{A}$ *is not an* m.i.f. *in* $M[G]$.

Proof. We first show that it is sufficient to consider the case when $I = \omega$. For, in general, fix $I_0 \subset I$ with $(|I_0| = \omega)^M$. Then $M[G \cap \mathrm{Fn}(I_0, 2)] \subset M[G]$, and $M[G \cap \mathrm{Fn}(I_0, 2)] = M[H]$ for some H which is $\mathrm{Fn}(\omega, 2)$-generic over M.

When $I = \omega$, $\bigcup G : \omega \to 2$ is the characteristic function of a set $C \subset \omega$, where $C = \{n \in \omega : \bigcup G(n) = 1\}$. We shall show that for every infinite $B \subset \omega$, if $B \in M$, then $B \cap C$ and $B \setminus C$ are both infinite. It will then follow that if $\mathscr{A} \in M$ and $\mathscr{A}$ is an i.f., then $\mathscr{A} \cup \{C\}$ is an i.f. also, so that $\mathscr{A}$ will not be an m.i.f. in $M[G]$.

To see that $B \cap C$ is infinite, observe that for each $m \in \omega$,

$$\{p : \exists n > m\, (p(n) = 1 \wedge n \in B)\}$$

is dense and in M. Likewise, to show that $B \setminus C$ is infinite, use

$$\{p : \exists n > m\, (p(n) = 0 \wedge n \in B)\}. \quad \square$$

2.6. Theorem. *Suppose that in* M, κ *is an infinite cardinal and* $\kappa^\omega = \kappa$. *Let* G *be* $\mathrm{Fn}(\kappa, 2)$*-generic over* M. *Then in* $M[G]$, $2^\omega = \kappa$ *and every* m.i.f. *has cardinality* κ.

Proof. It is immediate from VII 5.14 that $2^\omega = \kappa$ in $M[G]$ (since $\mathrm{Fn}(\kappa, 2)$ and $\mathrm{Fn}(\kappa \times \omega, 2)$ are isomorphic in M). Suppose that $\mathscr{A} \in M[G]$ is an i.f. of size $\lambda < \kappa$; we shall show that $\mathscr{A}$ is not maximal.

In $M[G]$, let $\mathscr{A} = \{A_\xi : \xi < \lambda\}$, and let

$$X = \{\langle \xi, n \rangle : n \in A_\xi\} \subset \lambda \times \omega.$$

By Lemma 2.2, there is an $I_0 \subset \kappa$, such that $I_0 \in M$, $(|I_0| \le \lambda)^M$, and $X \in M[G_0]$, where $G_0 = G \cap \mathrm{Fn}(I_0, 2)$. Let $I_1 = \kappa \setminus I_0$ and $G_1 = G \cap \mathrm{Fn}(I_1, 2)$. By Theorem 2.1, $M[G] = M[G_0][G_1]$, and G_1 is $\mathrm{Fn}(I_1, 2)$-generic over $M[G_0]$. Since I_1 is infinite and $\mathscr{A} \in M[G_0]$, Lemma 2.5 applied to $M[G_0]$ and I_1 implies that $\mathscr{A}$ is not maximal in $M[G_0][G_1]$. $\square$

§3. The independence of Kurepa's Hypothesis

We have seen that Con(ZFC) implies Con(ZFC + KH), since KH is true in **L** (see VI 5.2 and II 7.11). Also, KH may be made true in a generic extension of any c.t.m. M (see VII Exercise H19). We have also seen that KH is true in a Lévy extension of M (see VII 8.9), but that requires a strongly inaccessible cardinal in M.

We shall now give a proof, due to Silver, that if M contains a strongly inaccessible cardinal, then $\neg$KH is true in some generic extension of M. Here, the inaccessible is necessary, since $\neg$KH implies that ω_2 is inaccessible in **L** (see VII Exercise B8), so that we shall have

$$\mathrm{Con}(\mathrm{ZFC} + \neg\mathrm{KH}) \leftrightarrow \mathrm{Con}\big(\mathrm{ZFC} + \exists\kappa\,(\kappa \text{ is strongly inaccessible})\big).$$

Silver's p.o. is exactly like the Lévy p.o., except that we use countable rather than finite partial functions. We first present the elementary facts about such extensions; this is almost verbatim like the treatment of $\mathrm{Lv}(\kappa)$ in VII §8.

3.1. DEFINITION.

$$\mathrm{Lv}'(\kappa) = \{p : |p| \le \omega \wedge p \text{ is a function } \wedge \mathrm{dom}(p) \subset \kappa \times \omega_1 \wedge \forall \langle\alpha, \xi\rangle \in \mathrm{dom}(p)\,(p(\alpha, \xi) \in \alpha)\}.$$

$\mathrm{Lv}'(\kappa)$ is ordered by reverse inclusion. □

3.2. LEMMA. *$\mathrm{Lv}'(\kappa)$ is ω_1-closed. If κ is strongly inaccessible, then $\mathrm{Lv}'(\kappa)$ has the κ-c.c.*

PROOF. κ-c.c. is proved by a Δ-system argument, exactly as for $\mathrm{Lv}(\kappa)$ (see VII 8.7). □

3.3. LEMMA. *Suppose that κ is strongly inaccessible in M and G is $\mathrm{Lv}'(\kappa)$-generic over M. Then $\omega_1^{M[G]} = \omega_1^M$ and $\omega_2^{M[G]} = \kappa$.*

PROOF. ω_1-closed (in M) p.o.'s preserve ω_1 (see VII 6.15) and κ-c.c. p.o.'s preserve cardinals $\ge\kappa$ (see VII 6.9), so ω_1 and κ are cardinals of $M[G]$. However, as in VII 8.8, the map $(\bigcup G)_\alpha$ (defined by $(\bigcup G)_\alpha(\xi) = \bigcup G(\alpha, \xi)$) is a map from ω_1^M onto α whenever $0 < \alpha < \kappa$. □

Now, suppose T is an ω_1-tree of M. T may well be a Kurepa tree in M, but in M, it has at most $(2^{\omega_1})^M$ paths, and $(2^{\omega_1})^M < \kappa = \omega_2^{M[G]}$. The following lemma, which is the crux of Silver's proof, shows that T obtains no new paths in $M[G]$. Hence, T will not be a Kurepa tree in $M[G]$.

3.4. LEMMA. *Suppose that in M, T is an ω_1-tree and $\mathbb{P}$ is an ω_1-closed* p.o. *Let G be $\mathbb{P}$-generic over M, and suppose that in $M[G]$, C is a path through T. Then $C \in M$.*

PROOF. Assume not. We shall derive a contradiction. Let $\mathscr{C} = \mathscr{P}(T)^M = \mathscr{P}(T) \cap M$; then $C \notin \mathscr{C}$. Fix τ such that $C = \tau_G$, and fix $p \in G$ such that

$$p \Vdash [(\tau \text{ is a path through } \check{T}) \wedge \tau \notin \check{\mathscr{C}}]. \tag{1}$$

From now on, we shall work completely within M and derive a contradiction directly from (1).

For $\alpha < \omega_1$ and $q \in \mathbb{P}$, we use the notation $q \parallel \tau_\alpha$ to abbreviate the statement

$$\exists b \in \mathrm{Lev}_\alpha(T)\,(q \Vdash \check{b} \in \tau).$$

The symbol $\parallel$ is read "decides". Recall that a path contains precisely one element from every level of T. We are thinking (informally) of τ_α as being the element on the path named by τ on level α. Since τ_α may depend on the generic filter, q may or may not decide τ_α. On the one hand,

$$p \Vdash \exists x\,(x \in (\mathrm{Lev}_\alpha(T))\check{} \wedge x \in \tau),$$

so for $\alpha < \omega_1$,

$$\{q \leq p\colon q \parallel \tau_\alpha\} \text{ is dense below } p \tag{2}$$

(see VII 3.7(d)). On the other hand,

$$q \leq p \rightarrow \exists \alpha \neg (q \parallel \tau_\alpha). \tag{3}$$

To prove (3), assume $\forall \alpha (q \parallel \tau_\alpha)$. Fix $b_\alpha \in \mathrm{Lev}_\alpha(T)$ such that $q \Vdash (\check{b}_\alpha \in \tau)$, and let $B = \{b_\alpha\colon \alpha < \omega_1\}$. Then $q \Vdash \tau = \check{B}$, so $q \Vdash \tau \in \check{\mathscr{C}}$, which is impossible since $p \Vdash \tau \notin \check{\mathscr{C}}$.

Furthermore,

$$[(q \leq p) \wedge (q \parallel \tau_\beta) \wedge (\alpha < \beta)] \rightarrow (q \parallel \tau_\alpha). \tag{4}$$

To prove (4), let $b \in \mathrm{Lev}_\beta(T)$ with $q \Vdash (\check{b} \in \tau)$. Since $q \Vdash (\tau$ is a path), $q \Vdash (\check{a} \in \tau)$, where a is the element below b at level α.

By (3) and (4), if $q \leq p$, then $\{\alpha\colon q \parallel \tau_\alpha\}$ is a proper initial segment of ω_1. Let

$$\delta(q) = \{\alpha\colon q \parallel \tau_\alpha\} = \min\{\alpha\colon \neg (q \parallel \tau_\alpha)\}.$$

If $\alpha \geq \delta(q)$, then $\neg (q \parallel \tau_\alpha)$, which should imply that more than one possibility for τ_α is consistent with q. More formally,

$$(q \leq p \wedge \alpha \geq \delta(q)) \rightarrow \exists q_0, q_1 \leq q\, \exists a_0, a_1 \in \mathrm{Lev}_\alpha(T)$$
$$[(q_0 \Vdash \check{a}_0 \in \tau) \wedge (q_1 \Vdash \check{a}_1 \in \tau) \wedge a_0 \neq a_1]. \tag{5}$$

To prove (5), we first apply (2) and fix a $q_0 \leq q$ and an $a_0 \in \mathrm{Lev}_\alpha(T)$, such that $q_0 \Vdash (\check{a}_0 \in \tau)$. Since $\neg(q \parallel \tau_\alpha)$, $\neg(q \Vdash \check{a}_0 \in \tau)$, so fix $r \leq q$ with $r \Vdash \check{a}_0 \notin \tau$. Applying (2) again, fix $q_1 \leq r$ and $a_1 \in \mathrm{Lev}_\alpha(T)$, such that $q_1 \Vdash (\check{a}_1 \in \tau)$. Since $q_1 \Vdash (a_0 \notin \tau)$, $a_0 \neq a_1$.

We now apply (5) ω times to get a tree (of height ω) of extensions p. Thus, define p_s and a_s for $s \in 2^{<\omega}$, together with α_n for $n \in \omega$ so that

(a) $\alpha_0 < \alpha_1 < \alpha_2 < \cdots$.

(b) $\alpha_n \geq \max\{\delta(p_s)\colon \mathrm{dom}(s) = n\}$.

(c) $p_{\langle\,\rangle} = p$ (where $\langle\,\rangle = 0$, the empty sequence).

(d) $a_{s^\frown\langle 0\rangle} \neq a_{s^\frown\langle 1\rangle}$, and, for $i = 0, 1$, $p_{s^\frown\langle i\rangle} \leq p_s$, $a_{s^\frown\langle i\rangle} \in \mathrm{Lev}_{\alpha_n}(T)$, and $p_{s^\frown\langle i\rangle} \Vdash (\check{a}_{s^\frown\langle i\rangle} \in \tau)$.

Let $\gamma = \sup\{\alpha_n\colon n < \omega\}$. We show now that $|\mathrm{Lev}_\gamma(T)| \geq 2^\omega$, a contradiction. If $f \in {}^\omega 2$, we have

$$p \geq p_{\langle f(0)\rangle} \geq p_{\langle f(0), f(1)\rangle} \geq \cdots.$$

$\mathbb{P}$ is ω_1-closed, so fix q_f so that $\forall n (q_f \leq p_{f \restriction n})$, and then fix $r_f \leq q_f$ and $b_f \in \mathrm{Lev}_\gamma(T)$ so that $r_f \Vdash (\check{b}_f \in \tau)$. Since also $\forall n\, (r_f \Vdash \check{a}_{f\restriction n} \in \tau)$, $\forall n\, (a_{f\restriction n} < b_f)$. If $f \neq g$, then $\exists n\, (a_{f\restriction n} \neq a_{g\restriction n})$, so $b_f \neq b_g$. Thus, the b_f for $f \in 2^\omega$ are distinct, contradicting $|\mathrm{Lev}_\gamma(T)| \leq \omega$. □

If in Lemma 3.4, $M[G]$ contains a map from ω_1^M onto $(2^{\omega_1})^M$, then T cannot be a Kurepa tree in $M[G]$. However, this does not imply that $\neg$KH holds in $M[G]$, since a Kurepa tree T in $M[G]$ need not be in M. The reason that KH holds in an extension by $\mathrm{Lv}'(\kappa)$ is that we can view the extension as an iteration,

$$M \subset M[G_0] \subset M[G_0][G_1] = M[G],$$

where $T \in M[G_0]$, G_0 is a "small" piece of G, and G_1 collapses the 2^{ω_1} of $M[G_0]$. We may then apply Lemma 3.4 to $M[G_0]$ to show that T is not Kurepa in $M[G]$. The details of the argument now resemble 2.1 and 2.2.

3.5. Theorem. *In M, let κ be strongly inaccessible and let $\mathbb{P} = \mathrm{Lv}'(\kappa)$. Let G be $\mathbb{P}$-generic over M. Then $\neg$KH holds in $M[G]$.*

Proof. In $M[G]$: let T be an ω_1-tree. We show that T has $\leq\omega_1$ paths. T is isomorphic to a tree of the form $\langle \omega_1, W\rangle$ (so W is some tree ordering of the countable ordinals). Since isomorphic trees have the same number of paths, we may assume that $T = \langle \omega_1, W\rangle$.

Now, $W \subset (\omega_1^M \times \omega_1^M)$, so $W = \sigma_G$ for some nice name σ for a subset of $(\omega_1^M \times \omega_1^M)\check{}$.

In M:

$$\sigma = \bigcup \{\{\check{s}\} \times A_s\colon s \in \omega_1 \times \omega_1\},$$

where each A_s is an antichain in $\mathbb{P}$. Since $\mathbb{P}$ has the κ-c.c., each $|A_s| < \kappa$. Since κ is regular, fix a cardinal $\theta < \kappa$, such that $\theta \geq \omega_1$, and

$$\forall s \in \omega_1 \times \omega_1 \, \forall p \in A_s (\text{dom}(p) \subset \theta \times \omega_1).$$

Let

$$\mathbb{P}_0 = \{p \in \mathbb{P}: \text{dom}(p) \subset \theta \times \omega_1\},$$

and

$$\mathbb{P}_1 = \{p \in \mathbb{P}: \text{dom}(p) \subset (\kappa \smallsetminus \theta) \times \omega_1\};$$

then, as in Theorem 2.1, $\mathbb{P}_0 \times \mathbb{P}_1$ is isomorphic to $\mathbb{P}$.

Let $G_0 = G \cap \mathbb{P}_0$ and $G_1 = G \cap \mathbb{P}_1$. As in Theorem 2.1, $M[G] = M[G_0][G_1]$, G_0 is $\mathbb{P}_0$-generic over M, and G_1 is $\mathbb{P}_1$-generic over $M[G_0]$. As in Lemma 2.2, $T \in M[G_0]$.

$\mathbb{P}_1$ is countably closed in M. Furthermore, if $\langle p_n : n \in \omega \rangle \in M[G_0]$ is a sequence of elements of $\mathbb{P}_1$, then $\langle p_n : n \in \omega \rangle \in M$ since $\mathbb{P}_0$ is countably closed in M. It follows that $\mathbb{P}_1$ remains countably closed in $M[G_0]$. Thus, applying Lemma 3.4 to $M[G_0]$ and $\mathbb{P}_1$, every path through T in $M[G_0][G_1] = M[G]$ is already in $M[G_0]$. Thus, if we let $\lambda = (2^{\omega_1})^{M[G_0]}$, we have that

$$(|\{C: C \text{ is a path through } T\}| \leq \lambda)^{M[G]}.$$

Since $\omega_2^{M[G]} = \kappa$, we shall be done if we show that $\lambda < \kappa$.

Every subset of ω_1 in $M[G_0]$ is τ_{G_0} for some nice name τ for a subset of $\check{\omega}_1$. In M, $|\mathbb{P}_0| \leq \theta^\omega = \theta_1$ so $\mathbb{P}_0$ has at most $2^{\theta_1} = \theta_2$ antichains, so there are at most $\theta_2^{\omega_1} = \theta_3$ such nice names. Thus, $\lambda \leq \theta_3$; but κ is strongly inaccessible in M, so $\theta_3 < \kappa$. $\square$

§4. Easton forcing

In this section, we show how to modify the powers of infinitely many regular cardinals at once. We first review the situation for 2 cardinals (see VII 6.18) in terms of our results on products. In our discussion here, we shall replace the orders of the form $\text{Fn}(\kappa \times \lambda, 2, \lambda)$ used in VII §6 by the isomorphic $\text{Fn}(\kappa, 2, \lambda)$, which are simpler to write down and yield the same extensions (see VII 7.6 and discussion following).

Say M satisfies GCH, and we wish to find an extension satisfying $2^{\omega_1} = \omega_4 \wedge 2^{\omega_2} = \omega_5$. In the spirit of VII, we may let $\mathbb{P}_2 = (\text{Fn}(\omega_5, 2, \omega_2))^M$, and let G_2 be $\mathbb{P}_2$-generic over M; then $M[G_2]$ satisfies $2^\omega = \omega_1 \wedge 2^{\omega_1} = \omega_2 \wedge 2^{\omega_2} = \omega_5$. We now let $\mathbb{P}_1 = (\text{Fn}(\omega_4, 2, \omega_1))^{M[G_2]}$, and let G_1 be $\mathbb{P}_1$-generic over $M[G_1]$; then $M[G_2][G_1]$ satisfies $2^{\omega_1} = \omega_4 \wedge 2^{\omega_2} = \omega_5$.

Now $\mathbb{P}_1$ is also $(\mathrm{Fn}(\omega_4, 2, \omega_1))^M$, since M and $M[G_2]$ have the same objects of size ω. Thus, $\mathbb{P}_1 \in M$, $G = G_1 \times G_2$ is $\mathbb{P}_1 \times \mathbb{P}_2$-generic over M, and $M[G_2][G_1] = M[G]$ (by Theorem 1.4). We may thus think of our iterated extension as a one-step extension. It is also true (by Theorem 1.4 again) that $M[G] = M[G_1][G_2]$ and G_2 is $\mathbb{P}_2$-generic over $M[G_1]$. However, $\mathbb{P}_2$ is not $(\mathrm{Fn}(\omega_5, 2, \omega_2))^{M[G_1]}$, since $M[G_1]$ has new partial functions of size ω_1. In fact, forcing over $M[G_1]$ with $\mathrm{Fn}(\omega_5, 2, \omega_2)^{M[G_1]}$ collapses ω_3 and ω_4 (see VII Exercise G3), whereas forcing over $M[G_1]$ with $\mathbb{P}_2$ preserves cardinals, since $M[G_1][G_2] = M[G_2][G_1]$, and we have seen that M, $M[G_2]$, and $M[G_2][G_1]$ all have the same cardinals. Thus, although forcing with a product can be considered to be an iteration in either order, in the case of $\mathbb{P}_1 \times \mathbb{P}_2$ it is more "natural" to view the iteration as first adding ω_5 generic subsets of ω_2 in the "usual" way to get $M[G_2]$, and then forcing over $M[G_2]$ to add ω_4 generic subsets of ω_1 in the "usual" way.

Now, suppose we want a model of $\forall n \in \omega\,(2^{\omega_n} = \omega_{n+3})$. It may not be clear how to accomplish this by an infinite chain of extensions of M, but if we try to generalize our product construction, the most obvious thing works. Let $\mathbb{P}_n = (\mathrm{Fn}(\omega_{n+3}, 2, \omega_n))^M$, and let $\mathbb{P}$ be the infinite cartesian product, $(\prod_{n\in\omega} \mathbb{P}_n)^M$. We shall see that if G is $\mathbb{P}$-generic over M, then $M[G]$ satisfies $\forall n \in \omega\,(2^{\omega_n} = \omega_{n+3})$. We may think, informally, of $\mathbb{P}$ as producing an iterated forcing extension adding ω_{n+3} generic subsets of ω_n, taking the largest n first (whatever that means).

We shall now proceed with our formal development of Easton forcing, which will show that cardinal exponentiation on the regular cardinals can be anything not "obviously false".

4.1. Definition. (1) An *index function* is a function E, such that $\mathrm{dom}(E)$ is a set of regular cardinals.

(2) An *Easton index function* is an index function E, such that

(a) For all $\kappa \in \mathrm{dom}(E)$, $E(\kappa)$ is a cardinal and $\mathrm{cf}(E(\kappa)) > \kappa$, and

(b) $\forall \kappa, \kappa' \in \mathrm{dom}(E)\,(\kappa < \kappa' \rightarrow E(\kappa) \leq E(\kappa'))$.

(3) If E is an index function, $\mathbb{P}(E)$ is the set of functions p, such that

(a) $\mathrm{dom}(p) = \mathrm{dom}(E)$.

(b) $\forall \kappa \in \mathrm{dom}(p)\,(p(\kappa) \in \mathrm{Fn}(E(\kappa), 2, \kappa))$.

(c) For every regular λ (not necessarily in $\mathrm{dom}(E)$),

$$|\{\kappa \in \lambda \cap \mathrm{dom}(E) \colon p(\kappa) \neq 0\}| < \lambda.$$

$\mathbb{P}(E)$ is ordered co-ordinatewise:

$$p \leq p' \leftrightarrow \forall \kappa \in \mathrm{dom}(E)\,(p'(\kappa) \subset p(\kappa)).$$

The $\mathbb{1}$ of $\mathbb{P}(E)$ is of the element such that for each $\kappa \in \mathrm{dom}(E)$, $\mathbb{1}(\kappa) = 0$ (i.e., $\mathbb{1}(\kappa)$ is the $\mathbb{1}$ of $\mathrm{Fn}(E(\kappa), 2, \kappa)$). $\square$

We shall show, intuitively, that it is consistent for the powers of regular cardinals to be given by any Easton index function—that is, anything at all subject to the restrictions in (a) and (b) of (2), which embody the facts that $\mathrm{cf}(2^\kappa) > \kappa$ (König's Lemma, I 10.41), and $\kappa < \kappa' \to 2^\kappa \le 2^{\kappa'}$ (since $\mathscr{P}(\kappa) \subset \mathscr{P}(\kappa')$). Formally, we shall show that if E is any index function in M, and M satisfies GCH, then $(\mathbb{P}(E))^M$ preserves cardinals. Furthermore, if E is an Easton index function in M, then $(\mathbb{P}(E))^M$ makes

$$\forall \kappa \in \mathrm{dom}(E)\,(2^\kappa = E(\kappa))$$

true in $M[G]$.

If clause (c) were missing from the definition of $\mathbb{P}(E)$, then $\mathbb{P}(E)$ would be precisely

$$\prod \{\mathrm{Fn}(E(\kappa), 2, \kappa) \colon \kappa \in \mathrm{dom}(E)\},$$

ordered co-ordinatewise. Since $\mathrm{dom}(E)$ contains only cardinals, clause (c) only says something new when λ is weakly inaccessible, in which case it requires $p(\kappa)$ to be the condition $\mathbb{1}$ $(=0)$ for all but $<\lambda$ cardinals κ below λ.

Actually, our arguments never deal with $\mathbb{P}(E)$ as an infinite product. Rather, as in §§2, 3, what is important is that $\mathbb{P}(E)$ may be viewed as a product of two factors in many ways.

4.2. Definition. If E is an index function, then $E_\lambda^+ = E \restriction \{\kappa \colon \kappa > \lambda\}$ and $E_\lambda^- = E \restriction \{\kappa \colon \kappa \le \lambda\}$. $\square$

4.3. Lemma. *If E is an index function and λ is any cardinal, then $\mathbb{P}(E)$ is isomorphic to $\mathbb{P}(E_\lambda^+) \times \mathbb{P}(E_\lambda^-)$.* $\square$

A number of our arguments will use the fact that for any λ, we may use the results of §1 to view the extension by $\mathbb{P}(E)$ as an iterated extension—first by $\mathbb{P}(E_\lambda^+)$ and then by $\mathbb{P}(E_\lambda^-)$. There are two special cases worth noting: one when $\mathrm{dom}(E) \cap \lambda^+ = 0$ and the other when $\mathrm{dom}(E) \subset \lambda^+$. Although these special cases do not require the theory of products, they will be covered by the general argument and need not be considered separately. For example, if $\mathrm{dom}(E) \cap \lambda^+ = 0$, then $E = E_\lambda^+$, so $\mathbb{P}(E) = \mathbb{P}(E_\lambda^+)$, while $E_\lambda^- = 0$, whence $\mathbb{P}(E_\lambda^-) = \{0\}$, the 1-element order; thus, $\mathbb{P}(E)$ is trivially isomorphic to $\mathbb{P}(E_\lambda^-) \times \mathbb{P}(E_\lambda^+)$.

The importance of having clause (c) in the definition of $\mathbb{P}(E)$ is that it enables us to show that $\mathbb{P}(E_\lambda^-)$ has the λ^+-c.c.

4.4. Lemma. *If E is an index function, $\mathrm{dom}(E) \subset \lambda^+$, $2^{<\lambda} = \lambda$, and λ is regular, then $\mathbb{P}(E)$ has the λ^+-c.c.*

Proof. For $p \in \mathbb{P}(E)$, let

$$d(p) = \bigcup \{\{\kappa\} \times \mathrm{dom}(p(\kappa)) \colon \kappa \in \mathrm{dom}(E)\}.$$

By clause (c) of 4.1(3), $|d(p)| < \lambda$. Suppose $p_\alpha \in \mathbb{P}(E)$ for $\alpha < \lambda^+$. Since $2^{<\lambda} = \lambda$, the Δ-system lemma (II 1.6) says that there is an $X \subset \lambda^+$ such that $|X| = \lambda^+$ and the $d(p_\alpha)$ for $\alpha \in X$ form a Δ-system with some root r. Since $|r| < \lambda$, $2^{|r|} \le \lambda$, so there is a $Y \subset X$ such that $|Y| = \lambda^+$, and

$$\forall \alpha, \beta \in Y \, \forall \langle \kappa, i \rangle \in r \, (p_\alpha(\kappa)(i) = p_\beta(\kappa)(i));$$

then the p_α for $\alpha \in Y$ are all compatible. □

In Lemma 4.4, if $\mathrm{dom}(E) \subset \lambda$, we cannot conclude, even under GCH, that $\mathbb{P}(E)$ has the λ-c.c. For λ inaccessible, this is true iff λ is also Mahlo (Exercise J4).

4.5. LEMMA. *If E is an index function and $\mathrm{dom}(E) \cap \lambda^+ = 0$, then $\mathbb{P}(E)$ is λ^+-closed.* □

4.6. LEMMA. *Suppose that in M, E is an index function, $\mathbb{P} = \mathbb{P}(E)$, and GCH holds; then $\mathbb{P}$ preserves cofinalities (and hence cardinals).*

PROOF. If not, then there is a G which is $\mathbb{P}$-generic over M and a $\theta > \omega$ which is regular in M and singular in $M[G]$. Let $\lambda = \mathrm{cf}(\theta)^{M[G]} < \theta$. λ is regular in $M[G]$ and hence in M.

In M, let $\mathbb{P}_0 = \mathbb{P}(E_\lambda^-)$ and $\mathbb{P}_1 = \mathbb{P}(E_\lambda^+)$. Let $i_0\colon \mathbb{P}_0 \to \mathbb{P}_0 \times \mathbb{P}_1$ and $i_1 : \mathbb{P}_1 \to \mathbb{P}_0 \times \mathbb{P}_1$ be the usual embeddings (see 1.1), and let $j : \mathbb{P}_0 \times \mathbb{P}_1 \to \mathbb{P}$ be an isomorphism.

Let $G_0 = i_0^{-1}(j^{-1}(G))$ and $G_1 = i_1^{-1}(j^{-1}(G))$. $j^{-1}(G)$ is $\mathbb{P}_0 \times \mathbb{P}_1$-generic over M and $M[G] = M[j^{-1}(G)]$ (see VII 7.6), so G_1 is $\mathbb{P}_1$-generic over M, G_0 is $\mathbb{P}_0$-generic over $M[G_1]$, and $M[G] = M[G_1][G_0]$ (applying 1.3 and 1.4 to $j^{-1}(G)$).

In $M[G] = M[G_1][G_0]$, let f be a cofinal map from λ into θ. Since $(\mathbb{P}_1$ is λ^+-closed$)^M$, $(2^{<\lambda} = \lambda)^{M[G_1]}$, so by Lemma 4.4 applied within $M[G_1]$, $(\mathbb{P}_0$ has the λ^+-c.c.$)^{M[G_1]}$. Thus, there is an $F \in M[G_1]$ such that $F\colon \lambda \to \mathscr{P}(\theta)$, $\forall \alpha < \lambda\, (f(\alpha) \in F(\alpha))$, and $\forall \alpha < \lambda\, (|F(\alpha)| \le \lambda)^{M[G_1]}$ (see VII 6.8). But since $(\mathbb{P}_1$ is λ^+-closed$)^M$, $F \in M$ and $\forall \alpha < \lambda\, (|F(\alpha)| \le \lambda)^M$.

But now in M, $\bigcup_{\alpha<\lambda} F(\alpha)$ has cardinality $\le \lambda$ and is cofinal in θ, contradicting that θ is regular in M. □

4.7. THEOREM. *Suppose that in M, E is an Easton index function, $\mathbb{P} = \mathbb{P}(E)$, and GCH holds; then $\mathbb{P}$ preserves cofinalities (and hence cardinals) and, if G is $\mathbb{P}$-generic over M, then in $M[G]$:*

$$\forall \kappa \in \mathrm{dom}(E)\, (2^\kappa = E(\kappa)).$$

More generally, in $M[G]$, if θ is any infinite cardinal, let

$$E'(\theta) = \max(\theta^+, \sup\{E(\kappa)\colon \kappa \in \mathrm{dom}(E) \wedge \kappa \le \theta\}),$$

and let $E^*(\theta) = E'(\theta)$ *if* $\mathrm{cf}(E'(\theta)) > \theta$, *and* $E^*(\theta) = (E'(\theta))^+$ *otherwise; then* $2^\theta = E^*(\theta)$.

PROOF. Preservation of cofinalities was given in Lemma 4.6, and this implies that the definitions of E' and E^* are absolute for M, $M[G]$.

By our assumptions on E (see Definition 4.1(2)), $E^*(\kappa) = E(\kappa)$ for $\kappa \in \mathrm{dom}(E)$. Now fix any infinite cardinal θ of M. We shall show $2^\theta = E^*(\theta)$ holds in $M[G]$.

To show $2^\theta \geq E^*(\theta)$, it will be sufficient, by König's Lemma, to show that $2^\kappa \geq E(\kappa)$ for $\kappa \in \mathrm{dom}(E)$. To see that $2^\kappa \geq E(\kappa)$ holds in $M[G]$, we observe that the factor $\mathrm{Fn}(E(\kappa), 2, \kappa)^M$ in $\mathbb{P}$ forces an $E(\kappa)$-sequence of distinct subsets of κ to be added. More formally, in M let f be any 1–1 map from $E(\kappa) \times \kappa$ into $E(\kappa)$. In $M[G]$, let, for $\alpha < E(\kappa)$.

$$A_\alpha = \{\xi < \kappa\colon \exists p \in G\,(p(\kappa)(f(\alpha, \xi)) = 1)\};$$

then $\langle A_\alpha\colon \alpha < E(\kappa)\rangle \in M[G]$, and $\alpha \neq \beta \rightarrow A_\alpha \neq A_\beta$.

To see that $2^\theta \leq E^*(\theta)$ in $M[G]$, let $\lambda = \mathrm{cf}(\theta)^M = \mathrm{cf}(\theta)^{M[G]}$ and proceed as in the proof of Lemma 4.6. Within M, $\mathbb{P}$ is isomorphic to $\mathbb{P}_0 \times \mathbb{P}_1$, where $\mathbb{P}_0 = \mathbb{P}(E_\lambda^-)$ and $\mathbb{P}_1 = \mathbb{P}(E_\lambda^+)$; also, $\mathbb{P}_0$ has the λ^+-c.c. and $\mathbb{P}_1$ is λ^+-closed. Furthermore, $M[G] = M[G_1][G_0]$, where G_1 is $\mathbb{P}_1$-generic over M and G_0 is $\mathbb{P}_0$-generic over $M[G_1]$.

We consider first the special case when $\lambda = \theta$. Within M, $|\mathbb{P}_0| \leq E^*(\lambda)$, since for each $\kappa \leq \lambda$ in $\mathrm{dom}(E_\lambda^-)$,

$$|\mathrm{Fn}(E(\kappa), 2, \kappa)| \leq |\mathrm{Fn}(E^*(\lambda), 2, \kappa)| \leq E^*(\lambda)^\lambda,$$

whence $|\mathbb{P}_0| \leq (E^*(\lambda)^\lambda)^\lambda$; but $E^*(\lambda)^\lambda = E^*(\lambda)$ since $\mathrm{cf}(E^*(\lambda)) > \lambda$ and GCH holds. Since ($\mathbb{P}_1$ is λ^+-closed)^M, it is still true in $M[G_1]$ that $E^*(\lambda)^\lambda = E^*(\lambda)$, $\mathbb{P}_0$ has the λ^+-c.c., and $|\mathbb{P}_0| \leq E^*(\lambda)$. So, in $M[G_1]$, there are at most $(E^*(\lambda)^\lambda)^\lambda = E^*(\lambda)$ nice $\mathbb{P}_0$-names for subsets of λ. Thus, $2^\lambda \leq E^*(\lambda)$ in $M[G_1][G_0]$.

We now know that in $M[G]$, $2^\theta = E^*(\theta)$ in the special case when θ is regular.

To handle the general case, we first check that $E^*(\theta)^\lambda = E^*(\theta)$ in $M[G]$. If $f \in M[G] \cap {}^\lambda(E^*(\theta))$, then, since ($\mathbb{P}_0$ has the λ^+-c.c.)$^{M[G_1]}$, there is an F such that

$$F \in {}^{\lambda \times \lambda}(E^*(\theta)) \cap M[G_1], \tag{1}$$

and

$$\forall \alpha < \lambda\, \exists \beta < \lambda\, (f(\alpha) = F(\alpha, \beta)) \tag{2}$$

(see VII 6.8). Since ($\mathbb{P}_1$ is λ^+-closed)^M, any F satisfying (1) is in fact in M, so by GCH in M, there are only $E^*(\theta)$ such F. For each such F, the set of

$f \in M[G]$ satisfying (2) has size $\lambda^\lambda = E^*(\lambda) \le E^*(\theta)$ in $M[G]$ as we have just seen (since λ is regular in $M[G]$). Thus, $(|^\lambda E^*(\theta)| \le E^*(\theta))^{M[G]}$.

Now, argue completely within $M[G]$, and assume $\lambda < \theta$, so θ is singular. Let B be the set of bounded subsets of θ. We know that for regular $\delta < \theta$, $|\mathscr{P}(\delta)| = E^*(\delta) \le E^*(\theta)$, and the regular cardinals are cofinal in θ, so $|B| \le E^*(\theta)$. But we can map ${}^\lambda B$ onto $\mathscr{P}(\theta)$ by sending g to $\bigcup\{g(\alpha)\colon \alpha < \lambda\}$, so $2^\theta \le E^*(\theta)^\lambda = E^*(\theta)$. □

4.8. Corollary. Con(ZFC) *implies* $\text{Con}(\text{ZFC} + \forall n \in \omega (2^{\omega_n} = \omega_{n+3}) + \forall \kappa \ge \omega_\omega (2^\kappa = \kappa^+))$. □

By Theorem 4.7, if $\lambda \notin \text{dom}(E)$, then 2^λ is as small as possible in $M[G]$, so if we want cardinal exponentiation to "jump" at λ, we must put λ in $\text{dom}(E)$. For particular regular λ, this is no problem but suppose we want to prove the consistency of

$$\forall \kappa (\kappa \text{ regular} \rightarrow 2^\kappa = \kappa^{+++}).$$

Then the "obvious" $\mathbb{P}$ is a proper class of M. Although forcing with proper classes does not in general create models of ZFC, it does in the case of the Easton extension. For more details, see [Easton 1970] or Exercises G1–G5.

If λ is singular in M, then λ cannot be in $\text{dom}(E)$, and there are reasons in principle why it is difficult to make exponentiation jump at λ. Forgetting about models, let λ be singular and let $f(\lambda) = \sup(\{2^\kappa\colon \kappa < \lambda\})$. One of the following two cases must hold.

Case I: $\text{cf}(f(\lambda)) \ne \text{cf}(\lambda)$. Then $\exists \kappa_0 < \kappa\, \forall \kappa (\kappa_0 \le \kappa < \lambda \rightarrow 2^\kappa = 2^{\kappa_0})$, so in fact $\text{cf}(f(\lambda)) > \lambda$. In this case it is easily seen (Exercise H1) that $2^\lambda = f(\lambda)$.

Case II: $\text{cf}(f(\lambda)) = \text{cf}(\lambda)$. Then $2^\lambda \ge (f(\lambda))^+$.

By Theorem 4.7, $2^\lambda = (f(\lambda))^+$ in any Easton extension of a model of GCH. Furthermore, by results of Jensen, it is impossible to prove the consistency of $2^\lambda > (f(\lambda))^+$ by forcing. More precisely, consider the assertions

$$\forall \lambda (\lambda \text{ singular} \wedge \text{cf}(f(\lambda)) = \text{cf}(\lambda) \rightarrow 2^\lambda = (f(\lambda))^+), \qquad (*)$$

and

$$\forall X \subset \mathbf{ON} (|X| \ge \omega_1 \rightarrow \exists Y \subset \mathbf{ON} (X \subset Y \wedge |X| = |Y| \wedge Y \in \mathbf{L})). \qquad \text{(CL)}$$

It is easily seen (Exercise H2) that (CL) → (∗). (CL) is not a theorem of ZFC, but by Jensen's Covering Lemma, ¬(CL) implies a statement known as "$0^\#$ exists" (see [Drake 1974], [Shoenfield 1971b], or [Silver 1973]), which in turn implies that all uncountable cardinals are inaccessible (and also Mahlo and weakly compact) in **L**. Furthermore, "$0^\#$ exists" is false (and hence (CL) and (∗) are true) in every forcing extension of a model of $\mathbf{V} = \mathbf{L}$.

$0^\#$ had been introduced by Silver and Solovay, who showed that "$0^\#$ exists" follows from the existence of a measurable (or even Ramsey) cardinal. It is immediate from the definition of $0^\#$ that $\neg$(CL) is in fact equivalent to "$0^\#$ exists".

By results of Prikry and Silver, $\neg(*)$ can be made true in a generic extension of a model with a supercompact cardinal, and Magidor [1900] showed that if M has slightly more than a huge cardinal, then one can even get

$$\forall n < \omega\,(2^{\omega_n} = \omega_{n+1}) \wedge 2^{\omega_\omega} = \omega_{\omega+2}$$

holding in some generic extension of M. "Supercompact" and "huge" are both large cardinal properties (much bigger than measurable), with huge much bigger than supercompact in the sense that if κ is huge then

$$\{\lambda < \kappa\colon (\lambda \text{ is supercompact})^{R(\kappa)}\}$$

is stationary in κ. The existence of such cardinals is not known to be inconsistent. For more details, see [Solovay–Reinhardt–Kanamori 1978].

At singular cardinals of cofinality $>\omega$, some forms of (∗) are provable in ZFC. For example, Silver has shown that if $\omega < \mathrm{cf}(\lambda) < \lambda$ and

$$\forall \kappa < \lambda\,(2^\kappa = \kappa^+),$$

then $2^\lambda = \lambda^+$. For more details, see [Galvin–Hajnal 1975] or Exercises H3–H6.

§5. General iterated forcing

We now consider the situation where $\mathbb{P}$ is a p.o. in M, G is $\mathbb{P}$-generic over M, and $\mathbb{Q}$ is a p.o. in $M[G]$. We wish to find a p.o. $\mathbb{R} \in M$ such that a generic extension of M by $\mathbb{R}$ is the same as a generic extension of $M[G]$ by $\mathbb{Q}$. If $\mathbb{Q} \in M$, then we may take $\mathbb{R} = \mathbb{P} \times \mathbb{Q}$, but if $\mathbb{Q} \notin M$, then $\mathbb{P} \times \mathbb{Q} \notin M$. To handle the general situation of $\mathbb{Q} \in M[G]$, we shall build $\mathbb{R}$ in M out of $\mathbb{P}$ and a $\mathbb{P}$-name in M for $\mathbb{Q}$.

We may simplify the construction of $\mathbb{R}$ somewhat by observing that we need not consider all possible $\mathbb{P}$-names. First, we may assume that $\mathbb{Q} = \langle \alpha, \leq_{\mathbb{Q}}, 0\rangle$ (i.e., $\mathbb{Q}$ is a partial order of some ordinal α, with largest element 0), since in any case $\mathbb{Q}$ is isomorphic in $M[G]$ to such a p.o., and isomorphic p.o.'s yield identical generic extensions. Next, suppose $\leq_{\mathbb{Q}} = \tau_G$. Let $\phi(x)$ abbreviate the statement,

$$x \text{ partially orders } \check{\alpha} \text{ with largest element } \check{0}.$$

It need not be true that $\mathbb{1} \Vdash \phi(\tau)$, but

$$\mathbb{1} \Vdash \exists x\,(\phi(x) \wedge (\phi(\tau) \rightarrow x = \tau)),$$

so by the maximal principle (VII 8.2), there is a name σ, such that

$$\mathbb{1} \Vdash \phi(\sigma) \wedge (\phi(\tau) \rightarrow \sigma = \tau);$$

then $\mathbb{1} \Vdash \phi(\sigma)$ and, since $\phi(\tau_G)$ is true in $M[G]$, $\sigma_G = \tau_G = \leq_{\mathbb{Q}}$.

Actually, the fact that the domain of $\leq_{\mathbb{Q}}$ is an ordinal does not lead to any simplification, but the fact that we may restrict ourselves to names which are forced by $\mathbb{1}_{\mathbb{P}}$ to be a p.o., with a specified name for the largest element, does. We thus take only these latter restrictions as part of the official definition.

5.1. DEFINITION. If $\mathbb{P}$ is a p.o. in M, a $\mathbb{P}$-name for a p.o. is a triple of $\mathbb{P}$-names, $\langle \pi, \pi', \pi'' \rangle \in M$, such that $\pi'' \in \mathrm{dom}(\pi)$ and

$$\mathbb{1}_{\mathbb{P}} \Vdash_{\mathbb{P}} [(\pi'' \in \pi) \wedge (\pi' \text{ is a partial order of } \pi \text{ with largest element } \pi'')].$$

We often write π for $\langle \pi, \pi', \pi'' \rangle$, $\leq_\pi$ for π', and $\mathbb{1}_\pi$ or $\mathbb{1}$ for π''. □

5.2. DEFINITION. If $\mathbb{P}$ is a p.o. in M and π (i.e., $\langle \pi, \leq_\pi, \mathbb{1}_\pi \rangle$) is a $\mathbb{P}$-name for a p.o., then $\mathbb{P} * \pi$ is the p.o. whose base set is

$$\{\langle p, \tau \rangle : p \in \mathbb{P} \wedge \tau \in \mathrm{dom}(\pi) \wedge p \Vdash \tau \in \pi\}.$$

In $\mathbb{P} * \pi$, we define $\langle p, \tau \rangle \leq \langle q, \sigma \rangle$ iff

$$p \leq_{\mathbb{P}} q \wedge p \Vdash \tau \leq_\pi \sigma,$$

and we set $\mathbb{1}_{\mathbb{P}*\pi} = \langle \mathbb{1}_{\mathbb{P}}, \mathbb{1}_\pi \rangle$. Define $i: \mathbb{P} \rightarrow \mathbb{P} * \pi$ by $i(p) = \langle p, \mathbb{1}_\pi \rangle$. □

It is easily checked that $\mathbb{P} * \pi$ is a p.o. as claimed.

The product construction in §1 may be considered a special case of $\mathbb{P} * \pi$. Thus, if $\mathbb{Q}$ is a p.o. in M and $\pi = \check{\mathbb{Q}}$, then $\mathbb{P} * \pi$ is isomorphic to $\mathbb{P} \times \mathbb{Q}$.

As another example, suppose $\mathbb{P}$ preserves cardinals, G is $\mathbb{P}$-generic over M, and $\mathbb{Q} = (\mathrm{Fn}(\kappa, 2, \omega_1))^{M[G]}$. Although we may represent $\mathbb{Q}$ by a partial order of some ordinal, it is more natural to let $\pi = S \times \{\mathbb{1}_{\mathbb{P}}\}$, where S is defined in M by

$$S = \{\tau : (\mathbb{1} \Vdash_{\mathbb{P}} \tau \in \mathrm{Fn}(\check{\kappa}, \check{2}, \check{\omega}_1)) \wedge \mathrm{dom}(\tau) \subset \mathrm{dom}((\kappa \times 2)^{\check{}})\};$$

then $\mathbb{1}_\pi$ is the name 0. In this case, there will be names $\tau \neq 0$ in $S = \mathrm{dom}(\pi)$ such that $p \Vdash \tau = 0$ for some p, in which case $\langle p, \tau \rangle \leq \langle p, \mathbb{1}_\pi \rangle \leq \langle p, \tau \rangle$ even though $\langle p, \tau \rangle \neq \langle p, \mathbb{1}_\pi \rangle$. This is no problem, since we have not required our p.o.'s to be partial orders in the strict sense ($p \leq p' \leq p$ need not imply $p = p'$; see II 2.1). Although all p.o.'s occurring "naturally" in proofs from MA and in ordinary forcing constructions (as in VII) are p.o.'s in the strict sense, this need not be the case for "naturally" occurring p.o.'s in iterated forcing arguments.

5.3. LEMMA. *In the notation of Definition* 5.2:

(a) $\forall p, p' \in \mathbb{P}\,(p \leq p' \leftrightarrow \langle p, \mathbb{1}\rangle \leq \langle p', \mathbb{1}\rangle)$.

(b) $i(\mathbb{1}_{\mathbb{P}}) = \mathbb{1}_{\mathbb{P}*\pi}$.

(c) $\forall \langle p, \tau\rangle, \langle p', \tau'\rangle \in \mathbb{P} * \pi(p \perp p' \to \langle p, \tau\rangle \perp \langle p', \tau'\rangle)$.

(d) $\forall \langle p, \tau\rangle \in \mathbb{P} * \pi\, \forall p' \in \mathbb{P}(p \perp p' \leftrightarrow \langle p, \tau\rangle \perp \langle p', \mathbb{1}\rangle)$.

(e) $\forall p, p' \in \mathbb{P}(p \perp p' \leftrightarrow i(p) \perp i(p'))$.

(f) $i : \mathbb{P} \to \mathbb{P} * \pi$ *is a complete embedding.*

PROOF. (a)–(c) are immediate from the definitions, and (d) from left to right is a special case of (c). For (d) from right to left, if p'' is a common extension of p and p', then $\langle p'', \tau\rangle$ is a common extension of $\langle p, \tau\rangle$ and $\langle p', \mathbb{1}\rangle$. (e) is the special case of (d) when $\tau = \mathbb{1}$, and (f) follows from (a), (e), and (d); (d) implies that p is a reduction of $\langle p, \tau\rangle$ to $\mathbb{P}$. □

As in the case of products, the main result on $\mathbb{P}*\pi$ is that extending M by $\mathbb{P} * \pi$ is equivalent to extending by $\mathbb{P}$ and then by π (actually, by the p.o. named by π). Unlike the case for products, there is no symmetry between $\mathbb{P}$ and π. Thus, $\mathbb{P} \times \mathbb{Q}$ and $\mathbb{Q} \times \mathbb{P}$ are isomorphic, but $\pi * \mathbb{P}$ is not defined, and it is meaningless to attempt to extend by π first.

5.4. DEFINITION. In the notation of Definition 5.2, if G is $\mathbb{P}$-generic over M and $H \subset \pi_G$, then

$$G * H = \{\langle p, \tau\rangle \in \mathbb{P} * \pi : p \in G \wedge \tau_G \in H\}. \qquad \square$$

5.5. THEOREM. *Assume that* $\mathbb{P}$ *is a* p.o. *in* M *and* π *is a* $\mathbb{P}$*-name for a* p.o. *Let* K *be* $\mathbb{P} * \pi$*-generic over* M. *Let* $G = i^{-1}(K)$, *and let*

$$H = \{\tau_G : \tau \in \mathrm{dom}(\pi) \wedge \exists q(\langle q, \tau\rangle \in K)\}.$$

Then G *is* $\mathbb{P}$*-generic over* M, H *is* π_G*-generic over* $M[G]$, $K = G * H$, *and* $M[K] = M[G][H]$.

PROOF. Genericity of G follows from the completeness of i.

To prove genericity of H, we first check that H is a filter. Assume $\tau_G \in H$ and $\tau_G \leq \sigma_G$, where $\sigma_G \in \pi_G$ and $\sigma \in \mathrm{dom}(\pi)$; we show $\sigma_G \in H$. Fix q with $\langle q, \tau\rangle \in K$, and fix $p \in G$ such that $p \Vdash \tau \leq \sigma$. Let $\langle r, \rho\rangle \in K$ be such that $\langle r, \rho\rangle \leq \langle p, \mathbb{1}\rangle$ and $\langle r, \rho\rangle \leq \langle q, \tau\rangle$; then $r \Vdash \rho \leq \tau$ and $r \Vdash \tau \leq \sigma$ (since $r \leq p$), so $r \Vdash \rho \leq \sigma$. Thus, $\langle r, \rho\rangle \leq \langle r, \sigma\rangle$, so $\langle r, \sigma\rangle \in K$, so $\sigma_G \in H$. We omit the easy proof that any two elements of H have a common extension in H.

To prove that H meets all the dense sets of $M[G]$, we mimic the proof of (1) → (2) in Theorem 1.4. Fix $D \in M[G]$ such that D is dense in π_G. Now fix $\delta \in M^{\mathbb{P}}$ such that $D = \delta_G$ and fix $p \in G$ such that

$$p \Vdash_{\mathbb{P}} (\delta \text{ is dense in } \pi).$$

Now let

$$D' = \{\langle q, \tau \rangle \in \mathbb{P} * \pi \colon q \Vdash \tau \in \delta\}.$$

It is easily seen that D' is dense below $\langle p, \mathbb{1} \rangle$, so fix $\langle q, \tau \rangle \in K \cap D'$. Then $\tau_G \in H \cap D$.

If $\langle p, \tau \rangle \in K$, then $p \in G$ and $\tau_G \in H$ by the definitions of G and H, so $\langle p, \tau \rangle \in G * H$; thus $K \subset G * H$. If $\langle p, \tau \rangle \in G * H$, then $p \in G$, $\tau_G \in H$, and $\langle p, \tau \rangle \in \mathbb{P} * \pi$. Thus, $\langle p, \mathbb{1} \rangle \in K$, and $\langle q, \tau \rangle \in K$ for some q. Let $\langle r, \sigma \rangle$ be a common extension of $\langle p, \mathbb{1} \rangle$ and $\langle q, \tau \rangle$ in K; then $r \leq p$ and $r \Vdash \sigma \leq \tau$, so $\langle r, \sigma \rangle \leq \langle p, \tau \rangle$, so $\langle p, \tau \rangle \in K$. Thus $G * H \subset K$.

Now, $G, H \in M[K]$, so $M[G][H] \subset M[K]$. Also, $K = G * H \in M[G][H]$, so $M[K] \subset M[G][H]$. □

The converse to Theorem 5.5 says that given a $\mathbb{P}$-generic G and a π_G-generic H, $G * H$ is $\mathbb{P} * \pi$-generic. Since we do not plan to use it, we leave it as Exercise J15. In the case of products, the converse was proved in Theorem 1.4, but in fact was never used in the applications in §§2–4.

When we prove the consistency of MA + ¬CH, it will be important to know that iterating c.c.c. p.o.'s results in a c.c.c. p.o. First, we prove lemma

5.6. Lemma. *Assume that in M, κ is regular, $\mathbb{P}$ is κ-c.c., σ is a $\mathbb{P}$-name, and*

$$\mathbb{1} \Vdash (\sigma \subset \check{\kappa} \wedge |\sigma| < \check{\kappa}).$$

Then for some $\beta < \kappa$, $\mathbb{1} \Vdash (\sigma \subset \check{\beta})$.

Proof. In M, let

$$E = \{\alpha < \kappa;\ \exists p \in \mathbb{P}\, (p \Vdash (\check{\alpha} = \sup(\sigma)))\},$$

and, for $\alpha \in E$, pick $p_\alpha \in \mathbb{P}$ such that $p_\alpha \Vdash (\check{\alpha} = \sup(\sigma))$. Still in M, $\{p_\alpha \colon \alpha \in E\}$ is an antichain in $\mathbb{P}$, so by κ-c.c., $|E| < \kappa$. Fix $\beta < \kappa$ such that $E \subset \beta$.

Whenever G is $\mathbb{P}$-generic over M, κ remains regular in $M[G]$ by κ-c.c. (see VII 6.9), and $|\sigma_G| < \kappa$ in $M[G]$; so if $\alpha = \sup(\sigma_G)$, then $\alpha < \kappa$. Since $\exists p \in G\, (p \Vdash \check{\alpha} = \sup(\sigma))$, $\alpha \in E$, so $\sigma \subset B$.

Thus, $\mathbb{1} \Vdash \sigma \subset \check{\beta}$. □

5.7. Lemma. *Assume that $\mathbb{P}$ is a p.o. in M, π is a $\mathbb{P}$-name for a p.o., κ is a regular cardinal of M, $(\mathbb{P}$ is κ-c.c.$)^M$, and $\mathbb{1} \Vdash \pi$ is $\check{\kappa}$-c.c. Then $(\mathbb{P} * \pi$ is κ-c.c.$)^M$.*

Proof. If not, let $\{\langle p_\xi, \tau_\xi\rangle : \xi < \kappa\}$ be an antichain in $\mathbb{P} * \pi$ in M. Let $\sigma = \{\langle \check{\xi}, p_\xi\rangle : \xi < \kappa\}$; then $\sigma \in M^{\mathbb{P}}$, and $\mathbb{1} \Vdash \sigma \subset \check{\kappa}$.

Let G be $\mathbb{P}$-generic over M, then $\sigma_G = \{\xi < \kappa : p_\xi \in G\}$. We claim that in $M[G]$, the $(\tau_\xi)_G$ for $\xi \in \sigma_G$ are pairwise incompatible in π_G. To see this, suppose $\xi \neq \eta$, $p_\xi \in G$, $p_\eta \in G$, and $(\tau_\xi)_G$, $(\tau_\eta)_G$ had a common extension in π_G; then there would be a $q \in G$ and a $\rho \in \mathrm{dom}(\pi)$ with $q \Vdash \rho \leq_\pi \tau_\xi$ and $q \Vdash \rho \leq_\pi \tau_\eta$. Since G is a filter, we may assume $q \leq p_\xi$ and $q \leq p_\eta$. But then $\langle q, \rho\rangle$ would be a common extension of $\langle p_\xi, \tau_\xi\rangle$ and $\langle p_\eta, \tau_\eta\rangle$ in $\mathbb{P}^* \pi$.

Since $\mathbb{1} \Vdash (\pi \text{ is } \check{\kappa}\text{-c.c.})$, $(|\sigma_G| < \kappa)^{M[G]}$ whenever G is $\mathbb{P}$-generic over M, so $\mathbb{1} \Vdash |\sigma| < \check{\kappa}$. By Lemma 5.6, fix $\beta < \kappa$ so that $\mathbb{1} \Vdash \sigma \subset \check{\beta}$. But $p_\beta \Vdash \check{\beta} \in \sigma$, a contradiction. □

For a converse to Lemma 5.7, see Exercise C2.

As a special case of 5.7, suppose $\mathbb{Q} \in M$ and $\pi = \check{\mathbb{Q}}$. Then Lemma 5.7 says that if $\mathbb{P}$ is c.c.c. in M and $\mathbb{1} \Vdash \check{\mathbb{Q}}$ is c.c.c., then $\mathbb{P} \times \mathbb{Q}$ is c.c.c. in M. Lemma 5.7 does *not* say that the product of c.c.c. p.o.'s is c.c.c., since $\mathbb{Q}$ could be c.c.c. in M but fail to be c.c.c. in some $\mathbb{P}$-generic extension of M. c.c.c. is *not* absolute. See Exercise C4 for more on this.

We now turn to consider α-stage iterated forcing for α any ordinal. For $\alpha = 1$, we have ordinary forcing, and for $\alpha = 2$ we have iterations of the form $\mathbb{P} * \pi$ just discussed. For $\alpha = 3$, we start with an arbitrary p.o. $\mathbb{P}_1 \in M$ and a $\mathbb{P}_1$-name π_1 for a p.o., and let $\mathbb{P}_2 = \mathbb{P}_1 * \pi_1$. Now let π_2 be a $\mathbb{P}_2$-name for a p.o. and let $\mathbb{P}_3 = \mathbb{P}_2 * \pi_2$; then every element of $\mathbb{P}_3$ is of the form $\langle\langle \rho_0, \rho_1\rangle, \rho_2\rangle$, where $\rho_0 \in \mathbb{P}_1$, $\rho_0 \Vdash (\rho_1 \in \pi_1)$, and $\langle \rho_0, \rho_1\rangle \Vdash \rho_2 \in \pi_2$. One may likewise repeat this procedure n times for any $n \in \omega$ to form an n-stage iteration $\mathbb{P}_n$ for any $n \in \omega$. There is really nothing new here, but we shall, in our official definition (5.8) make a few trivial changes in the interest of notational simplicity. We shall consider elements of $\mathbb{P}_n$ to be n-tuples $\langle \rho_0, \rho_1, \ldots, \rho_{n-1}\rangle$ rather than the typographically more cumbersome $\langle\langle \cdots \langle \rho_0, \rho_1\rangle, \rho_2\rangle \cdots \rho_{n-2}\rangle\, \rho_{n-1}\rangle$. So elements of $\mathbb{P}_n$ are sequences of length n; i.e., functions with domain n. $\mathbb{P}_0$ was not defined above, but in conformity with our notation, we should take $\mathbb{P}_0 = \{0\}$, the one element order (since 0 is the empty sequence); thus, a 0-stage extension of M is M itself. The sequence of p.o.'s, $\mathbb{P}_0, \mathbb{P}_1, \ldots, \mathbb{P}_n$, has the property that if $p = \langle \rho_0, \rho_1, \ldots, \rho_{n-1}\rangle \in \mathbb{P}_n$, then each $p \restriction i = \langle \rho_0, \ldots, \rho_{i-1}\rangle$ is in $\mathbb{P}_i$ (for $i < n$), and $p \restriction i \Vdash_{\mathbb{P}_i} \rho_i \in \pi_i$, where π_i is a $\mathbb{P}_i$-name. In particular, π_0 is a $\mathbb{P}_0$-name, which must be a $\check{\mathbb{Q}}$ for some $\mathbb{Q} \in M$; thus $\mathbb{P}_1$ is not actually arbitrary as we originally had, but $\mathbb{P}_1$ is arbitrary up to isomorphism, since $\mathbb{Q}$ is arbitrary.

We must now decide what to do at limits. Suppose we have $\mathbb{P}_n$ for $n \in \omega$ constructed as above and we wish to define $\mathbb{P}_\omega$. One possibility (the *full limit*) is to take all sequences $p = \langle \rho_n : n \in \omega\rangle$ such that each $p \restriction n \in \mathbb{P}_n$. Another possibility (*finite support* iteration) is to require in addition that $\rho_i = \mathbb{1}$ for all but finitely many $i \in \omega$; this is the iteration relevant to proving

MA + $\neg$CH consistent. If γ is an uncountable limit, a third possibility in defining $\mathbb{P}_\gamma$ is to take all sequences $\langle \rho_\xi : \xi < \gamma \rangle$ of countable support. We may handle all such iterations simultaneously by demanding that at limit stages, the supports lie in some ideal (see II Definition 6.2).

5.8. DEFINITION. In M, suppose α is any ordinal $\mathscr{I} \subset \mathscr{P}(\alpha)$, $\mathscr{I}$ is an ideal on $\alpha + 1$, and $\mathscr{I}$ contains all finite subsets of α. An *α-stage iterated forcing construction with supports in $\mathscr{I}$* is an object in M of the form,

$$\langle\langle\langle \mathbb{P}_\xi, \leq_{\mathbb{P}_\xi}, \mathbb{1}_{\mathbb{P}_\xi}\rangle : \xi \leq \alpha\rangle, \qquad \langle\langle \pi_\xi, \leq_{\pi_\xi}, \mathbb{1}_{\pi_\xi}\rangle : \xi < \alpha\rangle\rangle,$$

which satisfies the following conditions: Each $\langle \mathbb{P}_\xi, \leq_{\mathbb{P}_\xi}, \mathbb{1}_{\mathbb{P}_\xi}\rangle$ is a p.o. Elements of $\mathbb{P}_\xi$ are all sequences of length ξ. If $p \in \mathbb{P}_\eta$ and $\xi < \eta$, then $p \restriction \xi \in \mathbb{P}_\xi$. Each $\langle \pi_\xi, \leq_{\pi_\xi}, \mathbb{1}_{\pi_\xi}\rangle$ is a $\mathbb{P}_\xi$-name for a p.o. If $\langle \rho_\mu : \mu < \xi \rangle \in \mathbb{P}_\xi$, then each $\rho_\mu \in \mathrm{dom}(\pi_\mu)$. $\mathbb{1}_{\mathbb{P}_\xi}$ is the sequence $\langle \rho_\mu : \mu < \xi \rangle$ such that each $\rho_\mu = \mathbb{1}_{\pi_\mu}$. Define $\mathrm{supt}(\langle \rho_\mu : \mu < \xi \rangle) = \{\mu < \xi : \rho_\mu \neq \mathbb{1}_{\pi_\mu}\}$. We demand further that the construction satisfy:

(1) *Basis.* $\mathbb{P}_0 = \{0\}$.

(2) *Successors.* If $p = \langle \rho_\mu : \mu \leq \xi \rangle$, then $p \in \mathbb{P}_{\xi+1}$ iff $p \restriction \xi \in \mathbb{P}_\xi$, $\rho_\xi \in \mathrm{dom}(\pi_\xi)$, and $p \restriction \xi \Vdash_{\mathbb{P}_\xi} (\rho_\xi \in \pi_\xi)$. If also $p' = \langle \rho'_\mu : \mu \leq \xi \rangle$, then $p \leq p'$ iff $p \restriction \xi \leq p' \restriction \xi$ and $p \restriction \xi \Vdash (\rho_\xi \leq \rho'_\xi)$.

(3) *Limits.* If η is a limit ordinal and $p = \langle \rho_\mu : \mu < \eta \rangle$, then $p \in \mathbb{P}_\eta$ iff

$$\forall \xi < \eta\, (p \restriction \xi \in \mathbb{P}_\xi) \wedge \mathrm{supt}(p) \in \mathscr{I}.$$

If $p, p' \in \mathbb{P}_\eta$, then $p \leq p'$ iff $\forall \xi < \eta\, (p \restriction \xi \leq p' \restriction \xi)$. $\square$

As usual, we shall often suppress the subscripts on the various $\mathbb{1}$'s, $\Vdash$'s, and $\leq$'s, and refer to $\mathbb{P}_\xi$ and π_ξ rather than $\langle \mathbb{P}_\xi, \leq_{\mathbb{P}_\xi}, \mathbb{1}_{\mathbb{P}_\xi}\rangle$ and $\langle \pi_\xi, \leq_{\pi_\xi}, \mathbb{1}_{\pi_\xi}\rangle$.

5.9. DEFINITION. In Definition 5.8, we say the iteration is of *finite support* iff $\mathscr{I} = \{X \subset \alpha : |X| < \omega\}$, and of *countable support* iff $\mathscr{I} = \{X \subset \alpha : (|X| \leq \omega)\}^M$. Iteration with *full limits* means $\mathscr{I} = (\mathscr{P}(\alpha))^M$. $\square$

Of course, if $\alpha < \omega$, then these three notions are the same. We have said in Definition 5.8 that $\mathscr{I}$ is an ideal on $\alpha + 1$ for the technical reason that $\mathscr{P}(\alpha)$ is not really a (proper) ideal on α.

Since $\mathscr{I}$ is an ideal and contains finite sets, $p \in \mathbb{P}_\xi \to \mathrm{supt}(p) \in \mathscr{I}$ for all ξ, even though this was part of the official definition only when ξ was a limit.

Definition 5.8 should be viewed as a prescription for building an iterated forcing construction. In applications we use some bookkeeping device to ensure that the π_ξ run through names for all p.o.'s of a specified type (e.g., all c.c.c. p.o.'s in the MA construction). Given a recipe for choosing the π_ξ, Definition 5.8 tells us how to inductively build the $\mathbb{P}_\xi$. Observe that Clause

(2) makes $\mathbb{P}_{\xi+1}$ isomorphic to $\mathbb{P}_\xi * \pi_\xi$, while Clause (3) uses $\mathscr{I}$ to handle the limits.

Generalizing the i: $\mathbb{P} \to \mathbb{P} * \pi$ from two-stage iterations, we have the following.

5.10. **Definition.** In the notation of Definition 5.8, if $\xi \leq \eta \leq \alpha$, define $i_{\xi\eta} : \mathbb{P}_\xi \to \mathbb{P}_\eta$ so that $i_{\xi\eta}(p)$ is the $p' \in \mathbb{P}_\eta$, such that $p' \restriction \xi = p$ and $p'(\mu) = \mathbb{1}_{\pi_\mu}$ for $\xi \leq \mu < \eta$. □

Of course, it must be checked that p' as defined really is in $\mathbb{P}_\eta$, but this is easily done by induction on η.

The following easy lemma summarizes the abstract order-theoretic properties of our iterated forcing constructions.

5.11. **Lemma.** *In the notation of Definitions* 5.8 *and* 5.10, *assume that* $\xi \leq \eta \leq \zeta \leq \alpha$.

(a) $i_{\xi\zeta} = i_{\eta\zeta} \circ i_{\xi\eta}$.
(b) $i_{\xi\eta}(\mathbb{1}_{\mathbb{P}_\xi}) = \mathbb{1}_{\mathbb{P}_\eta}$.
(c) $\forall p, p' \in \mathbb{P}_\eta (p \leq p' \to p \restriction \xi \leq p' \restriction \xi)$.
(d) $\forall p, p' \in \mathbb{P}_\xi (p \leq p' \leftrightarrow i_{\xi\eta}(p) \leq i_{\xi\eta}(p'))$.
(e) $\forall p, p' \in \mathbb{P}_\eta (p \restriction \xi \perp p' \restriction \xi \to p \perp p')$.
(f) $\forall p, p' \in \mathbb{P}_\eta [\text{supt}(p) \cap \text{supt}(p') \subset \xi \to (p \restriction \xi \perp p' \restriction \xi \leftrightarrow p \perp p')]$.
(g) $\forall p, p' \in \mathbb{P}_\xi (p \perp p' \leftrightarrow i_{\xi\eta}(p) \perp i_{\xi\eta}(p'))$.
(h) $i_{\xi\eta}$ *is a complete embedding.*

Proof. (a) and (b) are immediate from the definitions. (c) and (d) may be proved by induction on η. For (e), if p'' were a common extension of p and p', then by (c), $p'' \restriction \xi$ would be a common extension of $p \restriction \xi$ and $p' \restriction \xi$.

Half of (f) is contained in (e). For the other half, assume

$$\text{supt}(p) \cap \text{supt}(p') \subset \xi,$$

and assume $p \restriction \xi$ and $p' \restriction \xi$ are compatible; let $p'' \in \mathbb{P}_\xi$ be a common extension. Define $p^* = \langle \rho^*_\mu : \mu < \eta \rangle$ so that $p^* \restriction \xi = p'' \restriction \xi$; and, for $\xi \leq \mu < \eta$, ρ^*_μ is ρ_μ if $\mu \in \text{supt}(p)$, ρ'_μ if $\mu \in \text{supt}(p')$, and $\mathbb{1}$ if $\mu \notin \text{supt}(p) \cup \text{supt}(p')$. Now prove, by induction on γ, for $\xi \leq \gamma \leq \eta$, that

$$(p^* \restriction \gamma \in \mathbb{P}_\gamma) \wedge (p^* \restriction \gamma \leq p \restriction \gamma) \wedge (p^* \restriction \gamma \leq p' \restriction \gamma).$$

Setting $\gamma = \eta$, we have that p and p' are compatible.

(g) is a special case of (f), and (h) follows from (d), (g), and (f). (f) is used to show that, given $p \in \mathbb{P}_\eta$, $p \restriction \xi$ is a reduction of p to $\mathbb{P}_\xi$. □

(f) of Lemma 5.11 also allows us to use a Δ-system argument on the supports to show that the c.c.c. is preserved in finite support iteration. This will be of vital importance in the MA construction in §6.

5.12. LEMMA. *Assume that in* M: κ *is regular* $>\omega$,

$$\langle\langle \mathbb{P}_\xi\colon \xi \le \alpha\rangle, \langle \pi_\xi\colon \xi < \alpha\rangle\rangle$$

is an α*-stage finite support iterated forcing construction and for each* $\xi < \alpha$,

$$\mathbb{1} \Vdash_{\mathbb{P}_\xi} (\pi_\xi \text{ is } \check{\kappa}\text{-c.c.}).$$

Then for each ξ, $\mathbb{P}_\xi$ *is* κ*-c.c. in* M.

PROOF. By induction on ξ. For successor steps, this is just Lemma 5.7, since $\mathbb{P}_{\xi+1}$ is isomorphic to $\mathbb{P}_\xi * \pi_\xi$. Now, assume ξ is a limit, and suppose that for $\zeta < \xi$, $\mathbb{P}_\zeta$ has the κ-c.c. in M, but that in M, $\{p^\beta\colon \beta < \kappa\}$ is an antichain in $\mathbb{P}_\xi$. By passing to a sub-collection of size κ, we may assume that $\{\mathrm{supt}(p^\beta)\colon \beta < \kappa\}$ forms a Δ-system with some root r. Fix $\zeta < \xi$ with $r \subset \zeta$; then by Lemma 5.11(f), $\{p^\beta \restriction \zeta\colon \beta < \kappa\}$ would be an antichain in $\mathbb{P}_\zeta$. □

As a converse to Lemma 5.12, if for $n < \omega$, $\mathbb{1} \Vdash (\pi_n$ is not c.c.c.), then not only does $\mathbb{P}_\omega$ fail to be c.c.c., but in fact it collapses ω_1^M (see Exercise E1). This indicates that if one wishes to do cardinal-preserving iterations with non-c.c.c. p.o.'s, something other than finite supports is required. In some cases, countable supports works (see §7). It might be thought that for countable supports, Lemma 5.12 still holds if we take $\kappa = \omega_2$ and assume CH in M, in analogy with the ω_2-c.c. of $\mathrm{Fn}(I, 2, \omega_1)$. However, although one may apply the Δ-system lemma precisely as in 5.12, the proof breaks down when the root is cofinal in ξ, and in fact there is a counter-example with $\xi = \omega$ (see Exercises E5 and E6).

Since $i_{\xi\eta}$ is a complete embedding of $\mathbb{P}_\xi$ as a sub-order of $\mathbb{P}_\eta$, it would be possible to identify $\langle \mathbb{P}_\xi\colon \xi \le \alpha\rangle$ with an ascending chain under $\subset_c$. We do not choose to make such an identification here, as it is just as easy to work with the $i_{\xi\eta}$. However, we do need a lemma saying that extending by $\mathbb{P}_\alpha$ gives us an ascending chain of models, formed by iterating extensions by the π_ξ.

5.13. LEMMA. *Assume that in* M,

$$\langle\langle \mathbb{P}_\xi\colon \xi \le \alpha\rangle, \langle \pi_\xi\colon \xi < \alpha\rangle\rangle$$

is an α*-stage iterated forcing construction. Let* G *be* $\mathbb{P}_\alpha$*-generic over* M. *For each* $\xi \le \alpha$, *let* $G_\xi = i_{\xi\alpha}^{-1}(G)$. *Then* G_ξ *is* P_ξ*-generic over* M *and*

$$\xi \le \eta \to M[G_\xi] \subset M[G_\eta].$$

Let $\mathbb{Q}_\xi = \mathrm{val}(\pi_\xi, G_\xi)$, *and let*

$$H_\xi = \{\mathrm{val}(\rho, G_\xi) \colon \rho \in \mathrm{dom}(\pi_\xi) \wedge \exists p\ ((p^\frown\langle\rho\rangle) \in G_{\xi+1})\},$$

then each $H_\xi \in M[G_{\xi+1}]$ *and* H_ξ *is* $\mathbb{Q}_\xi$*-generic over* $M[G_\xi]$.

PROOF. G_ξ is generic because $i_{\xi\alpha}$ is a complete embedding. Since $G_\xi = i_{\xi\eta}^{-1}(G_\eta)$, $M[G_\xi] \subset M[G_\eta]$ (see VII 7.5). Genericity of H_ξ follows from Theorem 5.5 and the isomorphism between $\mathbb{P}_\xi * \pi_\xi$ and $\mathbb{P}_{\xi+1}$. □

It is also true that each $M[G_\xi] = M[\langle H_\mu \colon \mu < \xi\rangle]$; see Exercise J10.

If we think of the $\mathbb{P}_\xi$ as an ascending chain, then for finite support iteration, limit $\mathbb{P}_\eta$ are just the union of the preceding $\mathbb{P}_\xi$; in our notation $\mathbb{P}_\eta = \bigcup_{\xi<\eta} i''_{\xi\eta}\mathbb{P}_\xi$. It is *not* true, even in very simple cases, that

$$M[G_\eta] = \bigcup_{\xi<\eta} M[G_\xi],$$

but "small" sets in $M[G_\eta]$ do appear in some earlier stage. More precisely, the following holds.

5.14. LEMMA. *Assume that in* M, α *is a limit ordinal,*

$$\langle\langle \mathbb{P}_\xi \colon \xi \le \alpha\rangle, \langle \pi_\xi \colon \xi < \alpha\rangle\rangle$$

is an α*-stage iterated forcing construction with supports in* $\mathscr{I}$, *and each element of* $\mathscr{I}$ *is bounded in* α. *Suppose* G *is* $\mathbb{P}_\alpha$*-generic over* M, $S \in M$, $X \subset S$, $X \in M[G]$, *and* $(|S| < \mathrm{cf}(\alpha))^{M[G]}$. *Then for some* $\eta < \alpha$, $X \in M[i_{\eta\alpha}^{-1}(G)]$.

PROOF. Let σ be a $\mathbb{P}_\alpha$-name such that $X = \sigma_G$. Then for $s \in S$, $s \in X$ iff $\exists p \in G(p \Vdash_{\mathbb{P}_\alpha} \check{s} \in \sigma)$. By our assumption on $\mathscr{I}$, $\mathbb{P}_\alpha = \bigcup_{\xi<\alpha} i''_{\xi\alpha}\mathbb{P}_\xi$ and $G = \bigcup_{\xi<\alpha} i''_{\xi\alpha}G_\xi$, where $G_\xi = i_{\xi\alpha}^{-1}G$.

In $M[G]$, for each $s \in X$, fix $\xi_s < \alpha$ such that $\exists p \in G_\xi\,(i_{\xi\alpha}(p) \Vdash_{\mathbb{P}_\alpha} \check{s} \in \sigma)$. Let $\eta = \sup\{\xi_s \colon s \in X\}$. Since $|X| \le |S| < \mathrm{cf}(\alpha)$ in $M[G]$, $\eta < \alpha$.

Now, $X = \{s \in S \colon \exists p \in G_\eta\,(i_{\eta\alpha}(p) \Vdash_{\mathbb{P}_\alpha} \check{s} \in \sigma)$. Since $\Vdash_{\mathbb{P}_\alpha}$ is defined in M, $X \in M[G_\eta]$. □

We conclude this section with some further remarks, none of which are needed for the MA proof in §6.

The simplest examples of iterated forcing occur when each π_ξ is of the form $\check{\mathbb{Q}}_\xi$ for some p.o. $\mathbb{Q}_\xi \in M$ (there should really be a subscript on the ˇ, since it depends on the $\mathbb{1}$ of $\mathbb{P}_\xi$). Although such examples are more easily handled by the theory of products, it is amusing to see how the product construction of §§2–4 fit into the framework discussed here. Thus, the Cohen p.o., $\mathrm{Fn}(\alpha, 2)$, is isomorphic to iterating the 3-element order, $\mathrm{Fn}(1, 2)$, α times with finite supports (i.e., each $\mathbb{Q}_\xi = \mathrm{Fn}(1, 2)$). For the usual Lévy collapse, Lv(α) (see VII 8.6), take $\mathbb{Q}_\xi = \text{Fn}(\omega, \xi)$, the usual collapsing order for ξ, and again use finite supports. For the Lévy collapse with countable conditions (see Definition 3.1), take $\mathbb{Q}_\xi = \text{Fn}(\omega_1, \xi, \omega_1)^M$, and use countable supports.

For Easton forcing (see §4), let E be an index function, with $\text{dom}(E) = \{\kappa_\xi \colon \xi < \alpha\}$ in increasing enumeration. Then $\mathbb{P}(E)$ is isomorphic to the iteration $\mathbb{P}_\alpha$ obtained by taking (in M) $\mathbb{Q}_\xi = \text{Fn}(E(\kappa_\xi), 2, \kappa_\xi)$, and iterating with the ideal

$$\mathscr{I} = \{X \subset \alpha \colon \forall\lambda\,(\lambda \text{ regular} \rightarrow |\lambda \cap \{\kappa_\xi \colon \xi \in X\}| < \lambda)\}.$$

Observe that $\pi_\xi = \check{\mathbb{Q}}_\xi$ is the name for the $\text{Fn}(E(\kappa_\xi), 2, \kappa_\xi)$ of M. So-called *reverse* or *backwards* Easton forcing is obtained by instead taking π_ξ to be a name for the $\text{Fn}(E(\xi), 2, \kappa_\xi)$ of the extension by $\mathbb{P}_\xi$; the term *forwards* Easton forcing denotes the extension $\mathbb{P}(E)$ of §4 (although perhaps the terminology should be reversed). Reverse Easton forcing is slightly harder to handle, since it really does require the material in this section, and it will often collapse cardinals even when M satisfies GCH. Its importance is that it was used by Silver to obtain consistency results on violations of GCH at weakly compact and larger cardinals. Such results are known not to be obtainable via forwards Easton forcing; see Exercises I1–I9.

We now discuss some ways in which finite support iterations are "canonical", whereas countable support and more complicated iterations fail to be. These remarks will not aid the reader in understanding finite support iterations, but may serve to prevent errors when dealing with other iterations.

First, there are many names for the same p.o. If π and π' are both $\mathbb{P}$-names for p.o.'s, we take $\mathbb{1} \Vdash \pi = \pi'$ to mean that $\mathbb{1} \Vdash (\mathbb{1}_\pi = \mathbb{1}_{\pi'} \wedge \leq_\pi = \leq_{\pi'})$ also. If $\mathbb{1} \Vdash \pi = \pi'$, then $\mathbb{P} * \pi$ and $\mathbb{P} * \pi'$ need *not* be isomorphic (see Exercise K1), although they have isomorphic completions—i.e., they may both be densely embedded into the same complete Boolean algebra. In fact, if we write $\mathbb{P} \sim \mathbb{Q}$ to mean that $\mathbb{P}$ and $\mathbb{Q}$ have isomorphic completions, then $\mathbb{1} \Vdash \pi \sim \pi'$ implies that $\mathbb{P} * \pi \sim \mathbb{P} * \pi'$ in M. The same result holds for infinite iterations as well, *provided* the iteration is with finite supports (see Exercise K8). One can find π_n $(n \in \omega)$ such that each π_n is forced by $\mathbb{1}$ to equal $\text{Fn}(\omega, 2)$, but the countable support iteration of the π_n is not $\sim$ to the countable support iteration of ω copies of $(\text{Fn}(\omega, 2))\check{}$ (see Exercise K3).

Finite support iteration is also canonical when we pass to completions. Thus let $\mathscr{B}_\xi$ be the completion of $\mathbb{P}_\xi$. With finite supports, $\mathscr{B}_\eta$, for limit η, is the completion of $\bigcup_{\xi<\eta} \mathscr{B}_\xi$ (see Exercise K2). With countable supports, $\mathscr{B}_\omega$ is not definable at all from the chain $\langle \mathscr{B}_n \colon n < \omega \rangle$; i.e., the definition of $\mathscr{B}_\omega$ really does depend on the way the $\mathbb{P}_n$ were constructed (see Exercise K3).

The iterated forcing discussed in this section is not the most general kind possible. Indeed, any sequence $\langle \mathbb{P}_\xi \colon \xi \leq \alpha \rangle$ in M, together with complete embeddings $i_{\xi\eta}: \mathbb{P}_\xi \to \mathbb{P}_\eta$ which commute ($i_{\xi\zeta} = i_{\eta\zeta} \circ i_{\xi\eta}$) may be thought of as an iterated forcing construction, and if G is $\mathbb{P}_\alpha$-generic over M, then $\langle M[i_{\xi\alpha}^{-1}(G)]: \xi \leq \alpha\rangle$ forms an ascending chain of generic extensions of M. In this generality nothing interesting can be proved about the iteration beyond what is obvious from the properties of complete embeddings. However, there are important specific examples of such iterations which do not fall into our framework here; the best-known such example is iterated random real forcing (Exercise K6). If the sequence has the property that for each limit η, $\bigcup_{\xi<\eta} i''_{\xi\eta}\mathbb{P}_\xi$ is dense in $\mathbb{P}_\eta$, then one can identify the iteration as a finite support iteration in our sense, with appropriately defined π_ξ (see Exercise K5).

§6. The consistency of MA + ¬CH

Suppose that κ is a regular cardinal of M and we wish to construct a generic extension of M satisfying MA and $2^\omega = \kappa$. Naively, we iterate forcing κ times, forming a chain

$$M = M_0 \subset M_1 \subset \cdots \subset M_\xi \subset \cdots \subset M_\kappa.$$

Given M_ξ, we choose a c.c.c. p.o. $\mathbb{Q}_\xi \in M_\xi$, and $M_{\xi+1}$ will be $M_\xi[G_\xi]$, where G_ξ is $\mathbb{Q}_\xi$-generic over M_ξ. If $\lambda < \kappa$, we verify MA(λ) in M_κ as follows: Assume that in M_κ, $\mathbb{Q}$ is a c.c.c. p.o., $|\mathbb{Q}| \leq \lambda$, and $\mathcal{D}$ is a family of $\leq\lambda$ dense sets in $\mathbb{Q}$. Since $\mathcal{D}$ and $\mathbb{Q}$ are objects of size $<\kappa$, they will be in M_η for some $\eta < \kappa$. We arrange our construction so that $\mathbb{Q} = \mathbb{Q}_\xi$ for some $\xi \geq \eta$. Since $\mathcal{D} \in M_\eta \subset M_\xi$, G_ξ will be a filter in $\mathbb{Q}$ intersecting all the dense sets in $\mathcal{D}$.

In the rigorous treatment, we replace the naive approach with the results of §5, so instead of building a chain of models, we are building a chain, $\langle \mathbb{P}_\xi: \xi \leq \kappa\rangle$ of p.o.'s in M. We replace the $\mathbb{Q}_\xi$ by a $\mathbb{P}$-name, π_ξ, for a p.o. The basic lemmas on such iterations have already been proved in §5, so it remains only to show how, by the right bookkeeping devise, the π_ξ can be made to range over names for all relevant c.c.c. p.o.'s. Of course, *all* c.c.c. p.o.'s form a proper class, so we first prove a preliminary lemma stating that only a set of p.o.'s need be considered.

6.1. Definition. pord(λ, W) abbreviates the statement that W is a c.c.c. partial order of the set λ, with largest element 0. □

6.2. Lemma. *For any infinite cardinal λ, the following are equivalent:*

(1) MA(λ).

(2) *Whenever* pord(λ, W) *and $\mathcal{D}$ is a family of $\leq\lambda$ dense sets in $\langle\lambda, W\rangle$ (i.e., each $D \in \mathcal{D}$ is a subset of λ and is dense in the order W), there is a filter in $\langle\lambda, W\rangle$ intersecting each $D \in \mathcal{D}$.*

PROOF. (1) → (2) is obvious, so we assume (2) and try to prove MA(λ). We already know that MA(λ) is equivalent to MA(λ) restricted to partial orders of size $\leq\lambda$ (see II 3.1), so fix a c.c.c. partial order $\langle \mathbb{Q}, \leq \rangle$ with $|\mathbb{Q}| \leq \lambda$, and a family $\mathscr{D}$ of $\leq\lambda$ dense sets in $\mathbb{Q}$. We shall be done if we find a filter $G \subset \mathbb{Q}$ intersecing each $D \in \mathscr{D}$.

Now we did not require in II that partial orders had largest elements and $|\mathbb{Q}|$ could be $<\lambda$. To remedy both these problems let S be a set of size λ disjoint from $\mathbb{Q}$, and extend the order $\leq$ on $\mathbb{Q}$ to $\mathbb{Q} \cup S$ by putting $p < s$ for all $p \in \mathbb{Q}$ and $s \in S$, and $s \leq t$ for all $s, t \in S$. So any element of S is a largest element of $\mathbb{Q} \cup S$, and the $D \in \mathscr{D}$ are still dense in $\mathbb{Q} \cup S$. If $\mathbb{1}$ is any element of S, then $\langle \mathbb{Q} \cup S, \leq, \mathbb{1} \rangle$ is isomorphic to a p.o. of the form $\langle \lambda, W, 0 \rangle$, so (2) applies to give us a filter $G \subset \mathbb{Q} \cup S$ intersecting each $D \in \mathscr{D}$. Then $G \cap \mathbb{Q}$ is a filter in $\mathbb{Q}$ intersecting each $D \in \mathscr{D}$. □

We now wish to show that if κ is a cardinal of M, then there is a c.c.c. extension of M satisfying $2^\omega = \kappa$ and MA. Since MA implies 2^ω is regular (see II 2.19), we must require κ to be regular in M. Since MA implies $\forall \lambda < 2^\omega (2^\lambda = 2^\omega)$ (see II 2.18), we shall also need $2^{<\kappa} = \kappa$ in M. And, of course, κ cannot be ω.

6.3. THEOREM. *Assume that in M, $\kappa \geq \omega_1$, κ is regular and $2^{<\kappa} = \kappa$. Then there is a $\mathbb{P} \in M$ such that ($\mathbb{P}$ is* c.c.c.$)^M$ *and whenever G is $\mathbb{P}$-generic over M, $M[G]$ satisfies* MA *and $2^\omega = \kappa$.*

PROOF. To start with, fix a function, f, in M from κ onto $\kappa \times \kappa$ such that

$$\forall \xi, \eta, \gamma < \kappa (f(\xi) = \langle \eta, \gamma \rangle \rightarrow \eta \leq \xi).$$

(For example let g map κ onto $\kappa \times \kappa \times \kappa$; if $g(\xi) = \langle \eta, \gamma, \delta \rangle$, let $f(\xi)$ be $\langle \eta, \gamma \rangle$ if $\eta \leq \xi$ and $\langle 0, 0 \rangle$ if $\eta > \xi$.) f will be used for bookkeeping, as we shall explain later.

We now construct, in M, a finite support iteration of the form

$$\langle \langle \langle \mathbb{P}_\xi, \leq_{\mathbb{P}_\xi}, \mathbb{1}_{\mathbb{P}_\xi} \rangle : \xi \leq \kappa \rangle, \langle \langle \check{\lambda}_\xi, \sigma_\xi, \check{0} \rangle : \xi < \kappa \rangle \rangle.$$

We are following the notation of Definition 5.8 in exhibiting the p.o.'s, together with their orders and $\mathbb{1}$'s in gory detail. In the case of the $\mathbb{P}_\xi$, we shall now, as usual, write $\leq$ for $\leq_{\mathbb{P}_\xi}$, $\mathbb{1}$ for $\mathbb{1}_{\mathbb{P}_\xi}$, and $\mathbb{P}_\xi$ for $\langle \mathbb{P}_\xi, \leq, \mathbb{1} \rangle$. However, the π_ξ of Definition 5.8 will always be a $\mathbb{P}_\xi$-name for a p.o. whose domain is an ordinal, λ_ξ, with largest element 0, so the name for the order is important; we use σ_ξ rather than $\leq_{\pi_\xi}$.

We shall ensure that in our construction, for each $\xi < \kappa$,

(1) $\mathbb{1} \Vdash_{\mathbb{P}_\xi} \text{pord}(\check{\lambda}_\xi, \sigma_\xi)$, and

(2) $\lambda_\xi < \kappa$.

By (1), $\mathbb{P}_\xi$, for each $\xi \leq \kappa$, will be c.c.c. in M (see Lemma 5.12). By (2), an easy induction on ξ shows that in M, $|\mathbb{P}_\xi| < \kappa$ for $\xi < \kappa$ and $|\mathbb{P}_\kappa| \le \kappa$; this will be important for computing the number of nice $\mathbb{P}_\xi$-names; it is in such computations that $2^{<\kappa} = \kappa$ is used.

As one such computation, observe that in M there are at most $(\kappa^\omega)^\omega = \kappa$ nice $\mathbb{P}_\kappa$-names for subsets of $\check{\omega}$, so whenever G is $\mathbb{P}_\kappa$-generic over M, $2^\omega \le \kappa$ holds an $M[G]$.

For another such computation, observe that for each $\xi < \kappa$ and $\lambda < \kappa$, there are in M at most $(\kappa^\omega)^\lambda = \kappa$ nice $\mathbb{P}_\xi$-names for subsets of $(\lambda \times \lambda)\check{}$. Using this fact, we now describe how to choose the σ_ξ and λ_ξ.

Given each $\mathbb{P}_\xi$, pick, in M, an enumeration, $\langle\langle \lambda_\gamma^\xi, \sigma_\gamma^\xi\rangle: \gamma < \kappa\rangle$, of all pairs $\langle \lambda, \sigma\rangle$ such that $\lambda < \kappa$, λ is a cardinal, and σ is a nice $\mathbb{P}_\xi$-name for a subset of $(\lambda \times \lambda)\check{}$. We would like, at stage ξ, to consider all the σ_γ^ξ, but of course we can only use one at a time, so we put them off until later, which means that at stage ξ, we shall use a σ_γ^η for some $\eta \le \xi$. It is here that we use our function f to do the bookkeeping and make sure that each name gets considered eventually.

Let $f(\xi) = \langle \eta, \gamma\rangle$. Since $\eta \le \xi$, the $\mathbb{P}_\eta$-name, σ_γ^η has been defined. Let σ be the $\mathbb{P}_\xi$-name, $i_{\eta\xi*}(\sigma_\gamma^\eta)$, and let $\lambda = \lambda_\gamma^\eta$. We set $\lambda_\xi = \lambda$. We are tempted to set $\sigma_\xi = \sigma$, but σ may not be a name for a c.c.c. p.o. Instead, let σ_ξ be a $\mathbb{P}_\xi$-name such that

$$\mathbb{1} \Vdash_{\mathbb{P}_\xi} (\mathrm{pord}(\check{\lambda}, \sigma_\xi) \wedge [\mathrm{pord}(\check{\lambda}, \sigma) \rightarrow \sigma_\xi = \sigma]). \qquad (*)$$

To see that there is such a σ_ξ, note that it is theorem of ZFC that $\exists W\, \mathrm{pord}(\lambda, W)$ (e.g., take $W = \lambda \times \lambda$), so

$$\mathbb{1} \Vdash \exists W (\mathrm{pord}(\check{\lambda}, W) \wedge [\mathrm{pord}(\check{\lambda}, \sigma) \rightarrow W = \sigma]).$$

Thus, σ_ξ exists by the maximal principle (VII 8.2).

This specifies the construction. Now, set $\mathbb{P} = \mathbb{P}_\kappa$ and let G be $\mathbb{P}$-generic over M. We shall check that for each cardinal $\lambda < \kappa$ of M, MA(λ) holds in $M[G]$. This will imply $2^\omega \ge \kappa$ in $M[G]$. Since we have already seen that $2^\omega \le \kappa$, we shall have $2^\omega = \kappa$ and MA in $M[G]$.

In $M[G]$, assume that $\lambda < \kappa$, λ is a cardinal, $W \subset \lambda \times \lambda$, $\mathrm{pord}(\lambda, W)$ holds, and $\mathscr{D}$ is a family of $\le\lambda$ dense sets in $\langle \lambda, W\rangle$. We shall find a filter in $\langle \lambda, W\rangle$ intersecting each dense set in $\mathscr{D}$.

For $\xi \le \kappa$, let $G_\xi = i_{\xi\kappa}^{-1}(G)$. We first note that there is some $\eta < \kappa$ such that W and $\mathscr{D}$ are in $M[G_\eta]$. For W, this is immediate from Lemma 5.14. For $\mathscr{D}$, let $\mathscr{D} = \{D_\mu: \mu < \lambda\}$, and apply 5.14 to $\{\langle \mu, \nu\rangle \in \lambda \times \lambda: \nu \in D_\mu\}$.

Now, fix $\eta < \kappa$ such that $\mathscr{D}$, $W \in M[G_\eta]$. Then, fix $\gamma < \kappa$ so that $W = \mathrm{val}(\sigma_\gamma^\eta, G_\eta)$. Finally, fix $\xi \ge \eta$ so that $f(\xi) = \langle \eta, \gamma\rangle$. Let $\sigma = i_{\eta\xi*}(\sigma_\gamma^\eta)$. Then also $W = \mathrm{val}(\sigma, G_\xi)$ (see VII 7.13).

Next, note that $\mathrm{pord}(\lambda, W)$ holds in $M[G_\xi]$ as well as in $M[G]$. Indeed, most of pord is absolute for all transitive models of ZFC. "c.c.c." is not in general absolute; however, if W failed to be c.c.c. in $M[G_\xi]$, there would, in $M[G_\xi]$, be an uncountable antichain in the order W, and this antichain would remain uncountable in $M[G]$ since cardinals are preserved in these extensions.

By (*) above,

$$\mathbb{1} \Vdash [\text{pord}(\check{\lambda}, \sigma) \rightarrow \sigma_\xi = \sigma],$$

so $W = \text{val}(\sigma_\xi, G_\xi)$; hence $M[G_{\xi+1}]$ contains a filter H_ξ which is $\langle \lambda, W \rangle$-generic over $M[G_\xi]$ (see Lemma 5.13). Since $\mathscr{D} \in M[G_\eta] \subset M[G_\xi]$, H_ξ intersects all the dense sets in $\mathscr{D}$. □

As a minor technical point, the reader may wonder how we justify "picking" the enumeration

$$\langle \langle \lambda_\gamma^\xi, \sigma_\gamma^\xi \rangle : \gamma < \kappa \rangle,$$

since AC in M does not imply that we have a definable way of assigning, to each set of size κ, a well-ordering in type κ. To answer this objection, fix in M, at the outset, a well-ordering, $\lhd$, of the set $R(\kappa^+)$; then, given $\mathbb{P}_\xi$, let

$$\langle \langle \lambda_\gamma^\xi, \sigma_\gamma^\xi \rangle : \gamma < \kappa \rangle$$

be the $\lhd$-first such enumeration in type κ. Likewise, σ_ξ is chosen to be the $\lhd$-first $\mathbb{P}$-name satisfying (*). It is easily checked by induction that the whole construction stays within $R(\kappa^+)^M$.

6.4. Corollary. $\text{Con(ZFC)} \rightarrow \text{Con}(\text{ZFC} + \text{MA} + 2^\omega = \omega_{83})$.

Proof. Start with M satisfying GCH. □

6.5. Corollary. Con(ZFC + $\exists$ *a weakly inaccessible cardinal*) *implies* Con(ZFC + MA + 2^ω *is weakly inaccessible*).

Proof. Start with κ strongly inaccessible in M (see VII 5.16). □

§7. Countable support iterations

Iterated forcing arguments using other than finite supports occur frequently in the literature, although no one "quotable" axiom, like MA, stands out, and the various arguments take on a more ad hoc character. As a sample of what happens, we consider countable supports.

Since countable support extensions will be c.c.c. only in trivial cases, we must look for some other properties if we wish cardinals to be preserved. The easiest (but not the only) way of doing this is to force with p.o.'s which have the ω_2-c.c. and are ω_1-closed. We now look at whether these properties are preserved in iterations.

First, consider the property of being ω_1-closed. This property can fail to be preserved in countable support iterations (see Exercise E2), but it is preserved if the names for the p.o.'s satisfy an additional condition.

7.1. DEFINITION. If π is a $\mathbb{P}$-name for a p.o., π is *full for $\searrow$ ω-sequences* iff whenever $p \in \mathbb{P}$, $\rho_n \in \mathrm{dom}(\pi)\,(n \in \omega)$, and

$$p \Vdash ((\rho_n \in \pi) \wedge (\rho_{n+1} \leq \rho_n))$$

for each n, then there is a $\sigma \in \mathrm{dom}(\pi)$ such that $p \Vdash \sigma \in \pi$ and $p \Vdash (\sigma \leq \rho_n)$ for each n. $\square$

It is easily seen that if $\mathbb{P}$ is ω_1-closed and π is full for $\searrow$ ω-sequences, then $\mathbb{1} \Vdash_{\mathbb{P}} (\pi$ is ω_1-closed$)$. It is also easily seen (Exercise E3) that any finite iteration by ω_1-closed p.o.'s results in an ω_1-closed p.o. However, for infinite iterations this can fail (Exercise E2) unless the names used have the stronger property of being full for $\searrow$ ω-sequences.

7.2. LEMMA. *In M, let*

$$\langle\langle \mathbb{P}_\xi \colon \xi \leq \alpha\rangle, \langle \pi_\xi \colon \xi < \alpha\rangle\rangle$$

be a countable support iterated forcing construction (see §5), *and suppose that for each ξ, the $\mathbb{P}_\xi$-name π_ξ is full for $\searrow$ ω-sequences. Then $\mathbb{P}_\alpha$ is ω_1-closed in M.*

PROOF. In M, let $p^n \in \mathbb{P}_\alpha\,(n < \omega)$ with $p^{n+1} \leq p^n$ for each n. We shall find a p^ω with $p^\omega \leq p^n$ for each n. Say $p^n = \langle \rho^n_\mu \colon \mu < \alpha\rangle$; p^ω will be $\langle \rho^\omega_\mu \colon \mu < \alpha\rangle$. We shall define the ρ^ω_μ by recursion on μ, and check inductively that for each $\xi \leq \alpha$,

$$p^\omega \restriction \xi = \langle \rho^\omega_\mu \colon \mu < \xi\rangle \in \mathbb{P}_\xi \quad \text{and} \quad \forall n \in \omega\,(p^\omega \restriction \xi \leq p^n \restriction \xi). \tag{$*$}$$

We shall also have that

$$\mathrm{supt}(p^\omega) \subset \textstyle\bigcup_n \mathrm{supt}(p^n);$$

thus, if ξ is a limit and we have $(*)$ for all $\eta < \xi$, then $(*)$ easily follows for ξ.

Now, assume that $\xi < \alpha$ and we have defined $p^\omega \restriction \xi$ so that $(*)$ holds. We show how to define ρ^ω_ξ (and hence $p^\omega \restriction (\xi + 1)$). For each n, we have $p^{n+1} \restriction (\xi + 1) \leq p^n \restriction (\xi + 1)$, so by the definition of $\leq$ on $\mathbb{P}_{\xi+1}$, we have

$$p^{n+1} \restriction \xi \Vdash (\rho^n_\xi \in \pi_\xi \wedge \rho^{n+1}_\xi \leq \rho^n_\xi),$$

so, by (*) for ξ,

$$p^\omega \restriction \xi \Vdash (\rho_\xi^n \in \pi_\xi \wedge \rho_\xi^{n+1} \leq \rho_\xi^n).$$

Since π_ξ is full for $\rightsquigarrow$ ω-sequences, we may choose ρ_ξ^ω so that

$$p^\omega \restriction \xi \Vdash (\rho_\xi^\omega \in \pi_\xi) \quad \text{and} \quad p^\omega \restriction \xi \Vdash (\rho_\xi^\omega \leq \rho_\xi^n)$$

for each n. This choice makes (*) hold at $\xi + 1$. In addition, if $\xi \notin \bigcup_n \operatorname{supt}(p^n)$, so that each $\rho_\xi^n = \mathbb{1}$, then set $\rho_\xi^\omega = \mathbb{1}$ also, so that $\xi \notin \operatorname{supt}(p^\omega)$. □

Observe that for a $\mathbb{P}$-name π, being full for $\rightsquigarrow$ ω-sequences is a property of the name π, not of the p.o. named. If $\mathbb{1} \Vdash_{\mathbb{P}} (\pi$ is ω_1-closed), there is a canonical procedure for producing a $\mathbb{P}$-name π' such that $\mathbb{1} \Vdash (\pi = \pi')$ and π' is full for $\rightsquigarrow$ ω-sequences (see Exercise D1 and D2).

It might now be tempting to start with GCH in M, and perform a countable support iteration with ω_1-closed ω_2-c.c. p.o.'s, using names which are full for ω-sequences, thereby obtaining a model of CH + $2^{\omega_1} > \omega_2$ + MA_1, where MA_1 asserts that whenever $\mathbb{P}$ is an ω_1-closed ω_2-c.c. p.o. and $\mathscr{D}$ is a family of $<2^{\omega_1}$ dense sets in $\mathbb{P}$, there is a filter in $\mathbb{P}$ intersecting all the dense sets in $\mathscr{D}$. The details would be very much like those for the MA proof in §6. Unfortunately, however, it is *not* true that the ω_2-c.c. is preserved under such iterations (Exercise E6). In fact, although MA_1 follows easily from $\neg$CH or $2^{\omega_1} = \omega_2$ (Exercise J2), it is unknown whether CH + $2^{\omega_1} > \omega_2$ + MA_1 is consistent.[1] Certain weakened versions of MA_1, obtained by putting additional restrictions on $\mathbb{P}$, have been shown consistent by Baumgartner, Laver, and Shelah. In the absence of any definitive word on the consistency of the "right" axiom MA_1, we content ourselves here with showing how to prove the consistency of one of the consequences of MA_1. Other arguments of this type are given in the Exercises. All these arguments involve checking that the ω_2-c.c. is preserved when iterating with the specific p.o.'s needed.

Assume CH and $2^{\omega_1} > \omega_2$, and consider the possible cardinalities of a maximal almost disjoint family (m.a.d.f.) of subsets of ω_1 (see II 1.1). If $\lambda \leq \omega$, then any partition of ω_1 into λ uncountable pieces is a m.a.d.f. of size λ, and there is (by CH) a m.a.d.f. of size 2^{ω_1} (see II 1.3). There is no m.a.d.f. of size ω_1 (see II 1.2), so it remains to discuss m.a.d.f.'s of size λ where $\omega_1 < \lambda < 2^{\omega_1}$.

The consistency with CH + $2^{\omega_1} > \omega_2$ of the existence of a m.a.d.f. of size ω_2 is easily established by modifying the proof (in Theorem 2.3) of the consistency with $\neg$CH of a m.a.d.f. of ω_1 subsets of ω (see Exercise A8). Likewise, CH + MA_1 implies that there is no m.a.d.f. of subsets of ω_1 of size λ if $\omega_1 < \lambda < 2^{\omega_1}$; this is done by an obvious modification of the anal-

[1](Added in proof.) It is not, by a recent result of Stanley and Shelah.

ogous proof from MA (see II 2.16). Although we cannot settle the consistency of CH + MA_1 + $2^{\omega_1} > \omega_2$, we shall show, by iterating the almost disjoint sets partial order, that CH + $2^{\omega_1} > \omega_2$ is consistent with the non-existence of a m.a.d.f. of size strictly between ω_1 and 2^{ω_1}.

In analogy with II 2.7, we make the following definition.

7.3. DEFINITION. Let $\mathscr{A} \subset \mathscr{P}(\omega_1)$. The *almost disjoint sets* p.o. $\mathbb{Q}_{\mathscr{A}}$ is

$$\{\langle s, F\rangle \colon s \subset \omega_1 \wedge |s| < \omega_1 \wedge F \subset \mathscr{A} \wedge |F| < \omega_1\},$$

where $\langle s', F'\rangle \leq \langle s, F\rangle$ iff

$$s \subset s' \wedge F \subset F' \wedge \forall x \in F\,(x \cap s' \subset s).$$

$\mathbb{1} = \langle 0, 0\rangle$. □

7.4. LEMMA. *In* $\mathbb{Q}_{\mathscr{A}}$, $\langle s_1, F_1\rangle$ *and* $\langle s_2, F_2\rangle$ *are compatible iff*

$$\forall x \in F_1\,(x \cap s_2 \subset s_1) \wedge \forall x \in F_2\,(x \cap s_1 \subset s_2),$$

in which case $\langle s_1 \cup s_2, F_1 \cup F_2\rangle$ *is a common extension.* □

7.5. LEMMA. $\mathbb{Q}_{\mathscr{A}}$ *is* ω_1*-closed and, under* CH, *is* ω_2-c.c. □

7.6. LEMMA. *In* M, *assume that* $\mathscr{A} \subset \mathscr{P}(\omega_1)$, $\mathbb{Q} = \mathbb{Q}_{\mathscr{A}}$, *and*

$$\forall F \subset \mathscr{A}\,(|F| < \omega_1 \rightarrow |\omega_1 \setminus \bigcup F| = \omega_1). \qquad (*)$$

Let G *be* $\mathbb{Q}$*-generic over* M *and let*

$$d = \bigcup \{s \colon \exists F\,(\langle s, F\rangle \in G)\}.$$

Then in $M[G]$, $|d| = \omega_1$ *and* $\forall x \in \mathscr{A}\,(|d \cap x| < \omega_1)$.

PROOF. Since $\mathbb{Q}$ is ω_1-closed in M, $\omega_1^M = \omega_1^{M[G]}$. $(*)$ is used to show that for each $\delta < \omega_1^M$,

$$\{\langle s, F\rangle \colon \sup(s) > \delta\}$$

is dense, so that d is unbounded in ω_1^M. □

In particular, $(*)$ will hold if in M, $\mathscr{A}$ is an almost disjoint family of size $\geq \omega_1$, so such an $\mathscr{A}$ will fail to be a m.a.d.f. in $M[G]$. It thus looks hopeful that we can iterate forcing by p.o.'s like $\mathbb{Q}_{\mathscr{A}}$ to produce a model with no small m.a.d.f.'s. Since countable support iteration is sensitive to the particular names used for p.o.'s, we specify that next.

7.7. DEFINITION. Assume that $\mathbb{P} \in M$, ($\mathbb{P}$ is ω_1-closed)^M, and

$$\mathbb{1} \Vdash_{\mathbb{P}} (\tau \subset \mathscr{P}(\omega_1)).$$

The *standard* $\mathbb{P}$-name for $\mathbb{Q}_\tau$ is $\langle \pi, \leq_\pi, \mathbb{1}_\pi \rangle$, where

$$\pi = \{\langle \operatorname{op}(\check{s}, \sigma), \mathbb{1}_{\mathbb{P}} \rangle : s \subset \omega_1 \wedge |s| < \omega_1 \wedge \mathbb{1} \Vdash (\sigma \subset \tau \wedge |\sigma| < \omega_1) \wedge \sigma \text{ is a nice } \mathbb{P}\text{-name for a subset of } \tau\}.$$

$\mathbb{1}_\pi = \operatorname{op}(\check{0}, \check{0})$. □

Here, op is the invariant name for the ordered pair (see VII 2.16). We did not specify $\leq_\pi$ since its exact nature as a name is unimportant; $\leq_\pi$ can be any $\mathbb{P}$-name which is forced by $\mathbb{1}_{\mathbb{P}}$ to be the correct (as in Definition 7.3) partial order on $\mathbb{Q}_\tau$.

7.8. LEMMA. *In M, let $\mathbb{P}$, τ, and π be as in Definition 7.7. Let G be $\mathbb{P}$-generic over M, and let $\mathscr{A} = \tau_G$. Then in $M[G]$, $\pi_G = \mathbb{Q}_{\mathscr{A}}$.*

PROOF. Clearly $\pi_G \subset \mathbb{Q}_{\mathscr{A}}$. Let $\langle s, F \rangle \in \mathbb{Q}$. Since $\mathbb{P}$ is ω_1-closed in M, $s \in M$. Let σ be such that $F = \sigma_G$. It may not be the case that $\mathbb{1} \Vdash (\sigma \subset \tau \wedge |\sigma| < \omega_1)$, but by the maximal principle there is a σ' such that

$$\mathbb{1} \Vdash [[(\sigma \subset \tau \wedge |\sigma| < \omega_1) \rightarrow \sigma' = \sigma] \wedge [\neg(\sigma \subset \tau \wedge |\sigma| < \omega_1) \rightarrow \sigma' = 0]],$$

and then there is a nice $\mathbb{P}$-name σ'' for a subset of τ such that $\mathbb{1} \Vdash \sigma' = \sigma''$. Then $\langle s, F \rangle = \operatorname{val}(\operatorname{op}(\check{s}, \sigma''), G)$ and $\langle \operatorname{op}(s, \sigma''), \mathbb{1} \rangle \in \pi$. Thus, $\mathbb{Q}_{\mathscr{A}} \subset \pi_G$. □

7.9. LEMMA. *In M, let $\mathbb{P}$, τ, and π be as in Definition 7.7. Then π is full for $\searrow\omega$-sequences.*

PROOF. If for each n,

$$p \Vdash \langle \check{s}_{n+1}, \sigma_{n+1} \rangle \leq \langle \check{s}_n, \sigma_n \rangle,$$

then $s_n \subset s_{n+1}$ for each n. Let $s = \bigcup_n s_n$, and let σ be a nice $\mathbb{P}$-name such that $\mathbb{1} \Vdash \sigma = \bigcup_n \sigma_n$ (more formally, $\mathbb{1} \Vdash \sigma = \bigcup \rho$, where

$$\rho = \{\langle \sigma_n, \mathbb{1} \rangle : n \in \omega\}).$$

Then $p \Vdash \langle \check{s}, \sigma \rangle \leq \langle \check{s}_n, \sigma \rangle$ for each n. □

It might have seemed more natural, in the definition of π, to use arbitrary nice $\mathbb{P}$-names for countable subsets of ω_1, instead of just names of the form $\check{s}$. Indeed, Lemmas 7.8 and 7.9 would go through with this alternate definition, and we would not need to require $\mathbb{P}$ to be ω_1-closed in M. The restriction to names of the form $\check{s}$ is useful, however, in verifying that the ω_2-c.c. is preserved in transfinite iterations. It is true that the ω_2-c.c. would also be preserved if the alternate definition were used; but to show this, one first must prove that the standard iteration is densely embedded in the alternate iteration (Exercise D5).

7.10. LEMMA. *In M, assume CH, and let*

$$\langle\langle \mathbb{P}_\xi\colon \xi \le \alpha\rangle, \langle \pi_\xi\colon \xi < \alpha\rangle\rangle$$

be a countable support iterated forcing construction. For each ξ, let τ_ξ be a $\mathbb{P}_\xi$-name such that $\mathbb{1} \Vdash \tau_\xi \subset \mathscr{P}(\omega_1)$, and assume that π_ξ is the standard $\mathbb{P}_\xi$-name for $\mathbb{Q}_{\tau_\xi}$. Then $\mathbb{P}_\alpha$ has the ω_2-c.c.

PROOF. In M, suppose $p^\gamma \in \mathbb{P}_\alpha$ for $\gamma < \omega_2$. By CH, the Δ-system lemma implies that there is an $X \subset \omega_2$ with $|X| = \omega_2$ such that $\{\text{supt}(p^\gamma)\colon \gamma \in X\}$ forms a Δ-system with some root r. Let $p^\gamma = \langle \rho_\xi^\gamma\colon \xi < \alpha\rangle$, and let $\rho_\xi^\gamma = \text{op}(\check{s}_\xi^\gamma, \sigma_\xi^\gamma)$. By CH, there is a $Y \subset X$ with $|Y| = \omega_2$ such that for all $\xi \in r$, the s_ξ^γ for $\gamma \in Y$ are all the same; say $s_\xi^\gamma = s_\xi$ for $\xi \in r$ and $\gamma \in Y$. But then the p^γ for $\gamma \in Y$ are pairwise compatible; to see this, observe that if $\gamma, \delta \in Y$, then p^γ, p^δ have as a common extension $\langle \rho_\xi\colon \xi < \alpha\rangle$, where ρ_ξ is

(a) ρ_ξ^γ if $\xi \notin \text{supt}(p^\delta)$,
(b) ρ_ξ^δ if $\xi \notin \text{supt}(p^\gamma)$, and
(c) $\text{op}(\check{s}_\xi, \sigma_\xi)$ if $\xi \in r$,

where σ_ξ satisfies $\mathbb{1} \Vdash \sigma_\xi = \sigma_\xi^\gamma \cup \sigma_\xi^\delta$. □

It is now straightforward to obtain our result on m.a.d.f.'s. We state explicitly what we are proving consistent.

7.11. DEFINITION. P_1 is the statement that whenever $\mathscr{A} \subset \mathscr{P}(\omega_1)$, $|\mathscr{A}| < 2^{\omega_1}$, and

$$\forall F \subset \mathscr{A}\,(|F| < \omega_1 \rightarrow |\omega_1 \setminus \bigcup F| = \omega_1),$$

then

$$\exists d \subset \omega_1\,(|d| = \omega_1 \wedge \forall x \in \mathscr{A}\,(|d \cap x| < \omega_1)). \quad \square$$

7.12. LEMMA. *P_1 implies that if $\mathscr{A}$ is a* m.a.d.f. *and $|\mathscr{A}| \ge \omega_1$, then $|\mathscr{A}| = 2^{\omega_1}$.* □

7.13. THEOREM. *In M, assume* CH, *and assume that $\kappa > \omega_1$, κ is regular, and $2^{<\kappa} = \kappa$; then there is a $\mathbb{P} \in M$ such that $\mathbb{P}$ is ω_1-closed and ω_2-*c.c. *in M, and whenever G is $\mathbb{P}$-generic over M, $2^{\omega_1} = \kappa$ and P_1 hold in $M[G]$.*

PROOF. As in the MA construction of §6, build in M an iterated forcing construction

$$\langle\langle \mathbb{P}_\xi\colon \xi < \check{\kappa}\rangle, \langle \pi_\xi\colon \xi \le \kappa\rangle\rangle,$$

but now use countable supports. Given $\mathbb{P}_\xi$, list all names for relevant subsets of $\mathscr{P}(\omega_1)$; for example, let $\langle \sigma_\gamma^\xi\colon \gamma < \kappa\rangle$ enumerate all $\mathbb{P}_\xi$-names σ

such that for some $\lambda < \kappa$, σ is a nice name for a subset of $(\lambda \times \omega_1)\check{}$, and let τ_γ^ξ be such that

$$\mathbb{1} \Vdash \tau_\gamma^\xi = \{x \subset \omega_1 \colon \exists \mu < \check{\kappa}\,(x = \{\nu \colon \langle \mu, \nu \rangle \in \sigma_\gamma^\xi\})\}.$$

Let $f: \kappa \to \kappa \times \kappa$ be a bookkeeping function as in §6. If $f(\xi) = \langle \eta, \gamma \rangle$, let $\tau_\xi = i_{\eta\xi*}(\tau_\gamma^\eta)$ and let π_ξ be the standard $\mathbb{P}_\xi$-name for $\mathbb{Q}_{\tau_\xi}$. By Lemma 7.10, $\mathbb{P} = \mathbb{P}_\kappa$ has the ω_2-c.c. in M.

As in the MA proof, $\mathbb{P}$ preserves cardinals, and $2^{\omega_1} \leq \kappa$ in $M[G]$. Furthermore, in $M[G]$ the statement of P_1 holds for all $\mathscr{A} \subset \mathscr{P}(\omega_1)$ with $|\mathscr{A}| < \kappa$; since this implies $2^{\omega_1} \geq \kappa$, we have $2^{\omega_1} = \kappa$ and P_1. $\square$

A strengthening of P_1, known as SL_1 (Solovay's Lemma), says that if we also have $\mathscr{C} \subset \mathscr{P}(\omega_1)$, with $|\mathscr{C}| < 2^{\omega_1}$ and

$$\forall y \in \mathscr{C}\, \forall F \subset \mathscr{A}\, (|F| < \omega_1 \to |y \smallsetminus \bigcup F| = \omega_1),$$

then there is a $d \subset \omega_1$ such that $\forall x \in \mathscr{A}\,(|d \cap x| < \omega_1)$ and $\forall y \in \mathscr{C}\,(|d \cap y| = \omega_1)$. The analogue, SL_0, on ω, holds under MA (see II 2.15). One may verify directly that the $M[G]$ of Theorem 7.13 in fact satisfies SL_1. We do not bother to do this, however, since by a result of van Douwen $P_1 \leftrightarrow \mathrm{SL}_1$ under CH (see Exercise A2).

EXERCISES

A. Almost disjoint families, etc.

SL_θ (Solovay's Lemma) is the statement that whenever $\mathscr{A}, \mathscr{C} \subset \mathscr{P}(\theta)$ with $|\mathscr{A}|, |\mathscr{C}| < 2^\theta$ and

$$\forall y \in \mathscr{C}\, \forall F \subset \mathscr{A}\,(|F| < \theta \to |y \smallsetminus \bigcup F| = \theta),$$

then

$$\exists d \subset \theta\,(\forall x \in \mathscr{A}\,(|d \cap x| < \theta) \wedge \forall y \in \mathscr{C}\,(|d \cap y| = \theta)).$$

P_θ is the special case of SL_θ when $\mathscr{C} = \{\theta\}$. Then $\mathrm{MA} \to \mathrm{SL}_\omega$ (see II 2.15), and P_{ω_1} is consistent with $\mathrm{CH} \wedge 2^{\omega_1} > \omega_2$ (see Theorem 7.13).

(A1) Show that if θ is regular and $2^\theta = \theta^+$, then SL_θ holds.

(A2) (van Douwen) Show that if $2^{<\theta} = \theta$ and θ is regular, then $\mathrm{SL}_\theta \leftrightarrow P_\theta$. *Hint*. Let $S = \{s \subset \theta \colon |s| < \theta\}$. If $a \in \mathscr{A}$, let $a' = \{s \in S \colon s \cap a \neq 0\}$. If $c \in \mathscr{C}$ let $c' = \{s \in S \colon s \cap c = 0\}$. If $\alpha < \theta$, let $e'_\alpha = \{s \in S \colon s \cap \alpha \neq 0\}$. Apply P_θ on S to the a', c', and e'_α, to get a $d' \subset S$ with $|d'| = \theta$, and let $d = \bigcup d'$.

DF_θ (dominating functions) is the statement that whenever $\mathscr{F} \subset {}^\theta\theta$ with $|\mathscr{F}| < 2^\theta$, there is a $g \in {}^\theta\theta$ such that

$$\forall f \in \mathscr{F} \left(\left|\{\alpha < \theta\colon g(\alpha) \leq f(\alpha)\}\right| < \theta\right).$$

(A3) Show that if θ is regular, then $P_\theta \rightarrow \mathrm{DF}_\theta$. *Hint*. Apply P_θ on $\theta \times \theta$. $\mathscr{A}$ contains sets like $\{\langle\alpha, \beta\rangle\colon \beta \leq f(\alpha)\}$ and $\alpha \times \theta$. Observe that elements of $\mathscr{F}$ may be assumed to be increasing.

(A4) Show that $\mathrm{DF}_\omega \wedge \neg P_\omega$ is consistent. *Hint*. Let M satisfy $\mathrm{CH} \wedge 2^{\omega_1} = \omega_3$. Iterate c.c.c. forcing with finite supports ω_2 times. At each stage, add a function from ω to ω which almost dominates all functions in the previous model (using the p.o. of II Exercise 8). This produces a model of $\mathrm{DF}_\omega \wedge 2^\omega = \omega_2 \wedge 2^{\omega_1} = \omega_3$. But $P_\omega \rightarrow \mathrm{SL}_\omega \rightarrow \forall\kappa < 2^\omega\,(2^\kappa = 2^\omega)$; see II 2.18.

(A5) Generalize Exercise A4 to larger cardinals. Thus, assume that in M, θ is regular and $2^{<\theta} = \theta$. Find a θ-closed, θ^+-c.c. extension of M in which $\mathrm{DF}_\theta \wedge \neg P_\theta$ holds. *Hint*. First make $2^\theta = \theta^+$ and $2^{(\theta^+)} = \theta^{+++}$. Then iterate with $<\theta$ supports.

If $\mathscr{A} \subset \mathscr{P}(\theta)$, $\mathscr{A}$ is an independent family (i.f.) iff whenever $m, n \in \omega$ and $a_1, \ldots, a_m, b_1, \ldots, b_n$ are distinct members of $\mathscr{A}$,

$$|a_1 \cap \cdots \cap a_m \cap (\theta \smallsetminus b_1) \cap \cdots \cap (\theta \smallsetminus b_n)| = \theta.$$

$\mathscr{A}$ is a m.i.f. iff $\mathscr{A}$ is an i.f. and is maximal with respect to that property.

(A6) (Hausdorff) Show that there is an i.f. of size 2^θ. *Remark*. Unlike the corresponding statement for almost disjoint families, no assumption on cardinal arithmetic is needed (see Exercise B4). *Hint*. Let

$$I = \{\langle s, \mathscr{A}\rangle\colon s \subset \theta \wedge |s| < \omega \wedge \mathscr{A} \subset \mathscr{P}(s)\}.$$

For $X \subset \theta$, let $X' = \{\langle s, \mathscr{A}\rangle \in I\colon X \cap s \in \mathscr{A}\}$. Then $\{X'\colon X \in \mathscr{P}(\theta)\}$ is an independent family on I. *Remark*. For $\theta = \omega$ or 2^ω, this result is due to Fichtenholtz and Kantorovicz.

(A7) (R. Price) Assume P_θ, $2^{<\theta} = \theta$, and θ is regular. Show that every m.i.f. has size 2^θ. *Hint*. This is like Exercise A2. Now, for $a \in \mathscr{A}$, let

$$a' = \{s \in S\colon s \cap a = 0\}.$$

Pick $s_\beta \in d'$ $(\beta < \theta)$ so that the s_β are disjoint. Let $d = \bigcup\{s_{\beta+1}\colon \beta < \theta\}$. Then $\mathscr{A} \cup \{d\}$ is an i.f.

(A8) In M, assume θ is regular, $2^{<\theta} = \theta$, $2^\theta = \theta^+ < \kappa$, and $\kappa^\theta = \kappa$. Let

$M[G]$ be an extension via $\mathrm{Fn}(\kappa, 2, \theta)$. Show that in $M[G]$, $2^\theta = \kappa$, every m.i.f. has size κ, but there is a m.a.d.f. of size θ^+ (so that P_θ fails). *Hint.* See Theorems 2.3 and 2.6.

Let $\mathscr{U}$ be an ultrafilter on ω. $\mathscr{A} \subset \mathscr{U}$ *generates* $\mathscr{U}$ iff $\mathscr{U}$ is the unique ultrafilter extending $\mathscr{A}$. The *character* of $\mathscr{U}$, $\chi(\mathscr{U})$, is

$$\min(\{|\mathscr{A}|: \mathscr{A} \subset \mathscr{U} \wedge \mathscr{A} \text{ generates } \mathscr{U}\}).$$

(A9) Show that if $\mathscr{U}$ is non-principal, then $\omega_1 \leq \chi(\mathscr{U}) \leq 2^\omega$. Assuming P_ω, show that $\chi(\mathscr{U}) = 2^\omega$. *Hint.* If $\mathscr{A} \subset \mathscr{U}$, apply P_ω to $\{\omega \setminus a: a \in \mathscr{A}\}$.

(A10) Show that it is consistent with $2^\omega > \omega_1$ that there is a non-principal $\mathscr{U}$ with $\chi(\mathscr{U}) = \omega_1$. *Hint.* Let M satisfy $2^\omega > \omega_1$. Iterate c.c.c. forcing with finite supports ω_1 times to form M_α ($\alpha \leq \omega_1$). $\mathscr{U}_\alpha$ will be an ultrafilter on $\mathscr{P}(\omega)$ in M_α, and $\alpha < \beta \to \mathscr{U}_\alpha \subset \mathscr{U}_\beta$. $M_{\alpha+1}$ is obtained by adjoining to M_α an $a_\alpha \subset \omega$ with

$$\forall x \in \mathscr{U}_\alpha (|a_\alpha \setminus x| < \omega).$$

$\mathscr{U}_{\alpha+1} \in M_{\alpha+1}$ is any extension of $\mathscr{U}_\alpha \cup \{a_\alpha\}$. $\mathscr{U}_{\omega_1} \in M_{\omega_1}$ is generated by $\{a_\alpha: \alpha < \omega_1\}$.

(A11) (Pospíšil) Show that there is a $\mathscr{U}$ with $\chi(\mathscr{U}) = 2^\omega$. *Hint* (Hajnal–Juhász). Let $\mathscr{A}$ be an i.f. of size 2^ω. Let $\mathscr{U}$ be such that $\mathscr{A} \subset \mathscr{U}$ but

$$\forall \mathscr{B} \subset \mathscr{A} \, (|\mathscr{B}| \geq \omega \to (\bigcap \mathscr{B}) \notin \mathscr{U}).$$

*(A12) Assume that in M, I is infinite and $\mathscr{A} = \{a_i: i \in I\}$ is an i.f. Show that in a c.c.c. extension, one can adjoin a $d \subset \omega$ such that $\mathscr{A} \cup \{d\}$ is an i.f., but $\mathscr{A} \cup \{x, d\}$ is not an i.f. whenever $x \in \mathscr{P}(\omega) \cap M \setminus \mathscr{A}$. *Hint.* In M, let $\mathscr{B}$ be the regular open algebra of 2^I, and let $b_i \in \mathscr{B}$ be the basic clopen set, $\{f \in 2^I: f(i) = 1\}$. Still in M, let $h : \mathscr{P}(\omega) \to \mathscr{B}$ be an algebra homomorphism with $h(a_i) = b_i$. Adjoin d to be almost disjoint from every set in the kernel of h.

(A13) Show that it is consistent with $2^\omega > \omega_1$ that there is a m.i.f. of size ω_1. *Hint.* This is like Exercise A11, but now, along with the M_α one builds an i.f. of size ω_1. $M_{\alpha+1}$ is obtained from M_α by applying Exercise A12.

(A14) Show that in M there is a m.a.d.f. $\mathscr{A} \subset \mathscr{P}(\omega)$, such that whenever N is a model of ZFC with $M \subset N$ and $\mathscr{P}(\omega) \cap M \neq \mathscr{P}(\omega) \cap N$, $\mathscr{A}$ is not maximal in N. *Remark.* Hence, some care must be taken in choosing the $\mathscr{A}$ in Theorem 2.3. *Hint.* Obtain $\mathscr{A}$ by extending a perfect set of a.d. elements of $\mathscr{P}(\omega)$.

(A15) Let $\mathscr{A} \subset \mathscr{P}(\omega)$ be a m.i.f. Show that $|\mathscr{A}| \geq \omega_1$ and that MA implies $|\mathscr{A}| = 2^\omega$.

B. Applications of the Erdös–Rado Theorem

(B1) $[I]^2 = \{s \subset I: |s| = 2\}$. Suppose $|I| \geq (2^\kappa)^+$ and $f : [I]^2 \to \kappa$. Show that there is an $H \subset I$ such that $|H| \geq \kappa^+$ and f is constant on $[H]^2$. *Hint.* Suppose there is no such H. By closing under κ-ary functions, find a $J \subset I$ such that $|J| \leq 2^\kappa$ and

$$\forall X \subset J \, \forall i \in I \smallsetminus J \, [|X| \leq \kappa \to \exists j \in J \smallsetminus X \, \forall x \in X \, [f(\{x, j\}) = f(\{x, i\})]]. \tag{$*$}$$

Fix $i \in I \smallsetminus J$. For $\alpha < \kappa$, let $K_\alpha = \{x \in J: f(\{x, i\}) = \alpha\}$. Let M_α be any subset of K_α such that f has the constant value α on $[M_\alpha]^2$ and M_α is maximal with respect to that property. Let $X = \bigcup_\alpha M_\alpha$, and apply $(*)$ to obtain $j \in J \smallsetminus X$. If $f(\{i, j\}) = \alpha$, then $M_\alpha \cup \{j\}$ contradicts maximality of M_α. *Remark.* H is called *homogeneous* for f. This result is a special case of the Erdös–Rado Theorem. For more on such partition theorems, see [Erdös–Hajnal–Máté–Rado 1900] or [Kunen 1977]. The proof here is due to Simpson.

(B2) (Sierpiński) Show that in Exercise B1, if $|I|$ is only 2^κ then there is an $f: [I]^2 \to 2$ with no homogenous set of size κ^+. *Hint.* Let $I = {}^\kappa 2$, let W well-order I, and let $<$ be the lexicographic order on I. Define $f(\{i, j\}) = 1$ iff W and $<$ agree on $\{i, j\}$.

(B3) Show that in Exercise 1, if $|I|$ is only 2^κ, then there is an $f : [I]^2 \to \kappa$ with no homogeneous set of size 3. *Hint.* Let $I = {}^\kappa 2$. Let $f(\{i, j\})$ be the least α such that $i(\alpha) \neq j(\alpha)$.

(B4) (Baumgartner) Show that the existence of a family of 2^{ω_1} almost disjoint subsets of ω_1 cannot be proved in ZFC. *Remark.* Under CH there must be such a family (see II 1.3). *Hint.* Let M satisfy GCH and let $\mathbb{P}$ be c.c.c. in M. If p forces that $\tau : \omega_3 \to \mathscr{P}(\omega_1)$ enumerates an almost disjoint family, define in M an $f : [\omega_3]^2 \to \omega_1$ such that p forces $\tau(i) \cap \tau(j) \subset f(\{i, j\})$ whenever $\{i, j\} \in [\omega_3]^2$ (see Lemma 5.6). Applying Exercise B1 with $\kappa = \omega_1$ yields a contradiction.

(B5) Show that the existence of a family of 2^{ω_1} almost disjoint subsets of ω_1 can neither be proved nor refuted in ZFC + $2^\omega = 2^{\omega_1} = 2^{\omega_2} = \omega_3$, *Hint.* For one half of this, see Exercise B4. For the other half, let M satisfy CH plus $2^{\omega_1} = 2^{\omega_2} = \omega_3$, and force with $\mathrm{Fn}(\omega_3, 2)$.

(B6) Show that if $\mathbb{P}$ and $\mathbb{Q}$ are c.c.c. then $\mathbb{P} \times \mathbb{Q}$ is $(2^\omega)^+$-c.c. Likewise, show the product of two c.c.c. topological spaces has the $(2^\omega)^+$-c.c. *Hint.* If $I \subset \mathbb{P} \times \mathbb{Q}$ is an antichain, define $f : [I]^2 \to 2$ so that $f(\{\langle p, q \rangle, \langle p', q' \rangle\}) = 0$ iff $p \perp p'$. *Remark.* Can one replace $(2^\omega)^+$ by 2^ω? See Exercises C1, C6, and C8.

(B7) Let $\mathbb{P}_\alpha$ $(\alpha < \kappa)$ be p.o.'s. The *countable support product*, $\mathbb{Q}$, is the set of $q \in \prod_\alpha \mathbb{P}_\alpha$ such that $\{\alpha\colon q(\alpha) \neq \mathbb{1}\}$ is countable. If each $\mathbb{P}_\alpha$ has c.c.c., show that $\mathbb{Q}$ has the $(2^\omega)^+$-c.c. *Hint.* First apply the Δ-system lemma, and then the Erdös–Rado Theorem. *Remark.* If $\langle \mathbb{P}_n\colon n \leq \omega \rangle$ is a countable support iterated forcing construction (see §5) and each $\mathbb{P}_n$ is c.c.c., there is *no* bound on the chain condition on $\mathbb{P}_\omega$ (see Exercises E5, E6).

C. Chain conditions in products

(C1) Show that the following are equivalent:

(a) There are c.c.c. $\mathbb{P}$, $\mathbb{Q}$ such that $\mathbb{P} \times \mathbb{Q}$ is not κ-c.c.

(b) There are c.c.c. topological spaces X, Y such that $X \times Y$ is not κ-c.c.

(c) There are c.c.c. compact Hausdorff spaces X, Y such that $X \times Y$ is not κ-c.c.

Hint. See II §3. *Remark.* So MA + $\neg$CH implies that a product of two c.c.c. p.o.'s is c.c.c. (see II 2.24).

(C2) Let $\mathbb{P}$, $\mathbb{Q}$ be c.c.c. p.o.'s. Show that $\mathbb{P} \times \mathbb{Q}$ is c.c.c. iff $\mathbb{1}_{\mathbb{P}} \Vdash \check{\mathbb{Q}}$ is c.c.c. *Hint.* One direction is Lemma 5.7. For the other, if p forces that $\tau : \check{\omega}_1 \to \check{\mathbb{Q}}$ enumerates an antichain, let $p_\xi \leq p$ be such that $p_\xi \Vdash \tau(\check{\xi}) = \check{q}_\xi$. Then $\{\langle p_\xi, q_\xi \rangle\colon \xi < \omega_1\}$ is an antichain in $\mathbb{P} \times \mathbb{Q}$. *Remark.* Formally, the statement of this exercise should be relativized to M. However, it makes sense in $\mathbf{V}$, by replacing $\Vdash$ with $\Vdash^*$; see the discussion in VII §9 of Approach 2a.

(C3) Generalize Exercise C2 to show that in Lemma 5.7, $\mathbb{1} \Vdash (\pi$ is $\check{\kappa}$-c.c.) iff $\mathbb{P} * \pi$ is κ-c.c.

(C4) Obtain two different proofs that $\neg$SH implies that there are c.c.c. $\mathbb{P}$, $\mathbb{Q}$ with $\mathbb{P} \times \mathbb{Q}$ not c.c.c. *Hint 1.* Let $\mathbb{P} = \mathbb{Q}$ come from inverting an ω_1-Suslin tree T (see VII 8.4). Then $\mathbb{1} \Vdash (\check{T}$ is not Suslin), so $\mathbb{1} \Vdash \mathbb{P}$ is not c.c.c.), so apply Exercise C2.
Hint 2. Apply Exercise C1 and II 4.3.

(C5) Let $C \subset [I]^2$ (see Exercise B1). C is called a *graph* on I (elements of I are the "points"; $\{i, j\} \in C$ "means" that there is a "line" connecting i to j).

$\mathbb{P}(C)$, or $\mathbb{P}(C, I)$, is

$$\{s \subset I\colon [s]^2 \subset C \wedge |s| < \omega\}.$$

$\mathbb{P}(C)$ is ordered by reverse inclusion. Show that $\mathbb{P}(C, I) \times \mathbb{P}([I]^2 \setminus C, I)$ is not $|I|$-c.c. *Hint*. The $\langle\{i\}, \{i\}\rangle$ form an antichain.

(C6) (Fleissner) Let κ be an infinite cardinal of M, and let G be $\mathrm{Fn}(\kappa, 2)$-generic over M. Show that in $M[G]$, there are c.c.c. $\mathbb{P}$, $\mathbb{Q}$ with $\mathbb{P} \times \mathbb{Q}$ not κ-c.c. *Hint*. Instead, let G be $\mathrm{Fn}([\kappa]^2, 2)$-generic over M, and let $C = \{\{i, j\}\colon \bigcup G(\{i, j\}) = 0\}$. Suppose that p forces that $\tau : \check{\omega}_1 \to \mathbb{P}(C)$ enumerates an antichain. Let $p_\xi \leq p$, with $p_\xi \Vdash \tau(\check{\xi}) = \check{s}_\xi$. Assume (extending p_ξ if necessary), that $\mathrm{dom}(p_\xi) = [t_\xi]^2$, with $s_\xi \subset t_\xi$. Now, apply the Δ-system lemma to the t_ξ. *Remark*. This result was inspired by Exercise C7.

(C7) (Roitman) Let G be $\mathrm{Fn}(\omega, 2)$-generic over M. Show that in $M[G]$ there are c.c.c. $\mathbb{P}$, $\mathbb{Q}$ with $\mathbb{P} \times \mathbb{Q}$ not c.c.c. *Hint*. In M, fix $f_\xi : \xi \to \omega$ with f_ξ 1–1. In $M[G]$, define $C \subset [\omega_1]^2$ so that if $\eta < \xi$, then $\{\eta, \xi\} \in C$ iff $(\bigcup G)(f_\xi(\eta)) = 1$. *Remark*. This result was inspired by Exercise C8. It shows that MA is false in $M[G]$ unless CH holds in M.

*(C8) (Laver) Show that CH implies that there are c.c.c. $\mathbb{P}$, $\mathbb{Q}$ with $\mathbb{P} \times \mathbb{Q}$ not c.c.c. *Hint*. [Galvin 1900] Find a $C \subset [\omega_1]^2$ such that $\mathbb{P}(C)$ and $\mathbb{P}([\omega_1]^2 \setminus C)$ are both c.c.c.

(C9) Show that there are c.c.c. p.o.'s $\mathbb{Q}_m$ $(m \in \omega)$ such that for each $n \in \omega$, $\prod_{m<n} \mathbb{Q}_m$ has c.c.c., but $\prod_{m<\omega} \mathbb{Q}_m$ is not 2^ω-c.c. *Remark*. $\prod_{m<\omega} \mathbb{Q}_m$ is $(2^\omega)^+$-c.c. by Exercise B7. Also, the finite support product (which corresponds to the Tychonov product in topology) is c.c.c. by a Δ-system argument; see II 1.9. *Hint*. Each $\mathbb{Q}_m$ is $\mathrm{Fn}(1, 2)$.

(C10) Assume M satisfies CH, and let κ be an infinite cardinal of M. Let G be $\mathrm{Fn}(\kappa, 2, \omega_1)$-generic over M. Show that in $M[G]$ there are ω_1-closed p.o.'s, $\mathbb{Q}_m$ $(m < \omega)$ such that $\prod_{m<n} \mathbb{Q}_m$ has the ω_2-c.c. for each $n < \omega$ but $\prod_{m<\omega} \mathbb{Q}_m$ is not κ-c.c. *Hint*. Instead, let G be $\mathrm{Fn}([\kappa]^2, \omega, \omega_1)$-generic. Let $C_n = \{\{i, j\}\colon \bigcup G(\{i, j\}) \neq n\}$. Let $\mathbb{Q}_n = \{s \subset \kappa\colon [s]^2 \subset C_n \wedge |s| \leq \omega\}$. Now argue as in Exercise C6.

D. Full names

(D1) Let $\langle \pi, \leq_\pi, \mathbb{1}_\pi \rangle$ be a $\mathbb{P}$-name for a p.o. π is called *full* iff $\pi = \mathrm{dom}(\pi) \times \{\mathbb{1}_{\mathbb{P}}\}$ and whenever $p \in \mathbb{P}$, $\tau \in M^{\mathbb{P}}$, and $p \Vdash (\tau \in \pi)$, there is a $\sigma \in \mathrm{dom}(\pi)$ such that $p \Vdash (\tau = \sigma)$. Show that if $\langle \pi, \leq_\pi, \mathbb{1}_\pi \rangle$ is any $\mathbb{P}$-name for a p.o., there is a full $\mathbb{P}$-name $\langle \pi', \leq_\pi, \mathbb{1}_\pi \rangle$ such that $\mathbb{1} \Vdash (\pi = \pi')$. *Hint.* $\mathrm{dom}(\pi')$ is a suitably large subset of $\{\tau \in M^{\mathbb{P}} : \mathbb{1} \Vdash (\tau \in \pi)\}$. $\leq_\pi$ and $\mathbb{1}_\pi$ are unchanged and have nothing to do with this exercise; the same works for any $\mathbb{P}$-name π such that $\mathbb{1} \Vdash (\pi \neq 0)$. *Remark.* Iterating forcing with full names simplifies certain constructions. For example, $\mathbb{P} * \pi$ is, as a set, $\mathbb{P} \times \mathrm{dom}(\pi)$ if π is full.

(D2) Let $\langle \pi, \leq_\pi, \mathbb{1}_\pi \rangle$ be a $\mathbb{P}$-name for a p.o. Show that if π is full and $\mathbb{1} \Vdash (\pi$ is ω_1-closed), then π is full for $\rightsquigarrow \omega$-sequences (Definition 7.1).

*(D3) (Laver, Miller) Let

$$\langle \langle \mathbb{P}_\xi : \xi \leq \alpha \rangle, \langle \pi_\xi : \xi < \alpha \rangle \rangle$$

be a countable support iteration, and suppose that for each ξ, $\mathbb{1} \Vdash (\pi_\xi$ is c.c.c.), and π_ξ is a full name. Show that $\mathbb{P}_\alpha$ does not collapse ω_1. *Hint.* Assume this holds for all $\beta < \alpha$. If $s \subset \alpha$, and $p, p' \in \mathbb{P}_\alpha$, then $p' \leq_s p$ means that $p' \leq p$ and for all $\mu \in s$, $\rho'_\mu = \rho_\mu$. By induction on $|s|$, show that for all finite s, if $p \Vdash \tau \in \omega_1$, then there is a $\beta < \omega_1$, and a $p' \leq_s p$ such that $p' \Vdash \tau \in \beta$. If

$$p_0 \geq_{s_0} p_1 \geq_{s_1} p_2 \geq \cdots,$$

$$s_0 \subseteq s_1 \subseteq s_2 \subseteq, \ldots,$$

and $\bigcup_n s_n = \bigcup_n \mathrm{supt}(p_n)$, then there is a p_ω such that $\forall n\, (p_\omega \leq p_n)$. *Remark.* By Exercise E4, the assumption that the π_ξ are full cannot be dropped.

(D4) In Exercise D3, suppose that in M, α is strongly inaccessible and each π_ξ is a full $\mathbb{P}_\xi$ name for $\mathrm{Fn}(\omega, 2)$. Let G be $\mathbb{P}_\alpha$-generic over M. Show that in $M[G]$, $\omega_2 = \alpha = 2^\omega$. *Hint.* Each π_ξ "adds" an $f_\xi : \omega \to 2$, so $2^\omega \geq \alpha$. Let $f(\xi, n) = f_\xi(n)$. Fix $\eta < \alpha$. If $\xi < \eta$, then there is a $\mu < \omega_1$ such that

$$\forall n \in \omega \left(f_\xi(n) = f(\eta + \mu + n, 0) \right)$$

(since the π_μ are full, so there is a $\rho \in \mathrm{dom}(\pi_\mu)$ such that $\mathbb{1}$ forces $\rho(0)$ to be $f_\xi(n)$); thus, sending ξ into the least such μ defined a 1–1 map from η into ω_1.

(D5) In Lemma 7.10, show that the same extension would be obtained if full names are used.

E. Some pathologies

In this group,

$$\langle \langle \mathbb{P}_\xi : \xi \leq \omega \rangle, \langle \pi_\xi : \xi < \omega \rangle \rangle$$

denotes an iterated forcing construction in M.

(E1) Assume that $\mathbb{P}_\omega$ is defined by finite supports and for each $n < \omega$, $\mathbb{1} \Vdash_{\mathbb{P}_n} (\pi_n \text{ is not c.c.c.})$. Show that $\mathbb{P}_\omega$ collapses ω_1. *Hint*. Let

$$\mathbb{1} \Vdash (\sigma_n \text{ is a maximal antichain in } \pi_n \text{ and } \tau_n \text{ maps } \sigma_n \text{ onto } \omega_1).$$

If G is $\mathbb{P}_\omega$-generic over M, let G_n and H_n be as in Lemma 5.13. In $M[G]$, let $H_n \cap \mathrm{val}(\sigma_n, G_n) = \{q_n\}$, and $f(n) = (\mathrm{val}(\tau_n, G_n))(q_n)$. Then f maps ω onto ω_1^M.

(E2) Find an example where $\mathbb{P}_\omega$ is determined by countable supports, and for each n, $\mathbb{1} \Vdash (\pi_n \text{ is } \omega_1\text{-closed})$, but $\mathbb{P}_\omega$ collapses ω_1. *Hint*. π_n will be forced by $\mathbb{1}$ to be dense in $\mathrm{Fn}(\omega_1, \omega, \omega_1)$, and

$$\mathrm{dom}(\pi_n) \subset \{\check{s}\colon s \in \mathrm{Fn}(\omega_1, \omega, \omega_1) \wedge \mathrm{dom}(s) \in \omega_1\}.$$

Think of π_n as adding $f_n : \omega_1 \to \omega$. If $0 < \mathrm{dom}(s) = \gamma + k$, with γ a limit ordinal, then s will be in $\mathrm{val}(\pi_{n+1}, G_{n+1})$ iff $f_n(\gamma + \omega) = k$; but $0(=\mathbb{1})$ is always in $\mathrm{val}(\pi_{n+1}, G_{n+1})$. Then the iteration is essentially finite support, so Exercise E1 applies. *Remark*. By Lemma 7.2, this cannot happen if full names are used.

(E3) If for each n, $\mathbb{1} \Vdash (\pi_n \text{ is } \omega_1\text{-closed})$, then each $\mathbb{P}_n$ is ω_1-closed.

(E4) Let $\mathbb{P}_\omega$ be defined by countable supports, and let each π_n be $(\mathrm{Fn}(\omega, 2))\check{}$. Show that $\mathbb{P}_\omega$ collapses ω_1. *Remark*. By Exercise D3, this cannot happen if full names are used. *Hint*. $\mathbb{P}_\omega$ is isomorphic (in M) to $(\mathrm{Fn}(\omega, 2))^\omega$. Think of $\mathbb{P}_\omega$ as adding $f : \omega \times \omega \to 2$. Any element of $({}^\omega 2)^M$ can be coded by (or, is recursive in) f.

(E5) Let κ be any cardinal of M. Find an example where $\mathbb{P}_\omega$ is defined using countable supports, and for each $n < \omega$, $\mathbb{P}_n$ has the c.c.c. but $\mathbb{P}_\omega$ does not have the κ-c.c. *Hint*. See Exercise C9. π_0 is a name for $\mathrm{Fn}(\kappa, 2)$, and π_{n+1} is a name for $\mathrm{Fn}(1, 2)$. *Remark*. Unlike the pathologies in Exercises E2 and E4, this one will not go away if full names are used.

(E6) Assume CH in M. Modify E5 to make each $\mathbb{P}_n$ ω_1-closed and ω_2-c.c. *Hint*. See Exercise C10.

F. Kurepa trees

(F1) Assume $\theta < \kappa$, θ is regular, and κ is strongly inaccessible. Let $\mathrm{Lv}(\kappa, \theta)$ be the p.o. which adds a map from θ onto each ordinal less than κ using partial functions of size $<\theta$. Thus, $\mathrm{Lv}(\kappa, \omega_1)$ is $\mathrm{Lv}'(\kappa)$ (Definition 3.1). Show that $\mathrm{Lv}(\kappa, \theta)$ is θ-closed and κ-c.c.

(F2) Assume that in M, $\theta \neq \omega$ and θ is not strongly inaccessible. Let G be $\mathrm{Lv}(\kappa, \theta)^M$-generic over M. Show that in $M[G]$, $\neg\theta - \mathrm{KH}$ (i.e., there are no θ-Kurepa trees).

(F3) Assume that in M, $\kappa_1 < \kappa_2 < \kappa_3 < \cdots$, where $\kappa_1 = \omega_1$ and κ_n is strongly inaccessible for $n > 1$; let $\mathbb{P} = \prod_n \mathrm{Lv}(\kappa_{n+1}, \kappa_n)$. Let G be $\mathbb{P}$-generic over M. Show that in $M[G]$, $\kappa_n = \omega_n$, $\neg\omega_1$-KH, but ω_n-KH for $n > 1$. *Hint.* The complete binary κ_n-tree of M is a Kurepa tree in $M[G]$.

(F4) In Exercise F3, define $\mathbb{P}$ instead by iterating ω times with countable supports. Now, π_n is a name for the $\mathrm{Lv}(\kappa_{n+1}, \kappa_n)$ in the $\mathbb{P}_n$-extension. Let G be $\mathbb{P}$-generic over M. Show that in $M[G]$, $\kappa_n = \omega_n$ and $\neg\omega_n$-KH for $n \geq 1$.

(F5) Show the consistency of GCH plus

$$\forall n \geq 1\,(\omega_n\text{-KH} \leftrightarrow n \text{ is even}),$$

assuming the consistency of infinitely many inaccessible cardinals. *Hint.* It is easy to add Kurepa trees to a model (VII Exercise H19). *Remark.* The inaccessibles are necessary (VII Exercise B9).

(F6) Show that it is consistent with CH $\wedge$ $2^\omega = \omega_3$ that there is an ω_1-Kurepa tree with exactly ω_2 paths.

G. Proper class forcing

A *class* of M is an element of $\mathscr{D}(M)$ (VI Definition 1.1). If $\mathbb{P}$ and $\leq$ are classes of M, a filter G is generic iff $G \cap D \neq 0$ whenever D is dense and is a class of M. $M^{\mathbb{P}}$ and $M[G]$ are defined verbatim as in VII 2.6 and 2.8; so every $\mathbb{P}$-name of M is still an element of M.

(G1) Verify that the results of VII §§2, 3 still hold for proper class forcing.

(G2) Show that $M[G]$ may fail to satisfy ZF. *Hint.* Try $\mathrm{Fn}(\omega, \mathrm{o}(M))$.

(G3) Let E be a class of M such that (E is an index function)^M in the sense of Definition 4.1. Modify 4.1(3)(a) so that $\mathbb{P}(E)$ is a class of M (if $p \in \mathbb{P}(E)$, then $p \in M$ and $\mathrm{dom}(p) \subset \mathrm{dom}(E)$; $p(\kappa)$ is assigned the value 0 if $\kappa \notin \mathrm{dom}(p)$). Let G be $\mathbb{P}$-generic over M. Show that $M[G]$ satisfies ZFC. *Hint.* This involves identifying $\mathbb{P}(E)$ with $\mathbb{P}(E_\lambda^+) \times \mathbb{P}(E_\lambda^-)$ for various λ.

(G4) Show the consistency of $\forall\kappa\,(\kappa \text{ regular} \rightarrow 2^\kappa = \kappa^{+++})$.

(G5) Use forcing to prove

$$\text{Con(ZFC)} \to \text{Con(ZFC + GCH)}.$$

Hint. Recall that forcing with $\text{Fn}(\omega_1, 2, \omega_1)$ makes CH true (VII Exercise G4).

H. Exponentiation of singular cardinals

Let λ be singular. $f(\lambda)$, (CL), and $(*)$ are defined after Corollary 4.8.

(H1) Show that either $\text{cf}(f(\lambda)) = \text{cf}(\lambda)$, or $\text{cf}(f(\lambda)) > \lambda$. If $\text{cf}(f(\lambda)) > \lambda$, show that $2^\lambda = f(\lambda)$.

(H2) Show that (CL) $\to (*)$.

For the rest of this section, we assume throughout that $\lambda = \omega_{\omega_1}$ and $\forall \kappa < \lambda\, (2^\kappa = \kappa^+)$, and we derive the result, due to Silver, that $2^\lambda = \lambda^+$. The proof here is due to Baumgartner, Jensen, and Prikry. See [Galvin–Hajnal 1975] for more on this. Exercise H3 is a generalization of II 1.3, and Exercises H4 and H5 are from [Erdös–Hajnal–Milner 1968].

An *eventually different* (e.d.) family is an $\mathscr{F} \subset \lambda^{\omega_1}$ such that

$$\forall f, g \in \mathscr{F}\, (f \neq g \to |\{\alpha: f(\alpha) = g(\alpha)\}| \leq \omega).$$

(H3) Show that there is an e.d. $\mathscr{F}$ such that $|\mathscr{F}| = 2^\lambda$ and

$$\forall f \in \mathscr{F}\; \forall \alpha < \omega_1\, (f(\alpha) < \omega_{\alpha+1}).$$

Hint. If $X \subset \lambda$, define $f_X(\alpha) = X \cap \omega_\alpha \in \mathscr{P}(\omega_\alpha)$.

(H4) Show that if $\mathscr{F}$ is e.d. and

$$\forall f \in \mathscr{F}\, [\{\alpha: f(\alpha) < \omega_\alpha\} \text{ is stationary}],$$

then $|\mathscr{F}| \leq \lambda$. *Hint.* Use the pressing-down lemma.

(H5) Show that if $\mathscr{F}$ is e.d. and $g: \omega_1 \to \omega_{\omega_1}$ is such that $\forall \alpha\, (g(\alpha) < \omega_{\alpha+1})$, and

$$\forall f \in \mathscr{F}\, [\{\alpha: f(\alpha) < g(\alpha)\} \text{ is stationary}],$$

then $|\mathscr{F}| \leq \lambda$.

(H6) Show that $2^\lambda = \lambda^+$. *Hint.* Partially order the $\mathscr{F}$ of Exercise H3 by setting

$$f < g \quad \text{iff} \quad \{\alpha: g(\alpha) \leq f(\alpha)\} \text{ is non-stationary},$$

and let $\lhd$ be any total order of $\mathscr{F}$ extending $<$. By Exercise H5, every element of $\mathscr{F}$ has $\leq\lambda$ predecessors under $\lhd$.

I. Weakly compact cardinals

The exercises in this section (except for I8) assume familiarity with the characterization that κ is weakly compact (w.c.) iff κ is strongly inaccessible and Π_1^1 indescribable; see [Barwise 1975] or [Drake 1974]. Thus, if κ is w.c., $A \subset R(\kappa)$, ϕ is a Π_1^1 sentence, and

$$\langle R(\kappa), \in, A\rangle \models \phi,$$

then there is an inaccessible $\lambda < \kappa$ such that

$$\langle R(\lambda), \in, A \cap R(\lambda)\rangle \models \phi.$$

(I1) Show that if κ is w.c., then $(\kappa \text{ is w.c.})^{\mathbf{L}}$. *Hint.* If ϕ is Π_1^1, then there is a Π_1^1 ψ such that

$$R(\kappa) \models \psi \leftrightarrow (R(\kappa) \models \phi)^{\mathbf{L}}.$$

ψ uses its unbounded universal quantifiers to quantify over codes for the $L(\alpha)$ for $\kappa < \alpha < \kappa^+$. *Remark.* Hence, $\text{Con}(\text{ZFC} + \exists\kappa\,(\kappa \text{ w.c.}))$ implies $\text{Con}(\text{ZFC} + \text{GCH} + \exists\kappa\,(\kappa \text{ w.c.}))$. This may also be proved by forcing (Exercise I3).

(I2) In M, let κ be w.c. and E an index function with $\kappa \notin \text{dom}(E)$ and, for all $\alpha \in \kappa \cap \text{dom}(E)$, α is a successor cardinal and $|E(\alpha)| < \kappa$. Let G be $\mathbb{P}(E)$-generic over M. Show that κ remains w.c. in $M[G]$. *Remark.* This may fail if $\kappa \in \text{dom}(E)$; see Exercise I5. *Hint.* Assume $\text{dom}(E) \subset \kappa$. Let ϕ be Π_1^1. If $\mathbb{1} \Vdash_{\mathbb{P}(E)} (R(\kappa) \models \phi)$, then by Π_1^1 indescribability in M, there is an inaccessible $\lambda < \kappa$ such that $\mathbb{1} \Vdash_{\mathbb{P}(E\restriction\lambda)} (R(\lambda) \models \phi)$. Since $\lambda \notin \text{dom}(E)$, this implies $\mathbb{1} \Vdash_{\mathbb{P}(E)} (R(\lambda) \models \phi)$.

(I3) Show $\text{Con}(\text{ZFC} + \exists\kappa\,(\kappa \text{ w.c.}))$ implies $\text{Con}(\text{ZFC} + \text{GCH} + \exists\kappa\,(\kappa \text{ w.c.}))$ by using forcing. *Hint.* Combine Exercises I2 and G5.

(I4) In M, let κ be w.c. and $|\mathbb{P}| < \kappa$. Let G be $\mathbb{P}$-generic over M. Show that κ is w.c. in $M[G]$. *Remark.* Thus, GCH may fail in various ways below κ.

(I5) Let κ be w.c., $X \subset \kappa$, and assume that $\forall\alpha < \kappa\,(X \cap \alpha \in \mathbf{L})$. Show that $X \in \mathbf{L}$. *Hint.* This is like Exercise I1. *Remark.* Hence, if $\mathbf{V} = \mathbf{L}$ in the ground model, forcing with $\text{Fn}(I, 2, \kappa)$ destroys weak compactness, although it preserves Mahloness (VII Exercise H4). Thus, to show the consistency of $\exists\kappa\,(\kappa \text{ w.c.} \wedge 2^\kappa > \kappa^+)$, a new idea is needed (see Exercise I10).

(I6) In M, assume κ is regular, $\kappa > \omega$, and $\mathbb{P}$ is κ-closed. Let ϕ be Π^1_1, and $A \subset \kappa$. Let G be $\mathbb{P}$-generic over M. Show that

$$(\langle \kappa, <, A\rangle \models \phi)^M \quad \text{iff} \quad (\langle \kappa, <, A\rangle \models \phi)^{M[G]}.$$

Hint. For the hard direction, say ψ is first-order and

$$\mathbb{1} \Vdash (\tau \subset \check{\kappa} \wedge \langle \check{\kappa}, <, \check{A}, \tau\rangle \models \psi).$$

In M, inductively find $p_\xi \in \mathbb{P}$, $\gamma_\xi < \kappa$, and $t_\xi \subset \gamma_\xi$ such that

$$\xi < \eta \rightarrow (p_\eta \leq p_\xi \wedge \gamma_\xi < \gamma_\eta \wedge t_\xi = t_\eta \cap \gamma_\xi)$$

and

$$p_\xi \Vdash (\tau \cap \check{\gamma}_\xi = \check{t}_\xi \wedge \langle \check{\gamma}_\xi, <, \check{A} \cap \check{\gamma}_\xi, \check{t}_\xi\rangle \prec \langle \check{\kappa}, <, \check{A}, \tau\rangle).$$

Let $T = \bigcup_\xi t_\xi$. Then in M,

$$\langle \kappa, <, A, T\rangle \models \psi.$$

Remark 1. When $\kappa = \omega$, this is still true by the Kleene analysis of Π^1_1 properties, and in fact, by the Shoenfield Absoluteness Theorem (see [Barwise 1975]), even holds when ϕ is Π^1_2.
Remark 2. When κ is strongly inaccessible in M, we may replace $\langle \kappa, <\rangle$ by $\langle R(\kappa), \in\rangle$.
Remark 3. As a special case of this exercise, if T is a κ-Suslin tree in M, T remains Suslin in $M[G]$.

(I7) Use Exercise I6 to show, using standard Easton forcing, the consistency of a weakly compact κ such that $2^\lambda > \lambda^+$ for all regular $\lambda < \kappa$. *Hint.* Exercise I6 removes the requirement in Exercise I2 that elements of $\kappa \cap \mathrm{dom}(E)$ are successor cardinals

(I8) In M, let E be an index function, with $\mathrm{dom}(E) = \{\lambda_\xi : \xi < \alpha\}$ in increasing enumeration. The reverse Easton extension is obtained by letting π_ξ be a full $\mathbb{P}_\xi$-name for the $\mathrm{Fn}(E(\lambda_\xi), 2, \lambda_\xi)$ of the $\mathbb{P}_\xi$-extension; the ideal of supports is, as with standard Easton forcing

$$\{X \subset \alpha : \forall\theta(\theta \text{ regular} \rightarrow |\theta \cap \{\lambda_\xi : \xi \in X\}| < \theta)\}.$$

Give an example where M satisfies GCH but $\mathbb{P}_\alpha$ collapses ω_2. Show that if each λ_ξ is strongly inaccessible in M and $|E(\lambda_\xi)| < \lambda_{\xi+1}$, then $\mathbb{P}_\alpha$ preserves cardinals.

(I9) (Silver) In Exercise I8, assume that in M, κ is w.c., $\mathrm{dom}(E)$ is the set of strong inaccessibles $\leq\kappa$, and each $E(\lambda) = \lambda$. Let G be $\mathbb{P}_\kappa$-generic over M. Show that κ remains w.c. in $M[G]$. *Hint.* Let Γ_λ be the name for the generic function from $\lambda \to 2$. Suppose ϕ is Π^1_1 and

$$\mathbb{1} \Vdash (\langle R(\kappa), \in \Gamma_\kappa \rangle \models \phi);$$

then there is an inaccessible $\lambda < \kappa$ such that

$$\mathbb{1} \Vdash (\langle R(\lambda), \in, \Gamma_\lambda \rangle \models \phi).$$

But in *reverse* Easton forcing, there is a p such that $p \Vdash \Gamma_\kappa \restriction \lambda = \Gamma_\lambda$, so

$$p \Vdash (\langle R(\lambda), \in, \Gamma_\kappa \cap R(\lambda) \rangle \models \phi).$$

(I10) (Silver) Show

$$\operatorname{Con}(\text{ZFC} + \exists\kappa\,(\kappa \text{ is w.c.})) \to \operatorname{Con}(\text{ZFC} + \exists\kappa\,(\kappa \text{ is w.c.} \wedge 2^\kappa > \kappa^+)). \quad (*)$$

Hint. Modify Exercise I9 by setting $E(\lambda) = \lambda$ for $\lambda < \kappa$, but $E(\kappa) = \kappa^{++}$. Exercise I9 still applies, since any subset of κ in $M[G]$ lies in an extension of the form of Exercise I9. *Remark.* Silver also established $(*)$ with "supercompact" replacing "w.c."; see [Menas 1973]. However, ZFC plus the existence of a measurable cardinal κ with $2^\kappa > \kappa^+$ is consistency-wise stronger than ZFC plus the existence of two (or many) measurable cardinals [Kunen 1971].

J. Miscellaneous results

(J1) Suppose $\mathbb{P} \in M$ and $\mathbb{P}$ is non-atomic, and let $G \subset \mathbb{P}$. Show that $G \times G$ is not $\mathbb{P} \times \mathbb{P}$-generic over M.

(J2) Show that MA_1 (see the discussion after Lemma 7.2) follows from $\neg\text{CH} \vee 2^{\omega_1} = \omega_2$. *Hint.* Under $\neg$CH, every relevant $\mathbb{P}$ has an atom.

(J3) Let

$$\langle\langle \mathbb{P}_\xi \colon \xi \le \alpha \rangle, \langle \pi_\xi \colon \xi < \alpha \rangle\rangle$$

be a finite support iterated forcing construction in M, and let G be $\mathbb{P}_\alpha$-generic over M. Assume that $\alpha \ge \omega$ and for each ξ, $\mathbb{1} \Vdash (\pi_\xi$ is non-atomic). Show that $M[G]$ contains an H which is $\operatorname{Fn}(\omega, 2)$-generic over M. *Hint.* If the π_n are full names, let $\tau_n, \sigma_n \in \operatorname{dom}(\pi_n)$ with

$$\mathbb{1} \Vdash (\tau_n \in \pi_n \wedge \sigma_n \in \pi_n \wedge \tau_n \perp \sigma_n),$$

and let $p \in H$ iff for all $n \in \operatorname{dom}(p)$, $\mathbb{1}^\frown\langle \tau_{nG} \rangle \in G_{n+1}$.

(J4) Let λ be strongly inaccessible, and let E be an index function (Definition 4.1) such that $\operatorname{dom}(E)$ is an unbounded subset of λ and each $E(\kappa) \ne 0$.

Show that $\mathbb{P}(E)$ has the λ-c.c. iff λ is Mahlo. *Hint.* If λ is Mahlo and for $\xi < \lambda$, $a_\xi \subset \lambda$ and $|a_\xi \cap \theta| < \theta$ for all regular $\theta \leq \lambda$, then for some $X \subset \lambda$, $|X| = \lambda$ and $\{a_\xi : \xi \in X\}$ forms a Δ-system. To show this, let $f(\xi) = \sup(\xi \cap a_\xi)$, and apply the pressing-down lemma on $\{\xi < \lambda : \xi \text{ is regular}\}$.

(J5) Show the consistency with GCH of $\diamondsuit \wedge \neg \diamondsuit^+$. *Hint.* Use $\mathrm{Fn}(\omega_2, 2, \omega_1)$. For $\diamondsuit$, see VII 8.3. For $\neg \diamondsuit^+$ (and in fact $\neg \diamondsuit^*$), observe that adding a generic map from ω_1 into 2 destroys a $\diamondsuit^*$-sequence.

(J6) Assume that $\mathbb{P}$ is countably closed in M, and that $M[G]$ contains a subset of ω_1 which is not in M. Show that no sequence $\langle \mathscr{A}_\alpha : \alpha < \omega_1 \rangle \in M$ is a $\diamondsuit^*$-sequence in $M[G]$. Use this to show that $\neg \mathrm{KH} \wedge \diamondsuit \wedge \neg \diamondsuit^*$ holds in the model constructed in §3. *Hint* (for the first part). This is like 3.4.

(J7) Consider the question, "Does $\diamondsuit^+$ hold in the extension of M by $\mathrm{Fn}(\omega_1, 2, \omega_1)^M$?" Show that the answer is "yes" for some and "no" for other models of ZFC + GCH.

(J8) Let $\mathbb{P}_\xi$ ($\xi \leq \omega_1$) be p.o.'s such that $\xi < \eta \rightarrow \mathbb{P}_\xi \subset_c \mathbb{P}_\eta$, and assume that $\mathbb{P}_\xi$ is c.c.c. for $\xi < \omega_1$. Assume also that for *each* limit $\eta \leq \omega_1$ $\bigcup_{\xi<\eta} \mathbb{P}_\xi$ is dense in $\mathbb{P}_\eta$. Show that $\mathbb{P}_{\omega_1}$ is c.c.c.

(J9) Show that Exercise J8 can fail if we assume only that $\bigcup_{\xi<\omega_1} \mathbb{P}_\xi$ is dense in $\mathbb{P}_{\omega_1}$.

(J10) In Lemma 5.13, show that for each ξ,

$$M[G_\xi] = M[\langle H_\mu : \mu < \xi \rangle];$$

i.e., $M[G_\xi]$ is the least c.t.m. N such that $M \subset N$ and $\langle H_\mu : \mu < \xi \rangle \in N$.

(J11) [McAloon 1971] Show that $\mathrm{ZFC} + \mathrm{GCH} + \mathbf{V} = \mathbf{OD} + \mathbf{V} \neq \mathbf{L}$ is consistent. *Hint.* In M, assume $\mathbf{V} = \mathbf{L}$, and fix κ with $\kappa = \omega_\kappa$. For some $E \subset \kappa$, $M[G]$ will be $M[\langle f_\alpha : \alpha \in E \rangle]$, where $f_\alpha : \omega_{\alpha+1} \rightarrow 2$ will be $\mathrm{Fn}(\omega_{\alpha+1}, 2, \omega_{\alpha+1})$-generic over M. E will not be in M, but E will be ordinal-definable in $M[G]$, since in $M[G]$,

$$\alpha \in E \leftrightarrow \exists h\, [h : \omega_{\alpha+1} \rightarrow 2 \wedge h \notin \mathbf{L} \wedge \forall \gamma < \omega_{\alpha+1} (h \restriction \gamma \in \mathbf{L})].$$

Furthermore, arrange E so that $\langle f_\alpha : \alpha \in E \rangle$ is coded by E; for example, say $0 \in E$, and make $f_0(\mu) = 1$ iff $(\mu + 1) \in E$ for all $\mu < \omega_1$. $M[G]$ can be viewed as a κ-stage iteration, but the p.o. can easily be defined (as it was by McAloon) without reference to iterated forcing.

(J12) (McAloon) Show that $\mathbf{V} = \mathbf{OD}$ is consistent with $\neg$CH (or with MA + $\neg$CH, etc.).

(J13) Assuming the consistency of the existence of an inaccessible, show that it is consistent that there is a strongly inaccessible κ and an $a \subset \omega$ such that $a \in \mathbf{OD}$ but $(a \notin \mathbf{OD})^{R(\kappa)}$.

(J14) Show that it is consistent with ZFC + GCH that $(\neg\mathrm{CH})^{\mathbf{HOD}}$ and $(\mathbf{V} \neq \mathbf{OD})^{\mathbf{HOD}}$. *Hint.* See VII Exercise E5.

(J15) Assume that $\mathbb{P} \in M$, π is a $\mathbb{P}$-name for a p.o., G is $\mathbb{P}$-generic over M, and H is π_G-generic over $M[G]$. Show that $G * H$ is $\mathbb{P} * \pi$-generic over M. *Remark.* This is a converse to Theorem 5.5.

K. Boolean-valued models

By $\mathscr{B}(\mathbb{P})$ we denote the completion of the p.o. $\mathbb{P}$. $\mathbb{P} \sim \mathbb{Q}$ means that $\mathscr{B}(\mathbb{P})$ and $\mathscr{B}(\mathbb{Q})$ are isomorphic.

(K1) Suppose π and π' are $\mathbb{P}$-names for p.o.'s, and $\mathbb{1} \Vdash (\pi \sim \pi')$. Show that in M, $\mathbb{P} * \pi \sim \mathbb{P} * \pi'$. But give an example to show that even if $\mathbb{1} \Vdash (\pi = \pi')$, it need not be true that $\mathbb{P} * \pi$ is isomorphic to $\mathbb{P} * \pi'$.

(K2) Suppose that in M,

$$\langle\langle \mathbb{P}_\xi \colon \xi \leq \alpha \rangle, \langle \pi_\xi \colon \xi < \alpha \rangle\rangle$$

is a finite support iterated forcing construction. Show that in M, the $i_{\xi\eta}$ induce complete embeddings, $j_{\xi\eta} : \mathscr{B}(\mathbb{P}_\xi) \to \mathscr{B}(\mathbb{P}_\eta)$. If η is a limit, show that $\mathscr{B}(\mathbb{P}_\eta)$ is the completion of $\bigcup_{\xi<\eta} j''_{\xi\eta}\mathscr{B}(\mathbb{P}_\xi)$.

(K3) In M, let

$$\langle\langle \mathbb{P}_\xi \colon \xi \leq \omega \rangle, \langle \pi_\xi \colon \xi < \omega \rangle\rangle$$

be a countable support iterated forcing construction, where each π_n is $(\mathrm{Fn}(\omega, 2))\check{}$. Let $\mathbb{P}'_\xi$ be constructed by letting π'_n be a full P'_n-name for $\mathrm{Fn}(\omega, 2)$. Show that in M, $\mathbb{P}_\omega \not\sim \mathbb{P}'_\omega$. *Hint.* See Exercises D3, E4. *Remark.* But by Exercise K1, each $\mathbb{P}_n \sim \mathbb{P}'_n$, and it is easily seen that if $f_n : \mathscr{B}(\mathbb{P}_n) \to \mathscr{B}(\mathbb{P}'_n)$ are the "natural" isomorphisms, then the diagrams

$$\begin{array}{ccc} \mathscr{B}(\mathbb{P}_n) & \xrightarrow{j_{nm}} & \mathscr{B}(\mathbb{P}_m) \\ \downarrow{\scriptstyle f_n} & & \downarrow{\scriptstyle f_m} \\ \mathscr{B}(\mathbb{P}'_n) & \xrightarrow{j_{nm}} & \mathscr{B}(\mathbb{P}'_m) \end{array}$$

commute. Thus, $\mathscr{B}(\mathbb{P}_\omega)$ is not determined from the chain, $\langle \mathscr{B}(\mathbb{P}_n)\colon n \in \omega\rangle$. However, if full names are used, this pathology disappears (Exercise K4).

(K4) In M, let

$$\langle\langle \mathbb{P}_\xi\colon \xi \le \alpha\rangle, \langle \pi_\xi\colon \xi < \alpha\rangle\rangle$$

be an iterated forcing construction with supports in some ideal $\mathscr{I}$. Let $\mathscr{B}_\xi = \mathscr{B}(\mathbb{P}_\xi)$ and let $j_{\xi\eta}$ be as in Exercise K2. Assume that each π_ξ is a full $\mathbb{P}_\xi$-name for a p.o. Show that at a limit γ, $\mathscr{B}_\gamma$ is determined from the chain $\langle \mathscr{B}_\xi\colon \xi < \gamma\rangle$ as follows. If $\xi \le \eta$ and $b \in \mathscr{B}_\eta$, let

$$h_{\eta\xi}(b) = \bigwedge\{a \in \mathscr{B}_\xi\colon b \le j_{\xi\eta}(a)\}.$$

Then $\mathscr{B}_\gamma$ is the completion of the p.o. whose elements are sequences, $\langle b_\xi\colon \xi < \gamma\rangle$ such that $\xi < \eta \to b_\xi = h_{\eta\xi}(b_\eta)$, $\{\xi\colon b_{\xi+1} \ne j_{\xi,\xi+1}(b_\xi)\} \in \mathscr{I}$, and $b_\eta = \bigwedge_{\xi<\eta} j_{\xi\eta}(b_\xi)$ for limit η.

(K5) In M, let $\langle \mathbb{Q}_\xi\colon \xi \le \alpha\rangle$ be a sequence of p.o.'s, and suppose complete embeddings, $j_{\xi\eta}: \mathbb{Q}_\xi \to \mathbb{Q}_\eta$ are given which commute ($\xi < \eta < \zeta \to i_{\xi\zeta} = i_{\eta\zeta} \circ i_{\xi\eta}$). Assume that for limit η, $\bigcup_{\xi<\eta} i''_{\xi\eta}\mathbb{P}_\xi$ is dense in $\mathbb{Q}_\eta$. Show that there is an α-stage finite support iterated forcing construction,

$$\langle\langle \mathbb{P}_\xi\colon \xi \le \alpha\rangle, \langle \pi_\xi\colon \xi < \alpha\rangle\rangle$$

such that each $\mathbb{P}_\xi \sim \mathbb{Q}_\xi$. *Hint.* See VII Exercises D3–D5.

(K6) In M, let $\mathscr{B}_\xi$ be the measure algebra of 2^ξ (see VII Exercises H21–H23), and let $j_{\xi\eta}: \mathscr{B}_\xi \to \mathscr{B}_\eta$ be induced by projection of 2^η onto 2^ξ. Show that there is *no* iterated forcing construction (with any supports), such that $\mathscr{B}_\xi \sim \mathbb{P}_\xi$ for all $\xi \le \omega$. *Remark.* i.e., this iteration is not obtainable by the method of §5.

(K7) Verify the following treatment of 2-stage iterations in Boolean-valued models. If in M, $\mathscr{B}$ is a complete Boolean algebra (c.B.a.), $[\![\pi \text{ is a cBa}]\!] = \mathbb{1}_{\mathscr{B}}$, and π is full, then $\mathscr{B} * \pi$ may be defined as the cBa where elements are those of $\text{dom}(\pi)$. $\sigma_1 \vee \sigma_2$ is the $\sigma \in \text{dom}(\pi)$ such that $[\![\sigma = \sigma_1 \vee \sigma_2]\!] = \mathbb{1}$. A complete embedding $i\colon \mathscr{B} \to \mathscr{B} * \pi$ is defined by letting $i(b)$ be the $\sigma \in \text{dom}(\pi)$ such that

$$([\![\sigma = \mathbb{1}]\!] = b) \wedge ([\![\sigma = \mathbb{0}]\!] = b').$$

If $\mathscr{A}$ is another cBa of M and $\mathscr{B} \subset_c \mathscr{A}$, define $\mathscr{A}/\mathscr{B}$ to be the $\mathscr{B}$-name π for $\mathscr{A}/\tau$, where τ is the $\mathscr{B}$-valued ideal defined by

$$[\![\check{a} \in \tau]\!] = \left(\bigwedge\{b \in \mathscr{B}\colon a \le b\}\right)'.$$

(See also VII, Exercises D3–D5; our notation is slightly different here). Given $\mathscr{B}$ and $\mathscr{A}$, $\mathscr{A}$ is isomorphic in M to $\mathscr{B} * (\mathscr{A}/\mathscr{B})$. Given $\mathscr{B}$ and π, $\mathbb{1}_{\mathscr{B}}$ forces that π is isomorphic to $(\mathscr{B} * \pi)/\mathscr{B}$.

(K8) State and prove a rigorous version of the following: if $\langle \mathbb{P}_\xi : \xi \leq \alpha \rangle$ and $\langle \mathbb{P}'_\xi : \xi \leq \alpha \rangle$ are iterated forcing constructions using the same ideal of supports, $\mathbb{1} \Vdash (\pi_\xi \sim \pi'_\xi)$ for each ξ, and either the supports are finite or all the names π_ξ, π'_ξ are full, then $\mathbb{P}_\alpha \sim \mathbb{P}'_\alpha$. Observe that the statement $\mathbb{1} \Vdash (\pi_\xi \sim \pi'_\xi)$ does not really make sense. Exercise K3 is a counter example to this exercise if the names are not full and the supports are countable.

BIBLIOGRAPHY

BARWISE, J.

[1975] *Admissible Sets and Structures* (Springer, Berlin).

[1977] (editor) *Handbook of Mathematical Logic* (North-Holland, Amsterdam).

BAUMGARTNER, J. E.

[1970] Decompositions and embeddings of trees, *Notices of the American Mathematical Society* **17**, 967.

[1975] Inefability properties of cardinals I, in: HAJNAL et al. [1975] pp. 109–130.

BELL, J. R.

[1977] *Boolean-valued Models and Independence Proofs in Set Theory* (Clarendon Press, Oxford).

CHANG, C.C. and KEISLER, H. J.

[1973]-*Model Theory* (North-Holland, Amsterdam).

COHEN, P. J.

[1963] The independence of the continuum hypothesis I, *Proceedings of the National Academy of Science USA* **50**, 1143–1148.

[1964] The independence of the continuum hypothesis II. *Proceedings of the National Academy of Sciences USA* **51**, 105–110.

[1966] *Set Theory and the Continuum Hypothesis* (Benjamin, New York).

DEVLIN, K. J.

[1973] *Aspects of Constructibility*, Lecture Notes in Mathematics, Volume 354 (Springer, Berlin).

DEVLIN, K. J. and JENSEN, R.

[1900] *Marginalia to a theorem of Silver*, to appear.

DEVLIN, K. J. and JOHNSBRÅTEN, H.

[1974] *The Souslin Problem*, Lecture Notes in Mathematics, Volume 405 (Springer, Berlin).

DRAKE, F. R.

[1974] *Set Theory; an Introduction to Large Cardinals* (North-Holland, Amsterdam).

EASTON, W. B.

[1970] Powers of regular cardinals, *Annals of Mathematical Logic* **1**, 139–178.

ENDERTON, H. B.

[1972] *A Mathematical Introduction to Logic* (Academic Press, New York).

[1977] *Elements of Set Theory* (Academic Press, New York).

ERDÖS, P., HAJNAL, A., MÁTÉ, A. and RADO, R.

[1900] *Combinatorial Set Theory: Partition Relations for Cardinals*, to appear.

ERDÖS, P., HAJNAL, A. and MILNER, E. C.

[1968] On sets of almost disjoint subsets of a set, Acta *Mathematica Academiae Scientiarum Hungaricae* **19**, 209–218.

FRAENKEL, A. A., BAR-HILLEL, Y. and LÉVY, A.

[1973] *Foundations of Set Theory* (North-Holland, Amsterdam).

GALVIN, F.
[1900] Chain conditions and products, *Fundamenta Mathematicae* to appear.
GALVIN, F. and HAJNAL, A.
[1975] Inequalities for cardinal powers, *Annals of Mathematics* **101,** 491–498.
GANDY, R. O. and YATES, C. M. E. (editors)
[1971] Logic Colloquium '69 (North-Holland, Amsterdam).
GÖDEL, K.
[1940] *The Consistency of the Axiom of Choice and of the Generalized Continuum Hypothesis with the Axioms of Set Theory*, Annals of Mathematical Studies, Volume 3 (Princeton University Press, Princeton, NJ).
GREGORY, J.
[1976] Higher Souslin trees and the generalized continuum hypothesis, *Journal of Symbolic Logic* **41,** 663–671.
HAJNAL, A., RADO, R. and SÓS, V. T. (editors)
[1975] *Infinite and Finite Sets* (North-Holland, Amsterdam).
HALMOS, P. R.
[1960] *Naive Set Theory* (Van Nostrand Reinhold, New York).
[1963] *Lectures on Boolean Algebras* (Van Nostrand Reinhold, New York).
JECH, T. J.
[1973] *The Axiom of Choice* (North-Holland, Amsterdam).
KARP, C. R.
[1964] *Languages with Expressions of Infinite Length* (North-Holland, Amsterdam).
KEISLER, H. J.
[1973] Forcing and the omitting types theorem, in: MORLEY [1973] pp. 96–133.
KEISLER, H. J. AND MORLEY, M.
[1968] Elementa:y extensions of models of set theory, *Israel Journal of Mathematics* **6,** 49–65.
KEISLER, H. J. and SILVER, J. H.
[1971] End extensions of models of set theory, *Proceedings of Symposia in Pure Mathematics* **13,** 177–187.
KELLEY, J. L.
[1955] *General Topology* (Van Nostrand Reinhold, New York).
KLEENE, S. C.
[1952] *Introduction to Metamathematics* (North-Holland, Amsterdam).
KREISEL, J. and KRIVINE, J. L.
[1967] *Elements of Mathematical Logic* (North-Holland, Amsterdam).
KUNEN, K.
[1971] On the GCH at measurable cardinals, in: GANDY and YATES [1971] pp. 107–110.
[1977] Combinatorics, in: BARWISE [1977] pp. 371–401.
[1978] Saturated ideals, *Journal of Symbolic Logic* **43,** 65–76.
[1900] On random and Cohen reals, to appear.
KUREPA, D.
[1936] L'hypothése de ramification, *Comptes Rendus Hebdomaires des Séances de l'Academie des Sciences de Paris, Série* **202,** 185–187.
LAVER, R. and SHELAH, S.
[1900] *The* $\aleph_2$*-Souslin Hypothesis*, to appear.
MAGIDOR, M.
[1900] On the singular cardinals problem II, to appear.
MARTIN, D. A.
[1977] *Descriptive set theory: projective sets*, in: BARWISE [1977] pp. 783–815.
MARTIN, D. A. and SOLOVAY, R. M.
[1970] Internal Cohen extensions, *Annals of Mathematical Logic* **2,** 143–178.

McAloon, K.
[1971] Consistency results about ordinal definability, *Annals of Mathematical Logic* **2**, 449–467.
Menas, T. K.
[1973] On strong compactness and supercompactness, Ph.D. thesis, California.
Mitchell, W.
[1972] Aronszajn trees and the independence of the transfer property, *Annals of Mathematical Logic* **5**, 21–46.
Morley, M. (editor)
[1973] *Studies in Model Theory* (Mathematical Association of America, Buffalo, NY).
Moschovakis, Y. N.
[1980] *Descriptive Set Theory: A Foundational Approach* (North-Holland, Amsterdam).
Myhill, J. and Scott, D.
[1971] Ordinal definability, *Proceedings of Symposia in Pure Mathematics* **13**, 271–278.
Prikry, K. L.
[1970] Changing measurable into accessible cardinals, *Dissertationes Mathematicae* (Rozprawy Matematyczne) **68**, 1–55.
Quine, W. V. O.
[1937] New foundations for mathematical logic, *The American Mathematical Monthly* **44**, 70–80.
[1951] *Mathematical Logic*, (Revised edition; first edition 1940) (Harvard University Press, Cambridge, MA).
Rasiowa, H. and Sikorski, R.
[1963] *Mathematics of Metamathematics*, Monografie Matematyczne, Volume 41 (Polska Akademia Nauk, Warsaw).
Rosser, J. B.
[1953] *Logic for Mathematicians* (McGraw-Hill, New York).
[1969] *Simplified Independence Proofs* (Academic Press, New York).
Rubin, H. and Rubin, J.
[1963] *Equivalents of the Axiom of Choice* (North-Holland, Amsterdam).
Rudin, M. E.
[1977] Martin's axiom, in: Barwise [1977] pp. 491–501.
Sacks, G. E.
[1971] Forcing with perfect closed sets, *Proceedings of Symposia in Pure Mathematics* **13**, 331–355.
Shoenfield, J. R.
[1954] A relative consistency proof, *Journal of Symbolic Logic* **19**, 21–28.
[1967] *Mathematical Logic* (Addison-Wesley, Reading, MA).
[1971a] Unramified forcing. *Proceedings of Symposia in Pure Mathematics* **13**, 357–381.
[1971b] Measurable cardinals, in: Gandy and Yates [1971] pp. 19–49.
[1975] Martin's axiom, *The American Mathematical Monthly* **82**, 610–617...
Silver, J. H.
[1973] The bearing of large cardinals on constructibility, in: Morley [1973] pp. 158–182.
Solovay, R. M.
[1970] A model of set theory in which every set of reals is Lebesgue measurable, *Annals of Mathematics* **92**, 1–56.
[1971] Real-valued measurable cardinals, *Proceedings of Symposia in Pure Mathematics* **13**, 397–428.
Solovay, R. M., Reinhardt, W. N. and Kanamori, A.
[1978] Strong axioms of infinity and elementary embeddings, *Annals of Mathematical Logic* **13**, 73–116.

SPECKER, E.
[1953] The axiom of choice in Quine's "New foundations for mathematical logic", *Proceedings of the National Academy of Sciences USA* **39**, 972–975.
SUSLIN, M.
[1920] PROBLÈME 3, *Fundamenta Mathematicae* **1**, 223.
WANG, H.
[1949] On Zermelo's and von Neumann's axioms for set theory, *Proceedings of the National Academy of Sciences USA* **35**, 150–155.

INDEX OF SPECIAL SYMBOLS

GENERAL INDEX

Printed and bound by CPI Group (UK) Ltd, Croydon, CR0 4YY
03/10/2024
01040430-0001